Lecture Notes in Physics

The Lecture Notes in Physics

The series Lecture Notes in Physics (LNP), founded in 1969, reports new developments in physics research and teaching – quickly and informally, but with a high quality and the explicit aim to summarize and communicate current knowledge in an accessible way. Books published in this series are conceived as bridging material between advanced graduate textbooks and the forefront of research and to serve three purposes:

- to be a compact and modern up-to-date source of reference on a well-defined topic
- to serve as an accessible introduction to the field to postgraduate students and nonspecialist researchers from related areas
- to be a source of advanced teaching material for specialized seminars, courses and schools

Both monographs and multi-author volumes will be considered for publication. Edited volumes should, however, consist of a very limited number of contributions only. Proceedings will not be considered for LNP.

Volumes published in LNP are disseminated both in print and in electronic formats, the electronic archive being available at springerlink.com. The series content is indexed, abstracted and referenced by many abstracting and information services, bibliographic networks, subscription agencies, library networks, and consortia.

Proposals should be sent to a member of the Editorial Board, or directly to the managing editor at Springer:

Christian Caron
Springer Heidelberg
Physics Editorial Department I
Tiergartenstrasse 17
69121 Heidelberg / Germany
christian.caron@springer.com

Ion-Olimpiu Stamatescu
Erhard Seiler (Eds.)

Approaches to Fundamental Physics

An Assessment of Current Theoretical Ideas

Springer

Editors

Ion-Olimpiu Stamatescu
Forschungsstätte der Evangelischen
Studiengemeinschaft (FESt)
Schmeilweg 5
69118 Heidelberg, Germany
and
Institut für Theoretische Physik
Universität Heidelberg
Philosophenweg 16
69120 Heidelberg, Germany
stamates@thphys.uni-heidelberg.de

Erhard Seiler
Max-Planck-Institut für Physik
Werner-Heisenberg-Institut
80805 München, Germany
ehs@mppmu.mpg.de

I.-O. Stamatescu and E. Seiler (Eds.), *Approaches to Fundamental Physics*, Lect. Notes Phys. 721 (Springer, Berlin Heidelberg 2007), DOI 10.1007/978-3-540-71117-9

ISSN 0075-8450
ISBN 978-3-642-09021-9 e-ISBN 978-3-540-71117-9

Springer is a part of Springer Science+Business Media
springer.com

Cover design: eStudio Calamar S.L., F. Steinen-Broo, Pau/Girona, Spain

Preface

This book represents in the first place the desire of the authors of the various contributions to enter a discussion about the research landscape of present-day fundamental theoretical physics. It documents their attempt, out of their highly specialized scientific positions, to find a way of communicating about methods, achievements, and promises of the different approaches which shape the development of this field. It is therefore also an attempt to bring out the connections between these approaches, and present them not as disjoint ventures but rather as facets of a common quest for understanding.

Whether in competition to each other or in collaboration, the 'many-fold ways' of contemporary physics are characterized by a number of exciting findings (and questions) which appear more and more interrelated. Moreover, in the historical development of science, the steadily arriving new empirical information partly supports, partly contradicts the existing theories, and partly brings forth unexpected results forcing a total reorientation upon us. If we are lucky, the beginning of this century may prove to be as grand as that of the last one.

It is not an easy task in a situation so much in movement and in which various approaches strive for completion, to promote a constructive interaction between these and to achieve a level of mutual understanding on which such an interaction can be fruitful. Nearly all of the authors contributing to this book have been participating in a working group dedicated exactly to this task; this group met in many sessions over several years. This book is to a large extent the result of these discussions.

The support of the authors' home institutions was of course important for this project, but one institution has to be singled out for making this book possible: this is FESt, Heidelberg (Forschungsstätte der Evangelischen Studiengemeinschaft – Protestant Institute for Interdisciplinary Research).

FESt has a long tradition in bringing together interdisciplinary working groups. In particular, it has cultivated the dialogue between the natural sciences, philosophy, theology, and the life sciences – but also projects inside one discipline which involve discussion across the specialized fields and aim at a more general understanding of fundamental questions pertaining to this discipline. Our work has constituted a FESt project belonging to this class.

The intention of working groups at FESt typically is not only to present the differing perspectives but also to compare them and to find relations which could be fruitful for the fields involved. To achieve this goal, numerous group sessions are required and FESt provides hereto a unique scientific and organizational environment. This has been extremely useful for our project and we are very grateful to FESt for its support of our work as well as its continuous interest and confidence in it.

We appreciate very much the interest of Springer-Verlag in promoting the interdisciplinary exchange of information at the level of specialists. We thank Wolf Beiglböck for excellent advice and assistance in the completion of the book and the Springer team for dedicated editorial and publishing work.

Hans-Günter Dosch
Jürgen Ehlers
Klaus Fredenhagen
Domenico Giulini
Claus Kiefer
Oliver Lauscher
Jan Louis
Thomas Mohaupt
Hermann Nicolai
Kasper Peeters
Karl-Henning Rehren
Martin Reuter
Michael G. Schmidt
Erhard Seiler
Ion-Olimpiu Stamatescu
Norbert Straumann
Stefan Theisen
Thomas Thiemann

Heidelberg, September 2006

Contents

Part I Introduction

Part II Elementary Particle Theory

Part III Quantum Field Theory

Part VI String Theory

Part VII Cosmology

Part I

Introduction

Introduction – The Many-Fold Way of Contemporary High Energy Theoretical Physics

E. Seiler[1] and I.-O. Stamatescu[2]

[1] Max-Planck-Institut für Physik (Werner-Heisenberg-Institut), 80805 München, Germany
ehs@mppmu.mpg.de

[2] Forschungsstätte der Evangelischen Studiengemeinschaft (FESt), Schmeilweg 5, 69118 Heidelberg, Germany
and
Institut für Theoretische Physik, Universität Heidelberg, Philosophenweg 16, 69120 Heidelberg, Germany
stamates@thphys.uni-heidelberg.de

This book is trying to give an introductory account of the paradigms, methods and models of contemporary fundamental physics. One goal is to bring out the interconnections between the different subjects, which should not be considered as disjoint pieces of knowledge. Another goal is to consider them in the perspective of the quest for the physics of tomorrow. The term 'assessment' in the subtitle of our book is not meant as a comparative judgment but as a recognition of the state of the art. This also means that achievements, problems and promises will be touched in the discussion, as well as relations and cross-references.

The chapters in this volume are written in a style that is not very technical and should be intelligible by a graduate student looking for direction for his further studies and research. For established physicists they may help to remind them of the general context of research and may be an incentive to a look over the shoulder of the neighbor. The various chapters are written by authors who are workers in the respective fields and who are, unavoidably, of somewhat diverse character, also as far as the level of technicality is concerned. The following introduction is meant to sketch the frame in which these contributions are conceived, to offer some help in understanding the relationship between the different chapters and give the reader some guidance to their content.

This book is about the physics of the fundamental phenomena. This includes the physics of elementary particles, also known as high-energy physics, but also gravity and therefore the physics of space and time. The landscape of

E. Seiler and I.-O. Stamatescu: *Introduction – The Many-Fold Way of Contemporary High Energy Theoretical Physics*, Lect. Notes Phys. **721**, 3–18 (2007)
DOI 10.1007/978-3-540-71117-9_1

present day theoretical physics ranges from the standard model (of elementary particles) to the cosmological standard model, and the empirical information is at first interpreted in this conceptual framework (even though eventually it might require to go beyond it). The first and the last chapters of the book were chosen to indicate this span.

The term "fundamental" should be understood objectively. Physics research is a very broad enterprise and even if we restrict the view to the research not directly related to applications, fundamental phenomena make up only one among many directions: complex systems, laser physics, quantum information, solid state physics, atomic physics, nuclear physics, biophysics, astrophysics are only a few keywords to suggest the width of the research spectrum. The word "fundamental" implies in no way a judgment of importance. Sure enough, all the above fields introduce their own concepts and methods which allow genuine progress of our knowledge. On the other hand, any field of physics is dependent at a certain level on our understanding of the fundamental phenomena at this level. Laser physics or superconductor physics presupposes electrodynamics and quantum mechanics, nuclear physics is based on the interactions of the standard model (strong, weak and electromagnetic), solid state physics on statistical mechanics. One cannot say when and where new insights concerning the fundamental phenomena will enter other fields or, even more probable, form the basis of new ones, since this always has involved many other factors: quantum information and quantum computation, for instance, have arisen as important reasearch fields half a century after their quantum theoretical basis had been available.

The contemporary momentum in physics research appears to be reductionist unification in the physics of fundamental phenomena, and perfectionist diversification in the other fields. These can be seen as different components of the general research momentum, seemingly adequate each to the corresponding task, as suggested by the historical development. But even if committed to one or the other perspective, research has always been (willingly or unwillingly) critical enough to incessantly question the justification of the chosen approach and we can witness non-reductionist suggestions in the theory of fundamental phenomena as well as reductionist trends in, say, biophysics.

Finally we should note that in this book, experiment is not addressed directly, but only in the discussion of the empirical basis of the various theories. But this is by far not all that experimental physics is. In fact, the latter has its own momentum and task, which is not only to corroborate or falsify theories: it is by its independence that experiment can prompt the *new* in physical knowledge, and produce findings not "ordered" by any theory.[3] The restricted scope of this book does not allow a presentation of experimental

[3] "Who ordered that?" Nobel-prize winning physicist I. I. Rabi is said to have exclaimed over the discovery of the muon.

research, but the reader should be convinced that the latter stays in the background of all discussions.

We should also note that even in the restricted frame of the book there can be no claim of an even approximately exhaustive overview: many developments have not been described, or were only slightly touched. We do think, however, that we have collected here an essential part of the theoretical discussion, although the dynamics of the conceptual developments can hide many surprises.

1 Historical Remarks

Physics in the early 20th century saw two great revolutions: the development of the theories of relativity and quantum theory. Relativity actually involved two separate revolutions: special and general relativity.

The rest of the 20th century was largely concerned with working out the theories by building concrete models based on them, applying them to various physical problems and testing their predictions.

Soon it became clear that there are severe problems of compatibility between those theories; initially they referred to different regimes of physics but eventually the regions where they overlap could not be avoided, and the search for some more general theory combining and reconciling the theories valid in those different regimes could not be avoided.

The first such unification did not so much raise conceptual as technical problems: it was the unification of special relativity with quantum theory resulting in the highly successful structure of quantum field theory. Its success is typified by the extremely precise agreement between theory and experiment in quantum electrodynamics that began to emerge in the 1950s and is still being improved; this gave people confidence in the scheme of quantum field theory.

After this success, the story continued with the search for unification not so much of the theoretical frameworks of relativity and quantum theory but rather of the three different interactions that fit into the framework of (special) relativistic quantum field theory: the electromagnetic, weak and strong interactions. Unification between the first two was achieved with great success in the 1960s and 70s; the resulting electroweak theory has become a pillar of the standard model, which combines the electroweak theory with quantum chromodynamics (QCD) describing the strong interaction. The standard model is described in detail in the first chapter of Part II of this book. Models unifying all three non gravitational interactions, so-called grand unified theories (GUTs) were proposed soon after, but with less convincing success.

The theoretical basis for present day physics of fundamental phenomena consists of quantum field theory and general relativity. Their main ideas are presented in Parts III and IV of this book, respectively. However, a serious compatibility problem arose as people tried to bring gravity in the form of general relativity into the game. Intractable technical problems appeared, which had to do with the fact that any attempt to quantize general relativity introduces an intrinsic length scale (the 'Planck scale') that appears to make the interaction strength grow beyond all bounds as one goes to short distances or high energies. But even more serious is the conceptual clash between general relativity and any form of quantum theory: The main insight of Einstein's general relativity was the change of the role of space and time from a passive 'arena', in which physics takes place, to an active dynamical entity that is shaped by matter and acts back on it; but space-time remained a sharply defined classical object.

On the other hand, all interpretations of quantum theory and especially the measurement process, use space-time, and in particular time as something given, and even treat the future different from the past in such concepts as the 'reduction of the wave packet' (in the most common interpretation) or the 'splitting of worlds' (in the 'Many Worlds' interpretation). But anything of a dynamical nature in quantum theory also shows its typical non-classical behavior, described in somewhat simplistic terms as 'uncertainty', making uncertain the very arena in which the dynamical evolution of matter is to take place. Combining the ideas of quantum theory with those of general relativity leads unavoidably to fundamental conceptual difficulties and we think it is fair to say that they have not yet been resolved in any of the approaches.

But physicists are not easily deterred from trying the impossible: Various approaches to quantum gravity have been pursued with great vigor in the last few decades. On the one hand there are approaches that try to 'quantize' general relativity as a separate theory; these are described in Part V. On the other hand there is the even more ambitious project to construct a 'theory of everything' (TOE), describing all the forces of nature in a unified form. This has been the goal of string theory or M-theory, to be discussed in Part VI. Both these approaches have brought a wealth of new concepts and new views on the structure of space time and of matter.

If one is more modest, a lot can be learned by combining general relativity and quantum field theory in a less theoretically ambitious way by keeping gravity classical and therefore providing the arena for particle physics in the form of quantum field theory. This is pragmatically justified as long as the length scales involved are reasonably distinct. The astounding progress of physical cosmology in the last few decades was made possible by this pragmatic approach; the fact that many of its aspects are directly related to recent observations, makes this one of the most exciting areas of present-day physics (Part VII).

In the following sections we shall discuss some of these problems in more detail.

2 Systematic Considerations

2.1 Quantum Theory and Special Relativity

As mentioned above, the marriage of special relativity with quantum theory led to the structure of quantum field theory. This structure, though almost seventy years old, is still the most important paradigm for elementary particle physics. The general structure of quantum field theory, its status, concepts and their limitations, are discussed by K. Fredenhagen, K.-H. Rehren and E. Seiler (Part III).

There are different approaches, which can be labeled in short as 'axiomatic', 'constructive' and 'perturbative'. The purpose of the axiomatic approach is to gain structural insights and identify properties shared by all quantum field theories obeying the respective systems of axioms. For the phenomenological applications the perturbative approach is by far the most relevant one; its success depends on the method of renormalization of parameters, which removes the infinities that were present in the early, naive versions of the theory. Finally the constructive approach on the one hand tries to construct in a mathematically rigorous way quantum field theories satifying the axiom systems. On the other hand, in the form of lattice gauge theory it plays an important role in understanding the strong interactions, in particular the formation of hadrons as bound states and the very essential concept of the confinement of quarks inside the hadrons. Furthermore it opens the way to an application of the concept of renormalization in a non-perturbative and, in a certain sense, intuitive way, as integration over degrees of freedom which are irrelevant at a given scale.

The timeliness of the research in this field is also certified by the observation that many of the essential concepts – from renormalization group, to nonabelian gauge symmetry, confinement, Higgs mechanism – have been built in the course of time until recent days, and new conceptual developments, such as the so-called 'holographic principle', explicitly involve quantum field theory at the same time as quantum gravity and string theory.

The relation to other subjects such as general relativity (gravity) and string theory is discussed briefly in the mentioned chapter; in particular, it contains a discussion of quantum field theory in curved space-times, considered as fixed backgrounds, neglecting the back-reaction of the fields on space-time. This pragmatic approach has seen much progress in recent years.

The other aspect of quantum field theory is its application to describe high energy physics. Part I of this book deals with the practical ('phenomenological') use of quantum field theory: first H.G. Dosch describes the so-called standard model of elementary particle physics. This model is extremely successful in giving a quantitative account of all known particles and their interactions. In fact it is so successful that many physicists are desperately hoping for some disagreement with experiment to show some hints of 'new physics'. One of the hopes is of course that the 'Large Hadron Collider' (LHC), which should become operational next year at CERN in Geneva, will show such deviations from the standard model.

The interpretation and parametrization of such deviations (if they occur) requires models or theories that go beyond the standard model, since experimental data can never be interpreted without a theory. The chapter by M. Schmidt gives an overview of some of the ideas in this direction that are currently under consideration. All of them predict additional particles which so far have not been observed; the appearance of such particles at the LHC, it is expected, would help to narrow down the possibilities of such extended theories.

There are other reasons why physicists are not ready to accept the standard model as the last word on elementary particle physics: Cosmology, as discussed in Part VII, seems to require the existence of additional particles which do not have electromagnetic interactions (so-called dark matter) and moreover the mysterious 'dark energy'. It is hoped widely that the LHC will also shed some light on the question of dark matter by discovering some of the particles that might constitute it.

Finally there is a philosophical and esthetic reason for the search of a more fundamental theory: the standard model has at least 19 parameters, whose values should be explained in a truly fundamental theory. String theory, at least in the earlier stages of its development, seemed to offer the hope to determine some or all of these parameters; but lately there has been a shift away from this goal in (part of) the string theory community (see Part VI), where those parameters are now considered as contingent or environmental, roughly like the distance of the earth from the sun. But this view is by no means generally accepted even among string theorists; for most physicists the search for a theory explaining all or at least most of the free parameters remains on the agenda as a central goal of fundamental research.

To sum up the situation regarding the unification of special relativity with quantum theory, it can be said that it has been understood conceptually within the axiomatic approach and made practically useful by renormalized perturbation theory and numerical lattice gauge theory. But there are open mathematical problems: mathematically rigorous constructions of realistic quantum field theories, obeying one of the axiomatic schemes, have not been accomplished. This is the reason why the Clay Mathematics Institute offered a prize of one million dollars for a mathematically rigorous construction of a simplified version of quantum chromodynamics with the right physical properties.

2.2 General Relativity

As mentioned above, the crucial insight of Einstein's theory of gravitation known as general relativity (GRT) is that space-time no longer serves as a passive arena in which events take place, particles scatter, are created and annihilated, fields propagate, but rather all matter (every field) acts back on space-time, shaping it as it evolves.

Space-time becomes a dynamical object, not fundamentally distinct from matter. The classical theory of general relativity has been extremely successful in describing the world on a macroscopic scale up to farthest reaches of the observable universe. It even plays a role in mundane applications such as navigation systems in cars, which are based on the global positioning system (GPS). Without the use of general relativity, the GPS would accumulate an error of about 10 kilometers per day.

The chapter by J. Ehlers (Part IV) gives a concise introduction into the concepts and structure of classical general relativity. One of the characteristic features of that theory, which is invoked frequently, is the so-called 'general covariance' or 'diffeomorphism invariance'. Superficially this just means that one can use whatever coordinates or frames of reference one likes to describe the joint evolution of matter and space-time in formally the same way. But on closer inspection this statement may turn out to be, depending on how one interprets it, empty or false. In fact it is quite subtle to give a precise and correct meaning to the statement of general covariance or background independence, as it is sometimes called. This difficult issue is discussed in depth by D. Giulini (Part IV).

Beyond leading to the description of novel phenomena, such as the bending of light rays or the existence of black holes, there is one outstanding interest of general relativity: it provides space-time solutions which provide the basis for models of the universe. This leads directly to the discussion in Part VII. Cosmology is the scene for the collaboration (though not unification) of our most evolved theories: quantum field theory and general relativity. It is in fact a very fruitful scene, since new concepts at the interface of classical and quantum physics have been developed here and a great amount of empirical data has been obtained to guide the theoretical development.

2.3 Quantum Theory and General Relativity

The existence of quantum matter and the fact that this matter acts on space-time seems to make it unavoidable to assign quantum nature also to space-time itself. But, as said before, this leads to extremely hard technical as well as conceptual problems. On the other hand, the quantum nature of space-time, whatever this means precisely, should only become relevant at energy scales of the order of the Planck energy, which is 16 orders of magnitude above the highest accelerator energies. So a pragmatic approach is just to ignore the problem of unifying gravity with the other interactions. An even more extreme standpoint has been taken by the famous physicist Freeman Dyson[4]: he argued that the 'division of physics into separate theories for large and

[4] in his review of Brian Greene's bestseller 'Fabric of the Cosmos' (New York Review of Books, May 13, 2004).

small' is acceptable and a unification not necessary. However, most physicists disagree with this point of view, and the chapter by C. Kiefer (first chapter of Part V) explains why.

Before entering into the dangerous waters of quantum gravity, one can study a useful domain in which matter is treated quantum mechanically, but as far as its effect on space-time is concerned, only classical, large-scale properties of matter are considered. This is the regime where modern astrophysics and physical cosmology have their place; this has been an extremely active domain of research in the last decades. The beauty of this field is, as mentioned before, that it shows a very strong interplay between observations and theory, so theoretical predictions can actually be checked and have been checked with impressive success, using the satellite data on the cosmic microwave background. A discussion of some of the central aspects of modern cosmology is contained in the chapter by N. Straumann (Part VII); this chapter emphasizes in particular the problem of the so-called 'dark energy' or 'cosmological constant', which according to astronomical observations seems to pervade our universe.

Another preliminary way to join general relativity and quantum theory is the treatment of a general relativistic space-time as a fixed background arena for quantum field theory, neglecting the back-reaction of the quantum fields on space-time. This should be appropriate under certain circumstances, such as the situation where few particles (or particles of low density) are described in gravitational fields of large objects such as stars, galaxies, or even the universe as a whole; as remarked, this subject is discussed in Part III.

The really hard problem of quantum gravity is the subject of Part V; Part VI, which deals with string theory, could also be subsumed under this heading. This subject may seem to take a disproportionally large fraction of this book; this is so because of its fundamental importance as well as its difficulty, both technically and conceptually. This question has therefore been the focus of a large part of modern physics research.

The fundamental difficulty of a marriage between quantum field theory and general relativity, as alluded to before, lies in the totally different roles played by space-time, and time in particular, in the two frameworks. Any quantum theory treats and needs time as an external parameter, in order to give an interpretation in terms of measurement results. In general relativity, space-time is shaped by the evolution of matter, hence if matter behaves quantum mechanically, so will space-time. This fact leads almost unavoidably to such concepts as the quantum state or wave function of the universe, which would elevate the Schrödinger cat paradox to cosmic dimensions. In its standard interpretation quantum theory needs the concept of measurement, and it is hard to see what this would mean for the universe as a whole, therefore the interpretation of a wave function of the universe remains murky. Many researchers therefore are drawn to a 'many worlds' (better: many observers) interpretation which, again, is not free of conceptual problems.

In spite of these unresolved difficulties, it is legitimate to go ahead and try to construct something like a quantum field theory of gravity (or even of all interactions) and postpone the problems of interpretation to a later day. A rather direct approach is the so-called 'canonical quantization' of gravity, whose principles are described in the chapter by C. Kiefer and D. Giulini (Part V).

The starting point is that classical general relativity can be cast into the form of canonical field theory, in which the dynamics takes place in some phase space parametrized by coordinates and momenta; these then can be subjected to canonical quantization, the procedure that was so successful in non-relativistic quantum mechanics.

The situation is complicated by the way in which the classical system is *constrained* due to the general covariance of Einstein's equations. While such contraints already occur in gauge theories, such as the ones occurring in the standard model, here the situation is more serious: the Hamiltonian that should generate the evolution of the system is just a combination of constraints. This leads, after quantization, to the peculiar situation that, unlike in 'normal' quantum systems, physical states ('wave functions') have to be annihilated by the Hamiltonian. So there appears to be no evolution with respect to an external, given time. Of course this makes sense, because general relativity does not contain such an external time. Upon closer inspection, however, it seems possible to recover something like an evolution with respect to an 'intrinsic time'. The issues related to the ultraviolet problems (i.e. perturbative non-renormalizability) of canonical quantum gravity are not discussed here; they are addressed in different ways in the following three chapters (Part V).

After the discussion of the general ideas of canonical quantum gravity by Kiefer and Giulini, H. Nicolai and K. Peeters give an introductory account to so-called loop and spin foam quantum gravity. Loop quantum gravity is an elaboration of the canonical approach discussed before, whereas the spin foam formulation of quantum gravity is trying to avoid the different treatment of space and time inherent in that approach. This presentation is given by 'outsiders' to the subject, i.e. physicists who mostly worked on other subjects (strings in this case) but studied the loop and spinfoam approaches, to understand its advantages as well as its problems. One advantage of this 'outside' view may be the pedagogical style of this 'brief guide for beginners', as the authors call it. The presentation by Nicolai and Peeters also raises some critical questions about the prospects of the enterprise; some of these questions are addressed or answered in the following chapter by Thomas Thiemann. Reading both chapters should make it possible to form an educated opinion about the loop approach.

T. Thiemann then gives a moderately technical account of loop quantum gravity. This chapter is written by an 'insider', that is a physicist who has intensely worked on this subject. As remarked, the approach is an elaboration of the canonical approach discussed before, striving for mathematical rigor. Partly this has become possible by the introduction of more appropriate canonical variables (the 'Ashtekar variables'). The word 'loop' in this approach refers to the fact that (at least on the kinematical level) the basic coordinates are parallel transporters along curves, and the corresponding momenta are 'electric fluxes' through two-surfaces bordered by closed curves. A special feature of the approach is the appearance of a non-separable (i.e. having uncountably many dimensions) 'kinematical Hilbert space' which is supposed to collapse to a separable one (as is physically desirable) by imposing the constraints.

The main virtues of loop quantum gravity may be listed as first of all background independence, secondly existence of length, area and volume operators with discrete spectra, and finally the possibility to couple other field theories ('matter') to this form of quantum gravity. The first property means that no given, prescribed space-time geometry is present, in accordance with the crucial property of classical general relativity stressed repeatedly. The second one is interpreted as a sign that at distances of the order of the Planck length the usual continuous manifold structure of space-time disappears (but questions of interpretation of these quantized space-time structures remain). The discreteness at the Planck scale also offers hope for an effective physical cutoff in other, non-gravitational theories, which can be coupled to loop quantum gravity. The great difficulty of this approach is to understand the emergence of a classical space-time, as we experience it, at distances large compared to the Planck length.

A totally different approach has been taken by O. Lauscher and M. Reuter (also in Part V). Again the goal is to quantize gravity 'in isolation' and to overcome the main technical obstacle, the alleged nonrenormalizability of the theory due to the presence of a coupling constant with positive length dimension (given by the Planck length). The idea, in short, is that this problem is entirely due to the conventional treatment, which is based on perturbation expansion in the coupling constant. It has been known for a long time that in quantum field theory perturbatively non-renormalizable models may turn out to be renormalizable, once treated non-perturbatively. Steven Weinberg has coined the term 'asymptotic safety' for this phenomenon and it is the thesis of the chapter that this is indeed what happens in quantum gravity. Since nobody can actually solve the theory exactly, the authors collect evidence in favor of this scenario from approximations which are distinct from the usual perturbative ones.

2.4 String Theory

The most ambitious approach to quantum gravity is the enterprise variously known as 'String Theory', 'Superstring Theory' or 'M-Theory'. In Part VI J. Louis, T. Mohaupt and S. Theisen give an overview over this vast subject. We will call it generally 'String Theory' here, like these authors do.

String theory has a peculiar history: it started out as a theory of the strong interaction around 1970, going into hibernation with the advent of quantum chromodynamics as the part of the standard model describing the strong interaction, and re-emerged in the mid 1980s as a 'theory of everything', that is all interactions including gravity. This came about because the originally unwanted massless spin two particle appearing in string theory was identified with the graviton (hence the saying that string theory implies gravity) and the realization in 1984 that there was a version in which all anomalies canceled and apparently a theory free of ultraviolet divergencies emerged.

Ever since then, string theory has been the most popular area of fundamental research, attracting a huge number of young and talented theoreticians as well as the support of many influential senior physicists and being in particular shaped by Edward Witten, who is recognized as the leading figure in present-day mathematical physics. Like the ancient Greek hero Proteus, string theory has gone through many metamorphoses. Originally it was really considered to be a theory that replaced the points appearing as formal arguments of fields by extended strings (this is often not quite correctly phrased as the replacement of 'point particles' by strings), whereas later it was sprouting 'branes', that is submanifolds of various other dimensions, and then it was even discovered that it was 'dual' to an 11-dimensional supergravity (a quantum field theory). The discovery of various dualities between different versions of string theory and that field theory was considered as a major breakthrough, since it suggested the existence of a unifying theory, dubbed 'M-Theory' by Witten, behind all this.

The physical results expected from the theory also evolved over time: initially it was hoped that one could eventually predict in a more or less unique way the standard model (or some extension of it) as a low energy approximation. This hope was not fulfilled and today the currently dominating view is that it has an incomprehensibly large number (10^{500} is often quoted) of 'vacua', each corresponding to a world with different physics, making the parameters of, say, the standard model, merely contingent or accidental facts of the universe we are living in, much like the distances of the planets from the sun.

String theory is not a closed theoretical structure with fixed concepts and axioms, but an evolving enterprise; somebody even proposed to define it simply as follows: 'String Theory is what string theorists do'. The chapter by Louis, Mohaupt and Theisen describes the evolution of the theory methodically, but the different steps described roughly follow the historical development.

One aspect of string theory is that it led to strong interaction between mathematicians and physicists. Its influence on mathematics can be seen by the frequent appearance of the name 'Witten' in various mathematical contexts, such as the 'Seiberg-Witten' functional or the 'Gromov-Witten invariants' (for instance, in the work of the 2006 Fields medal winner A. Okounkov) or, most importantly, by the awarding of the Fields medal to Witten himself in 1990.

One criticism that is leveled against string theory as a proposed theory of quantum gravity is its dependence on an unquantized background geometry, serving again as the arena in which the dynamics unfolds. String theory replies that this is only apparent, since the split between the classical background and the quantized fluctuations around it is arbitrary; while this seems at first sight to imply that the topology of the background is still fixed, there are some proposals how string theory might provide a context for fluctuations of topology as well. It is also hoped that full background independence should become manifest in a future 'String Field Theory'. The debate about this issue, mostly between loop quantum gravity and string theory can to some extent be followed in this book by comparing the chapters dealing with these subjects.

It is clear from this brief discussion that there is no unique current paradigm, but there are some competing and even conflicting paradigms that have to be explored much further, before a consensus may be reached. It is appropriate to stress at this point that, in spite of all diversity and even contradiction among the various approaches towards a fundamental theory of the future, as testified by this book, there is a broad agreement about the established physics in which such a theory has to be rooted. The development of any new theory must take into account the huge amount of accumulated empirical evidence, since the ultimate judge for a theory will always be the experiment. Any future theory also must retain contact with the present theories which successfully describe these empirical data, and build upon the conceptual base offered by these theories since – according to the experience we have until now – a superseding theory will indeed contain successful partial theories in some well-defined 'limit'.

As remarked, the subject of quantum gravity suffers from the problem that it is beyond any direct contact with experiment or observation now and will arguably remain so in the foreseeable future. Nevertheless it is to be hoped that eventually also Nature itself will be kind enough to help us decide. Until then we have to rely on exploring the internal consistency and predictive power of the different approaches and also try to stay aware of their mutual interdependence.

3 Conceptual Questions

One cannot be unaware of the interpretational and conceptual problems raised by the developments of modern physics. While these are not directly the matter of the normal physics research they find their way into the philosophy of science discussion – and color the books for the general public written by well-known physicists.

There are essentially two scenes in which these problems are raised: the forming of our concepts and the character of our knowledge with reference to reality.

In building up our concepts we normally proceed by extending older ones and redefining them in new theoretical schemes. So, for instance, we took the concept of particle from classical physics over to quantum mechanics and to quantum field theory while changing it in major ways. In doing this we increasingly departed from the classical intuition, which is strongly webbed in our everyday life. Most of the concepts of present-day physics are mathematically based, both in geometry (branes, loops) and in analysis (Lagrangians, Hilbert spaces, operators, group representations). It may be an interesting question to ask: what kind of general new intuitions, both physical and mathematical do we construct in this way?

One of the notions related to forming concepts is that of effective or approximate conceptual schemes. Let us consider, e.g. the concept of electron. We can mean by this the electron of classical electrodynamics, of quantum mechanics, or of quantum electrodynamics. To the extent we want to consider them to be related to each other we must use the notion of effective theory. In fact this notion is very powerful and allows us to unambiguously define lines of relationship: there is no need to look for some kind of similarity, what we need is to establish the procedure by which a well-defined approximation is realized – both mathematically and as the definition of a physical situation. So, for instance, we can speak of the classical electron as decohered quantum object: both the physical situation and the mathematical derivation are well defined. Another example is that of space and time: there are very different intuitions related to these concepts in the various theoretical schemes and the contact between them can be less based on following these intuitions but more on their binding in a fundamental vs effective setting (asymptotic flatness in general relativity models, for instance). An enlightening construction in this process is Wilson's renormalization group. Normally this construction shows a unique direction, from small to large scales, but it in fact is defined more generally in terms of identifying relevant degrees of freedom and averaging (or integrating) over the irrelevant ones.

The other scene for the discussion is the character of our knowledge. If we leave aside the 'postmodernist' views, and since the a priori stance of critical idealism is difficult to bring into agreement with modern physical knowledge, the main argument seems to go between some kind of positivist, empiricist or instrumentalist positions on the one hand, and some kind of realist or fundamentalist positions on the other hand. It may be interesting therefore to risk some brief comments on these issues.

Both kind of positions appear to have their advantages and disadvantages. To insist on empiricism and demand that physics only be concerned with relating and describing observations discards a lot of interpretational problems but fails to account for the progress of the scientific process. To assume, on the other hand, that we always have access to the 'real thing' cannot work, unless, may be, we mean this in a 'weak' sense and qualify this access in terms of effective and approximate concepts. So, for instance, the electron of classical electrodynamics, quantum mechanics and quantum electrodynamics cannot represent the same and therefore One real thing: Either we consider them as 'unfinished', with the real thing behind being only suggested asymptotically by them, or we assume that they do point to real 'manifestations' of this thing which however depend on a certain frame – e.g. scale. Then we can introduce a notion of continuity and progress, which is theoretically as well as phenomenologically well defined: a typical event picture, for instance, shows particle generation processes of quantum field theory, quantum mechanical interaction with atoms and decoherence, and classical electromagnetic interaction with external fields all interrelated and in one shot (see Fig. 1).

The advantages or disadvantages we have been speaking of do not seem to interfere with the dynamics of the physics research. A positivist, for instance, may not be particularly uncomfortable with the many parameters of the standard model, since for him reduction is not a question of explanation, but only one of optimization in the reproduction of observations. Hence reduction is only good if it allows better predictions (in that sense the Copernican model was, at the beginning, a failure). No new theoretical ansatz achieves this. But also for a realist, who might be more eager to take a risk for the sake of such criteria as simplicity, explanatory promises and faith in the existence of 'laws of nature' there is too much theoretical indefiniteness and too little empirical support for any particular ansatz going beyond the standard model to be convincing. Fortunately, however, there seems to be no way to improve the acknowledged problems of the standard model the Ptolemaic way and there is also the fundamental question of quantization of gravity which is both of theoretical and empirical significance (in as much as cosmology is). This raises enough uneasiness, independently of 'philosophical' position, to motivate the quest for a superior theory.

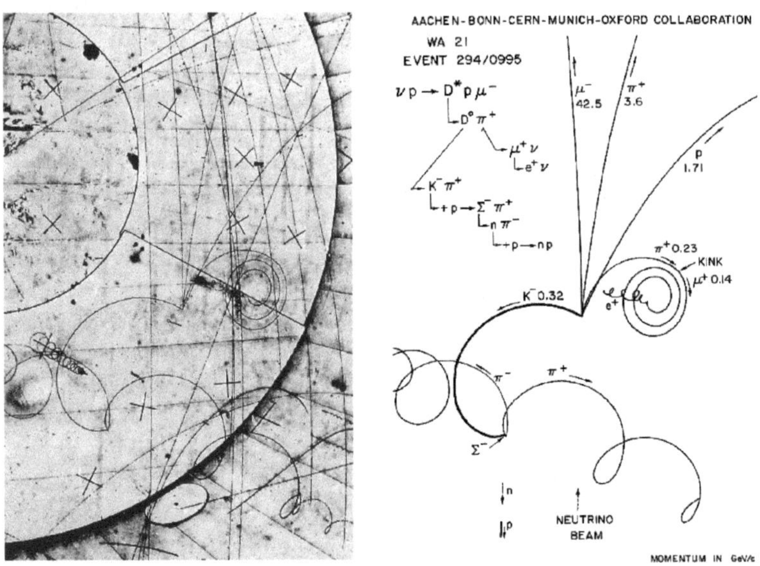

Fig. 1. Bubble chamber event: production and decay of a D* meson in a neutrino beam [CERN copyright; we thank CERN for the permission to publish this picture]

Another aspect of this opposition is the discussion on truth and justification. Roughly said, an instrumentalist perspective would stay with the justification concept, understood as internal and empirical consistency of theories (with post-diction and prediction), while a realistic perspective would ask for truth to be prompted by nature and incorporated in the theory. In the first case empirical tests support or contradict a certain justification scheme (in short, a theory), possibly asking for a new one – with hypothesis building being a qualified trial-and-error endeavor. In the second case one assumes that the hypotheses are 'conducted' or inspired by empirical and conceptual considerations and that the change in the justification (from one theory to the next one) captures an element of 'truth'. These seem to be just different ways of talking, but reflect in fact different positions: are our concepts just convenient but arbitrary instruments or do they follow some lines traced by nature?

Now both perspectives appear difficult to follow to the very end. The positivist attitude simply renounces of posing questions (as Born says, it does not deny the existence of a 'reality', but it states that it is meaningless to speak of it). Justification as introduced in this perspective appears insufficient since it always remains one step behind in the process of development of physical knowledge. If the scientific process is based only on justification one cannot explain why this process seems directed – and as a result, of course, continuity and directedness in this perspective is either denied or claimed to be only historically (culturally, socially) generated.

On the other hand, a 'strong' realistic hypothesis also fails, since it needs to accommodate contradictions. Such as, for instance, the clash between causality and a description with help of 'elements of reality' pointed at by the Einstein–Podolsky–Rosen argument (generally, to assume 'reality' for the concepts of quantum mechanics – e.g. for Hilbert space vectors – may be difficult to secure against non-locality). Therefore a truth concept in the strong sense is also problematic: it gets into trouble, it relies itself on metaphysical assumptions, and in fact shoots beyond its aim, namely to explain the features of the scientific process.

In fact, what we can only claim is that there seems to be evidence for some kind of continuity and directedness of the scientific process and that these features themselves have at least in part something to do with reality. This too can be contradicted, but one may also feel that there are some good arguments for this position. One class of arguments concern the evolution of theories, with the trends, inclusions etc. which can be found here, and the way our conceptual tools change and develop in this evolution – all indicating such directedness and not supporting sheer disconnectedness. The other class of arguments consider the alternatives, which, if followed to conclusion, all seem to lead to diverging plurality – at the best in the sense of 'one law for one phenomenon', at the worst in the postmodernist 'social determination' view.

In this connection it might be mentioned that some people would interpret the 'holographic principle' as an adequate picture of the knowledge interface between us and the world (recalling the platonic metaphor of the shadows on a screen). This may be an example of how physics tries to impinge on philosophy. The conclusion of this discussion is then that philosophical considerations are helpful from the point of view of understanding the world, but we should not feel compelled to hastily draw philosophical conclusions from physical conjectures, and this primarily for the sake of philosophy, not of physics.

Besides the above two questions – that of the forming of our concepts and of the character of our knowledge – there are some more pragmatic ones concerning the structure of our theories and which are especially relevant in the context of the contemporary high energy physics research. So, for instance, when do we speak of a theory, when do we consider to have it 'under control'? We may find different answers to this question in this book, and in fact we may ask which understanding of it is assumed by a theory we are developing. A more special question may be whether we must expect any relevant quantum theory to have a classical limit and whether we are able to find quantum theories not by quantization of a classical precursor. Still another question is: What impact on our understanding does the development of new methodologies have – e.g. numerical simulations in quantum field theory?

We included this discussion here to suggest to the reader that these may also be interesting questions to consider when reading this book.

Part II

Elementary Particle Theory

The Standard Model of Particle Physics

H. G. Dosch

Institut für Theoretische Physik, Universität Heidelberg,
Philosophenweg 16, 69120 Heidelberg, Germany
H.G.Dosch@thphys.uni-heidelberg.de

1 Introduction

Phenomenology of elementary particles is, since 1980, in an excellent state. Presumably there are more physicists who complain that the standard model is too good than those who complain that it is too bad. The reason for that paradoxical behaviour is that the excellent agreement between theory and experiment leaves little space for evidence of 'new physics'.

Before I give an outline of the standard model of particle physics, I shall shortly describe the development that lead to that model. This will be done in an woodcut-like and therefore oversimplifying manner. The reason for this historical introduction is threefold:

1. The historical development shows to what extend the present model meets the expectations of a theory of elementary particles.
2. I do not believe that we can learn from history, from history of science no more than from political history, but nevertheless history is the only arsenal we have of realized possibilities in science.
3. I think it is adequate to emphasize in this book, which is mainly focused on theoretical issues, the decisive role that experiment and especially the interaction between experiment and theory has played in the development of present-day particle physics.

2 The Development of the Standard Model

At the turn of the 19th to the 20th century two developments of physics were evident. Firstly, the field theory of electric phenomena, as conceived by Faraday and put in its final mathematical form by Maxwell, could not be considered as a branch of mechanics in the sense Euler had developed

H. G. Dosch: *The Standard Model of Particle Physics*, Lect. Notes Phys. **721**, 21–50 (2007)
DOI 10.1007/978-3-540-71117-9_2

mechanics of continua. Einstein even reversed the order: he took the symmetries of the Maxwell equations more serious than those of classical mechanics and he thereby modified the latter to relativistic mechanics. Secondly, around the same time there was evidence from statistical mechanics and atomic spectra that classical mechanics had to be modified essentially at the scale of atomic extensions, that is around a tenth of a nanometer. This first led to the 'old quantum mechanics' initiated by Planck in 1900 and essentially extended by Einstein and Bohr. The 'new quantum mechanics' was originated by Heisenberg in his paper on ON QUANTUM-THEORETICAL REINTERPRETATION OF KINEMATICAL AND MECHANICAL RELATIONS.[1] Not even two months after this paper was submitted, Born and Jordan formulated Heisenberg's ideas in a systematic way and at the end of their paper they made 'the attempt, to fit the laws of the electromagnetic field into the new theory'. They introduced matrices, that is non-commuting operators, not only for the mechanical observables, but also for the electric and magnetic field. The next essential step towards a realistic quantum electrodynamics was due to Dirac (1927). He could already rely on the interpretation given in a sequel to the paper of Born and Jordan, the famous 'Dreimännerarbeit' (three-men paper) of 1925, where also Heisenberg participated. Dirac used his approach based on analogies of quantum theory with higher mechanics and introduced annihilation and creation operators for photons. Since he had the full dynamics incorporated in his approach, he could give a dynamical derivation of the famous relation between the spontaneous and the induced emission coefficient, established by Einstein in 1916/17. Dirac was emphasizing the particle character of the electromagnetic radiation (photons), but in the same year Jordan and Klein, following in some respect Dirac's ideas, stressed the opposite, namely the field character of matter. Jordan also realized that for fermion fields the commutation relations had to be substituted by anti-commutation relations.

Two papers authored by Heisenberg and Pauli and published in 1929 can be regarded as the first papers having the essential ingredients of relativistic quantum field theory. They treated both the matter fields and of course the radiation field relativistically. For the matter field they used the relativistic wave equation found by Dirac, which shall be mentioned later several times. They used the canonical formalism of classical field theory for the quantization procedure, in analogy to the application of the canonical formalism of mechanics in establishing quantum mechanics. On their way they met a tremendous obstacle: as a consequence of the Maxwell equations the conjugate field of the electric potential is zero. This and other difficulties made the two silent for nearly a year, a very long period in a time where seminal papers were often separated by only a few weeks. The real breakthrough came when they realized the importance of gauge invariance in quantum theory, a feature first clearly recognized by H. Weyl and already stressed in his famous book GROUP THEORY AND QUANTUM MECHANICS, the first edition of

[1] For references, see remarks in the literature section.

which appeared in 1928. So the main ingredients of quantum field theory were found in a period of only four years and a few months after the first appearance of new quantum mechanics. But further progress was by no means easy. Heisenberg later remembers that in contrast to quantum mechanics quantum electrodynamics became never simple. The mood of the early 1930s is caught in his reminiscence:

> In 23 and 24 we knew that there were difficulties and we also had the feeling that we were quite close to the final solution of the difficulties. ...It was as if we were just before entering the harbor, while in this later period we were just going out into see again, i.e. all kinds of difficulties coming up.

I will not dwell on these difficulties mentioned, some of the most obstinate ones are discussed in the contribution by Fredenhagen et al. to this book. The outcome of the adventure on open sea was renormalized relativistic quantum field theory, which governed large parts of physics for the rest of the 20th century and is still going strong in the 21st.

Quantized field theory led in the sequel to a dichotomy with epistemological consequences. In the theoretical description the *field* concept is the fundamental one, but on the other hand all our knowledge comes from accelerated and detected *particles.* Only in perturbation theory is there a clear-cut relation between particles and fields: the field quanta are the (observed) particles.

Physics did not stop on the level of atoms. After the essential questions of atomic spectra had been clarified, nuclear physics entered the scene. The classical scattering experiments of Rutherford, Geiger and Marsden showed that the atoms had a nucleus which was extremely small as compared to the extension of the atom. The appropriate scale for the atom is the nanometer (10^{-9} m), that of the nucleus the femtometer (10^{-15} m). Elementary particles at the time were the electron and the proton, the nucleus of the Hydrogen atom. There were good reasons to believe that the nuclei of the other atoms were composite objects, their constituents being presumably protons and electrons. There was strong evidence for such a hypothesis: The mass of a nucleus was roughly an integer multiple of the mass of a proton and the charge was also a multiple of the charge of a proton, therefore the difference between the mass and charge number had to be explained by an extremely light negatively charged particle, just the typical properties of an electron. Furthermore the emission of electrons from a nucleus could be observed in the nuclear β-decay.

In β-decay there was, however, a serious problem. Chadwick and Ellis (1914–1927) had found that the electron spectrum in that decay was not discrete, as in the case of α-decay, but continuous. Furthermore there seemed for certain decays to be a problem with the relation between spin and statistics, if only one fermion was emitted. After Lise Meitner had, by her own experiment, convinced herself and Pauli of the correctness of the results of Chadwick and Ellis, Pauli found 'a desperate way out' from both problems: in β-decay not only an electron, but also another very light neutral fermion, later called neutrino, is emitted.[2]

Besides these problems there were also some other serious difficulties to reconcile the otherwise good phenomenological evidence of the picture of the nucleus with theoretical principles. So it was not far fetched to assume 'new physics' to set in at the scale of the atomic nucleus, that is at several femtometer. Increasing the resolution by a factor of a million, that is from millimeter to nanometer, had led from classical to quantum physics. Why should a further factor of a million, that is going from nanometer to femtometer, not also necessitate far-reaching modifications? The scale where new physics should set in was generally considered to be the classical electron radius $r_e = \alpha\hbar/(m_e c) \approx 2.8$ fm.

The doubts that quantum physics could not be applied to scales much smaller than the atomic ones even influenced the interpretation of experiments. It was not clear that one could trust results obtained by quantum electrodynamics, for instance for the energy loss of charged particles in matter, when the wavelength of the involved photons is of the order of a femtometer. It turned out soon, however, that such a transition to 'new physics' was not necessary and that quantum physics, as derived from atomic physics, also applied to nuclear physics. Several experimental and theoretical findings contributed to this insight. The α-decay of the nucleus was explained by Gamow (1928) as a quantum mechanical tunnel effect. Part of the theoretical problems of electrons inside a nucleus were solved through the discovery of the neutron by Chadwick (1932). It was immediately proposed (Heisenberg 1932) that the nucleus consisted of protons and neutrons rather than of protons and electrons.

In the same year local quantum field theory had its first spectacular triumph: the antiparticle of the electron, predicted by Dirac in 1928, was discovered in a cosmic ray experiment by Anderson. Though it was already predicted on the basis of local interaction in relativistic quantum mechanics, it is essentially a consequence of quantum field theory and can only be properly accounted for in a quantum field theoretical framework.

The neutrino hypothesis of Pauli was incorporated by Fermi in his quantum field theoretical description of β-decay (1933). In this theory the occurrence of creation and annihilation operators for fermions was essential.

Though the interaction strength was very weak, the theory had problems if one applied to it the procedures of perturbation theory used in quantum mechanics.[3] But on the other hand the lowest order (tree level) contributions

---

[2] This was communicated in an open letter to the 'radioactive ladies (L. Meitner was present) and gentlemen' at a meeting in December 1930.

[3] These corrections were first derived in a truly mechanistic field theory, namely the theory of sound by Rayleigh (1877), in quantum mechanics they were derived by Max Born.

of Fermi's theory were the basis for a very successful quantitative explanation of many observed decay spectra.

The success of quantum field theory in the description of β-decay, that is weak interactions, motivated Yukawa to develop a quantum field theory of nuclear forces (1935). In some sense it was closer to electrodynamics than to the Fermi's theory and it predicted as quantum of interaction the existence of a new kind of elementary particle, namely a massive particle with integer spin, which was first called mesotron, later π-meson. The mass (Compton wavelength) should be corresponding to the size of nuclei, that is several hundred electron masses. A particle of such a mass was indeed discovered by Neddermayer and Anderson (1937), it turned out later, however, that it could not be the particle wanted for the Yukawa theory.

The discovery of the neutron had another very important impact on theory: it initiated the concept of internal symmetries. Since the mass of the neutron differs from that of the proton by only about 1 permill, Heisenberg proposed immediately a symmetry between the two particles, later called nucleons. On the basis of results of nuclear spectroscopy and first precise measurements of cross sections of proton–proton scattering, this theory was finally developed into the theory of isospin symmetry (Condon, Kemmer, Wigner and others).

The particle predicted by Yukawa, later called π-meson, was discovered in 1947 by Powell and collaborators, shortly after it had been shown that the mesotron, the particle found by Neddermayer and Anderson ten years earlier, did not have the properties to mediate strong interactions. The situation of particle physics seemed in the middle of the 20th century to be in a similarly good state as at the end of that century, though the standard model of that time was completely different from the present one. The elementary particles were the proton, the neutron, the electron, the neutrino(s) and, as particles mediating the electromagnetic and strong interaction, the photon and the π-meson, respectively. To that came a a particle, which 'nobody had ordered', the muon, the former mesotron.

Quantum field theory turned out to be extremely successful. The problems occurring by just transposing the concepts of quantum mechanics to quantum field theory were solved by Dyson, Feynman, Gell-Mann, Schwinger and Tomunaga in renormalized perturbation theory of quantum electrodynamics (see 'Quantum Field Theory: Where We Are', by K. Fredenhagen et al.) and results were brilliantly confirmed by experiment (as they still are with increasing precision). Quantum field theory was also the basis for a treatment of weak and strong interactions, though there were some flaws: In weak interactions the qualitative results were impressive, but the renormalization programme, which was so successful in quantum electrodynamics, was not applicable without increasing the numbers of parameters indefinitely. In strong interactions, the problems were just the opposite. The field theory with pseudoscalar mesons was renormalizable, but the quantitative results of renormalized perturbation theory were by no means satisfactory. This was not unexpected, however, since the interaction constant between the nucleons and the π-mesons turned out to be several orders of magnitude larger than the electromagnetic coupling.

In contrast to today there were, however, strong signs that particle physics was more complex than the picture outlined above. This was inferred essentially from results of nuclear physics and confirmed by events produced by cosmic rays.

In 1947, Rochester and Butler discovered in a cloud chamber experiment traces with the topology of a V and which were called V-particles, today they are called *strange* particles. They were unstable but lived long enough to form traces in cloud chambers; their mass was definitely higher than that of a π-meson. Their unwanted presence could not be ignored by theoreticians for too long a time, especially since they were soon produced in large number in accelerator experiments. The development of accelerator and beam construction and of more and more refined detectors (e.g. bubble chambers) led soon to a true profusion of elementary particles which started a crisis for the whole field and initiated a search for new concepts. Since meson field theory did not lead to more than just qualitative results, G. Chew made the famous statement (1961):

> I do not wish to assert (as does Landau) that conventional field theory is necessarily wrong, but only that it is sterile with respect to strong interactions and that, like an old soldier, it is destined not to die but just to fade away.

In weak interactions there was, from a strictly phenomenological point of view, no need to look for new concepts. Experimentalists were looking for the field quantum of weak interactions, the so-called 'intermediate boson', but even if the search had been successful, the presence of an intermediate boson alone would not have solved the theoretical problem of non-renormalizability of the Fermi theory. Furthermore it was not clear if non-renormalizability was only a problem of weak interactions, since it was not known how strong interactions can influence weak interactions at small distances.

In strong interactions several lines of research, partially in parallel, partially in contradiction to each other were followed. All of them were motivated and inspired by quantum field theory, but none of them was willing to accept its full programme, namely to calculate observable quantities directly from a Lagrangian. They all tried to handle the problem of the ever increasing number of *elementary* particles:

1. In the theory of the analytic S-matrix one tried to eliminate the field concept from strong interactions and concentrate on properties of scattering matrix elements derived solely from conservation of probability. Though part of this programme had a strong effect against field theory (see the quotation of Chew above), many of the postulated analytic properties of the S-matrix were results obtained in the framework of local quantum field theory. The approach culminated in the concept of 'nuclear democracy', in which all observed strongly interacting particles and resonances

were treated on the same footing and were related through self-consistency conditions. This was the so-called *bootstrap programme*. The application of Regge's theory of potential scattering to high energy scattering and the use of dispersion relations in particle physics were an outcome of this programme. Another important consequence of the theory was a model developed by Veneziano: It showed duality, that is it related the high energy behaviour of the scattering-matrix elements to the resonance structure (poles) of the matrix elements. It eventually gave rise to string theory.

2. There was a strong emphasis on internal symmetries, motivated by the success of the isospin symmetry $SU(2)$ in the analysis of π-meson-nucleon scattering.
3. The discovery of 'several new particles' led already Fermi and Yang (1949) to speculate that not all of them were elementary. They therefore proposed, rather as an illustration of a possible programme than as a realistic model, to consider the π-meson as a bound state of a nucleon and an anti-nucleon. Though Fermi was coauthor of this paper, the idea was not enthusiastically embraced by the majority of the community. But the phenomenological evidence for the composite nature of strongly interacting particles grew with time. In Fig. 1 the hydrogen spectrum is compared with the spectrum of the nucleons, that is the particles and resonances with baryon number 1 and isospin 1/2. The search for a constituent picture of the strongly interacting particles led eventually to the phenomenologically very successful quark model of Gell-Mann and Zweig.

From the concepts mentioned above, only the bootstrap philosophy has disappeared. Regge theory is a prerequisite for the description of hadronic high energy scattering processes and it gave birth, through the Veneziano model, to string theory. Dispersion relations are not in the focus of present-day theoretical interest, but they are still an important tool in the analysis of strong interactions. The second and third point are cornerstones of the

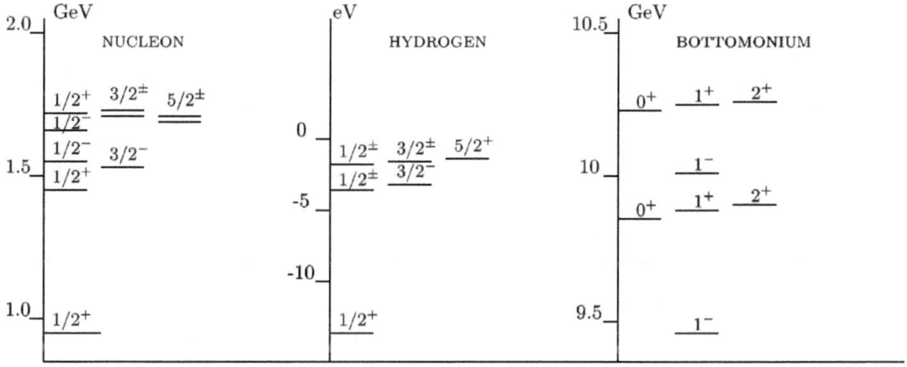

Fig. 1. Lowest lying states of the spectra of the nucleon, hydrogen and the heavy meson state bottomonium

present standard model, but it was a long and tedious way to incorporate these concepts into the frame of relativistic quantum field theory.

Before I come to a description of this standard model, I shall just quote a few important steps which lead to his final establishment.

An important step, though not recognized immediately as such, was the construction of a classical gauge field theory in which the gauge field transform under a non-Abelian symmetry group (Yang and Mills theory, 1954) (see 'Quantum Field Theory: Where We Are' by K. Fredenhagen et al.) It took, however, some time before this theory was formulated as a quantum field theory in the sense of a formal power series or, non-perturbatively, on a discrete set of space and time points (lattice).

In 1967 a theory of weak and electromagnetic interactions was proposed based on a classical Lagrangian, gauge invariant under $SU(2) \times U(1)$, with a mechanism for mass generation of the interaction quanta (massive gauge bosons). It led to the prediction of neutral weak currents, that is to reactions like $\bar{\nu}_\mu + e \to \bar{\nu}_\mu + e$. It also led, together with the experimentally confirmed absence of strangeness changing neutral currents, to the prediction of a new quantum number, besides isospin and strangeness, later called *charm* (GIM mechanism, after its inventors Glashow, Iliopoulos and Mainai). In 1971 the proof of renormalizability of the interaction based on the classical $SU(2)\times U(1)$ (electroweak) Lagrangian was finished ('t Hooft and Veltman).

Though the theory was now in a good shape, it was evidently not taken too seriously in the community. The search for neutral currents was only on position 8 in a priority list of 10 points of the relevant Gargamelle experiment. However, the experimentalists, who in an heroic effort found in 1972 three events of $\bar{\nu}_\mu + e$ scattering in 1.4 million pictures, write that they were motivated by the proof of renormalizability of the electroweak Lagrangian. This discovery of neutral currents opened the way for the general acceptance of the electroweak $SU(2) \times U(1)$ model. The Nobel prize was awarded to Glashow, Salam and Weinberg for their contribution to the theory of unification of weak and electromagnetic interactions in 1979, before the quanta of the weak interaction, the massive gauge bosons, were found experimentally in 1983 at CERN.

The development of the gauge theory of strong interactions was a bit slower and there the interplay between experiment and theory was even stronger.

In the deep inelastic scattering experiments at SLAC (1966 ff.) electrons with high energy were scattered off protons and especially reactions with high momentum transfer (more than 1 GeV^2) were analysed. It turned out that special features of these reactions were best described by a picture in which the proton consisted of a bunch of practically free constituents, the so-called 'partons'. A scheme of this picture is given in Fig. 2.

The detection of the heavy J/ψ-(1974) and Υ-(1976) meson and their resonances made it even more evident that a bound state picture of hadrons could explain many features (see Fig. 1, bottomonium).

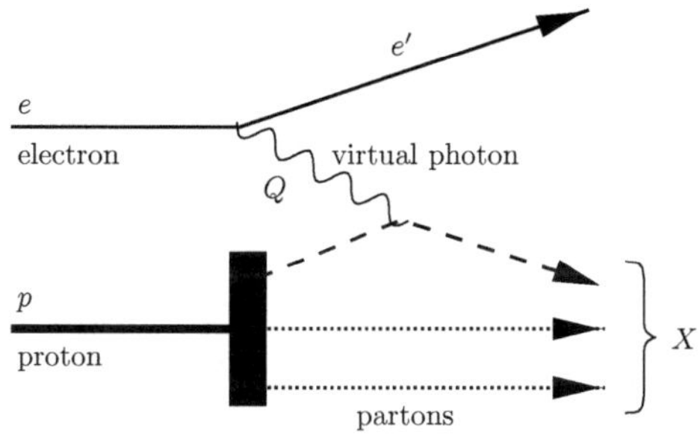

Fig. 2. Schematic description of the parton model in deep inelastic scattering. The virtual photon interacts with only one parton, the rest are not affected. This intuitive picture makes sense only in a reference frame where the momentum of the proton tends to infinity

In 1973, Fritzsch, Gell-Mann and Leutwyler proposed quantum chromodynamics (QCD) as the dynamical theory of strong interactions. It was a gauge theory based on unbroken $SU(3)$ (colour) symmetry. Its phenomenological basis was the success of two different approaches, namely current algebra and the quark model. It could explain extremely well the deep inelastic scattering experiments and, with some extra ingredients, the spectra of the J/ψ- and Υ- states.

Though an essential ingredient of the electroweak Lagrangian, the so-called 'Higgs boson', has not yet been found, past experience lets us believe that it will be detected in the next decade or so at the Large Hadron Collider in CERN. The only very clear-cut evidences that the standard model has to be modified in its present form are the neutrino oscillations which in the most favourable case would lead to a rather straightforward extension with 9 new parameters. Apart from this major and some minor *black clouds* there is a very nice blue sky over the model. In the next section it will be described more systematically.

3 Systematic Description of the Standard Model

In this section' I concentrate pragmatically on the phenomenological aspects of a particular realization of local quantum field theory, namely the local gauge theory of the symmetry $SU(2) \times U(1)$. For the more theoretical aspects I refer to the contribution 'Quantum Field Theory: Where We Are' by K. Fredenhagen et al.

3.1 Local Gauge Invariance and Fermionic Matter Fields

(see 'Quantum Field Theory: Where We Are' by K. Fredenhagen et al.)
Be $\mathcal{G}$ an unitary semi-simple Lie group with hermitian generators τ_i, $i = 1 \dots L$, that is any element of $\mathcal{G}$ can be expressed as $\exp[i \sum_{i=l}^{L} c_i \tau_i]$. Be $\{\psi(x)\}$

a N-tuple of fields which transforms according to a certain x-dependent representation $\mathbf{U}(x)$ of $\mathcal{G}$

$$\psi(x) \to \psi'(x) = \mathbf{U}(x)\psi(x) \ , \tag{1}$$

in components:

$$\psi'(x) = U_{\alpha\beta}(x)\psi_\beta(x) = \sum_{\beta'=1}^{N} \left(\exp[\sum_{i=1}^{L} ic_i(x)\tau_i] \right) \psi'_\beta(x) \ . \tag{2}$$

Terms of the form $\psi^\dagger\psi$ and powers of them are invariant under this local gauge transformation. Kinetic terms or derivative couplings in a Lagrangian, however, will not be invariant due to the x dependence of the transformation. In order to achieve gauge invariance one has to replace the gradient ∂_μ by the covariant derivative $\mathbf{D}^\mu$. It has the form:

$$\mathbf{D}_\mu = \partial_\mu \mathbf{1} + ig\mathbf{A}_\mu \tag{3}$$

with

$$\mathbf{A}_\mu \equiv \sum_{i=1}^{L} A^i_\mu(x)\tau_i \tag{4}$$

The operator valued vector field $\mathbf{A}_\mu$ transforms according to

$$\mathbf{A}_\mu(x) \to \mathbf{U}(x)\mathbf{A}_\mu(x)\mathbf{U}(x)^\dagger - i\mathbf{U}\partial_\mu\mathbf{U} \ . \tag{5}$$

The introduction of the covariant derivative fixes the interaction with all matter fields completely. The gauge g is sometimes included in the definition of the gauge fields.

A gauge invariant kinetic term for the gauge fields is given by:

$$\mathrm{Tr}\ \mathbf{F}_{\mu\nu}\mathbf{F}^{\mu\nu} \tag{6}$$

with the field tensor

$$\mathbf{F}_{\mu\nu} \equiv \frac{-i}{g^2}\mathbf{D}_\mu\mathbf{D}_\nu - \mathbf{D}_\nu\mathbf{D}_\mu \ . \tag{7}$$

The field tensor transforms homogeneously under the local gauge transformation:

$$\mathbf{F}_{\mu\nu} \to \mathbf{U}(x)\mathbf{F}_{\mu\nu}\mathbf{U}(x)\dagger \ . \tag{8}$$

A mass term for the gauge fields would violate gauge invariance.

For Abelian gauge groups the kinetic term is quadratic in the gauge potentials $\mathbf{A}_\mu$, for non-Abelian groups it contains also cubic and quartic terms; this is the origin of the interactions among the gauge bosons themselves.

3.2 Left- and Right-Handed Spinor Fields

The lowest dimensional non-trivial representations of the Lorentz group are the two-dimensional unitary inequivalent spinor representations. One of them acts on the so-called 'right handed' Weyl spinors, the other on the 'left handed' spinors. The names come from the helicity of the spinors. The helicity operator is the scalar product of the spin and the momentum operator and left-handed spinors are eigenstates of the helicity operator with negative eigenvalue, right-handed ones have a positive eigenvalue. Space reflection (parity transformation) transforms a left-handed into a right-handed Weyl spinor and vice versa. A mass term in the Lagrangian couples the right-handed to the left-handed spinor fields, it is therefore not possible to construct massive particles which transform as single-handed Weyl spinors. For massive fermions one has to consider the direct sum of a right- and a left-handed Weyl spinor, leading to the four spinor introduced by Dirac. The projectors of a four-spinor on the left- and right- handed parts are constructed with the four-dimensional γ_5-matrix:

$$P_L = \frac{1}{2}(1-\gamma_5) \qquad P_R = \frac{1}{2}(1+\gamma_5)\ . \tag{9}$$

If a four-spinor field is invariant under charge conjugation, it is called a Majorana field. This can only happen if the particles have quantum numbers which do not change under charge conjugation, that is they must necessarily be neutral. Four-spinor fields which are not invariant under charge conjugation are called Dirac fields.

3.3 Quantum Chromodyanamics, the Strong Interaction Sector

The fundamental fields of the hadrons, the strongly interacting particles, are six triplets of Dirac fields, the quark fields:

$$\psi_c^f,\ f=1\ldots 6,\ c=1\ldots 3\ . \tag{10}$$

The six quantum numbers f are called flavour, the three quantum numbers c are called colour. The conventional names for the flavours are *down*, *up*, *strange*, *charm*, *bottom*, and *top*. To these fields there corresponds no asymptotic particles.

The local gauge transformation inducing strong interactions is a three-dimensional unitary group acting on the colour triplets; it is called colour group, $SU(3)_c$. It has eight generators and correspondingly eight vector gauge fields, the gluon fields. The symmetry is unbroken, hence no mass term for the gluons occurs. With the notation introduced above the Lagrangian has the remarkably simple form:

$$\mathcal{L} = \mathrm{Tr}\ \mathbf{F}_{\mu\nu}\mathbf{F}^{\mu\nu} + i\sum_f \left(\bar{\psi}^f\gamma^\mu\mathbf{D}_\mu\psi^f - m_f\bar{\psi}^f\psi^f\right)\ . \tag{11}$$

The resulting quantum theory is called quantum chromodynamics (QCD), it is renormalizable. The problem of unphysical states, already apparent in quantum electrodynamics, becomes virulent in QCD and necessitates the explicit appearance of ghost fields (Fadeev–Popov ghosts) in covariant perturbation theory.

Already the pure gauge theory, that is the part of the Lagrangian not involving the quark fields, is an interacting theory. It contains no scale, since a mass term for the gauge fields is forbidden by gauge invariance. Nevertheless a scale is introduced by the necessity to regularize the theory. Since observable results should not depend on the choice of that renormalization scale μ, a scale dependence of the gauge coupling $\alpha_s \equiv g^2/(4\pi)$ is induced. By choosing a mass-independent renormalization scheme like $\overline{MS}$ this behaviour is not essentially influenced by the quark masses. The dependence of α_s on the scale is given by the β-function:

$$\mu \frac{\partial \alpha_s}{\partial \mu} = \beta'(\alpha_s) \tag{12}$$

The expansion of the β-function is known up to the three-loop level, the lowest (scheme independent) contribution is

$$\beta(\alpha_s) = \frac{-1}{2\pi}(11 - \tfrac{2}{3} n_f)\alpha_s^2 + O(\alpha_s^3) \,. \tag{13}$$

Here n_f is the number of active flavours, it is maximally 6 and therefore the β-function is negative and the theory is asymptotically free: if the renormalization scale is shifted to higher higher masses, the gauge coupling decreases with an inverse power of a logarithm. The perturbative theory seems to be safe for high mass scales, that is at short distances. There are, however, strong indications that this is not the case for low scales. This makes itself remarked already in renormalization group improved perturbation theory through the so-called 'infrared renormalons'. These difficulties are presumably closely related to a property called confinement: physical states are supposed to occur only as colour singlets, therefore there are no asymptotic fields corresponding to the quarks and the gluons. This confinement behaviour should be a dynamical consequence of the Lagrangian, but up to now a one-million-dollar award for a proof of confinement is still waiting for a winner.

Presently the problem can be tackled only in models or, numerically, in the lattice regularized version of QCD (see 'Quantum Field Theory: Where We Are' by K. Fredenhagen et al.). We shall come back to this question in the discussion on the merits and deficiencies of the standard model. One consequence of confinement is the lack of a natural scheme to define the quark masses. For the light quarks (d, u, s) the mass is normally quoted as the mass in the already mentioned $\overline{MS}$-scheme taken at a scale of 1 GeV (Older entries) or 2 GeV. For the heavy quarks (c, b, t) two schemes are usual: either the so-called 'pole mass' which is convenient for non-relativistic calculations or the $\overline{MS}$ scheme, with the mass value itself as scale. Though the pole mass is quite intuitive, it is plagued by the so-called 'renormalon ambiguities' and therefore the less intuitive $\overline{MS}$-mass is more appropriate for theoretical calculations.

Most probably related to confinement is the spontaneous breaking of a symmetry which is nearly present in the QCD-Lagrangian (11). The masses of the very light quarks u and d are indeed very small (see Table 1) and therefore it is a good approximation to put them to zero. In that case the Lagrangian is invariant under independent global two-dimensional unitary symmetry transformations for the left- and right-handed doublets of the u- and d-quark fields; this is the so-called 'chiral $SU(2) \times SU(2)$ symmetry'. Such a symmetry is not observed in the hadron spectrum, not even approximately;

Table 1. The elementary fields of the standard model. The charges are those of the particles, antiparticles have opposite charge. The masses of the quarks are Lagrangian masses in the $\overline{MS}$-scheme. The renormalization point for the light quarks (u, d, s) is 2 GeV, those of the heavy quarks the mass itself

Field	Spin	Charge	Mass	Width
			Leptons	
electron	1/2	-1	0.5109989 MeV	stable
muon	1/2	-1	105.63806 MeV	$3 \cdot 10^{-16}$ MeV
τ-lepton	1/2	-1	1776.99 ± 0.29 MeV	$2.3 \cdot 10^{-9}$MeV
e-neutrino	1/2	0	< 3 eV	
μ-neutrino	1/2	0	< 190 eV	
τ-neutrino	1/2	0	< 18.2 MeV	
			Gauge bosons of electroweak interactions	
photon	1	0	0	stable
$W^{\pm}$-boson	1	± 1	80.425 ± 0.039 GeV	2.118 ± 0.042 GeV
Z-boson	1	0	91.1876 ± 0.0021 GeV	2.4952 ± 0.0023 GeV
			Higgs boson	
H^0	0	0	> 114 GeV	
			Quarks	
d-quark	1/2	$-1/3$	5 to 8.5 MeV	
u-quark	1/2	2/3	1.5 to 4.5 MeV	stable
s-quark	1/2	$-1/3$	80 to 155 MeV	
c-quark	1/2	2/3	1.0 to 1.4 GeV	
b-quark	1/2	$-1/3$	4 to 4.5 GeV	
t-quark	1/2	2/3	174.4 ± 5 GeV	
			Gauge boson of strong interactions	
gluon	1	0	0	

therefore one concludes that it is broken sponataneously, that is by some vacuum expectation value. A consequence of this breaking of a global symmetry is the occurrence of massless bosons, the so-called 'Goldstone bosons'. Since the symmetry is not only broken spontaneously but also directly in the Lagrangian through the small but finite masses of the light quarks, the (pseudo-)Goldstone bosons are not massless but proportional to the quark masses. The perturbation theory of (pseudo-)Goldstone bosons based on a Lagrangian invariant under spontaneously broken chiral $SU(2) \times SU(2)$ is highly developed and has led to results in good agreement with experiment. This 'chiral perturbation theory' is, however, not renormalizable and therefore further and further refinement leads to more and more new parameters.

3.4 The Electroweak Sector

We first discuss the purely leptonic sector of the standard model. Since the impact of the recent results on neutrino oscillations on the standard model is still ambiguous, I start, against better knowledge, with the original form, that is with massless neutrinos. The matter content of the leptonic sector of the standard model consists of the three charged Dirac fields, namely those of the electron, the muon and the τ-lepton, and of the three chargeless fields of the corresponding neutrinos, which might be Dirac or Majorana fields. The gauge group of electroweak interactions is $SU(2) \times U(1)$. To the factor $SU(2)$ a weak isospin T and to the factor $U(1)$ a hypercharge Y is assigned. The electric charge Q is related to the third component of the weak isospin and to the hypercharge by

$$Q = T_3 + \frac{1}{2} Y \tag{14}$$

For the gauge transformations the lepton fields are divided into the right- and left-handed ones. The right-handed ones transform as singlets under the $SU(2)$ part of the electroweak group and have the hypercharge $Y = -2$. The left-handed charged leptons and the neutrinos form doublets under the transformation, they have the hypercharge $Y = -1$. We have thus three doublets of left-handed leptonic fermion fields with $Y = -1$:

$$\begin{pmatrix} \psi_{\nu_e} \\ \psi_e \end{pmatrix}_L , \quad \begin{pmatrix} \psi_{\nu_\mu} \\ \psi_\mu \end{pmatrix}_L , \quad \begin{pmatrix} \psi_{\nu_\tau} \\ \psi_\tau \end{pmatrix}_L \tag{15}$$

and three singlets of right-handed leptonic fermion fields with $Y = -2$:

$$(\psi_e)_R \, , \quad (\psi_\mu)_R \, , \quad (\psi_\tau)_R \tag{16}$$

The $SU(2) \times U(1)$ gauge group leads to four gauge bosons which are vector fields:

1. an iso-triplet, $A^i_\mu, i = 1, 2, 3$ for SU(2) and
2. a singlet B_μ for $U(1)$.

The covariant derivatives acting on the weak iso-doublets fields are

$$\mathbf{D}_\mu^L = \partial_\mu \mathbf{1} - \frac{1}{2} i g' Y \, B_\mu - \frac{1}{2} i g \, \mathbf{A}_\mu \tag{17}$$

Those acting on the iso-singlets are

$$D_\mu^R = \partial_\mu - \frac{1}{2} i g Y' \, B_\mu \, . \tag{18}$$

It was known for a long time that the weak interactions have a very short range. Therefore the gauge bosons of weak interactions must be massive. A direct mass term in the Lagrangian would violate gauge invariance and lead to a non-renormalizable theory, therefore the mass generation has to have more subtle reasons, see 'Quantum Field Theory: Where We Are' by K. Fredenhagen et al.. It can, indeed be achieved by an additional scalar field, the Higgs field, $\phi = (\phi^+, \phi^0)$; which couples as a doublet to the $SU(2)$ part of the electroweak gauge group, it has hypercharge $Y = 1$.

The pure Higgs part of the Lagrangian can have the most general potential compatible with renormalizability and stability:

$$V(\phi) = \mu^2 \phi^\dagger \phi + \lambda (\phi^\dagger \phi)^2 \tag{19}$$

with $\lambda > 0$.

For the case, that the quadratic term is negative, that is $\mu^2 < 0$, the minimal energy density occurs for a non-trivial field configuration at the expectation value of $\langle \phi^\dagger \phi \rangle$ given by

$$\langle \phi^\dagger \phi \rangle = -\frac{\mu^2}{2\lambda} \equiv \frac{1}{2} v^2 \tag{20}$$

This happens already on the classical level. If in the quantization procedure modulus *and* phase of the vacuum expectation value are fixed, the $SU(2)$ part of the symmetry is broken.

The Higgs doublet ϕ can be parameterized by the new fields $\xi_j, \quad j = 1, 2, 3$ and H:

$$\phi(x) = \exp \left[\frac{i}{2v} \sum_{j=1}^{3} \xi_j \tau_j \right] \begin{pmatrix} 0 \\ \frac{v + H(x)}{\sqrt{2}} \end{pmatrix} , \tag{21}$$

where τ^j are the generators of $SU(2)$.

The fields $\xi_j(x)$ can be 'gauged away' so that in the so-called 'unitary gauge' one has

$$\phi(x) = \begin{pmatrix} 0 \\ \frac{v + H(x)}{\sqrt{2}} \end{pmatrix} , \tag{22}$$

$H(x)$ is the field of the observable Higgs boson. The covariant derivative in the kinetic part of the Lagrangian [see (6)] containing this form of the Higgs doublet leads not only to gauge couplings of the Higgs to the gauge bosons, but also to terms quadratic in the gauge fields. They are multiplied by terms proportional to the expectation value of the Higgs field, v, and lead to masses of the gauge fields.

If the underlying symmetry is not gauged but global, a similar procedure leads to a so-called spontaneous symmetry breaking (see end of Sect. 3.3). In analogy, the Higgs mechanism described above is often also called 'spontaneous symmetry breaking', though no Goldstone bosons, characteristic for the degenerate symmetry-breaking vacuum states, are present.

There exist three massive vector gauge fields, two are charged, the W-bosons ($W^{\pm}$), one is uncharged, the Z-boson (Z^0). They are related to the original gauge fields by

$$\begin{aligned} W^{\pm}_{\mu} &= \frac{1}{\sqrt{2}}(A^{(1)}_{\mu} \pm iA^{(2)}_{\mu}) \\ Z_{\mu} &= \frac{1}{\sqrt{g^2 + (g')^2}}(-gA^{(3)}_{\mu} + g'B_{\mu}) \,. \end{aligned} \tag{23}$$

The masses of these gauge bosons are

$$M_W = \frac{1}{2}vg \,, \qquad M_Z = \frac{1}{2}v\sqrt{g^2 + (g')^2} \,. \tag{24}$$

The photon field is given by

$$A_{\mu} = \frac{1}{\sqrt{g^2 + (g')^2}}(g'A^{(3)}_{\mu} + gB_{\mu}) \,, \tag{25}$$

it is massless and couples only to charged particles.

It is convenient to introduce the weak mixing angle θ_W by

$$\tan\theta_W = \frac{g'}{g} \,. \tag{26}$$

Then one obtains the following relations for the electric elementary charge e and the weak Fermi coupling constant G_F:

$$e = g\sin\theta_W \,, \qquad G_F = \frac{g^2}{8M_W^2} = \frac{1}{2v^2} \,. \tag{27}$$

The electroweak mixing angle had been determined from neutrino scattering experiments, therefore one could already from these tree-level considerations predict the masses of the gauge bosons with enough precision to make a dedicated experiment at CERN; this led to their experimental detection.

Also the Fermion masses can be generated by the vacuum expectation value of the Higgs doublet by couplings of the Dirac fields to the Higgs doublet

(Yukawa coupling). The coupling is then proportional to the mass of the fermion divided by v.

In order to incorporate hadrons into the electroweak sector, one assigns to the right- and left-handed quarks different hypercharges Y. To all the left-handed quarks one assigns the hypercharge $Y = \frac{1}{3}$, to the right-handed *up, strange* and *top* quark $Y = \frac{4}{3}$, and to the *down, strange* and *bottom* quark $Y = -2/3$.

The eigenstates of the quark mass matrix in the strong interacting sector are not eigenstates of the weak interaction and it is usual to express the lower components of the doublets (d', s', b') as linear combinations of the mass eigenstates (d, s, b). The matrix relating them is the Cabbibo–Kobayashi–Maskawa (CKM) mass matrix:

$$\begin{pmatrix} d' \\ s' \\ b' \end{pmatrix} = \begin{pmatrix} V_{ud} & V_{us} & V_{ub} \\ V_{cd} & V_{cs} & V_{cb} \\ V_{td} & V_{ts} & V_{tb} \end{pmatrix} \cdot \begin{pmatrix} d \\ s \\ b \end{pmatrix} \tag{28}$$

In Table 1 all these fields are listed, and the CKM mass matrix is discussed in Sect. 4.2.

4 Achievements and Deficiencies of the Standard Model

4.1 Strong Interactions

If the renormalized gauge coupling constant of strong interactions, $\alpha_s(\mu)$, is small, perturbation theory can be used to calculate observable quantities. This happens if in a process at least one energy scale μ is large. A typical example of such a hard process is deep inelastic scattering. This is the reaction of highly energetic electrons and protons, where the large quantity is the momentum transfer of the electron to the hadron. It is schematically displayed in Fig. 2. It can be viewed as the scattering of a virtual photon, γ^*, on a proton, that is the reaction equation is

$$\gamma^* + p \rightarrow \text{anything} \tag{29}$$

The virtual photon carries the momentum transfer $(p_e' - p_e)$ from the electron to the proton. A process is called *deep inelastic* if the square of the four momentum transfer $Q^2 = -(p_e' - p_e)^2$ is at least 1 GeV/c and the invariant mass of the outcoming hadronic system, the 'anything' in (29), is at least 2 GeV/c^2. The 'wavelength' of the virtual photon is $\hbar/Q$, that is for high momentum transfer the wavelength is short and, as in an optical microscope, the resolution power is high.

The cross section for deep inelastic scattering divided by the momentum transfer Q^2 is the so-called 'structure function'. It depends in general on the momentum transfer Q and the invariant mass of the outgoing hadronic system, $W^2 = (p'_e - p_e + p)^2$, conveniently expressed through the dimensionless quantity $x = Q^2/(W^2 + m_p) \approx Q^2/W^2$. The structure function is

$$F(x, Q^2) \equiv \sigma_{\gamma^* p \to X}/Q^2 \tag{30}$$

In the naive parton model of deep inelastic scattering the hadron is viewed as consisting of non-interacting particles (partons); in that case the structure function depends only on the variable x. The function $F(x)$ can be viewed as the distribution of the longitudinal momentum fraction of the partons. This intuitive picture, however, applies only in the frame, in which the hadron has infinite longitudinal momentum.

In QCD the quarks and gluons can be viewed as partons. At short distances, the gauge coupling becomes weak (asymptotic freedom, see (12)) but does not vanish completely; this leads to a Q^2 dependence of the structure function $F(x, Q^2)$. If this function is known for a fixed value $Q^2 = Q_0^2$, perturbative QCD allows to calculate it for all values of Q^2, provided $\alpha_s(Q^2)$ is small enough to justify a perturbative expansion. This has been done in the enormous range:$10^{-5} < x < 1$ and $2 < Q^2/\mathrm{GeV}^2 < 10000$; the agreement between next-to-leading order calculations and experiment is very satisfactory. This can be seen from Fig. 3 where the values of the structure function measured by different experimental groups are shown. Also displayed are the

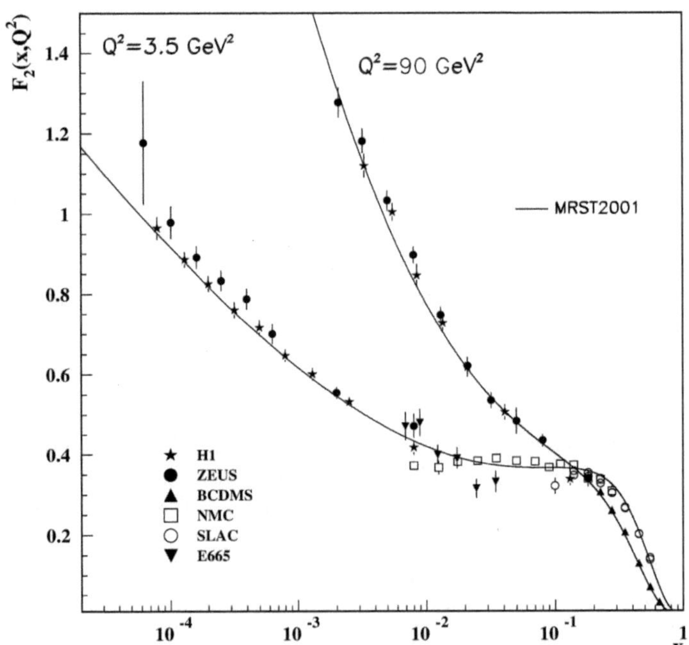

Fig. 3. Experimental points and theoretical curves for the structure function at two different values of the photon virtuality Q^2

curves obtained from perturbative QCD: one can be viewed as input, the other one as theoretical prediction.

For purists it should be noted that due to asymptotic freedom most probably the formal power series in α_s could be made convergent by resummation.

Another very successful application of perturbative QCD is jet physics. The annihilation of an electron and a positron into many highly energetic hadrons can be described perturbatively, that is on the quark-gluon level. In Fig. 4 the lowest order contribution to the annihilation of an electron and a positron into a quark, an antiquark and a gluon is depicted. The final particles in the perturbative amplitude, the quark, antiquark, and gluon cannot be observed as asymptotical particles, but under certain well-defined conditions some so-called 'infrared save' features of the reaction should survive in the real asymptotic, that is the hadronic final state. The final states of the perturbative calculation determine the axes of the hadronic jets.

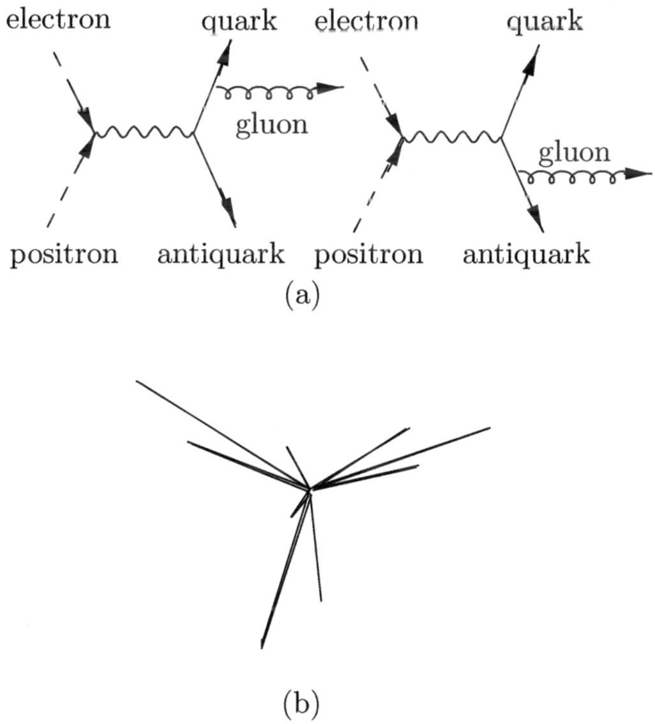

Fig. 4. Two- and three-jet events in electron–positron annihilation. (**a**) simplest graphs for three-jet events. Dashed lines represent electrons, wavy lines photons, solid lines quarks and coiled lines gluons. Quarks and gluons are treated here as if they were real particles. (**b**) Schematical picture of the observed three-jet events of DESY; the lines are traces of hadrons, the three-jet axes are clearly visible

The three-jet events played an important role for consolidation of QCD; very loosely speaking they can be considered as the existence proof of the gluon.

Another important domain of perturbative QCD is the spectroscopy and decay of hadrons composed of heavy quarks. Here the high quark masses set the hard scale which makes the relevant gauge coupling small. Especially in the effective theory of non-relativistic QCD (NRQCD) calculations can be performed in a very controlled way.

The running of the gauge coupling and the consistency of QCD in a large variety of applications can be inferred from Fig. 5. There the values of the strong gauge coupling α_s, obtained from different processes at different scales, are shown, left figure. In the right figure the values are rescaled to $\mu = M_Z$ using (12).

Impressive as the successes of perturbative QCD are, one has to note that the explained phenomena are rater hand-picked in order to be tractable. The most obvious phenomena of particle physics, as hadron spectra and hadron–hadron interactions at arbitrary scales, cannot be treated in it. Perturbative QCD allows for instance only to calculate the Q^2 dependence of the structure function of deep inelastic scattering, but the structure function itself at a given value of Q^2 has to be taken from experiment or from QCD-motivated models.

Unfortunately there is no analytic way to treat QCD non-perturbatively. In order to obtain quantitative non-perturbative results in an analytical calculation one has to rely on models, that is on more or less deformed or

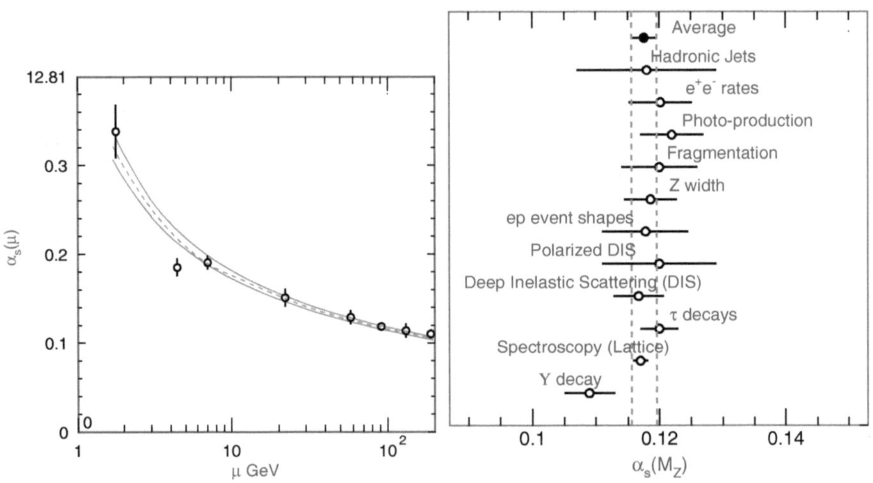

Fig. 5. The strong gauge coupling $\alpha_s(\mu)$ obtained from different reactions at different scales μ. In the right figure the values are rescaled with (12) to $\mu = M_Z = 91.2$ GeV. From Review of Particle Physics, W.-H. Yao et al., Particle Data Group, Journal of Physics G, Vol. 33, p 1 ff., Fig. 9.1 and 9.3

simplified versions of QCD, or one has to perform numerical calculations. A method to calculate certain matrix elements is based on a regularization of QCD on a space-time lattice (see 'Quantum Field Theory: Where We Are' by K. Fredenhagen et al.). It is possible to give such a regularization preserving gauge invariance, by assigning the gauge fields rather to links between lattice points than to lattice points itself.

The lattice-regularized version has two advantages: it is well defined for any finite lattice spacing and the functional integrals of continuum quantum field theory become high-dimensional 'ordinary' integrals, in general of Grassman or Haar-type. If one also transforms the field theory on the space-time continuum with Minkowski metric to a continuum with Euclidean metric, these high-dimensional integrals can be performed numerically using Monte-Carlo methods. The natural expansion parameter in lattice QCD is the inverse of the gauge coupling constant, therefore even analytic results can be obtained for finite lattice spacing. On the way to the continuum limit one has, however, to choose the lattice spacing smaller and smaller and therefore the mass scale becomes higher and higher and the inverse coupling larger and larger. Therefore there is no way to translate results of the strong coupling expansion reliably to continuum QCD. It is, however, comforting that numerical calculations show at small enough couplings a scaling behaviour which agrees with that obtained from perturbative QCD. This is by no means a proof, but a very good indication that in that case the numerical results are close to the ones of a continuum theory. It is also noteworthy that numerical results in general agree quite nicely with results obtained with analytic results of models which are less closely related to the QCD Lagrangian but respect Lorentz invariance.

The fast development of computer power and also the development of sophisticated analytic developments of lattice field theory has led to many interesting results. It should, however, be noted that the results of lattice calculations, which can be compared with experimental results, imply some more or less justified extrapolation and that there is no proof that the continuum limit exists at all.

The theoretical treatment of high energy reactions of hadrons is generally in a rather provisionary shape, depending largely on simplifying models. Some general features can qualitatively be understood, but there are practically no truly hard statements derived directly from the QCD Lagrangian.

The biggest success of non-perturbative QCD would of course be a non-perturbative continuum theory with asymptotic freedom and confinement.

4.2 Weak and Electromagnetic Interactions

The electroweak sector of the standard model is phenomenologically in a better shape, since there the gauge coupling constants are so small that all relevant questions can be answered in perturbation theory. However, since the strongly interacting fundamental fields are the quark fields and not the hadrons, and since experiments only yield information on hadronic amplitudes, the problems of strong interactions are also imported into the electroweak sector. The leptonic sector is free of these problems.

Electroweak theory made spectacular predictions, both qualitative and quantitative ones. In order to explain the observed absence of neutral flavour-changing currents due to the so-called 'GIM mechanism', the quarks have to occur in families. Therefore a (heavier) brother of the *strange* quark was predicted, it was followed, after discovery of *bottom* flavour, by the prediction of the *top* quark, forming with the *bottom* quark the third family. These additional quarks had another very satisfactory theoretical consequence. The purely leptonic sector of the standard model is plagued by the presence of axial anomalies which jeopardize renormalizability, since they violate the $SU(2)$ symmetry. The quantum corrections through internal quark loops with the above-mentioned three families cancel exactly the anomalies of the three leptons (e, μ, τ).

Another appealing feature of three quark families is the possibility to explain the observed CP-violation in the framework of the standard model. This would not be possible with less families.

Already on the tree level the model yielded rather precise values for the masses of the gauge bosons. The experimental input were the weak and electromagnetic gauge couplings, known for almost a century, and new data from neutral currents, determining the electroweak mixing angle (26) with sufficient precision. Perhaps even more spectacular was the prediction of the existence of the heavy *top* quark and determination of its mass (ca 170 GeV) from radiative corrections. This prediction can be compared with the theoretical prediction and the determination of the position of the planet Neptune by perturbations of the orbit of the Uranus. Effects of the *top* quark had been first observed at the Tevatron in 1995.

There exists a long and truly impressive list of very precise predictions, which have been measured with equally good precision at LEP. But most physicists see in the good agreement not so much a triumph of theory but rather resent tight constraints on new physics. In the standard model there is no place for new families with lightest members (presumably neutrinos) lighter than one half of the Z-mass. This follows from the excellent agreement between the theoretical and experimental values of the widths of the gauge standard model experiment bosons, displayed in Table 2 and Fig. 6.

Table 2. Theoretical prediction and experimental value of the Z-width

	Standard Model	Experiment
total Z^0-width/GeV	2.4961 ± 0.0012	2.4952 ± 0.0023
total $W^{\pm}$-width/GeV	2.0921 ± 0.0025	2.114 ± 0.0025

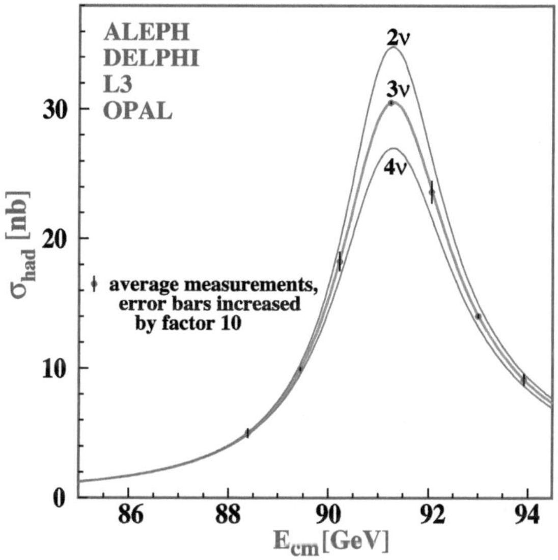

Fig. 6. The resonance curve of the Z-boson. In order to make the errors visible, they have been enlarged by a factor 10, they are in reality smaller than the thickness of the lines. From Review of Particle Physics, W.-H. Yao et al., Particle Data Group, Journal of Physics G, Vol. 33, p 1 ff., Fig. 40.8

Another excellent example for the little space left for new physics is the so-called 'ρ parameter' which is exactly one for the standard mass generating mechanism through one Higgs doublet but would deviate from one if other or more complicated scenarios would prevail. The experimental value (one σ) is

$$\rho_0 = 1.0012^{+0.0023}_{-0.0014} \tag{31}$$

There is, however, still one essential particle of the standard model not (yet) detected, the Higgs–boson. It is responsible for a key ingredient, namely mass generation of the gauge bosons of the electroweak theory. Unlike the *top* quark, which is a fermion, the influence of the Higgs mass, m_{H}, on radiative corrections is rather weak. Correspondingly the errors for the Higgs mass obtained from these corrections are large. On the 90% confidence level the newest limits (unpublished 2004) are

$$45 < m_H/\mathrm{GeV} < 183 \quad (260 \text{ for } 95\% \text{ CL}) \,. \tag{32}$$

The lower limit from direct experimental search (95% CL) is $m_{\mathrm{H}} > 114.4$ GeV.

A special feature of the quark content of the electroweak model is the mixing of the mass eigenstates in the weak families. It is expressed by the CKM-matrix, see (25). This matrix is unitary and after extraction of a trivial phase factor can be expressed through three real and one imaginarynumbers.

In the Wolfenstein parametrization it reads, neglecting terms $O(\lambda^4)$,

$$\begin{pmatrix} V_{ud} & V_{us} & V_{ub} \\ V_{cd} & V_{cs} & V_{cb} \\ V_{td} & V_{ts} & V_{tb} \end{pmatrix} = \begin{pmatrix} 1-\frac{\lambda^2}{2} & \lambda & A\lambda^3(\rho - i\eta) \\ -\lambda & 1-\frac{\lambda^2}{2} & A\lambda^2 \\ A\lambda^3(1-\rho-i\eta) & -A\lambda^2 & 1 \end{pmatrix} + O(\lambda^4) \tag{33}$$

The numerical values for the four parameters are (again neglecting higher terms in λ)

$$\begin{aligned} \lambda &= 0.2272 \pm 0.0010\,, \quad & A &= 0.818^{+0.007}_{-0.017}\,, \\ \rho &= 0.221^{+0.064}_{-0.028}\,, \quad & \eta &= 0.34^{+0.017}_{-0.045} \end{aligned} \tag{34}$$

The imaginary parameter $i\eta$ is responsible for CP-violation and all presently available data of particle physics are compatible with the violation induced by it. One can see that in the matrix representation (33) it enters only in the elements involving the *top* or *bottom* quark; this explains the smallness of CP-viaolation. Indeed, three families are necessary to allow an imaginary parameter leading to CP-violation in the frame of the standard model; soon stringent predictions in B-meson decays will be tested.

We now have a short look at the deficiencies of the standard model. Before coming to more ideological objections, we first discuss the most compelling evidence for a necessary modification, though it is not yet clear how and to which extent the model has to be altered. There are compelling indications that neutrinos have a finite mass and that, as for the quarks, the mass eigenstates are not the eigenstates of the $SU(2)$ doublets and therefore neutrinos of one flavour, e.g. electron neutrinos can oscillate into that of another flavour, for instance into μ-neutrinos.

The strong indications for these oscillations are as follows:

1. There is a long-standing problem of solar neutrinos. The sun emits less neutrinos than it is to be expected from the very well understood solar model. This lack of neutrinos has been observed over the full energy spectrum and therefore a true loss of neutrinos on the way from the sun to the earth is very probable. It had been an old suggestion that neutrino oscillations are responsible for that loss, since all experiments were only sensitive to electron neutrinos and blind to the other ones.
2. In a huge counting experiment (Super-Kamiokande) of neutrinos produced in the atmosphere, it turned out that there was a strong asymmetry between those neutrinos coming from above and from below the horizon. This asymmetry can best be explained by the hypothesis that a longer path leads to a loss of neutrinos due to oscillations.
3. From the KEK accelerator to the distant Kamioka counter (K2K) less neutrinos arrived than expected. Statistics is poor, but the result is in excellent agreement with the predictions from neutrino oscillations based on the results of the other experiments.
4. The most direct proof is the observation of a sizable flux of μ-neutrinos coming from the sun by the SNO collaboration. Since in the thermonuclear

reactions on the sun only electron neutrinos are generated, this can be considered as direct proof of neutrino oscillations, especially as the total number of observed neutrinos, e together with μ-neutrinos, agrees very well with the prediction of the solar models.

Since there are three different neutrinos, the mixing matrix is at least a three-by-three matrix, which is analogous to the CKM-matrix, except that it can contain two additional CP-violating phases. It is also possible that there exist additional neutrinos, which do not couple to the W- and Z-boson (so-called 'sterile neutrinos').

The direct upper bounds of the neutrino masses are given in Table 1; neutrino oscillations indicate mass differences of the order of 0.01 eV. It is plausible that this is the order of magnitude of the neutrino masses. As mentioned above, this must lead to a modification of the standard model. If the massive neutrinos can be incorporated in a simple extension of the present standard model, these small masses indicate that mass generation occurs through effects at short distances, e.g. through a very heavy additional Higgs boson; the value of the ρ parameter, see (31), which is very close to 1, points in the same direction.

A quite serious theoretical point, which speaks against the standard model as a possible final theory, is the group structure of the electroweak gauge group $SU(2) \times U(1)$. The $U(1)$ part leads to a gauge coupling which increases with increasing mass scale, since the β-function is positive for $U(1)$. One cannot conclude from perturbation theory alone that this rise increases indefinitely, but calculations on the lattice indicate that this is indeed the case. The only solution to that problem is that the coupling is zero, that is there is no $U(1)$ interaction at all, in contradiction to experiment (see 'Quantum Field Theory: Where We Are' by K. Fredenhagen et al.). This objection is however – at least partially – met in a modification discussed in the next section.

Unjustified from a formal point of view but nevertheless serious if one adopts a realistic view of the regularization procedure of quantum field theory is the question of the energy density of the vacuum. Generally speaking it makes no sense, to use the classical definition of the energy density in a quantized theory, since the latter involves products of fields at the same space-time point and therefore has in general to be regularized and to be renormalized. The energy density is in this procedure rather an input for the renormalization than a prediction of the theory. If one adopts, however, the view that the regularization is provided by the short distance behaviour of the theory, the natural choice for the regularization is a cutoff at the scale where new physics sets in. If one assumes that this happens at the Planck mass, one obtains the tremendous energy density $\rho_E = m_{\text{Planck}}^4 \approx 3 \cdot 10^{78}$ GeV fm^{-3} and even a cutoff at the modest scale of 1 TeV leads to the value $\rho_E \approx 10^{14}$ GeV fm^{-3}. In gravitational theory the energy density plays the role of a cosmological constant (see 'Dark Energy' by Straumann). The value for the energy density obtained from astronomical observations is $\rho_E \approx 5 \cdot 10^{-45}$ GeV fm^{-3}. This huge discrepancy must be of a major concern for all who have a realistic point of view for the regulatization procedure.

There is seemingly a way out, namely a supersymmetric field theory. In this theory the energy density of the vacuum vanishes, since the fermionic and bosonic contributions cancel each other. Supersymmetry, if it exists in nature at all, must however be broken at a scale of at least 1 TeV; this reduces the dicreapancy, but it remains still formidable.

More ideological are the following arguments, which have to to with simplicity and 'beauty'. They are not compelling in empirical science but have been (sometimes) fruitful in the past. The most often heard objection against the standard model is the large number of parameters occurring in it. There are indeed 18 or 19 free parameters even in the version with massless neutrinos:

- three gauge couplings, one for the colour gauge group $SU(3)$ of strong interactions, one for the $SU(2)$ part and one for the $U(1)$ part of the electroweak gauge group
- nine mass parameters for the three charged leptons and 6 quarks
- four parameters in the CKM quark mixing matrix
- two parameters in the Higgs part of the Lagrangian: the self-coupling λ and the vacuum expectation value v of the Higgs doublet To this one may add a 19th parameter:
- the so-called 'θ parameter'.[4]

Incorporation of neutrino mixing will at least add nine new free parameters.

5 Extrapolation to the Near Future

The purpose of this contribution is to give an account of the present experimentally well-established situation. Nevertheless there are already now some experimental hints which indicate a certain direction of development in the frame of local quantum field theory, see for this point especially the contribution Beyond the Standard Model' by M.G. Schmidt. An obvious way to reduce parameters is to embed the gauge groups of the standard model, that is $SU(3) \times SU(2) \times U(1)$ into a simple group, for instance $SU(5)$. In that case we have only one gauge coupling. Such a large symmetry is called a GUT (from grand unified theory) symmetry. This symmetry is supposed to be broken at a certain scale M_G, and all particles which do not occur in the standard model have masses larger than that scale. Since this scale is supposed to be very high, there is little hope to observe these new particles

[4] The θ term is an additional term in the pure gauge part of the Lagrangian which does not change the equations of motion but can lead to CP-violation.

directly. The standard model is an effective theory of the GUT, this means that in the model all degrees of freedom relevant at scales higher than M_{G} are absorbed in the renormalized parameters of the standard model. This is possible through the decoupling of the heavy particles – even inside loops – if the external scales are small compared to M_{G} and the theory is renormalizable. Renormalization group arguments based on the gauge group of the standard model may be applied and yield the observed, experimentally well-established scaling behaviour for the strong and the electroweak gauge couplings. If the scale comes, however, into the region of the GUT scale M_{G}, the heavy states can no longer be neglected and the different gauge couplings of the standard model must meet and follow together the renormalization group equation of the unified gauge group, e.g. $SU(5)$. In this way the grand unification scale M_{G} is constrained by the low-energy effective theory. Furthermore there is a stringent consistency condition for the effective theory since all three gauge couplings have to meet at the same point.

The GUT theory will generally lead to additional interactions in the standard model which from the point of view of the effective theory might not be renormalizable, but they are suppressed by powers of μ/M_{G}, where μ is a scale typical for the effective theory, that is $\mu \ll M_{\mathrm{G}}$. A nice example of an effective theory is the Fermi theory of weak interactions. It is an effective theory of the standard model at scales small compared to the mass of the gauge bosons W and Z. In it occurs the unrenormalizable four-fermion coupling, but the coupling constant G_{F} in front of it is proportional to M_W^{-2},

$$G_{\mathrm{F}} = \frac{\sqrt{2}g^2}{8M_W^2} \tag{35}$$

Important additional terms introduced by GUT are interactions leading to the decay of the proton into leptons and other non-baryonic states. These effects will be small since they are suppressed by powers of M_{G}. Nevertheless experimental limits on the proton decay in specific channels have already falsified a lot of proposals for grand unified theories. Present limits for important decay channels are, with 90% confidence level,

$$t_{p\to e^+\pi^0} \geq 1.6 \cdot 10^{33}\ \text{years} \quad t_{n\to e^+\pi^-} \geq 0.16 \cdot 10^{33}\ \text{years} \tag{36}$$

At the moment the most widely accepted extension of the standard model is a supersymmetric gauge theory. The supersymmetry is broken at a scale of about 1 TeV, that is a scale low compared to the Grand unification scale M_{G}. This model is called the supersymmetric standard model. The indications that this might indeed be the next standard model are

1. The supersymmetric standard model predicts a light Higgs, $m_{\mathrm{H}} < 170$ GeV. This is a value well inside the range of the LEP predictions.[5]

[5] These limits were much lower before the radiative corrections were calculated and before the lower limit of the Higgs mass was pushed to around 115 GeV.

2. All three gauge couplings meet nicely at a scale $M_G \approx 3 \cdot 10^{16}$ GeV.
3. This scale can lead to a proton lifetime compatible with present-day experimental bounds (but close to them).

The supersymmetric standard model is still a speculative model since up to now no particle predicted by supersymmetry has been observed, but the speculations are based on present-day experimental findings; they make also stringent predictions for the near future:

1. The light Higgs-boson should be observed at LHC.
2. The proton decay should be observed in the near future.
3. Supersymmetric partners of known particles should most probably seen at LHC.

If these predictions will be fulfilled, then the supersymmetric GUT will be the new standard model. Precision experiments could then indirectly explore the large desert from the TeV region to 10^{16} GeV.

6 Conclusion

The standard model is apparently the adequate description of subnuclear physics up to scales of the order of several hundred GeV. In the framework of renormalized perturbative local quantum field theory it gives an excellent quantitative description of electromagnetic and weak interactions. In strong interactions there are also many quantitative results, but a major deficiency is the lack of understanding of confinement. The success of models and numerical calculations in the lattice-regularized theory with finite lattice spacing indicate that the fundamental Lagrangian is the correct one, but a formal proof of a confining continuum limit is still missing.

All interactions can be derived from the postulate of local gauge invariance, at the moment the gauged group is $SU(3) \times SU(2) \times U(1)$. The matter content of the model is well structured, but masses and particle mixing introduce more than a dozen free parameters. Many people dislike this and they also want answers to questions like 'Why three families? Why integer and fractional charges?' Questions of this type are certainly legitimate and have sometimes led to real progress in physics.[6] But we should not forget that the principal questions of physics as a empirical science should be those of 'What' or 'How'.

At the moment there are several burning questions to be answered:

- Is there a Higgs boson and is it light?
- Does the proton decay and what is its lifetime?
- Are there supersymmetrie partners of the known particles with masses below 1 TeV?

[6] An example is the detection of ultraviolet radiation by J.W. Ritter (1801). He was guided by the principle of 'polarity' which was very popular among natural philosophers at that time.

These are not questions for the far future, the relevant experimental facilities exist or are already under construction. The answers to these questions will decide over the viability of the new supersymmetric standard model with a simple gauge group. The dream of a final theory may have disappeared, but the dream of a big leap forward seems quite realistic.

Literature

Since this short review concentrates on results, it would not be adequate to select only the original literature which is directly referred to in the text. I therefore only quote some books and reviews where the original literature can be found easily.

I mention especially the two bibliographic data bases of particle physics, where the newer and some of the classical older literature can be found: `http://www.slac.stanford.edu/spires.hep` and `http://www.arXiv.org`

For Sect. 2:

A great history of particle physics in the 20th century with an excellent bibliography (until 1985) is the book by Abraham Pais:

Abraham Pais. *Inward Bound.* Clarendon press, Oxford, 1986.

Interesting articles on the early history of particle physics and a chronological bibliography of the classical papers (until 1964) can be found in

International Colloqium on the History of Physics. *Journal de Physique*, 43, Colloque C-8, 1982. 1930–1960.

The development of quantum electrodynamics is described in

S.S. Schweber. *QED and the Men who Made It: Dyson, Feynman, Schwinger, and Tomonaga.* Princeton University Press, Princeton, 1994.

Proceedings of a conference on the rise of the standard model are published in

L. Hoddeson et al., eds., *The Rise of the Standard Model*, p. 299. Cambridge University Press, 1997.

For the remaining sections I warmly recommend the Review of Particle Properties (RPP), the 2006 edition appeared as

W.-M Yao et al. (Particle data Group), J. Phys. G: Nucl. Part. Phys. **33** (2006) 1.

It contains not only all particle data, but also very concise review articles with rather complete bibliographic data.

For this contribution, the following review articles are of particular interest:

- Quantum chromodynamics
- Electroweak model and constraints on new physics
- CKM quark mixing matrix

- Neutrino mass, mixing and flavor change
- Grand unified theories

The home page of the particle data group is `http://pdg.lbl.gov`

Some text books and readers treating the standard model or parts of it are, without the slightest claim on completeness:

C. Itzykson and J.-B. Zuber. *Quantum Field Theory.* MacGraw-Hill, New York, 1980.

O. Nachtmann. *Concepts and Phenomena of Particle Physics.* Springer, Berlin, Heidelberg, New York 19XX.

M.A. Shifman, ed., *At the Frontiers of Particle Physics, Handbook of QCD.* World Scientific, 2001.

Beyond the Standard Model

M. G. Schmidt

Institut für Theoretische Physik, Universität Heidelberg,
Philosophenweg 16, 69120 Heidelberg, Germany
M.G.Schmidt@thphys.uni-heidelberg.de

The well-founded cornerstones of our discussion are the classical (Einstein) theory of gravity, local relativistic quantum field theory (QFT), and elementary particle physics, today described so impressively by its so-called "Standard Model" (SM). "Well-founded" does not just imply mathematical elegance but most importantly a solid fundament of observational/experimental findings – the relevance of black holes and of the standard cosmological model confirmed by astrophysical observations; the spectacular successes of quantum electrodynamics (QED), e.g. for anomalous magnetic moments; non-abelian gauge theories and the three-quark-lepton generation structure in the SM explaining a huge body of data ("Rosenfeld table"). The big question remains how to raise a building with these cornerstones: are there further essential pieces still missing? Will it be one unifying building as suggested by the only theory ansatz with this claim – superstring theory?

This involves questions rather far away from our present experimental/observational possibilities. The Planck scale $M_{PL} \sim 10^{18}$ GeV of gravity, important in quantum gravity, and related scales in string theory are far, far above the "high energy" scale of present accelerators, being above, but still in the range of the electroweak scale $\sim 10^2$ GeV. Even the highest observed scale of cosmic rays $\sim 10^{10}$ GeV still is intermediate on the way to M_{PL}.

A century ago there was a similar problem how to connect microscopic and macroscopic physics. Its solution was one of the great successes of mankind. A deep understanding on one side was based on experimental access to microscopic physics – molecules, crystal lattices, nuclei, ..., on the other side on theoretical methods to develop "effective theories" for the macroscopic world, – gases, fluids, solid states. The appropriate language of the discussion raised above, today, is "Wilsonian" renormalization leading to "effective field theories" designed for a certain scale of observation. The description of a physical system in such a language changes if one considers the same system with varying resolution. Going to a weaker resolution, "to the infrared" (IR), finer details of the object are "integrated out". Perturbation theory in such effective theories is finite since Feynman loop integrals are cut off in momentum/energy

M. G. Schmidt: *Beyond the Standard Model*, Lect. Notes Phys. **721**, 51–57 (2007)
DOI 10.1007/978-3-540-71117-9_3

above the fixed scale of the theory. Adding non-renormalizable terms at very small distances (large momenta) is harmless since they vanish going to the IR region interesting for us. This explains the importance of renormalizable theories in the SM and why we can work quite successfully in this well-known area of QFT (well-known only if infrared properties like confinement are dealt with numerically; there still appear questions concerning its precise nature, and progress, e.g. using supersymmetric variants of quantum chromodynamics (QCD), is slow). The process of renormalization "group" by integrating out physics can strictly speaking not be inverted going to small distances – the "ultraviolet" (UV). Still in QCD the postulate that there is a simple asymptotic freedom behavior at small distances is consistent. All this can be made very concise in the path integral formulation ("sum over fields") of QFT and can be discussed even quantitatively in numerical studies of discretized theories ("lattice (gauge) theory").

Let us inspect questions of elementary particle physics which cannot be answered within the SM: certainly most prominent is the *unification* of the strong and electroweak *gauge forces*. Indeed continuing the "running" (effective) gauge couplings based on the particle content of the SM towards the UV, one observes some convergence to a common value at a scale of about 10^{14} GeV. Going to the minimal supersymmetric extension of the SM (MSSM) this is much improved resulting in a common value of about $M_{GU} \approx 10^{16}$ GeV. Postulating a grand unified gauge symmetry corresponding to a semisimple gauge group ($SU(5), SO(10)$) quarks and leptons are in common representations and related by gauge interactions. For a spontaneous breaking to the SM one has to introduce eventually further Higgs fields (or some non-local Wilson-loop operators) all in representations of these groups. This clearly leads beyond the SM. Massive neutrinos, now well established in experiments, can be considered still partly in the range of the SM, but their small mass naturally induces new scales $M_M \approx 10^7 - 10^{14}$ GeV via $m_\nu \sim m_D^2/M_M$ (where m_D is a typical charged lepton mass), the so-called "see-saw mechanism". The scale $M_{GU} \simeq 10^{16}$ GeV is still two orders of magnitude below the Planck scale M_{PL} of gravity but big enough to suppress the decay of protons and neutrons of our universe within its age of $\approx 10^{10}$ years.

Supersymmetry (SUSY) is essential in most of these model buildings. This is a symmetry between bosons and fermions, e.g. between a photon and a photino, an electron and its bosonic partner ("selectron"). In its gauged ("local") form it also changes gravity to supergravity (SUGRA). Supersymmetry is the only possible enlargement of the Ponicare group. In more practical terms it allows to continue the successful way to do calculations in the SM. The main point here is its ability to tame UV-divergences requiring an artificial fine tuning in the SM in order to preserve hierarchies between vastly different scales. This is why it allows to continue the successful calculations of the SM and why it is so popular now in phenomenologically minded circles though being a genuinely theoretical concept not confirmed in experiments up to now (but hopefully soon). Since there are certainly no observed super-partners with equal mass in nature, SUSY has to be broken in a "soft" way such that the above virtues are not destroyed. There are numerous attempts to model such soft breakings which should have spontaneous origin, but still this problem is not finally settled. The alternative to SUSY is a plethora of problems, mostly related to non-perturbative effects for strong coupling systems, and thus this is much less attractive – though maybe the future! If supersymmetry in connection with SM physics will not be detected in the next generation of experiments, this will have serious drawbacks for most of the theoretical developments in elementary particle physics in the last 30 years.

Modern *superstring* theory contains supersymmetric structures, if it is considered as a 2-dimensional conformal QFT, but it does not necessarily imply space-time supersymmetry, in particular not at the electroweak scale. Still the latter is a genuine ingredient. Unfortunately, even after marvelous detections of connections between the various types of string theories this approach is (still?) not a very concrete guide how to go beyond the SM. There is a vast number of possible string ground states ("vacua") – a "landscape" of string vacua – being on equal footing in our present understanding and there is no well-understood dynamics preferring one from the others. Even if one fixes to one of these vacua the calculational abilities to extract phenomenological information – how far the SM is obtained and where there are deviations – is quite limited. These days it is very popular to invoke the "anthropic principle", saying roughly that we are living in a particular world (vacuum), in one of these billions of other possibilities, because only there the structures could develop which made our universe so beautiful and sophisticated and which allow us to live and to do research as we do. Of course, this is not what a scientist likes to believe, who wants to explain and predict quantitatively physical phenomena and fundamental constants in our world from simple principles. However, like the improbable existence of our planet earth the choice of such an improbable "vacuum" state is a logical possibility. Since almost nothing is known about the dynamics leading to different vacua, just some counting of states has been done – these ideas are still very vague.

Still string theory already today is an invaluable source of inspiration, if it comes to constructing models beyond the SM: (i) Grand unification with some (semi)simple non-abelian gauge groups can be very genuinely realized (though it is not mandatory and) though "details" like the pattern of spontaneous breakings and of the generalized Higgs fields can only be attempted for very specific string vacua. (ii) The genuine role of supersymmetry and supergravity in string theory we have already mentioned. (iii) Maybe most interestingly more than four space-time dimensions are required in a very concise way in consistent string theories. These may show up "only" at the Planck scale M_{PL} and maybe in realizations of a grand unification of gauge interactions. These *higher dimensional worlds* at extremely small distances allow for beautiful GUT-constructions leading to structures like in the SM in the corresponding effective theory for light particles. Topological structures in the higher dimensions curled up at these small distances can be related to the observed quark-lepton generations. It is conceivable that all or some part of the quarks and leptons may exist in lower dimensional subdomains ("branes", "singular surfaces") in such a theory and that one can calculate their (Yukawa) interactions analyzing some intersections. This is all beset with mathematical problems and thus has some strong relations to modern mathematics. Still somewhere in this haystack one might find the needle some day. Of course, it would be amazing, if one could see these extra dimensions in accelerator experiments not involving gravity. Up to now these exclude such structures below energies of about 10^4 GeV. Substantial deviations from 4-dimensional gravity are still not excluded at distances in the sub-mm range. One can also try to construct models with extra dimensions, say one or two for simplicity. This is quite in the spirit of the old Kaluza–Klein approach except that now part of the fields are allowed to be based on lower dimensional submanifolds ("branes"), say our 4D world. These models are only vaguely related to string theory, though there is a trend these days to base them in string theory. For example, quantum anomaly cancellations, a basic point in string theory, still have to be checked "by hand". However, these models provide us with some insight concerning the embedding of the SM not seen in genuine string theory with all its complexity so easily, e.g. the reduction of Higgs multiplets to their doublet components appearing in the SM, the use of non-local Wilson lines made out of extra dimensional gauge field components for realizing Higgs breaking, the discussion of the famous quadratic divergences for the Higgs mass in this context and last but not least the unification of gauge forces below the Planck scale (not necessarily true for general string vacua).

Theoretical physicists determined to raise a unified (triangular?!) building with the cornerstones mentioned at the beginning cannot just concentrate on one edge. They have to jump back and forth trying to connect formal theories with experimental/observational facts. The dream to create a theory just with the criteria of elegance of principles never worked out (perhaps with the exception of Einstein's general relativity?). Thus the "bottom-up" approach – more modest, less brilliant might be necessary: to construct models partly realizing the ideas discussed above and to proceed with "trial and error".

Trying to look beyond the SM, observations related to very *early cosmology* may be very helpful. The physics of "the first three minutes" (S. Weinberg) after the big bang can be connected to high energy elementary particle physics observed with present accelerators and in underground observatories (neutrino physics) and well described by the SM with simple extensions in the case of neutrinos. The big bang model itself contains some problematic features, in particular the homogeneity over distances not connected by light signals. This is remedied by a period of exponential growth, the so-called *"inflation"*, at the end of which thermalization leads to the successful standard big bang model. Indeed, realization of inflation in quantum field theoretical models brings along quantum fluctuations which later on turn into classical inhomogeneities creating the observed large-scale structure including galaxies when the universe is getting neutral and transparent. Such fluctuations also show up in the thermal background radiation of the universe measured in recent years with incredible accuracy by satellite observatories (WMAP, etc.) and can be well explained by an inflationary period. The paradigm of inflation is not tied to a particular model, one just needs a period with an effective cosmological "constant" above the TeV region, maybe not much smaller than the Planck-mass scale. It is intriguing to speculate about a relation of this scale to grand unification. There is one common feature of all such models: they go beyond the SM! Thus they usually contain also particles which are too heavy to be seen in present experiments, but which are candidates for "(cold) *dark matter*", i.e. non-baryonic matter which is not luminous and forms invisible halos of the galaxies. This kind of matter should make $\sim 25\%$ of the matter of the universe (which in accordance with inflation has the critical density) as an analysis of the WMAP data suggests. Baryonic matter, the material of our stars and of gas nebula, is analyzed to amount to only $\sim 5\%$, and the remaining 70% of energy is vacuum energy, a cosmological constant, or alternatively, the energy of a field almost spacially constant in the observed universe, but changing in time – "*dark energy*" ("quintessence"). Again this cannot be explained by the SM. Another number to be explained is the baryon number of our universe, the remaining *baryon asymmetry* after recombination of baryon–antibaryon pairs in the cooling down universe $\eta = \frac{n_B}{n_\gamma} \sim 10^{-10}$ (with photon number density n_γ and baryon density n_B). This requires (with Sakharov) a violation of baryon number, Charge/Charge-Parity ("C", "CP") violation, and thermal non-equilibrium due to the expansion of the universe and/or a phase transition. These are conditions one finds in SM physics, but it turns out that again "Beyond the SM" physics is required to get realistic estimates. All this invites for model constructions, though again this allows for quite a few different realizations. Models trying to explain many or even all known observational facts then might contain the decisive clue how to go "beyond". The sound scientific principle "divide and impera" does not bless such a procedure, but unfortunately we cannot perform experiments with our universe. Approaching the Planck scale, of course, the big bang theory, even with inflation, loses its meaning and a deeper understanding of quantum gravity and of its connection to elementary particle physics is required. String theory, proposed to contain all the basic ingredients of a fundamental theory, offers some fascinating pictures, but is still far away from explaining early cosmology in a detailed dynamics. Loop quantum gravity does not have a compatible universal claim and is even less developed. In this situation a modest phenomenologically oriented procedure in going "beyond" seems to be appropriate. Unfortunately, these are sometimes painful exercises, most of them without long-term merits. One can only hope that a few of them will survive in a more complete theory.

Selected References

- **General relativity:**
  K. Ehlers contribution in this book
  R. Wald "General Relativity", Chicago Univ. Press 1984
  N. Straumann "General Relativity" with Applications to Astrophysics", Texts and Monographs in Physics, Springer-Verlag 2004
- **Elementary particle physics:**
  H.G. Dosch contribution in this book
  O. Nachtmann "Concepts and Phenomena of Particle Physics", Springer-Verlag
- **Quantum field theory:**
  K. Fredenhagen, K.-H. Rehren, E. Seiler contribution in this book
  C. Itzykson, I.B. Zuber "Quantum Field Theory", McGraw Hill
  M.E. Peskin, D.E. Schroeder "An introduction to QFT", Addison-Wesly
  M. Maggiore "A modern Introduction to QFT", Oxford Univ. Press
  M. Stone "The Physics of Quantum Fields", Springer-Verlag
- **Cosmology:**
  N. Straumann contribution in this book
  A.R. Liddle, D.H. Lyth "Cosmological Inflation and Large Scale Structure", Cambridge Univ. Press 2000
  V. Mukhanov "Physical Foundations of Cosmology", Cambridge Univ. Press, 2005
  P.D.B. Collins, A.D. Martin, E.J. Squires "Particle Physics and Cosmology", Wiley 1989
- **String theory:**
  J. Louis, S. Theisen contribution in this book and references quoted there
- **Extra dimensions and model building:**
  M. Quiros "New Ideas in Symmetry Breaking", Boulder 2002, Particle Physics and Cosmology 549-601, hep-ph/ 0302189
  R. Rattazzi "Cargese Lectures on Extra Dimensions" (hep-ph/ 0607055)
  I. Antoniadis "The Physics of Extra Dimensions", 3rd Aegean Summer School, Karfas, Greece, hep-ph/ 0512182
  H.P. Nilles "Five golden rules of superstring phenomenology", Boston 2004, Theories in Unification 264-274, hep-th/ 0410160

- **Supersymmetry:**

S.P. Martin	"A Supersymmetry Primer", Perspectives on Supersymmetry (ed. G.L.Kane) 1997, 1–98
H.E. Haber, G.L. Kane	"The search for supersymmetry: Probing physics beyond the Standard Model", Physic Reports 1985, 117–175

Part III

Quantum Field Theory

Quantum Field Theory: Where We Are

K. Fredenhagen[1], K.-H. Rehren,[2] and E. Seiler[3]

[1] II. Institut für Theoretische Physik, Universität Hamburg, 22761 Hamburg, Germany
klaus.fredenhagen@desy.de
[2] Institut für Theoretische Physik, Universität Göttingen, 37077 Göttingen, Germany
rehren@theorie.physik.uni-goe.de
[3] Max-Planck-Institut für Physik (Werner-Heisenberg-Institut), 80805 München, Germany
ehs@mppmu.mpg.de

1 Introduction

Quantum field theory (QFT) aims at the synthesis of quantum physics with the principles of classical field theory, in particular the principle of locality. Its main realm is the theory of elementary particles where it led to a far-reaching understanding of the structure of physics at subatomic scales with an often amazingly good agreement between theoretical predictions and experiments. Typical observables in QFT are current densities or energy flow densities which correspond to what is measured in particle physics detectors. The original aim of QFT was to compute expectation values and correlation functions of the observables, and to derive scattering cross sections in high-energy physics. In the course of development, QFT has widened its scope, notably towards the inclusion of gravitational interactions.

The purpose of this contribution is to take stock and to comment on the present status, the concepts and their limitations, and the successes and open problems of the various approaches to a relativistic quantum theory of elementary particles, with a hindsight to questions concerning quantum gravity and string theory.

Quantum field theory rests on two complementary pillars. The first is its broad arsenal of powerful modeling methods, both perturbative and constructive. These methods are based on the gauge principle, and have been tremendously successful especially for the modeling of all the interactions of the standard model of elementary particles. The perturbative treatment of the standard model and its renormalization, as well as lattice approximations of quantum chromodynamics (QCD), give enormous confidence into the basic correctness of our present understanding of quantum interactions. (For the impressive phenomenological support for the standard model, we refer to the

K. Fredenhagen et al.: *Quantum Field Theory: Where We Are*, Lect. Notes Phys. **721**, 61–87 (2007)
DOI 10.1007/978-3-540-71117-9_4

contribution by Dosch to this book.) Despite these successes, however, establishing the standard model (or part of it) as a mathematically complete and fully consistent quantum field theory remains an unsettled challenge, as will be explained in the sequel.

The second pillar of QFT are axiomatic approaches, putting the theory on a firm conceptual ground, which have been developed in order to understand the intrinsic features of a consistent QFT, irrespective of its construction. In these approaches, the focus is set on the fundamental physical principles which any QFT should obey, and their axiomatic formulation in terms of the observable features of a theory is addressed.

In fact, several such axiomatic approaches, which have been shown to be partially but not completely equivalent, are pursued. None of them indicates a *necessary* failure or inconsistency of the framework of QFT. (Of course, this does not mean that a realistic QFT should not include new Physics, say at the Planck scale, cf. Sect. 8.)

2 Axiomatic Approaches to QFT

Axiomatic QFT relies on the fact that the fundamental principles which every quantum field theoretical model should satisfy are very restrictive. On the one hand this is a great obstacle for the construction of models, on the other hand it allows to derive a lot of structural properties which a QFT necessarily has. They often can be tested experimentally, and they provide a criterion whether a construction of a model is acceptable.

The main principles are

- the superposition principle for quantum states, and the probabilistic interpretation of expectation values. These two principles together are implemented by the requirement that the state space is a Hilbert space, equipped with a positive definite inner product.
- the locality (or causality) principle. This principle expresses the absence of acausal influences. It requires the commutativity of quantum observables localized at acausal separation (and is expected to be warranted in the perturbative approach if the action functional that determines the interaction is a local function of the fields).

In addition, one may (and usually does) require

- covariance under spacetime symmetries (in particular, Lorentz invariance of the dynamics) and
- stability properties, such as the existence of a ground state (vacuum) or of thermal equilibrium states.

The critical discussion of these principles themselves ("axioms") is, of course, itself an issue of the axiomatic approaches. For a review, see [1]. Various axiomatic approaches (Wightman QFT, Euclidean QFT, Algebraic QFT) may differ in the technical way the principles are formulated. Several theorems establishing (partial) equivalences among these approaches have been established, such as the Osterwalder–Schrader reconstruction theorem [2] stating the precise prerequisites for the invertibility of the passage from real-time QFT to Euclidean QFT ("Wick rotation"), or the possibility to recover local fields from local algebras [3].

In the Wightman formulation, one postulates the existence of fields as operator-valued distributions defined on a common dense domain within a Hilbert space. The field operators should commute at space-like distance and satisfy a linear transformation law under the adjoint action of a unitary representation of the Poincaré group. Moreover, there should be a unique Poincaré invariant vacuum state which is a ground state for the energy operator. The assumption of local commutativity may be relaxed admitting anti-commutativity for fermionic fields. One may also relax the assumption of the vacuum vector, retaining only the positivity of the energy (unless one is interested in thermal states) in order to describe charged states; in the algebraic approach, such theories are most advantageously regarded as different representations (superselection sectors) of the same field algebra, originally defined in the vacuum representation, see below.

Due to the restrictive character of these principles, they typically are violated in intermediate steps of approximation schemes. One often has to introduce auxiliary fields without a direct physical meaning as observables.

As an illustration, consider the Dirac equation featuring a charged electron field coupled to the electromagnetic field:

$$i\gamma^\mu(\partial_\mu + ie\,A_\mu(x))\,\psi(x) = m\,\psi(x)\ . \tag{1}$$

The Fermi field $\psi(x)$ satisfies anti-commutation relations and can therefore not be an observable field strength subject to causality. The vector potential $A_\mu(x)$ is already in the classical theory not an observable. Related to the gauge arbitrariness, the vector field cannot be covariantly quantized on a Hilbert space with a probabilistic interpretation. (Other problems related with the promotion to QFT of classical field products, appearing in evolution equations such as (1), will be considered later.)

The general principles can therefore not be applied to the objects of basic relations such as (1). The principles rather apply to the physical sector of the theory where the typical fields are current and stress–energy densities or electromagnetic fields, such as

$$j^\mu = \bar\psi\gamma^\mu\psi, \qquad T^{\mu\nu} = T^{\mu\nu}(\psi, A), \qquad F_{\mu\nu} = \partial_\mu A_\nu - \partial_\nu A_\mu\ . \tag{2}$$

These fields, corresponding to observable quantities, should be well-defined in a QFT, admitting that the individual quantities on the right-hand sides of (2) turn out to be very ill-defined.

In this spirit, the axiomatic approaches focus directly on the *observable* aspects of a theory, which have an unambiguous and invariant physical meaning, and which should be computed in order to compare with experiment. They thus strive to develop analytic strategies to extract these quantities from a given theory. For example, the particle spectrum emerges in terms of poles in renormalized correlation functions, or in terms of the spectrum of the time evolution operator, rather than as an input in terms of a classical action. The Haag–Ruelle scattering theory showing how the space of scattering states (and its structure as a Fock space) is intrinsically encoded, and how cross sections are obtained as asymptotic limits of correlations, was one of the first successes.

The power of the axiomatic approach resides not least in the ability to derive structural relations among elements of the theory without the need to actually compute them in a model. These relations are recognized as necessary consequences of the axioms. The most familiar examples are the PCT theorem and the Spin-Statistics theorem, which arise from functional identities among the Wightman functions due to covariance, energy positivity and locality.

Another example is the discovery ("Doplicher–Haag–Roberts theory") of the coherence among the intrinsic data relating to the superselection structure (charge structure). To value this approach, it is important to note that *if one assumes* (as one usually does) the presence of unobservable charged fields in a theory, these will typically "create" charged states from the vacuum state Ω. As a specific example,

$$\Psi = \psi(f)\,\Omega \tag{3}$$

is an (electrically charged) fermionic state if $\psi(f) = \int d^4x f(x)\psi(x)$ is an (electrically charged) Fermi field smeared with some function f. These states cannot be created by observable fields such as those in (2), and their charge can be distinguished by looking at suitable characteristics of the state functional

$$\mathcal{O} \mapsto \omega_\Psi(\mathcal{O}) \equiv (\Psi\,,\,\mathcal{O}\,\Psi) \tag{4}$$

as the local observables $\mathcal{O}$ vary, e.g., when the charge operator Q is approximated by integrals over the charge density $j^0(x)$. States of different charge belong to inequivalent representations (superselection sectors) of the observables. The DHR theory provides the means to study charged sectors intrinsically, i.e. without the assumption of charged fields creating them.

More recently, the DHR theory culminated in the proof ("Doplicher–Roberts reconstruction") that the observables along with their charged representations in fact *determine* an algebra of charged unobservable fields transforming under a global symmetry group, which create charged sectors from the vacuum and among which the observables are the invariants under the symmetry [4]. Indeed, the presence of Fermi fields, although these do not correspond to observable quantities, can be inferred (and their conventional use can be justified) from the existence of fermionic representations of the bosonic fields of the theory.

At least the relevance of *global* symmetry as the origin of charged sectors has thus been derived from the physical principles of QFT. At the same time, the way how geometric properties of spacetime enter this analysis shows clearly why the analogous conclusion fails in low-dimensional QFT. Here, the charge structure turns out to be much richer, opening the way to a much broader symmetry concept beyond global symmetry groups.

In realistic models of QFT, the most important symmetry concept is that of *local* gauge groups, to which we devote a section of its own below. Unfortunately, local gauge symmetry is not covered by the DHR theory.

Axiomatic approaches also allow to investigate the infrared problem of theories containing electromagnetism. The infrared problem is due to the fact that the mathematical description of particle states as eigenstates of the mass operator

$$P_\mu P^\mu \Psi = m^2 \Psi \tag{5}$$

(which is the starting point of the Haag–Ruelle scattering theory) cannot be used for particles which carry an electric charge. It was proven under very general conditions [5] that electrically charged sectors contain no eigenstates of the mass operator. Instead it turns out to be physically more appropriate to use so-called "particle *weights*" rather than *states* which share many properties with the latter but are not normalizable [6].

A more pragmatic way out is the artificial introduction of a photon mass as a regulator. One computes the cross sections in the auxiliary theory and takes the limit of vanishing photon mass for suitable inclusive cross sections (where "soft" photons, i.e. photons below an arbitrary small, but finite energy in the final state, are not counted) at the very end. On the conceptual level, this method involves an exchange of limits. Namely, scattering theory in the sense of Haag and Ruelle amounts to look at distances which are large compared to the Compton wavelengths of the particles. The physically relevant limit for scattering of electrically charged particles should therefore be to perform first the limit for the photon mass and then to go to large distances. As was emphasized by Steinmann [7], it is doubtful whether the limits may be exchanged.

The section about axiomatic approaches should not be concluded without the remark that the complete construction of models fulfilling all required principles has been achieved with methods described in Sect. 6, although presently only in two- and three-dimensional spacetime (polynomial self-interactions of scalar fields, Yukawa interactions with Fermi fields).

Low-dimensional models are of interest as testing grounds for the algebraic methods and concepts of axiomatic approaches, and to explore the leeway left by the fundamental principles. Apart from that, since string theory can in some respect be regarded as (a ten-dimensional "target space re-interpretation" of) a conformal QFT in two dimensions, the exact control available for a wealth of these models could thus indirectly provide insight into higher-dimensional physics.

Conformally invariant theories in two dimensions have been constructed rigorously (and partially classified [8]) by methods of operator algebras, especially the theory of finite index subfactors [9]. It is here crucial that a "germ" of the theory is given, such as the subtheory of the stress-energy tensor field, and is verified to share certain algebraic features. Then any local and covariant QFT which contains this subtheory is strongly constrained, and can be constructed from certain data associated with the subtheory. Even if the "germ" (as is usually the case) can be realized as a subtheory of some auxiliary free field theory, the new theories thus constructed extend the relevant subtheory but not this auxiliary theory, and therefore cannot be considered as free theories in their turn.

Quite recently, a novel scheme for the construction of quantum field theories has been developed in a genuinely operator algebraic approach, which is not based on quantum fields and some classical counterpart, but on the relation between the localization of quantum observables and their interpretation in terms of scattering states. As a consequence of the phenomenon of vacuum polarization, this relation is subtle since interacting local fields can never create pure one-particle states from the vacuum. The basic new idea stems from modular theory (see below) by which geometric properties such as localization in causally independent regions and the action of Poincaré transformations can be coded into "modular data" of suitable algebras.

Although this is not the place to introduce modular theory [10] to a general audience, we wish to add a rough explanation. There is a mathematical theorem that the pair of a von Neumann algebra and a (cyclic and separating) Hilbert space vector determine an associated group of unitaries and an antiunitary involution, the "modular data", which have powerful algebraic and spectral properties. In the case of algebras of covariant quantum observables localized in a wedge region (any Poincaré transform of the region $|ct| < x_1$) and the vacuum vector, these properties allow to identify the modular data with a subgroup of the Poincaré group and the PCT conjugation. The joint data for several such wedge algebras generate the unitary representation of the full Poincaré group. Exploiting this algebraic coding of geometry in the opposite direction, it is in fact possible to construct a QFT by specifying a distinguished vector in a Hilbert space and a small number of von Neumann algebras, provided these are in a suitable "relative modular position" to each other to warrant the necessary relations among their modular data to generate the Poincaré group and ensure local commutativity and energy positivity.

This opens an entirely new road for the non-perturbative construction of QFT models [11]. As an example in two spacetime dimensions, algebras of putative observables localized in spacetime wedges can be constructed in terms of one-particle states. Observables with bounded localization are then obtained by taking intersections of wedge algebras. That this road indeed leads to the desired construction of interacting theories with a complete interpretation in terms of asymptotic particle states has been established [12] for a large class of two-dimensional models with factorizing scattering matrices. Even though – by lack of a priori knowledge of a nontrivial scattering matrix – a directly analogous program in four dimensions is not available, the approach shows the advantage of constructing observables with poorer localization properties in the first step, before identifying local observables as subalgebras of the latter.

3 The Gauge Principle

It happens very often that complicated structures can be more easily accessed by introducing redundant quantities. The extraction of the relevant information then requires a notion of equivalence. In fundamental physics it is the notion of a local interaction which forces the introduction of redundant structures. To ensure that the observable quantities do not influence each other at a distance (causality), one wants to describe their dynamics by field equations which involve only quantities at the same point. But it turns out that this is possible only by introducing auxiliary quantities, such as gauge potentials in electrodynamics. This difficulty already exists in classical field theory, and it complicates considerably the structure of classical general relativity.

Classical gauge theories describe the interaction of gauge fields (understood as connections of some principal bundle) and matter fields (described as sections in associated vector bundles). The interaction is formulated in terms of covariant derivatives and curvatures. (In this way, the rather marginal gauge symmetry of Maxwell's electrodynamics is turned into a paradigmatic symmetry principle determining the structure of interactions.) The combination

$$D_\mu = \partial_\mu + ieA_\mu(x) \tag{6}$$

providing the coupling between the fields in (1) is a covariant derivative which ensures that the equation is invariant under the abelian gauge transformation

$$\begin{aligned} \psi(x) &\mapsto \mathrm{e}^{ie\alpha(x)}\, \psi(x) \\ A_\mu(x) &\mapsto A_\mu(x) - \partial_\mu\alpha(x) \,, \end{aligned} \tag{7}$$

i.e. $D_\mu\psi(x)$ transforms in the same way as $\psi(x)$ itself, and the equation of motion (1) is preserved by the transformation. The electromagnetic field strength tensor $F_{\mu\nu}(x)$ in (2) is obtained through the commutator of two covariant derivatives, i.e. geometrically speaking, the curvature.

The presence of this group of automorphisms (7) of the bundle (gauge transformations) makes the description redundant, and only the space of orbits under the automorphism group corresponds to the relevant information.

Non-abelian gauge transformations generalize these structures by replacing the charged field by a multiplet and the complex phase in (7) by an element of a compact group in a matrix representation. Along with the appropriate generalization of the Maxwell–Dirac action of quantum electrodynamics, one arrives at a gauge covariant coupled system of equations of motion for the charged fields and the gauge potentials. It is considered as the triumph of the gauge principle that these equations successfully describe most of the dynamics and symmetries of standard model of elementary particles, provided one chooses $U(1)\times SU(2)\times SU(3)$ as the gauge group, and assigns appropriate representations ("quantum numbers") to the fermions. For a more detailed account of the gauge symmetry of the standard model, we refer to Dosch's contribution to this book.

In QFT, the very concept of gauge theories is strictly speaking not well defined, because of the singular character of pointlike localized quantities. These singularities are absent in the lattice approximation (see Sect. 6). There matter fields are attached to the lattice points, while gauge fields are, as parallel transporters, attached to the links between them.

In the continuum, perturbation theory is used to deal with these singularities (see Sect. 5). In the cause of gauge theories, additional auxiliary structure has to be invoked in order to be able to use the canonical formalism. Namely, the Cauchy problem in gauge theories is not well posed because of the ambiguities associated with time-dependent gauge transformations. Therefore one has to introduce a gauge-fixing term in the Lagrangean which makes the Cauchy problem well posed, and the so-called "ghost and antighost" fields which interact with the gauge field in such a way that the classical theory is equivalent to the original gauge theory. This auxiliary theory is quantized on a "kinematical Hilbert space" $\mathcal{H}$ which is not positive definite. The observables of the theory are then defined as the cohomology of the BRST transformation s which is an infinitesimal symmetry of the theory with $s^2 = 0$ (see, e.g., [13]). More precisely, s is a graded derivation implemented as the graded commutator with a charge operator q satisfying $q^2 = 0$, the observables are those local operators that commute with q:

$$q\,A = A\,q\;, \tag{8}$$

physical states are those annihilated by it:

$$q\,\Psi_{\mathrm{phys}} = 0\;, \tag{9}$$

and two physical states are equivalent if they differ by a state in the image of it:

$$\Psi_1 - \Psi_2 \;\in\; q\mathcal{H}\;. \tag{10}$$

The BRST method ensures that the equivalence classes of physical states form a positive-definite Hilbert space

$$\mathcal{H}_{\mathrm{phys}} = \mathrm{Ker}\; q \,/\, \mathrm{Im}\; q\;, \tag{11}$$

and the observables are well-defined operators on $\mathcal{H}_{\mathrm{phys}}$.

We will see in Sect. 5 that within perturbation theory, BRST gauge theories are distinguished by their good behaviour under renormalization.

It is not clear how the gauge principle should enter the axiomatic formulations. Namely, these approaches focus on the observables of a quantum system, while gauge fields are per se unobservable. Put differently, one should ask the question which observable features tell us that a QFT is a gauge theory. In the abelian case, there is of course the characteristic long-range nature of Gauss' law, but there is no obvious equivalent in the non-abelian case. Could there be, in principle, an alternative description of, say, QCD without gauge symmetry?

There are, of course, experimental hints towards the color symmetry, ranging from particle spectroscopy over total cross section enhancement factors to "jets" in high-energy scattering. In algebraic QFT, the counterpart of these observations is the analysis of the global charge structure of a theory, i.e. the structure of the space of states.

The DHR theory of superselection sectors is precisely an analysis of the charge structure entirely in terms of the algebra of observables. As we have seen, it leads to the derivation of a symmetry principle from the fundamental principles of QFT (see Sect. 2), but the result pertains to global symmetries only. The case of local gauge symmetries is still open. Yet, a local gauge theory without confinement should possess charged states in non-trivial representations of the gauge group. If the theory has confinement, but is asymptotically free, then its gauge group should become visible through the charge structure of an appropriate short-distance limit of the observables [1]. It is therefore expected that gauge symmetry, if it is present, is not an artefact of the perturbative description but an intrinsic property coded in algebraic relations among observables.

4 The Field Concept

It is the irony of quantum field theory that the very notion of a "quantum field" is not at all obvious. The field concept has been developed in classical physics as a means to replace the "action at a distance" by perfectly local interactions, mediated by the propagating field. Classical fields, such as the electromagnetic fields, can be observed and measured locally. On the other hand, in quantum field theory one usually interprets measurements in terms of particles. The fields used in the theory for the prediction of counting rates appear as (very useful, undoubtedly) theoretical constructs, imported from the classical theory. But what is their actual status in reality?

The conventional particle interpretation requires that a given state behaves like a multi-particle state at asymptotic times. A closer look shows that this feature may be expected only in certain circumstances, say, in a translationally invariant theory in states close to the vacuum. Once one leaves these situations, neither the concept of a vacuum (ground state of the energy) nor that of particles (eigenstates of the mass operator) keep a distinguished meaning, as may be exemplified by the occurrence of Hawking radiation, by the difficulties of a notion of particles in thermal states, and last but not least, in the infrared problem.

The field concept, on the other hand, keeps perfect sense in all known cases. Fields may be understood, generally speaking, as a means to interpret quantum theoretical objects in geometrical terms. In Minkowski space, they may assume the form of distributions whose values are affiliated to the algebras of local observables and which transform covariantly under Poincaré transformations. Here, the test function f plays the role of a "tag" which keeps track of the localization of the associated field operator $\varphi(f)$. In a generally covariant framework (see Sect. 8.1), fields can be viewed abstractly as natural transformations from the geometrically defined functor which associates test function spaces to spacetime manifolds, to the functor which associates to every spacetime its algebra of local observables [14].

On the mathematical side, the field concept leads to hard problems in the quantum theory. They are due to the quantum fluctuations of localized observables which diverge in the limit of pointlike localization. But in perturbation theory as well as in algebraic QFT one has learned to deal with these problems, the most difficult aspect being the replacement of ill-defined pointwise products by the operator product expansion.

In free field theory on Minkowski space, one associates to every particle a field which satisfies the field equation. While in this case, the use of the term "particle" for the associated field is perfectly adequate, the analogous practice for fields which appear in the classical equation of motion of *interacting* field theory is justified only in special cases. It may happen (this seems to be the case in asymptotically free theories) that in a short distance limit, the analogy to the particle–field correspondence of free field theory becomes meaningful. In theories which become free in the infrared limit, a similar phenomenon happens at large distances; then the scattering data can be directly interpreted in terms of these distinguished fields.

In general, however, besides the observable fields one uses a whole zoo of auxiliary fields which serve as a tool for formulating the theory as a quantization of a classical Lagrangean field theory. Such a formulation may not always exist nor must it be unique. In the functional ("path integral") approach to QFT, such auxiliary fields (which are not coupled to external sources) may be regarded as mere integration variables. The most powerful functional techniques involve deliberate changes in such variables (introduction of "ghost fields", BRST transformations or the renormalization program by successive integration over different energy scales). While this is by far the most successful way to construct models, at least in the sense of perturbation theory, the intrinsic physical significance of these auxiliary fields is unclear, and it would be misleading to think of them in terms of particles in a similar way as discussed before.

The delicacy of the field concept in quantum theory, contrasted with the clarity of the classical field concept, may be just one aspect of the more fundamental question: Is a quantum theory necessarily the quantization of a classical theory? Does it always have a classical limit (think of QCD, for the sake of definiteness), and can it be reconstructed from its classical limit?

5 The Perturbative Approach to QFT

The main approximative schemes for relativistic QFT are Perturbation Theory (or other expansions like the $1/N$ approximation) and lattice approximations of Euclidean functional integrals. All these approximations of QFT are based on the idea of "quantization of a classical field theory". Perturbation theory proceeds by producing a formal power series expansion in a coupling constant, hoped to be asymptotic to a QFT yet to be constructed, and therefore requires weak couplings; lattice approaches can in principle also treat strongly coupled regimes, using cluster expansions or Monte Carlo simulations; although numerical simulations of lattice QFT are limited to rather coarse lattices, aspects of the continuum and infinite volume limits can be studied. As far as comparisons are possible, there seems to be little doubt about the basic consistency among different approaches.

Our discussion in this section will mainly pertain to Perturbation Theory. This is a general scheme applicable to any QFT with a "free" dynamics perturbed by an "interaction" which is considered as a small correction. Locality requires the interaction to be described by a local density, called the interaction Lagrangean. Characteristic limitations to the scheme arise, however, through various sources which are described below.

First of all, there is the need to "renormalize" the single terms of the perturbative expansion. This is the procedure to fix the parameters of the theory to their physical values, thereby also avoiding any infinities that occur if one proceeds in the traditional way using "bare" parameters. One must demand that renormalization can be achieved without the introduction of infinitely many new parameters which would jeopardize the predictive power of the theory. This necessity restricts the admissible form of the interaction Lagrangean. Provided the polynomial order in the fields is limited (depending on the spacetime dimension, and on the spins of the fields involved), a simple "power counting" argument (controlling the behaviour of potentially divergent terms in terms of the momentum dependence of propagators and interactions) ensures renormalizability. For spins larger than 1, there are no interactions in four dimensions which are renormalizable by power counting. (This fact also prevents the direct incorporation of gravitational fields into the perturbative scheme.) In the presence of additional symmetries which ensure systematic cancellations of divergent terms, renormalizability might be shown in spite of a failure of the power counting criterium (but in supersymmetric perturbative gravity the desired effect seems to fail beyond the first few lowest orders).

For the theory of elementary particles, experiments have revealed the prime relevance of vector (spin 1) couplings, starting with parity violation in the weak interaction which could be explained by V–A but not by scalar and tensor couplings. The idea that vector couplings are mediated by vector fields lies at the basis of the standard model. For interactions involving massless vector fields, however, there is a conflict between locality, covariance and Hilbert space positivity, while massive vector fields do not possess couplings which are renormalizable by power counting. This is due to the fact that the necessary decoupling of modes which otherwise would give rise to states of negative norm changes the large-momentum behaviour of the propagator.

The only successful way to incorporate vector fields into a perturbative QFT is to treat them as gauge fields, with couplings which are necessarily gauge couplings (see Sect. 3). Thus, the gauge principle imposes itself through the inherent limitations of the perturbative scheme [15]. However, it brings about several new problems which have to be solved in turn: the unphysical degrees of freedom can be eliminated by cohomological methods ("BRST theory", see Sect. 3) which at the same time can be used to systematically control the preservation of gauge invariance. While gauge invariance forbids the introduction of explicit mass terms for the vector fields, masses can be generated by coupling to a Higgs field with "spontaneous symmetry breakdown" (see the next section for more details). That this can indeed be done in a way which keeps the theory renormalizable in spite of the bad power counting behaviour of massive propagators is one of the great achievements of the perturbative standard model.

In the process of renormalization there may appear "anomalies" which break symmetries present in the classical theory. While anomalies per se are not problematic (and may even be phenomenologically desirable), anomalies of the *gauge* symmetry will spoil the renormalizability. Their absence has therefore to be imposed as a consistency condition. In chiral gauge theories, it can be achieved by a suitable choice of representations of the gauge group (particle multiplets).

The circumstance that the "cascade of problems" outlined in the preceding paragraph can in fact be consistently overcome within the setting of perturbative QFT, and in excellent agreement with the phenomenology of High Energy Physics, gives enormous confidence in the basic correctness of the general framework. The standard model precisely exhausts the leeway admitted by the perturbative approach.

Besides the renormalization problems caused by ultraviolet singularities, perturbative QFT has infrared problems, when the free theory used as the starting point contains massless particles. In quantum electrodynamics (QED), the infrared problem can be traced to the computational use of particles with sharp masses which is illegitimate in the presence of massless particles (see Sect. 5).

A more severe kind of infrared problem arises in theories like QCD; here it is due to the fact that the fields (quarks and massless gluons) do not correspond to the massive particles (hadrons) presumably described by the full, non-perturbative theory. A fully consistent solution of these problems, i.e.

the confinement of hadronic constituents, can therefore not be expected in a perturbative treatment. If the confinement problem can be addressed at all, then only by non-perturbative methods (see the next section). However, effects like the deviations from naive scaling of hadronic structure functions have been successfully predicted by perturbative methods.

The infrared problems of perturbation theory may be circumvented by the use of interactions which are switched off outside some compact region of spacetime. This leads to the concept of causal perturbation theory which was developed by Epstein and Glaser [16] on the basis of previous ideas of Stückelberg and Bogoliubov. This approach is crucial for a consistent treatment of QFT on curved spacetimes. On Minkowski space it allows a perturbative construction of the algebra of observables. The infrared problem then is the physical question on the *states* of the theory, such as the existence of a ground state, the particle spectrum, thermal states etc.

Whether one considers the rationale for the gauge principle in the standard model outlined above (see also Sect. 3) to be logically cogent depends on the implicit expectations one imposes on the formal structure of a QFT. In any case, the standard model is by no means uniquely determined by these constraints. QED (given by the gauge group $U(1)$) and QCD (given by the gauge group $SU(3)$) are completely self-consistent subtheories (i.e., on the level of a formal perturbative expansion); the subtheory of electro weak interactions (given by the gauge group $U(1) \times SU(2)$ with parity-violating representations) is consistent provided the gauge anomalies are eliminated by suitable charged multiplets. The gauge groups themselves may be considered as free parameters of a model, as long as anomaly cancellation is possible.

The possibility of grand unification and/or supersymmetric extensions is an aesthetic feature of the standard model, for which, however, there is no fundamental physical need, nor is it required for reasons of mathematical consistency. QFT alone presumably cannot answer the question why there are so many "accidental" free parameters (notably the mass matrices or Yukawa couplings, according to the point of view) in the theory of fundamental interactions.

To conclude this section, we should point out that, as far as model building is concerned, the limitation to renormalizable interactions might be too narrow. There are perturbatively non-renormalizable model theories in which non-trivial fixed points have been established, meaning that the theories are non-perturbatively renormalizable [27].

6 The Constructive Approach to QFT

In spite of its tremendous numerical success, the perturbative scheme to evaluate QFT approximately suffers from a severe defect: it provides answers only in the form of formal, most likely *divergent* power series. The usual answer to this is that the series is an asymptotic expansion. But aside from the problem where to truncate the series in order to convert the formal power series into numbers, there is the fundamental question: asymptotic to what? There are well-known cases (such as the so-called "Φ_4^4 theory" of a self-interacting scalar field Φ in four spacetime dimensions) in which the perturbation expansion, according to the accumulated knowledge, is *not* an asymptotic expansion to any QFT and it may very well be that the most successful of all QFTs, quantum electrodynamics, also suffers from this disease.

The axiomatic approach, on the other hand, does not answer the question whether the axioms are not empty, i.e. whether any *non-trivial* QFTs satisfy them.

The constructive approach is in principle addressing both of these problems. On the one hand it attempts to show that the axiomatic framework of QFT is not empty, by mathematically constructing concrete non-trivial examples satisfying these axioms, and on the other hand it provides non-perturbative approximation schemes that are intimately related to the attempted mathematical constructions; the prime example are the lattice approximations to QFTs. Even where the goal of a mathematical construction of models satisfying all the axioms is not (yet) attained, this kind of approximative scheme differs in a fundamental way from the formal perturbative expansions: it produces approximate numbers which, if all goes right, *converge* to a limit that would be the desired construction.

The constructive approach (see for instance [17]) is based on a modification and generalization of Feynman's "sum over histories". The main modification is the transition from the indefinite Lorentz metric of Minkowski spacetime to a Euclidean metric; the return to the physical Lorentzian metric is expected to be manageable via the so-called "Osterwalder–Schrader reconstruction" [2] (see Sect. 2). The approach starts from a classical field theory, with dynamics specified by a Lagrangean. Formally one then proceeds by writing an ill-defined functional integral over all field configurations, weighted with a density given in terms of the classical action $S = \int \mathcal{L}\, dx$ depending on some fields collectively denoted by Φ; the expectation value of an "observable" $\mathcal{O}[\Phi]$ (a suitable function of the fields) would be given by

$$\langle \mathcal{O} \rangle = \frac{1}{Z} \int \mathcal{D}\Phi \; \mathcal{O}[\Phi] \; e^{-S[\Phi]} \, . \tag{12}$$

Here the symbol $\mathcal{D}\Phi$ is supposed to indicate a (non-existing) Lebesgue measure over the set of all field configurations Φ and Z a normalization constant.

To make mathematical sense of this, the theory first has to be "regularized" by introducing a finite spacetime volume and deleting or suppressing high frequencies (by an "ultraviolet cutoff"). The job of the constructive field theorist then consists of controlling the two limits of infinite volume ("thermodynamic limit") and restoring the high frequencies ("ultraviolet limit") by removing the cutoff; the latter can only be done if the parameters of the Lagrangean (and the observables of the theory) are made cutoff dependent in a suitable way – this procedure is the non-perturbative version of renormalization.

The constructive program has been completed only in spacetime dimensions less than four, but at least in these unrealistic cases it has shown that axiom systems such as Wightman's are not vacuous for interacting theories. In these low-dimensional cases it has also given a justification to the perturbative expansion by showing that it produces indeed an asymptotic expansion to the constructed QFTs.

A particularly useful way of introducing an ultraviolet cutoff consists in replacing the spacetime continuum by a discrete structure, a lattice. Together with the introduction of a finite spacetime volume one thereby reduces QFT to a finite dimensional "integral" (the quotation marks indicate that this "integral" is just some linear form for the fermionic degrees of freedom). In other words, QFT has been reduced to quadratures. The advantage of this is that QFT thereby becomes amenable to numerical evaluation; there is a whole industry of lattice field theory exploiting this fact, most notably in approximately evaluating the theory of strong interactions, QCD. The subject of lattice (gauge) field theory has been covered in detail in several books [18].

But the lattice approach is very important also for more fundamental reasons: it is the only known constructive approach to a non-perturbative definition of gauge field theories, which are the basis of the standard model. The constructive approach and the numerical procedures to extract infinite volume and continuum information from finite lattices are closely parallel:

Typically a lattice model produces its own dynamically generated scale ξ ("correlation length") which, unlike the lattice spacing, has a physical meaning. It may be defined – after the thermodynamic limit has been taken – by the exponential decay rate of suitable correlation functions, such as

$$\xi = -\lim_{n\to\infty} \frac{1}{|n|} \ln\langle \Phi(0)\Phi(n) \rangle \ , \tag{13}$$

where $\Phi(\cdot)$ stands for a field of the lattice theory and n is a tupel of integers labeling lattice points.

In a finite volume version, finite volume effects disappear exponentially fast, like $\exp(-L/\xi)$, with the size L of the volume. The thermodynamic limit can then be controlled numerically and often also mathematically, borrowing techniques from classical statistical mechanics.

The next step is to identify the dimensionless number ξ with a physical standard of length (e.g. some appropriate Compton wave length, say 1 fm), such that ξ lattice spacings equal 1 fm. The lattice points can then be relabelled by $x_i = (n_i/\xi)$ fm where the coordinates x_i have now acquired the dimension of length. Taking the lattice spacing to zero (i.e. taking the continuum limit) then amounts to sending the correlation length to infinity while keeping x_i fixed. The n-point correlation functions of a field in the continuum should therefore be defined as limits of the form

$$\langle \varphi(x_1)\dots\varphi(x_n) \rangle = \lim_{\xi\to\infty} \langle \Phi([x_1]\xi)\dots\Phi([x_n]\xi) \rangle \ Z(\xi)^{-\frac{n}{2}} \tag{14}$$

where $x = [x]$ fm and $\varphi(x)$ is the resulting continuum quantum field.

So the continuum limit requires to drive the parameters of the system (such as the coupling constants) to a point of divergent correlation length, i.e. a critical point in the language of statistical mechanics. $Z(\xi)$ is a "field strength renormalization" needed to prevent the limit from being 0.

This procedure makes it clear that the lattice spacing is a derived dynamical quantity proportional to $1/\xi$, not something to be specified beforehand. The inverse of the correlation length in the chosen physical units is the mass gap of the theory in physical units. The procedure of choosing the dynamically generated scale as the standard of length or mass leads generally to a phenomenon usually attributed to special features of perturbation theory: "dimensional transmutation". Let us explain this in a simple case, QCD with massless quarks: the only parameter of the lattice theory is the gauge coupling; since we find the continuum limit at the (presumably unique) critical point, this ceases to be an adjustable parameter. Instead we obtain a free scale parameter by the freedom of choosing a certain multiple of the correlation length as the standard of length (or a certain multiple of the inverse correlation length as the standard of mass). So we have traded a dimensionless parameter (the coupling constant) for a parameter with dimensions of a mass (e.g. the mass of the lightest particle).

Quite generally the particle spectrum of any QFT is extracted by looking at exponential decay rates of suitable correlation functions; when applied to QCD the lattice approach has been reasonably successful in reproducing the observed spectrum of baryons and mesons. It has also been successfully extended to the computation of weak decay matrix elements of hadrons. All this gives us confidence that QCD is indeed an appropriate description of the strong interactions.

On the side of mathematically rigorous construction, the success with gauge theories in four dimensions has been much more modest, even though some impressive work towards control of the continuum limit has been done by Balaban [19]. The problem is one of the seven "millenium problems" for whose solution the Clay Mathematics Institute has offered a prize of one million dollars [20].

There is another issue where the constructive approach via a spacetime lattice has helped to understand a fundamental property of the standard model: this is the "Higgs mechanism" of mass generation, whose phenomenological importance is made clear in the contribution by Dosch in this book. This mechanism comes into play when gauge fields are coupled to a scalar ("Higgs") field with a quartic self-interaction potential, symmetric around zero but with an orbit of minima away from zero.

Textbooks normally call this an instance of "spontaneous symmetry breaking", but this term is somewhat misleading and not really appropriate. On the lattice, gauge fixing is not necessary, and it is a general fact known as Elitzur's theorem [21] that it is not possible to spontaneously break local gauge invariance. The local gauge freedom prevents the occurrence of long range order since there is no energy penalty for locally "rotating" any fields. The Higgs mechanism appears instead as a conspiracy between the gauge fields and the fluctuations of the Higgs field along the orbit of its minimal potential energy, making the transverse components of the gauge field massive; no unphysical massless components remain in the spectrum [22, 23].

It has been pointed out by 't Hooft [24] that from this point of view there is no fundamental difference between confinement, as seen in QCD, and the Higgs mechanism, as it operates in the electroweak part of the standard model. The massive gauge bosons (the W and Z particles) may, for instance, be viewed as permanently bound combinations of bare gauge fields and Higgs constituents (not really particles), and similarly for the massive fermions, just as hadrons are viewed as permanently bound compounds of quarks and gluons. This point of view was worked out in more detail by Fröhlich, Morchio and Strocchi [25].

Once a gauge fixing is introduced, as is necessary in perturbation theory, in general it does no longer make sense to speak of *spontaneous* breaking of gauge invariance, since this invariance is broken explicitly. There are, however, classes of gauge fixings in which the global part of the symmetry remains intact. In these circumstances it remains meaningful to ask whether this global symmetry is spontaneously broken; this is then a problem to be studied with the methods of statistical mechanics. The answer, it turns out, depends both on the spacetime dimension and the precise form of the gauge fixing. In four dimensions, for an abelian model, only one particular gauge fixing (the so-called "Landau gauge") has been found to lead to spontaneous breaking of the remnant gauge symmetry [26]. From the point of view of physics, on the other hand, all versions, with different or no gauge fixing, should be equivalent; in this perspective spontaneous symmetry breaking is thus nothing but a gauge artefact.

7 Effective Quantum Field Theories

In applications one often encounters the term "effective field theory". We can distinguish three different meanings:

(1) The result of an exact renormalization group (RG) transformation applied to a QFT in the sense discussed before.
(2) An approximate QFT that is supposed to give a good approximation to a certain assumed QFT.
(3) A phenomenological theory that is not to be taken seriously beyond a certain energy; in this case it does not matter if the theory arises from a bona fide QFT by some approximation or by integrating out high-momentum modes.

The notion (1) is at least conceptually very clear. The idea is to start with an already constructed well-defined QFT and then to apply an exact "Renormalization Group step". This means that one performs the part of the functional integral (which has been made well-defined before) corresponding to the "hard" (i.e. high momentum, fast varying) part of the fields, formally

$$\exp(-S_{\text{eff}}[\Phi_{\text{soft}}]) = \frac{1}{Z} \int \mathcal{D}\Phi_{\text{hard}} \, \exp(-S_{\text{eff}}[\Phi_{\text{soft}} + \Phi_{\text{hard}}]) \tag{15}$$

and also performs some rescalings of fields and spacetime variables. The combination of the integration in (15) and this rescaling constitutes one renormalization group step. The resulting "effective theory" describes exactly the same physics as the original full theory when applied to the soft (low momentum, slowly varying) degrees of freedom. It is clear that this may be "effective", but it is not efficient because it requires control of the full theory before one can even start.

Of course, the RG step sketched above can be iterated; thereby one generates the semigroup usually called renormalization group.

A more useful variation of the RG idea is used in constructive QFT (see for instance [19, 27]). Here one starts with a regularized version of the theory, defined with a high-momentum cutoff; one then performs a number of RG steps as indicated above until one reaches a predefined "physical scale" leading to an effective low-energy theory still depending on the cutoff. In the final step one attempts to show that the effective low-energy theory has a limit as the cutoff is removed; this requires adjusting the parameters of the starting "bare action" such that the effect of the increasing number of successive renormalization group steps is essentially compensated.

The notion (2) is widely used to describe the low-energy physics of QCD (assumed to exist as a well-defined QFT even though this has not been shown so far). Specific examples are

- "Effective Chiral Theory" [28] to describe the interactions of the light pseudoscalar mesons,
- "Heavy Quark Effective Theory" (HQET) [29], in which the effect of the heavy (charmed, bottom and top) quarks is treated by expanding around the limit where their masses are infinite,
- "Nonrelativistic QCD" (NRQCD) [30, 31] used in particular for bound state problems of heavy quarks.

For an overview over various applications of these ideas see [32].

Examples for notion (3) are the old Fermi theory of weak interactions (before the electro-weak part of the standard model was known). A more modern example is presumably the standard model itself, because it contains the scalar self-interacting Higgs field which suffers from the presumed triviality of Φ_4^4 theories; the same applies to any other model involving Higgs fields. One often finds the words "something is only an effective theory"; this expresses the fact that the author(s) do not want to claim that their model corresponds to a true QFT.

8 Gravity

Given the state of affairs for the standard model of elementary particles, being comfortably well described by QFT as outlined in the previous sections, the "missing link" in our present conception of fundamental physics is the incorporation of the gravitational interaction into quantum physics (or vice versa).

For a review of classical gravity, we refer to the contribution by Ehlers to this book. Empirically, gravity is a theory valid at macroscopic scales only, and it is well known that, if extrapolated to very small scales (the Planck length), it becomes substantially incompatible with the quantum uncertainty principle ("quantum energy fluctuations forming virtual black holes"). This suggests that at small scales gravity needs modification, although one might as well argue conversely that at small scales gravity modifies quantum theory (by acting as a physical regulator for the UV problems of QFT, or possibly in a much more fundamental manner). The truth is not known, and one might expect that neither quantum theory nor gravity will "survive" unaffected in the ultimate theory.

Empirical evidence for this case is, of course, extremely poor due to the smallness of the Planck length. The most promising candidates for empirical evidence about effects of quantum gravity are astronomical observations of matter falling into black holes, cosmological remnants of the very early universe, or perhaps signals in accelerator experiments of "large extra dimensions", which in some theories are claimed to lead to an increase of the effective Planck length to a size accessible at accelerator energies. On the theoretical side, it is generally expected that black hole physics (Hawking radiation and Bekenstein entropy) represents the crucial point of contact. It appears very encouraging that both major approaches (string theory and canonical quantum gravity, see below), in spite of their great diversity, make more or less the same predictions on this issue. But it should be kept in mind that also Hawking radiation of black holes is far from being experimentally accessible.

The attempt to incorporate the gravitational interaction into quantum theory raises severe conceptual difficulties. Classical gravity being a field theory, QFT is expected to be the proper framework; but QFT takes for granted some fixed background spacetime determining the causal structure, as one of its very foundations, while spacetime should be a dynamical agent in gravity theory. This argument alone does not preclude the logical possibility of perturbative quantization of gravity around a fixed background, but on the other hand, the failure of all attempts so far which split the metric into a classical background part and a dynamical quantum part (cf. Sect. 2) should not be considered as a complete surprise or as a testimony against QFT.

On the other hand, the existing arguments against the quantization of gravity within a conventional QFT framework are not entirely conclusive. They are based on the most simple notion of renormalizability which demands that the renormalization flow closes within a finite space of *polynomial* couplings, thus giving rise to the limitation by power counting. It is conceivable, and there are indications that something in this way actually occurs (see the contribution by Lauscher and Reuter to this book), that a renormalization flow closes within a finite space of suitable *non-polynomial* Lagrangeans (which are present in classical Einstein gravity anyway). In this case, the renormalized theory also would contain only finitely many free parameters, and would have the same predictive power as a theory with polynomial interactions.

Taking the geometrical meaning of gravitational fields seriously, it is clear that the framework of QFT has to be substantially enlarged in order to accomodate a quantum theory of gravity. It is questionable whether this can be done by formal analogies between diffeomorphism invariance and gauge symmetry.

8.1 QFT on Gravitational Background Spacetime

An intermediate step on the way towards a theory of quantum gravity is a semiclassical treatment, where "matter" quantum fields are defined on classical curved spacetimes. This situation brings along severe technical and conceptual problems, since crucial tools of QFT in flat spacetime (energy–momentum conservation, Fourier transformation and analyticity, Wick rotation, particle interpretation of asymptotic scattering states) are no longer available due to the lack of spacetime symmetries.

Considerable progress in this direction has been made notably concerning the problem of the absence of a distinguished ground state (the vacuum). In globally hyperbolic spacetimes, the ground state can be substituted by a class of states (Hadamard states) which guarantee the same stability properties of quantum fields, and allow for a similar general set-up of causal perturbation theory as in flat space [33]. Of crucial importance is the incorporation of the principle of general covariance. It is realized as a covariant functor which associates to every globally hyperbolic spacetime its algebra of observables and which maps isometric embeddings of spacetimes to homomorphic embeddings of algebras. The interpretation of the theory is done in terms of covariant fields, which are mathematically defined as natural transformations from a geometrically defined functor which associates to every spacetime its test function space to the functor describing the QFT [14].

One may include into the set of quantum fields also the fluctuations of the metric field. One then has to impose the consistency condition that the result does not depend on the chosen split of the metric into a background field and a fluctuation field (this is essentially Einstein's equation in QFT). One may hope to obtain in this way reliable information on the "back reaction" of the energy of the quantum matter on the background. It remains, however, the bad power counting behaviour of quantum gravity which might point to limitations of the perturbative approach.

8.2 Non-commutative Spacetime

Taking into account the expectation that localization should be an operational concept which at very small scales is limited by the interference between quantum and gravitational effects, models of non-commutative spacetimes have been formulated which exhibit an intrinsic localization uncertainty. While these are definitely not more than crude models, in which gravity is not itself present but just motivates the localization uncertainty, it could be established that they are compatible with QFT; contrary to widespread hopes, however, the quantum structure of spacetime does not act as a "physical regulator" at the Planck scale for the ultraviolet behaviour of QFT [34].

8.3 Canonical Quantum Gravity

Other approaches to quantum gravity focus on the purely gravitational self-interaction. The most prominent ones, going under the name "Canonical Quantum Gravity", are built upon the geometric nature of classical gravity. In these approaches, the dynamics of three-dimensional (space-like) geometries is studied in a canonical framework. However, due to general covariance, the dynamics turns out to be strongly constrained, giving rise to severe complications (see the contribution by Giulini and Kiefer to this book.)

Within the general framework of canonical approaches, loop quantum gravity (LQG) has been pursued and developed furthest as a model for the structure of quantum spacetime [35] (see also the contributions by Thiemann and by Nicolai and Peeters to this book). It is asserted that the model can be supplemented by any kind of "conventional" matter (e.g. the standard model). It therefore denies every ambition towards a unified or unifying theory.

For these reasons, critical questions confronting the model with the requirements for a "true" theory of quantum gravity are more or less void. As for its intrinsic consistency and mathematical control, the model meets rather high standards, consolidating and improving previous attempts of canonical quantization of gravity.

The model predicts that geometric observables such as areas and volumes are quantized, with lowest eigenvalue spacings of the order of the Planck size. This feature appears most promising in that quantum deviations from classical geometry are derived as an output, with no classical ("background") geometry being used as an input.

On the other hand, one of the most serious flaws of LQG is the lack of understanding of its relation to gravity "as we know it", i.e. the construction of semiclassical states in which Einstein's general relativity at large scales is (at least in some asymptotic sense) restored.

Another, presumably related drawback of LQG (like any other model within the canonical approach to quantum gravity) is that in the physical Hilbert space, once it has been constructed, the Hamiltonian vanishes. Thus, the question of the nature of "time" evolution of the quantum gravitational states is presently poorly understood.

8.4 String Theory

A detailed discussion of successes and problems of string theory will be given in the contribution by Louis, Mohaupt and Theisen to this book. We will here restrain ourselves to some questions focussing on the intrinsic structure and the conceptual foundations of string theory, which appear quite natural to ask having in mind the benefits of axiomatic approaches in the case of QFT. Even if some of our questions might appear immodest, the theory being still under construction, they should be settled in some sense before string theory can be considered as a mature theory.

String theory is a quantum theory naturally including gravitational degrees of freedom in a unified manner along with "conventional" matter. Gravitons and other particles arise as different "zero modes" of strings which are the fundamental objects; higher-vibrational modes would correspond to undetected heavy particles (with masses far beyond accelerator energies). This fact is the prominent source of enthusiasm with the theory. (For a critical comparison of the achievements of string theory and of loop quantum gravity as candidates for the quantum theory of gravitation, see e.g. [36].)

The theory can successfully reproduce scattering cross sections for gravitons as they are expected in the lowest orders of perturbation theory with the Einstein–Hilbert action. In contrast to perturbation theory (cf. Sect. 5), the theory is believed to have a better UV behaviour due to the finite size of the string, but its alleged finiteness (or renormalizability) could not be established with the increasing understanding of higher-order contributions to string theory.

On the phenomenological side, it was hoped that a unified theory including the standard model of elementary particles would naturally emerge as an effective theory at low (compared to the Planck scale) energies, but these hopes were considerably reduced by an enormous number of possible "string vacua", destroying the predictive power of the theory.

String theory was originally formulated in a perturbative scheme, where spacetime appears just as a classical background. The dynamics of the string moving in this background is given by a two-dimensional conformal QFT (organizing its internal degrees of freedom), whose consistency requires the background to satisfy Einstein's equations. In the course of time it became clear that a consistent formulation of string theory has to take into account non-perturbative structures like duality symmetries, including the need to introduce higher-dimensional objects ("branes"). The presence of these classical objects is expected to be related to the question (although still far from answering it) of the quantum nature of spacetime itself [37].

Non-perturbative formulations of string theory are in the focus of most modern developments. Yet, the mathematical structure of non-perturbative string theory and the picture of spacetime and quantum gravity which emerges are at the present time not yet well understood beyond a huge body of heuristic imagination, based on the various duality symmetries of string theory and the "holographic principle" concerning the quantum degrees of freedom of general relativity. A most fascinating recent development is Maldacena's conjecture which states that non-perturbative string theory could be "equivalent" (in a sense involving duality) to a QFT, possibly even in four dimensions. The theory which started off to supersede QFT may in the end be equivalent to a QFT!

As a computational scheme, string theory is highly constrained and determined by its internal consistency. For this reason, it is often claimed to be a "unique" theory, hence it makes little sense to "axiomatize" string theory in a similar way as quantum field Theory was axiomatized (Sect. 2). Nevertheless, the justification of its computational rules deserves some critical scrutiny.

The central question is, which are the fundamental insights into the nature of physical laws (principles) that are implemented by string theory? Is string theory unique in doing so, or is it possibly only *one* consistent realization of the same principles? Accepted principles such as quantum uncertainty, locality and general relativity should be transcended by the new principles without recourse to (classical) notions outside the theory.

An important "message" from algebraic QFT is that the intrinsic invariant structure of the quantum observables are their algebraic relations such as local commutativity, rather than their description in terms of fields. (Neither the classical action nor Feynman diagrams are intrinsic; field equations and canonical commutation relations cannot even be maintained after quantization.) The "concrete" (Hilbert space) representations of these "abstract" algebraic relations determine the physical spectrum (masses, charges).

In this spirit, one would like to identify the intrinsic elements of string theory, and the structural relations which hold a priori among them. An intrinsic characterization would also turn claims such as the Maldacena conjecture into predictions that can be verified (or falsified).

It is generally agreed that a classical background manifold should not appear in an ultimate formulation of string theory. This is not only because the metric is expected to fluctuate, so that it is impossible to describe its expectation values in a particular state by a classical geometry. Since spacetime structures smaller than the string size cannot be probed, and hence cannot have an operational meaning, string theory is expected to produce a radically new concept of spacetime.

While string theory is an S-matrix theory, i.e. in a suitable limit it admits the computation of "on-shell" particle scattering amplitudes, "off-shell" string field theory has been rigorously constructed only without interactions [38]. The resulting theory may be viewed as a collection of infinitely many "ordinary" quantum fields, but their local commutativity cannot be ensured in a covariant way. The reason is that the constraints on the string degrees of freedom prevent the construction of sharply or only compactly localized observables on the physical (positive-definite) Hilbert space out of string fields defined on an indefinite space. In view of the previous remark, this conflict with the classical spacetime concept should not come as a surprise. The result underlines that defining localization in terms of a classical labelling of test functions is misleading. Instead, here is another instance where modular theory (see Sect. 2 and [10, 11]) can deploy its power: using Poincaré covariance, one can identify families of subalgebras which by their transformation and commutation properties behave like algebras of local observables localized in wedge regions. These algebraic properties should therefore be used as the definition of localization.

With interactions, the description in terms of an infinite tower of quantum fields is expected to survive, but the structure of the interactions (string corrections to the effective action) goes beyond the framework of local Lagrangean QFT. Correspondingly, string field theory (even in a regime where gravity can be neglected) is not expected to be a QFT in the sense of Sects. 2 or 5.

On the other hand, string theory exhibits a new fundamental symmetry called "duality". The Maldacena conjecture suggests that under a duality transformation, string theory could turn into a QFT. A clarification of the precise non-perturbative meaning of this conjecture is highly desirable, not least in view of the numerous and far-reaching implications drawn from it.

As an example, T-duality, relating vibrational and winding modes of a string, is a most characteristic symmetry of string theory. With the help of T-duality one can understand how a string fails to be able to probe certain singularities of a classical background [37]. Positing duality symmetry as an abstract fundamental symmetry is a promising candidate for an intrinsic structure of the theory which can be formulated without recourse to the classical picture of a string embedded into spacetime.

As for the intrinsic texture of string theory (assuming it to be a consistent theory), it would be desirable to understand in which sense its subtheories ("spacetime without matter", "QFT without Planck scale gravity") are separately consistent, or rather only effective theories obtained by a singular limit, which is regulated by the full theory.

While some of these questions might indeed rather reflect the authors' personal rooting in QFT (and also some lack of understanding of string theory), we think that they are urgent enough that expert string theorists should provide answers in order to legitimate string theory as a candidate for the Fundamental Unified Theory of all interactions.

9 Conclusions and Outlook

Whether the remaining gaps in the theory are merely of technical nature, or rather signal a fundamental shortcoming of QFT, is not known at present, and is by many researchers not considered as the most urgent question.

Instead, the prime concern at present is the clash between gravity and quantum theory, whose unification is considered as the (last) "missing link" in our conception of fundamental physics. There are promising candidate theories to achieve this ambitious goal, but none of them shares the same conceptual clarity as has been attained for QFT, nor are there empirical data available favouring or disfavouring either of them.

Unlike almost every historical precedent, the guiding principle at the frontiers of research in fundamental physics is therefore mainly intrinsic consistency, rather than empirical evidence. Every active researcher should be aware of the delicacy of such a situation.

It should be remarked that, while various lines of research presently pursued call basic notions such as geometry and symmetry into question, the basic rules of quantum theory are never challenged. One may be tempted to ascribe this fact to the solidity of our conceptual understanding of quantum physics, developed over several decades not least in the form of QFT.

Note to the References

There is a long list of standard textbooks on quantum field theory. The subsequent list of references leaves out most of them, as well as much of the "classical" research articles. Instead, it includes a number of less well-known articles, stressing some points which are relevant in our discussion but which do not belong to the common knowledge about quantum field theory.

References

1. D. Buchholz and R. Haag: The quest for understanding in relativistic quantum physics, J. Math. Phys. **41** (2000) 3674–3697.
2. K. Osterwalder and R. Schrader: Axioms for Euclidean Green's functions, Commun. Math. Phys. **31** (1973) 83; Commun. Math. Phys. **42** (1975) 281.
3. H. Bostelmann: Phase space properties and the short distance structure in quantum field theory, J. Math. Phys. **46** (2005) 052301.
4. S. Doplicher and J.E. Roberts: Why there is a field algebra with a compact gauge group describing the superselection structure in particle physics, Commun. Math. Phys. **131** (1990) 51–107.
5. D. Buchholz: The physical state space of quantum electrodynamics, Commun. Math. Phys. **85** (1982) 49.
6. D. Buchholz, M. Porrmann and U. Stein: Dirac versus Wigner: Towards a universal particle concept in local quantum field theory, Phys. Lett. **B 267** (1991) 377–381.
7. O. Steinmann: *Perturbative Quantum Electrodynamics and Axiomatic Field theory*, Springer-Verlag, Berlin etc. 2000.
8. Y. Kawahigashi and R. Longo: Classification of two-dimensional local conformal nets with $c < 1$ and 2-cohomology vanishing for tensor categories, Commun. Math. Phys. **244** (2004) 63–97.

9. R. Longo and K.-H. Rehren: Nets of subfactors, Rev. Math. Phys. **7** (1995) 567–598.
10. H.-J. Borchers: On revolutionizing quantum field theory with Tomita's modular theory, J. Math. Phys. **41** (2000) 3604–3673.
11. B. Schroer and H.-W. Wiesbrock: Modular constructions of quantum field theories with interactions, Rev. Math. Phys. **12** (2000) 301.
    H.-J. Borchers, D. Buchholz and B. Schroer: Polarization-free generators and the S-matrix, Commun. Math. Phys. **219** (2001) 125–140.
12. G. Lechner: An existence proof for interacting quantum field theories with a factorizing S-matrix, [arXiv:math-ph/0601022].
13. S. Weinberg: *The Quantum Theory of Fields*, Cambridge University Press, Cambridge, 1996.
14. R. Brunetti, K. Fredenhagen and R. Verch: The generally covariant locality principle: A new paradigm for local quantum physics, Commun. Math. Phys. **237** (2003) 31–68.
15. R. Stora: Local gauge groups in quantum field theory: Perturbative gauge theories, talk given at ESI workshop "Local Quantum Physics", Vienna (1997).
    D.R. Grigore: On the uniqueness of the non-Abelian gauge theories in Epstein-Glaser approach to renormalisation theory, Rom. J. Phys. **44** (1999) 853–913 [arXiv:hep-th/9806244].
    M. Duetsch and B. Schroer: Massive vector mesons and gauge theory, J. Phys. **A 33** (2000) 4317.
16. H. Epstein and V. Glaser: The role of locality in perturbation theory, Ann. Inst. H. Poincaré **A 19** (1973) 211.
17. J. Glimm and A. Jaffe: *Quantum Physics, a Functional Integral Point of View*, Springer-Verlag, Berlin etc. 1987.
18. E. Seiler: *Gauge Theories as a Problem of Constructive Quantum Field Theory and Statistical Mechanics*, Springer-Verlag, Berlin etc. 1982.
    M. Creutz: *Quarks, Gluons and Lattices*, Cambridge University Press, Cambridge 1983.
    H. J. Rothe: *Lattice Gauge Theories: An Introduction*, World Scientific, Singapore 1992;
    I. Montvay and G. Münster: *Quantum Fields on a Lattice*, Cambridge University Press, Cambridge 1994.
    J. Smit: *Introduction to Quantum Fields on a Lattice – A robust mate*, Cambridge University Press, Cambridge 2002.
19. T. Balaban: The large field renormalization operation for classical N-vector models, Commun. Math. Phys. **198** (1998) 493, and references therein.
20. A. M. Jaffe and E. Witten: Quantum Yang-Mills theory, The Millenium Problems, Official Problem Description, Clay Mathematics Institute 2000.
21. S. Elitzur: Impossibility of spontaneously breaking local symmteries, Phys. Rev. **D 12** (1975) 3978.
22. K. Osterwalder and E. Seiler: Gauge field theories on a lattice, Ann. Physics **110** (1978) 440.
23. E. Fradkin and S.H. Shenker: Phase diagrams of lattice gauge theories with Higgs fields, Phys. Rev. **D19** (1979) 3682.
24. G. 't Hooft: Why do we need local gauge invariance in theories with vector particles? An introduction, in: G. 't Hooft et al. (eds.), *Recent Developments in*

Gauge Theories, Proceedings NATO Advanced Study Institute, Cargèse 1979, Plenum, New York 1980.
25. J. Fröhlich, G. Morchio and F. Strocchi: Higgs phenomenon without symmtery breaking order parameter, Nucl. Phys. **B 190[FS3]** (1981) 553.
26. T. Kennedy and C. King: Spontaneous symmetry breakdown in the Abelian Higgs model, Commun. Math. Phys. **104** (1986) 327.
27. K. Gawedzki and A. Kupiainen: Renormalizing the nonrenormalizable, Phys. Rev. Lett. **54** (1985) 2191; Renormalization of a nonrenormalizable quantum field theory, Nucl. Phys. **B 262** (1985) 33; Gross-Neveu model through convergent perturbation expansions, Commun. Math. Phys. **102** (1985) 1.
28. J. Gasser and H. Leutwyler: Chiral perturbation theory: expansions in the mass of the strange quark, Nucl. Phys. **B 250** (1985) 465; Low-energy expansion of meson form factors, Nucl. Phys. **B 250** (1985) 517; $\eta \to 3\pi$ to one loop, Nucl. Phys. **B 250** (1985) 539.
29. E. Eichten and B. Hill: An effective field theory for the calculation of matrix elements involving heavy quarks, Phys. Lett. **B 234** (1990) 511; Static effective field theory: $1/m$ corrections, Phys. Lett. **B 243** (1990) 427.
30. W.E. Caswell and G.P. Lepage: Effective Lagrangians for bound state problems in QED, QCD, and other field theories, Phys. Lett. **B 167** (1986) 437.
31. G.T. Bodwin, E. Braaten and G.P. Lepage: Rigorous QCD analysis of inclusive annihilation and production of heavy quarkonium, Phys. Rev. **D 51** (1995) 1125; erratum Phys. Rev. **D 55** (1997) 5853.
32. A. Pich: Effective field theory (Les Houches lectures 1997), in: Probing the Standard Model of Particle Interactions, R. Gupta, A. Morel, E. de Rafael and F. David (eds.), Vol. 2, pp. 949–1049, North Holland, Amsterdam 1999 [arXiv:hep-ph/9806303].
33. S. Hollands and R.M. Wald: On the renormalization group in curved spacetime, Commun. Math. Phys. **237** (2003) 123–160.
34. D. Bahns, S. Doplicher, K. Fredenhagen and G. Piacitelli: On the unitarity problem in space/time noncommutative theories, Phys. Lett. **B 533** (2002) 178–181.
35. A. Ashtekar: Gravity and the quantum, New J. Phys. **7** (2005) 198.
36. L. Smolin: How far are we from the quantum theory of gravity? [arXiv:hep-th/0303185].
37. G. T. Horowitz: Spacetime in string theory, New J. Phys. **7** (2005) 201.
38. J. Dimock: Locality in free string field theory, J. Math. Phys. **41** (2000) 40; Ann. H. Poinc. **3** (2002) 613.

Part IV

General Relativity Theory

General Relativity

J. Ehlers

Max-Planck-Institut für Gravitationsphysik (Albert-Einstein-Institut), Am Mühlenberg 1, 14476 Golm, Germany
Juergen.Ehlers@aei.mpg.de

1 Introduction

This introduction is meant to indicate some properties of general relativity theory (GRT) which distinguish it from other branches of physics considered in this book, to relate it to other branches of physics and to mention some of its achievements and open problems. The subsequent chapters will give details.

1.1 GRT is the only empirically supported theory in which the spacetime structure is treated as dynamical, and not specified once and for all, independently of physical processes. Since the spacetime metric is interrelated to matter and field variables via field equations, the distinction between kinematics and dynamics is abolished in GRT.

Conceptually, the background independence must be seen as the principal achievement of GRT; it is, however, at the same time the main obstacle to overcome if GRT and quantum theory are to be united.

1.2 According to GRT the spacetime metric (and the connection and curvature derived from it) represents both the "metric" in the original sense – time, distance, causal order – and the gravitational inertial field; it unifies geometry, chronometry, gravity and inertia. (Einstein: "gravitational field and metric are manifestations of the same physical field".)

1.3 GRT may be viewed as encompassing in a coherent system all of *macroscopic*, phenomenological physics, from laboratory scales to cosmology.

1.4 So far, all physical theories, classical or quantum, employ a metric to represent matter or fields and their interactions. For this reason GRT is, in principle, a basic ingredient of physics even if gravitation is quantitatively negligible in many contexts. Since inertial mass is inseparable from active,

J. Ehlers: *General Relativity*, Lect. Notes Phys. **721**, 91–104 (2007)
DOI 10.1007/978-3-540-71117-9_5

gravity-producing mass, an ultimate understanding of mass can be expected only from a theory comprising inertia and gravity.

1.5 Mathematically, GRT is fairly well understood. Several physically interpreted exact solutions to its field equations, with and without matter, are known, as well as general existence and uniqueness theorems [1]. For complex realistic circumstances, perturbation schemes and numerical methods are available. There is, at least in principle, no interpretation problem.

1.6 The existence of a Lorentz metric, the most basic assumption of GRT, implies the approximate validity of special relativity theory (SRT) in spacetime regions which are small compared to the time and distance scale set by the curvature of spacetime. Even in neutron stars this scale is much larger than the scales relevant for the properties of bulk matter, atoms or nuclei. Therefore equations of state, cross sections, transport coefficients etc. derived from quantum theory can be incorporated into the classical matter models used in GRT in spite of the fact that these theories are in principle incompatible.

1.7 So far all experimental tests of GRT have supported the theory [2]. This concerns laboratory experiments which test the existence of a Lorentz metric or, equivalently, of local inertial frames; experiments with clocks, satellites and electromagnetic signals around the Earth and in the solar system, and the dynamics of binary pulsar systems including gravitational radiation damping. GRT has also been used increasingly to analysze and interprete astrophysical and cosmological phenomena. (Here one so far unexplained observation, the "pioneer anomaly" [3], deserves to be mentioned, which has been related tentatively to quintessence [4].)

1.8 At present, gravitational physics is one of the most active areas of research. Great efforts are being made to directly detect gravitational waves, with the prospect to open another window into the universe. Another goal is to find direct evidence supporting the assumption that the large concentrations of mass in the centers of galaxies are indeed black holes. High energy astrophysics offers additional challenges such as the explanation of gamma ray bursts. At a more conservative side, the investigation of gravitomagnetism, opened up by gravity probe B, might be mentioned. This shows that classical GRT is not a closed subject; compared to electrodynamics, gravitational physics has not yet reached the stage of Hertz's experiments.

1.9 The fundamental problem of unifying quantum theory and GRT is considered in other contributions to this book. Here I want to remark only that, in my view, "quantizing general relativity" is a rather inadequate way to address the problem. A unification presumably requires basic changes of quantum theory as well as of GRT, at least if the resulting theory is to remove infinites from both theories and to bring light to issues such as non-baryonic dark matter and dark energy.

2 Basic Assumptions of GRT

2.1 In GRT as well as in Newtonian physics and SRT, spacetime, the arena of directly perceivable phenomena, is represented as a connected real, 4-dimensional differentiable manifold M. This manifold is not generally identified with $\mathbf{R}^4$, however. M depends on the situation to be modelled; it can only be determined in connection with a solution to the field equation (see 3.5). M by itself has no physical meaning; it gets meaning only through fields defined on it.

The manifold M is assumed to carry a Lorentz metric $g_{\alpha\beta}$. This assumption guarantees that (i) SRT with its non-gravitational laws remains approximately valid locally even if gravitational fields are taken into account (2.2, 2.3); (ii) the connection $\Gamma^{\alpha}_{\beta\gamma}$ (or covariant derivative operator ∇_α) determined by $g_{\alpha\beta}$ provides a natural way to express the influence of gravity on "matter" (2.4 – 2.6); and (iii) the interaction between matter and gravity can be expressed via the curvature associated with $\Gamma^{\alpha}_{\beta\gamma}$ (2.7). Here "matter" is used to denote all physical entities besides $g_{\alpha\beta}$, i.e. everything which carries localizable energy and momentum.

The beauty of GRT is due to the fact that one mathematical object, the metric field $g_{\alpha\beta}$ and fields derived from it, provides all three aspects of gravity listed above, without the need to introduce additional structures.

On the other hand, the division of physical entities into the metric and "everything else" calls perhaps for a more democratic or, even better, a monistic structure which, however, apparently is not in sight.

2.2 Given a point ("event") x on a spacetime $(M, g_{\alpha\beta})$, there exists a coordinate system (x^α) on a neighbourhood N of x such that $x^\alpha = 0$ at x and, on N,

$$g_{\alpha\beta}(x^\varepsilon) = \eta_{\alpha\beta} + p_{\alpha\gamma\beta\delta}(x^\varepsilon)x^\gamma x^\delta \tag{1}$$

where $\eta_{\alpha\beta} = diag(1,1,1,-1)$, the functions p have the symmetries of the *curvature tensor* $R_{\alpha\gamma\beta\delta}$ associated with $g_{\alpha\beta}$, and $p_{\alpha\gamma\beta\delta}(0) = -\frac{1}{3}R_{\alpha\gamma\beta\delta}(0)$. For fixed x, such "normal" coordinates are unique up to Lorentz transformations.[1]

[1] The statements about normal coordinates are equivalent to a coordinate-independent fact: a neighbourhood of the zero vector of the tangent space at x can be mapped diffeomorphically onto a neighbourhood of x in M such that straight lines go into geodesics starting at x. In spite of their usefulness in GRT, (1) and (2) are rarely mentioned; I found it only in Pauli's relativity article in the Encyclopedia of Mathematics and Mathematical Sciences. The theorem is true for arbitrary dimensions and signatures; it holds if the metric is C^2. A proof has recently been given by B.G. Schmidt (unpublished) .

These coordinates satisfy

$$(at\ x) : g_{\alpha\beta} = \eta_{\alpha\beta}, g_{\alpha\beta,\gamma} = 0, g_{\alpha\beta,\gamma\delta} = \frac{2}{3} R_{\alpha(\gamma\delta)\beta} \tag{2}$$

Coordinates obeying the first line of (2) are called"locally inertial at x"; normal coordinates form a subclass of them.[2] The existence of normal coordinates indicates that any Lorentz metric can be approximated by the flat Minkowski metric $\eta_{\alpha\beta}$ in a region small compared to the curvature scale, $max|R_{\alpha\gamma\beta\delta}|^{-1/2}$, and (2) identifies the curvature tensor as a measure of an "intrinsic" gravitational field, i.e. one that cannot be "transformed away" by a coordinate change.

2.3 As a global restriction on physical spacetimes one assumes the manifold M to be orientable and $(M, g_{\alpha\beta})$ to be time-oriented. This last property means that there exists a continuous, never-vanishing timelike vector field which is said to point into the future. Timelike and lightlike vectors pointing into the same half of the null cone as that specified vector are then also called future pointing. These (rather weak) global restrictions are made to give meaning to the discrete symmetry operations T (time reversal) and P (parity), and to formulate local laws which presuppose a time-orientation such as the second law of thermodynamics, molecular chaos, or the quantum law for transition probabilities.

2.4 The existence of normal coordinates suggests the transfer of local physical laws from special to general relativity: formulate the law in SRT as a tensor equation with respect to inertial coordinates and substitute $g_{\alpha\beta}$, ∇_α for $\eta_{\alpha\beta}$, ∂_α, respectively, to obtain a tensorial GRT law. This law is seen to be identical to the original law at the origin x of any normal coordinate system, hence it will differ from its ancestor very little in a sufficiently small neighbourhood of any event x.

This rule is unambiguous if the SRT-law is algebraic or of first differential order. It provides a physical interpretation of the metric and the matter variables involved. The consistency of the laws so obtained is not implied by the rule itself.[3]

Simple consequences of this hypothesis are Einstein's generalized law of inertia freely; falling test particles have timelike geodesic world lines given by

---

[2] Given a timelike geodesic G, it is also possible to introduce local coordinates in a neighbourhood of G such that G is the "spatial origin", and such that the first two equations of (2) are valid on G. Such coordinates are "locally inertial on G" and represent Einstein's elevator better that those defined in the text. It is instructive to consider how "freely falling test masses" contained in a drag-free satellite realize, as precisely as possible, geodesics enclosed in a local inertial frame.

[3] Examples where difficulties arise have been discovered by H.A. Buchdal, G. Velo and D. Zwanziger. For discussion and refs. see, e.g., [5].

$$\ddot{x}^\alpha + \Gamma^\alpha_{\beta\gamma}\dot{x}^\beta\dot{x}^\gamma = 0 ; \tag{3}$$

ideal clocks measure proper time $\int |g_{\alpha\beta}dx^\alpha dx^\beta|^{1/2}$ along their (not necessarily geodesic) world line; light rays in vacuo correspond to lightlike geodesics.

The rule also supplies GRT-laws for classical matter models including kinetic theory and hydro-, elasto-, thermo- and electrodynamics. These matter models each contain an *energy–momentum tensor* $T^{\alpha\beta}$. The total energy–momentum tensor obeys, in agreement with the correspondence rule $SRT \to GRT$, the law

$$T^{\alpha\beta}{}_{;\beta} = 0 \tag{4}$$

which, because of the covariant derivative, is not a conservation law in the ordinary sense. This comes as no surprise since the gravitational field acts on matter. (See 3.8.)

The considerations of this section, which concern non-gravitational matter laws in gravitational fields, may be taken as an exact expression of (many formulations of) Einstein's heuristic "principle of equivalence".

2.5 Energy is usually asumed to be positive and to dominate stresses. Accordingly, $T^{\alpha\beta}$ is said to be energy dominated if its components with respect to any orthonormal basis satisfy

$$T^{00} \geq |T^{\alpha\beta}| \tag{5}$$

for all α, β.

Hawking [6] has shown: if (4) and (5) hold, and if $T^{\alpha\beta} = 0$ on a compact part S of a spacelike hypersurface, then $T^{\alpha\beta} = 0$ in the domain of dependence of S. Thus, matter obeying (4) and (5) cannot move faster than light into an empty region, since otherwise it could enter the domain of dependence of S from the outside of S. This result is remarkable, since (4) represents 4 equations for 10 unknowns; without (5) the conclusion does not hold.

2.6 In GRT the concept "free particle" is abandoned since all matter seems to be universally coupled to gravity. Accordingly the law of inertia is replaced in GRT by the geodesic law (3) to represent free fall. No concept of mass enters that law (or its predecessor, Galileo's law), though for historical reasons it is said to express the universal proportionality (or equality) of inertial and gravitational mass.

It follows from (3) that the relative position vector r^α of two infinitesimally close (in the sense of a variation), freely falling particles obeys the equation of *geodesic deviation*

$$\ddot{r}^\alpha = R^\alpha_{\beta\gamma\delta}\dot{x}^\beta\dot{x}^\gamma r^\delta \tag{6}$$

where the dot indicates covariant differentiation with respect to the proper time of one of the geodesics $\dot{x}^\alpha(\tau)$.

This equation characterizes the curvature tensor. It shows that the ordinary law of inertia, if expressed in terms of relative motions, holds, within the framework of Lorentzian spacetimes, if and only if the spacetime is flat, and it provides the interpretation of the curvature tensor as the gravitational tidal field.

2.7 So far, the assumptions which have been introduced hold in any "metric" theory of spacetime including SRT, since no field equation has been imposed on $g_{\alpha\beta}$.

To obtain a field equation relating $g_{\alpha\beta}$ to matter, Einstein assumed, in analogy to Poisson's law, an equation of the form

$$V^{\alpha\beta}(g_{..}, \partial g_{..}, \partial^2 g_{..}) = \kappa T^{\alpha\beta}$$

where the l.h.s. is a tensor-valued function depending on the arguments indicated; it is assumed to be linear in the second derivatives $g_{\alpha\beta,\gamma\delta}$.

Remarkably these assumptions determine $V^{\alpha\beta}$, as follows. Equation (2) shows: A function like $V^{\alpha\beta}$ can be expressed algebraically in terms of $g_{\alpha\beta}$ and $R_{\alpha\beta\gamma\delta}$ (specialize to the origin of normal coordinates), linearly in $R_{\alpha\beta\gamma\delta}$. Therefore $V^{\alpha\beta}$ must be a linear combination of $R^{\alpha\beta}$, $g^{\alpha\beta}$ and $Rg^{\alpha\beta}$ with constant coefficients. Hence, the looked-for field equation is equivalent to the "tracefree equation"

$$R_{\alpha\beta} - 1/4 g_{\alpha\beta} R = \kappa (T_{\alpha\beta} - 1/4 g_{\alpha\beta} T) \tag{7a}$$

and a relation involving the traces R, T.

Equation (7a), the contracted Bianchi identity, and (4) imply that $R + \kappa T$ is constant. Putting

$$R + \kappa T = 4\Lambda \tag{7b}$$

gives Einstein's gravitational field equation

$$R^{\alpha\beta} - 1/2 R g^{\alpha\beta} + \Lambda g^{\alpha\beta} = \kappa T^{\alpha\beta} \tag{7}$$

Equation (7a) may be considered as that part of the gravitational field equation which is independent of the "mechanical" or "matter" law (4), while (7b) expresses the compatibility of (7a) with (4). In this argument Λ appears as an integration constant.

In 1915, Einstein had assumed in addition that the Minkowski metric should satisfy the vacuum field equation. Then $\Lambda = 0$. In 1917 he added Λ to allow for a static model of the universe with pressureless matter. For a discussion of the present views on Λ in physics and cosmology, see Part VII.

By construction, the field equation (7) implies the energy–momentum law (4). Thus (7) accounts both for the inertia of matter and for its power to attract gravitationally. The constant $\kappa = 8\pi G c^{-4}$ is chosen such that for weak fields and slowly moving and weakly stressed matter, Newton's theory emerges as an approximation [7] (c = speed of light, G = Newtons's constant of gravity).

Just as in Newtonian gravity the Poisson equation contains the trace of the tidal field tensor, so in GRT (7) contains a "trace" $R^{\alpha\beta}$ of the curvature tensor.

3 General Comments on the Structure of GRT

3.1 The field equation (7) has physical meaning only if $T^{\alpha\beta}$ is specified; this specification always contains the metric. Mathematical studies often consider the vacuum case, $T^{\alpha\beta} = 0$, with or without Λ. Matter models studied in some detail include perfect fluids, electromagnetic fields, collisionless particle systems idealized by kinetic theory and, to a lesser extent, elastic bodies. In these cases the system of partial differential equations consisting of (7) and the relevant matter law admits a (locally) well-posed initial value problem.

A model of a physical system in GRT thus consists of a structure $(M, g_{\alpha\beta}, m)$, where m stands for matter variables. Two such models are physically equivalent if their underlying manifolds can be smoothly and bijectively mapped onto each other such that the fields $g_{\alpha\beta}, m$ of one model are mapped into those of the other one. Ideally, a particular model should be characterized by invariant properties. For example, Einsteins's static universe is characterized as the only static solution of (7) with pressureless matter, with the density ρ being the only independent invariant.

3.2 Contrary to appearance, the Einstein equation (7) does not imply matter to be the source that determines the gravitational potential $g_{\alpha\beta}$, for only (at least) the pair $(T^{\alpha\beta}, g_{\alpha\beta})$ describes matter, not $T^{\alpha\beta}$ by itself. Equation (7) states a mutual *inter-action* between metric and matter.

3.3 Equation (7) is incompatible with point particles as matter models. For static, stellar models the mass/radius ratio has an upper bound $c^2/2G$. The simplest "objects" of GRT which may be taken to replace mass points are black holes, see (4.5) below.

3.4 The tensors in (7) are symmetric. This follows from Einstein's assuptions stated in 2.6. The symmetry of the total energy–momentum tensor is, therefore, essentially a consequence of the assumption that gravity is completely represented by $g_{\alpha\beta}$ and fields derived from it. The same holds for the special kind of non-linearity ("self interaction") of the l.h.s. of (7).

3.5 A solution of (7) is usually constructed in some local coordinate system. Frequently the components $g_{\alpha\beta}$ in that system exhibit singularities. These may either be due to the choice of coordinates or to the existence of an intrinsic singularity. A solution can be considered as fully understood only if it has been maximally extended. A maximal solution may be free of singularities; otherwise its (suitably defined) boundary will be singular. The problems of finding maximal extensions and/or characterizing singularities are difficult; we know examples, but no general theorems.

3.6 The background independence, mentioned already in 1.1, is an important characteristic of GRT. Its meaning is not properly grasped by "general covariance", i.e. the possibility to formulate the laws such that arbitrary local coordinates may be used; that can be done for SRT as well as for Newton's theory. Rather, "absence of background" means that the laws of GRT, in contrast to those of Newtonian physics and SRT, do not presuppose the existence of an "absolute" spacetime structure which is specified categorically prior to dynamical laws and not influenced by physical processes.

In GRT the metric is said to be "dynamical". This involves two interrelated aspects: (i) a $g_{\alpha\beta}$-field is specifiable by independent initial data ("has degrees of freedom") which determine, together with matter data, its evolution (see Sect. 4.3), (ii) the $g_{\alpha\beta}$-field not only acts on matter as, e.g., via (4), but interacts with matter, (7).

The history of physics shows that some essential changes in the foundation of theories consisted in substituting dynamical structures for absolute ones. It appears to be generally accepted that a fundamental theory should be free of any background structure, i.e. its basic structures, not only those of spacetime, should be dynamical, not absolute ones. One might call this the "Mach–Einstein principle".

Historically, the formulation of a theory directly identified its absolute structures. Systematically and in general, it is difficult (if at all possible) to identify these structural elements of a theory unambiguously, especially since a theory may be based on different basic concepts. If, however, the variables and laws to be taken as basic are specified, the distinction absolute/dynamical is unambiguous, in my view.

The issue briefly considered here, and its relation to the principles of general covariance, general relativity and diffeomorphism invariance, is discussed carefully in the next chapter by D. Guilini. For a related discussion from a different viewpoint, see [8], Sects. 2.2.5 and 2.3.

3.7 The assumptions introduced in Sect. 2 are not independent. That light rays are given by lightlike geodesics, e.g., can be deduced from the generally covariant Maxwell equations, the geodesic law can be deduced from (4), (5), and a definition of "test particle", and dynamical clock models can be shown to exhibit proper time. It appears that GRT is semantically consistent, though a complete axiomatic has not been given.

3.8 As remarked in 2.3, (4) is not a conservation low; integration cannot transform it into a statement saying that the amount of energy contained in some finite volume changes only in accordance with a flux through the boundary. This fact cannot be remedied by adding to the matter energy tensor a gravitational energy tensor; according to GRT, such a tensor does not exist. The reason is simple: The *state* of a gravitational field, i.e. its Cauchy data, is given by some components of $g_{\alpha\beta}$ and its first partial derivatives. From these data at a point one cannot construct a tensor as required.

It *is* possible to find non–tensorial energy–momentum "complexes" which, added to $T^{\alpha\beta}$ (or to its densitized version), obey ordinary divergence equations in consequence of (7), and which give rise to non-tensorial integral conservation laws. Such complexes and laws are used in connection with approximation methods to express GRT-relations in familiar energy terms. It is possible, however, to describe all observable relations of GRT without such non-covariant tools.

In contrast to energy–momentum, scalar quantities like electric or baryonic charges do admit "decent" local and integral conservation laws since scalars at different events can be added unambiguously while vectors cannot.

4 Theoretical Developments, Achievements and Problems in GRT

4.1 Einstein's gravitational field equation (7) is the Euler–Lagrange equation associated with the action functional (with $c \equiv 1$)

$$A_D[g,m] = \int_D \left\{ \frac{1}{2\kappa}(R(g) - 2\Lambda) + L(g,m) \right\} dV \tag{8}$$

in which D denotes a compact domain of spacetime, g stands for the metric, m for matter variables and $dV = \sqrt{|det g_{\alpha\beta}|}d^4x$ is the invariant volume element of spacetime. Up to a divergence, the curvature scalar is the only invariant function of the metric and its derivatives (of any order) whose variational derivative is of second order in the metric. Up to a divergence, R is a quadratic form in the connection coefficients. The action density L of matter contains the metric and the connection, but not the curvature. The energy tensor is obtained as the variational derivative of L,

$$\frac{1}{2}T^{\alpha\beta} = \frac{1}{\sqrt{|g|}} \frac{\partial(\sqrt{|g|}L)}{\partial g_{\alpha\beta}} \tag{9}$$

Varying A with respect to g gives (7); varying it with respect to m gives the matter equations. These statements summarize the mathematical contents of Chap. 2 if the appropriate expressions for L are chosen.

For first-order Lagrangian field theories in SRT, the rule for generalizing them to GRT stated in 2.3 is equivalent to the simple device of substituting g and ∇ for η and ∂ *in the matter action density*. This prescription includes that no curvature term should be introduced into the matter action; this minimal coupling rule may be considered as a version of Einstein's equivalence principle. In this form the principle can be applied not only to the classical matter models mentioned in 2.3, but also to the formally classical, Lagrange-based standard model of particle physics. That requires, however, that spacetime is considered as the base manifold of a principal fibre bundle with structure group $U(1) \times SU(2) \times SU(3)$; see Part II.

The action (8) is also the starting point for Hamiltonian formulations of gravity, either in terms of metric variables or connection variables (see 4.2). These formulations make it possible to introduce *canonical variables* and to try canonical *quantization* of gravity.

4.2 In the standard model of particle physics, *principal connections* play the part of mediating interactions between massive particles. But although GRT was the first theory in which a connection appeared, besides objects related to linear representations of an underlying group, and the name "gauge" derives from Weyl's attempt to unify electromagnetism and gravitation, GRT is not a pure gauge theory since the gravitational connection $\Gamma^{\alpha}_{\beta\gamma}$ is not a basic field, but is derived from the metric. This is related to the fact that, in contrast to pure gauge theories, the points of the fibres of the SO(3,1) bundle over M are orthonormal frames of (M, g); the bundle space is said to be soldered to the base space. This special role of the spacetime connection shows up in the gravitational Lagrangian density $R = g^{\alpha\beta} R_{\alpha\beta}$, which is linear, not quadratic, in the curvature like that of Maxwell and Yang–Mills fields. It appears that this is another characteristic feature of gravity which distinguishes it from the other fundamental interactions. In gravity, the gauge potential $\Gamma^{\alpha}_{\beta\gamma}$ itself derives from a potential, the metric.

A historical remark: the procedure

$$g_{\alpha\beta} \rightarrow \Gamma^{\alpha}_{\beta\gamma} \rightarrow R^{\alpha}_{\beta\gamma\delta} \tag{10}$$

consisting of two non-linear steps of first differential order leads from a tensor via a connection to a tensor. The impossibility to form tensors from $g_{\alpha\beta}$ by differentiation without the intervention of a non-tensorial field was one of the obstacles Einstein had to overcome on his arduous way to his general theory of relativity. Connections as quantities independent of a metric were introduced only in 1918 by J.A. Schouten and H. Weyl after T. Levi-Civita's introduction of metric connections in 1917. While only the second step in (10) occurs in gauge theories, the first step is peculiar to gravity; it was a step which enabled Einstein to make the metric dynamical and to identify it with (a new kind of) gravitational potential.

4.3 An essential test for the viability of a classical field theory is whether its field equations admit a well-posed initial value problem. Solving the Cauchy problem requires identifying initial data and, thereby, states and degrees of freedom, as well as to determine the dependence of the evolved field on its data, i.e. the causal behaviour of the field.

Carrying out this analysis for Einsteins's field equation (7), without or with coupling to matter, turned out to be difficult for reasons which can all be traced back to diffeomorphism invariance. Here I report only the main results without technical details.

The equations split into two subsets. One of these imposes conditions, usually in the form of non-linear elliptic partial differential equations, on the initial data specified on a 3-dimensional Riemannian space (constraint equations). The free data for the gravitational field turn out to correspond to two degrees of freedom per space point, as in the case of ordinary electromagnetism. The second subset consists of the evolution equations. After imposition of coordinate conditions, these turn out to be hyperbolic, non-linear wave equations. For all matter models mentioned in this survey,[4] classical ones as well as Dirac and Yang–Mills fields, the outermost characteristics turn out to be lightlike hypersurfaxes, i.e. wave fronts propagating with fundamental speed c. This expresses *Einstein causality.*

Hyperbolicity means the following: the laws imply relations between the fields within finite domains of spacetime, relations which are not affected by the fields outside that domain. The laws, and data on a compact part S of space, uniquely determine the fields in the (finite) domain of dependence of S. This kind of determinism is fundamentally different from that of Laplace which requires data on the whole, infinite space at one instant.

A further important fact is that the first set is preserved under the evolution. Thus, later states of the field again satisfy the so-called "constraint equations". Finally, irrespective of coordinate conditions, the evolved field is determined by the data uniquely up to diffeomorphisms in the domain of dependence of the data.

Spacetimes determined by initial data are said to be globally hyperbolic; their manifolds M are products of a 3-manifold "space" and a 1-manifold "time". The initial value problem for the vacuum field equation with $\Lambda = 0$ has been used to prove the existence of global, singularity free, asymptotically flat spacetimes filled with gravitational radiation only. Such solutions arise from initial data close to trivial data giving flat spacetime. They describe how incoming gravitational radiation scatters on itself and propagates out again. Theorems about global solutions with Λ are also known.

4.4 An important task for any gravitation theory it the modelling of an isolated system such as a single star, the solar system or a binary star system far removed from other bodies. All quantitative tests of the field equation (7)

---

[4] For models of bulk matter such as fluids, equations of state have to be restricted, however, to exclude, e.g., superluminal sound waves. This holds in SRT already.

are based on approximate solutions to such spacetimes. In this subsection we put $\Lambda = 0$.

One expects the spacetime of an isolated system to resemble flat spacetime at large distance from the bodies. To express that asymptotic behaviour R. Penrose proposed to rescale the metric $g_{\alpha\beta} \to \Omega^2 g_{\alpha\beta}$ and to let Ω tend to zero at large physical distances such that one can attach a "boundary at infinity" where $\Omega = 0, \Omega_{,\alpha} \neq 0$. The boundary consists of ideal endpoints of outgoing and incoming light rays, respectively, and of spacelike infinity. Such spacetimes may contain outgoing and/or incoming gravitational and electromagnetic radiation. Some exact implications of the vacuum field equation about the asymptotic behaviour of such radiation have been derived, but the motion of bodies emitting radiation so far is the domain of analytical, post-Newtonian approximations and, increasingly, numerical relativity.

For asymptotically flat spacetimes a constant *total 4-momentum* at spacelike infinity has been defined, and a celebrated result says that it is future-directed timelike; so it has positive energy (except, of course, for Minkowski spacetime, where it is zero), provided $T^{\alpha\beta}$ is energy dominated. A total 4-momentum at null infinity whose (positive) energy decreases towards the future according to the amount of the outgoing radiation has also been defined.

4.5 The vacuum field equation with $\Lambda = 0$ has asymptotically flat, particle like solutions, *black holes*. Their stationary states are characterized by only three parameters, namely mass, angular momentum and charge. The outer part of a black hole spacetime, connected to infinity, is separated from an interior part by a horizon which acts as a one-way membrane: test particles and radiation can pass through from the outside only, not from the inside.

A *thermodynamic* of black holes has been elaborated [9]; its relation to statistical mechanics, quantum and string theories is a subject of current research.

In astrophysics, black holes are considered as objects which may form when a massive star collapses at the end of its thermonuclear evolution. They are also thought to exist at the centres of most galaxies. Efforts are being made to observe spectroscopic features characteristic of the geometry near a horizon.

4.6 *Light cones* not only govern the propagation of electromagnetic and gravitational radiation, but they determine causal relations, too. While in flat spacetime light cones are, apart from their vertices, smooth hypersurfaces, in curved spacetimes they have self-intersections and caustics. Observationally these geometric properties show up as the phenomena of *gravitational lensing* [10]. Distant galaxies, e.g., are observed in different images which differ in brightness and shape. Modelling such phenomena has become a useful tool in astronomy for determining the masses and mass distributions of the deflecting matter including dark matter. The successes of such modelling provide direct evidence for spacetime curvature. They support the light deflection measurements which followed, with ever increasing accuracy, the famous solar eclipse measurements of 1919.

4.7 Relativistic celestial mechanics which began with Einstein's perihelion paper of 1915 has been much developed since about 1980, in the form of post-Newtonian dynamics whose approximate equations of motion now include corrections of Newton's laws of order up to $(\frac{v}{c})^7$, where v is a typical relative speed, e.g., in a binary system.

This theory, or rather its first post-Newtonian version which includes only $(\frac{v}{c})^2$ corrections, has been used to test whether GRT-predictions of relations between observable parameters agree with real observations made on binary systems composed of neutron stars. So far, agreement prevails, which is very remarkable in view of the precision of the data. The predictions include the slowing down rates of the orbital periods, and the agreement with measured values has given indirect evidence for the existence of gravitational waves.

The higher-order approximations are applied tentatively to late stages of compact binaries when the components approach each other ever closer until they "plunge" together to form a single object, perhaps a black hole. Such processes are thought to emit bursts of gravitational radiation which might be detected by gravitational wave interferometers which have started to operate.

Selected References

1. Schmidt, B.G. (ed.), Einstein's Field Equations and Their Physical Applications, Lect. Notes Phys. 540, Springer 2000.
2. Will, C.M., Theory and Experiment in Gravitational Physics, CUP, 1993.
3. Anderson, J.D. et al., Study of the anomalous acceleration of Pioneer 10 and 11, Phys. Rev. D **65** 082004 (2002) and gr-qc/0104064, March 2005.
4. Mbelek, J.P., General Relativity and Quintessence Explain the Pioneer Anomaly, gr-qc/0402088, 2006.
5. Frauendiener, J., On the Velo-Zwanziger Phenomena, J. Phys. **A 36**, 8433, 2002.
6. Hawking, S.W. and Ellis, G.F.R., The Large Scale Structure of Space-Time, CUP, 1973.
7. Ehlers, J., Newtonian Limit of General Relativity, pp. 503–509 in Encyclop. of Mathem. Physics *3*, Francoise, J. J. et al. (eds.), Elsevier, Amsterdam, 2006.
8. Rovelli, C. Quantum Gravity, (Cambridge University Press, Cambridge, 2004).
9. Wald, R.M., Quantum Field Theory in Curved Spacetime and Black Hole Thermodynamics, University of Chicago Press, 1994.
10. Schneider, P., Ehlers, J. and Falco, E.E., Gravitational Lenses, Springer, 1992.

In addition, the following references – not cited in the text – provide perspective and detailed references:

Hawking, S.W. and Israel, W. (eds.) 300 Years of Gravitation, CUP, 1987.

Classical and Quantum Gravity, **16**, No. 12A, 1999.

Straumann, N., General Relativity, Springer, 2004.
Monas, L. and Diaz-Alonso J. (eds.), A Century of Relativity Physics, AIP Conference Proc., N.Y. 2006.

Remarks on the Notions of General Covariance and Background Independence

D. Giulini

Max-Planck-Institut für Gravitationsphysik, Am Mühlenberg 1,
14476 Golm/Potsdam
domenico.guiliniaei.mpg.de

1 Introduction

It is a widely shared opinion that *the* most outstanding and characteristic feature of general relativity is its manifest *background independence*. Accordingly, those pursuing the canonical quantization programme for general relativity see the fundamental virtue of their approach in precisely this preservation of 'background independence' (cf. Kiefer's and Thiemann's contributions). Indeed, there is no disagreement as to the background dependence of competing approaches, like the perturbative spacetime approach[1] (see the contribution by Lauscher and Reuter) or string theory (see the contribution by Louis, Mohaupt, and Theisen, in particular their Sect. 10). Accordingly, many string theorists would subscribe to the following research strategy:

> Seek to make progress by identifying the background structure in our theories and removing it, replacing it with relations which evolve subject to dynamical laws. ([18], p. 10).

But how can we reliably identify background structures?

There is another widely shared opinion according to which the principle of *general covariance* is devoid of any physical content. This was first forcefully argued for in 1917 by Erich Kretschmann [11] and almost immediately accepted by Einstein [20] (Vol. 7, Doc. 38, p. 39), who from then on seemed to have granted the principle of general covariance no more physical meaning than that of a formal heuristic concept.

From this it appears that it would not be a good idea to define 'background independence' via 'general covariance', for this would not result in a

[1] Usually referred to as the 'covariant approach', since perturbative expansions are made around a maximally symmetric spacetime, like Minkowski or DeSitter spacetime, and the theory is intended to manifestly keep covariance under this symmetry group (i.e. the Poincaré or the DeSitter group), not the diffeomorphism group!

D. Giulini: *Remarks on the Notions of General Covariance and Background Independence*, Lect. Notes Phys. **721**, 105–120 (2007)
DOI 10.1007/978-3-540-71117-9_6

physically meaningful selection principle that could effectively guide future research. What would be a better definition? 'Diffeomorphism invariance' is the most often quoted candidate. What precisely is the difference between general covariance and diffeomorphism invariance, and does the latter really improve on the situation? These are the questions to be discussed here. For related and partially complementary discussions, which also give more historical details, we refer to [14, 15] and [4] respectively.

As a historical remark we recall that Einstein quite clearly distinguished between the *principle of general relativity* (PGR) on one hand, and the *principle of general covariance* (PGC) on the other. He proposed that the formal PGC would imply (but not be equivalent to) the physical PGR. He therefore adopted the PGC as a heuristic principle, guiding our search for physically relevant equations. But how can this ever work if Kretschmann is right and hence PGC devoid of any physical content? Well, what Kretschmann precisely said was that *any* physical law can be rewritten in an equivalent but generally covariant form. Hence general covariance alone cannot rule out any physical law. Einstein maintained that it did if one considers the aspect of 'formal simplicity'. Only those expressions which are formally 'simple' after having been written in a generally covariant form should be considered as candidates for physical laws. Einstein clearly felt the lack for any good definition of formal 'simplicity', hence he recommended to experience it by comparing general relativity to a generally covariant formulation of Newtonian gravity (then not explicitly known to him), which was later given by Cartan [5, 6] and Friedrichs [9] and which did not turn out to be outrageously complicated, though perhaps somewhat unnatural. In any case, one undeniably feels that this state of affairs is not optimal.

2 Attempts to Define General Covariance and/or Background Independence

A serious attempt to clarify the situation was made by James Anderson [2, 3], who introduced the notion of *absolute structure* which here we propose to take synonymously with background independence. This attempt will be discussed in some detail below. Before doing this we need to clarify some other notions.

2.1 Laws of Motion: Covariance versus Invariance

We represent spacetime by a tuple (M, g), where M is a four-dimensional infinitely differentiable manifold and g a Lorentzian metric of signature $(+,-,-,-)$. The global topology of M is not restricted a priori, but for definiteness we shall assume a product-topology $\mathbb{R} \times S$ and think of the first factor as time and the second as space (meaning that g restricted to the tangent spaces of the submanifolds $S_t := \{t\} \times S$ is negative definite and positive definite along $\mathbb{R}_p := \mathbb{R} \times \{p\}$). Also, unless stated otherwise, the Lorentzian metric g is assumed to be at least twice continuously differentiable. We will generally not need to assume (M, g) to be geodesically complete.

Being a C^∞-manifold, M is endowed with a maximal atlas of coordinate functions on open domains in M with C^∞-transition functions on their mutual overlaps. Transition functions relabel the points that constitute M, which for the time being we think of as recognizable entities, as mathematicians do. (For physicists these points are mere 'potential events' and do not have an obvious individuality beyond an actual, yet unknown, event that realizes this potentiality.) Different from maps between coordinate charts are global diffeomorphisms on M, which are C^∞ maps $f : M \to M$ with C^∞ inverses $f^{-1} : M \to M$. Diffeomorphisms form a group (multiplication being composition) which we denote by Diff(M). Diffeomorphisms act (mostly, but not always, naturally) on geometric objects representing physical entities, like particles and fields.[2] The transformed geometric object has then to be considered a priori as a *different* object on the *same* manifold (which is not meant to imply that they are necessarily physically distinguishable in a specific theoretical context). This is sometimes called the 'active' interpretation of diffeomorphisms to which we will stick throughout.

Structures that obey equations of motion are, e.g., particles and fields. Classically, a structureless *particle* (no spin etc.) is mathematically represented by a map *into* spacetime:

$$\gamma : \mathbb{R} \to M \, , \tag{1}$$

such that the tangent vector field $\dot{\gamma}$ is everywhere timelike, i.e. $g(\dot{\gamma}, \dot{\gamma}) > 0$. Other structures that are also represented by maps *into* spacetime are strings, membranes, etc.

A *field* is defined by a map *from* spacetime, i.e.

$$\Phi : M \to V \tag{2}$$

where V is some vector space (or, slightly more general, affine space, to include connections). To keep the main argument simple we neglect more general situations where fields are sections in non-trivial vector bundles or non-linear target spaces.

Let γ collectively represent all structures given by maps into spacetime and Φ collectively all structures represented by maps from spacetime. Equations of motions usually take the general symbolic form

$$\mathcal{F}[\gamma, \Phi, \Sigma] = 0 \tag{3}$$

which should be read as equation for γ, Φ *given* Σ.

[2] For example, diffeomorphisms of M lift naturally to any bundle associated to the bundle of linear frames and hence act naturally on spaces of sections in those bundles. In particular, these include bundles of tensors of arbitrary ranks and density weights. On the other hand, there is no natural lift to, e.g., spinor bundles, which are associated to the bundle of *orthonormal* frames (which are only naturally acted upon by isometries, but not by arbitrary diffeomorphisms).

Σ represents some *non-dynamical* structures on M. Only if the value of Σ is prescribed do we have definite equations of motions for (γ, Φ). This is usually how equations of motions are presented in physics: solve (3) for (γ, Φ), *given* Σ. Here only (γ, Φ) represent physical 'degrees of freedom' of the theory to which alone observables refer (or out of which observables are to be constructed). By 'theory' we shall always understand, amongst other things, a definite specification of degrees of freedom and observables.

The group Diff(M) acts on the objects (γ, Φ) (here we restrict the fields to tensor fields for simplicity) as follows:

$$(f, \gamma) \to f \cdot \gamma := f \circ \gamma \qquad \text{for particles etc. ,} \tag{4a}$$

$$(f, \Phi) \to f \cdot \Phi := D(f_*) \circ \Phi \circ f^{-1} \qquad \text{for fields etc. ,} \tag{4b}$$

where D is the representation of $GL(4,\mathbb{R})$ carried by the fields. In addition, we require that the non-dynamical quantities Σ to be geometric objects, i.e. to support an action of the diffeomorphism group.

Definition 1. *Equation (3) is said to be* ***covariant*** *under the subgroup* $G \subseteq$ Diff(M) *iff*[3] *for all* $f \in G$

$$F[\gamma, \Phi, \Sigma] = 0 \Leftrightarrow F[f \cdot \gamma\, ,\, f \cdot \Phi\, ,\, f \cdot \Sigma] = 0\ . \tag{5}$$

Definition 2. *Equation (3) is said to be* ***invariant*** *under the subgroup* $G \subseteq$ Diff(M) *iff for all* $f \in G$

$$F[\gamma, \Phi, \Sigma] = 0 \Leftrightarrow F[f \cdot \gamma\, ,\, f \cdot \Phi\, ,\, \Sigma] = 0\ . \tag{6}$$

Note the difference: in Definition 2 the non-dynamical structures Σ are the same on both sides of the equation, whereas in Definition 1 they are allowed to be also transformed by $f \in$ Diff(M). Covariance merely requires the equation to 'live on the manifold', i.e. to be well defined in a differential-geometric sense, whereas an invariance is required to transform solutions to the equations of motions to solutions of the *very same* equation,[4] which is a much more restrictive condition.

As a simple example, consider the vacuum Maxwell equations on a fixed spacetime (Lorentzian manifold (M, g)):

$$dF = 0\ , \tag{7a}$$

$$d \star F = 0\ , \tag{7b}$$

[3] I use 'iff' as an abbreviation for 'if and only if'.

[4] In the mathematical literature this is called a symmetry (of the equation). We wish to avoid the term 'symmetry' here altogether because that – in our terminology – is reserved for a further distinction of invariances into *symmetries*, which change the physical state, and *redundancies* (gauge transformations) which do not change the physical state. Here we will not need this otherwise very important distinction.

where F denotes the 2-form of the electromagnetic field and d the exterior differential. The $\star$ denotes the (linear) 'Hodge duality' map, which in components reads

$$\star F_{\mu\nu} = \tfrac{1}{2}\varepsilon_{\mu\nu\alpha\beta}F^{\alpha\beta} \,, \tag{8}$$

and which depends on the background metric g through ε and the operation of raising indices: $F^{\alpha\beta} := g^{\alpha\mu}g^{\beta\nu}\,F_{\mu\nu}$.[5] The system (7) is clearly Diff(M)-covariant since it is written purely in terms of geometric structures on M and makes perfect sense as equation on M. In particular, given any diffeomorphisms f of M, we have that $f \cdot F$ satisfies (7a) iff F does. But it is *not* likewise true that $d \star F = 0$ implies $d \star f \cdot F = 0$. In fact, it may be shown[6] that this is true iff f is a conformal isometry of the background metric g, i.e. $f \cdot g = \lambda\, g$ for some positive real-valued function λ on M. Hence the system (7) is not Diff(M)-invariant but only G-invariant, where G is the conformal group of (M, g).

2.2 Triviality Pursuit

Covariance Trivialized (Kretschmann's Point)

Consider the ordinary 'non-relativistic' diffusion equation for the $\mathbb{R}$-valued field ϕ (giving the concentration density):

$$\partial_t \phi = \kappa \Delta \phi \,. \tag{9}$$

This does not look Lorentz covariant, let alone covariant under diffeomorphisms. This changes if it is rewritten in the following form

$$\{n^\mu \nabla_\mu - \kappa(n^\mu n^\nu - g^{\mu\nu})\nabla_\mu \nabla_\nu\}\,\phi = 0 \,. \tag{10}$$

Here $g^{\mu\nu}$ are the contravariant components of the spacetime metric (recall that we use the 'mostly minus' convention for its signature), ∇_μ is the associated Levi-Civita covariant derivative, and n^μ is a normalized covariant-constant timelike vector field which gives the preferred flow of time encoded in (9) (i.e. on scalar fields $\partial_t = n^\mu \nabla_\mu$). Equation (10) has the form (3) with no γ, $\Phi = \phi$, and $\Sigma = (g^{\mu\nu}, n^\mu)$ and is certainly diffeomorphism covariant in the sense of Definition 1. The largest invariance group – in the sense of Definition 2 – is given by that subgroup of Diff(M) whose elements stabilize the non-dynamical structures Σ. We write

$$\mathrm{Stab}_{\mathrm{Diff(M)}}(\Sigma) = \{f \in \mathrm{Diff(M)} \mid f \cdot \Sigma = \Sigma\} \tag{11}$$

---

[5] Note that in 3+1 dimensions this means that the $\star$ operation only depends on the conformal equivalence class of g, since $g^{\alpha\beta}g^{\gamma\delta}\sqrt{|\det\{g_{\mu\nu}\}|}$ is invariant under $g_{\mu\nu} \mapsto \Omega^2\, g_{\mu\nu}$. Accordingly, in this case, it is only the conformal equivalence class of g and not g itself that should be identified with Σ.

[6] This is true in 1+3 dimensions. In other dimensions higher than two, f must even be an isometry of g.

In our case, $\mathrm{Stab}_{\mathrm{Diff(M)}}(g)$ is the 10-parameter Poincaré group. In addition, f stabilizes n^μ if it is in the 7-parameter subgroup $\mathbb{R} \times E(3)$ of time translations and spatial Euclidean motions.

This example already shows (there will be more below) how to proceed in order to make any theory covariant under Diff(M). As already noted, Diff(M)-covariance merely requires the equation to be well defined in the sense of differential geometry, i.e. it should live on the manifold. It seems clear that any equation that has been written down in a special coordinate system on M (like (9)) can also be written in a Diff(M)-covariant way by introducing the coordinate system – or parts of it – as background geometric structure. This is, in more modern terms, the formal core of the critique put forward by Erich Kretschmann in 1917 [11].

Invariance Trivialized

Given that an equation of the form (3) is already G-covariant, we can equivalently express the condition of being G-invariant by

$$F[\gamma, \Phi, \Sigma] = 0 \Leftrightarrow F[\gamma, \Phi, f \cdot \Sigma] = 0, \quad \forall f \in G, \tag{12}$$

i.e. any solution of the equation parameterized by Σ is also a solution of the *different* equation parameterized by $f \cdot \Sigma$. Evidently, the more non-dynamical structures there are, the more difficult it is to satisfy (12). In generic situations it will only be satisfied if $G = \mathrm{Stab}_{\mathrm{Diff(M)}}(\Sigma)$. Hence, in distinction to the covariance group, increasing the amount of structures of the type Σ cannot enlarge the invariance group. The case of the largest possible invariance group deserves a special name:

Definition 3. *Equation (3) is called* ***diffeomorphism invariant*** *iff it allows* Diff(M) *as invariance group.*

In view of (12), the requirement of Diff(M)-invariance can be understood as a strong limit on the amount of non-dynamical structure Σ. Generically it seems to eliminate any Σ, i.e. the theory should contain no non-dynamical background fields whatsoever. Intuitively this is what background independence stands for. Conversely, any Diff(M)-covariant theory without non-dynamical fields is trivially Diff(M)-invariant. Hence it seems sensible to simply identify 'Diff(M)-invariance' and 'background independence', and this is what most working physicists seem to do.

But this turns out to be too simple. The origin of the difficulty lies in our distinction between dynamical and non-dynamical structures, which turns out not to be sufficiently sharp. Basically we just said that a structure (γ or Φ) was dynamical if it had no a priori prescribed values, but rather obeyed some equations of motion. We did not say what qualifies an equation as an 'equation of motion'. Can it just be *any* equation? If yes then we immediately object that there exists an obvious strategy to trivialize the requirement of Diff(M)-invariance: just let the values of Σ be determined by equations rather than by hand; in this way they formally become 'dynamical' variables and no non-dynamical quantities are left. Formally this corresponds to the replacement scheme

$$\Phi \mapsto \Phi' = (\Phi, \Sigma) , \tag{13a}$$

$$\Sigma \mapsto \Sigma' = \emptyset , \tag{13b}$$

so that invariance now becomes as trivial as the requirement of covariance.

More concretely, reconsider the examples (7) and (10) above. In the first case we now regard the spacetime metric g as 'dynamical' field for which we add the condition of flatness as 'equation of motion':

$$\mathbf{Riem}[g] = 0 , \tag{14}$$

where **Riem** denotes the Riemann tensor of (M, g). In the second case we regard g as well as the timelike vector field n as 'dynamical' and add (14) and the two equations

$$g(n, n) = c^2 , \tag{15a}$$

$$\nabla n = 0 . \tag{15b}$$

In this fashion we arrive at diffeomorphism invariant equations. But do they really represent the same theory as the one we originally started from? For example, are their solution spaces 'the same'? Naively the answer is clearly 'no', simply because the reformulated theory has – by construction – a much larger space of solutions. For any solution Φ of the original equations $F[\Phi, \Sigma] = 0$, where Σ is fixed, we now have the whole Diff(M)-orbit of solutions, $\{(f\cdot\Phi, f\cdot\Sigma) \mid f \in \text{Diff(M)}\}$ of the new equations, which treat Σ as dynamical variable. A bijective correspondence can only be established if the transformations f that act non-trivially on Σ (i.e. $f \notin \text{Stab}_{\text{Diff(M)}}(\Sigma)$) are declared to be *gauge transformations*, so that any two field configurations related by such an f are considered to be physically identical.

If this is done, the simple strategy outlined here suffices to (formally) trivialize the requirement of diffeomorphism invariance. Hence defining background independence as being simple diffeomorphism invariance would also render it a trivial requirement. How could we improve its definition so as to make it a useful notion? This is precisely what Anderson attempted in [3]. He noted the following peculiarities of the reformulation just given:

1. The new fields g or (g, n) obey an autonomous set of equations which does not involve the proper dynamical fields F or ϕ respectively. In contrast, the equations for the latter *do* involve g or (g, n). Physically speaking, the system whose states are parameterized by the new variables acts upon the system whose states are parameterized by F or ϕ, but not vice versa. An agent which dynamically acts but is not acted upon may well be called 'absolute' – in generalization of Newton's absolute space. Such an absolute agent should be eliminated.

2. The sector of solution space parameterized by g or (g, n) consists of a single diffeomorphism orbit. For example, this means that for any two solutions (ϕ, g, n) and (ϕ', g', n') of (10), (14), and (15) there exists a diffeomorphism f such that $(g', n') = (f \cdot g\,,\, f \cdot n)$. So 'up to diffeomorphisms' there exists only one solution in the (g, n) sector. This is far from true for ϕ: the two solutions ϕ and ϕ' are generally not related by a diffeomorphism. This difference just highlights the fact that the added variables really did not correspond to new degrees of freedom (they were never supposed to) because the added equations were chosen strong enough to maximally fix their values (up to diffeomorphisms).

A closer analysis shows that the first criterion is really too much dependent on the presentation to be generally useful as a necessary condition. Absolute structures will not always reveal their nature by obeying autonomous equations. The second criterion is more promising and actually entered the literature with some refinements as criterion for absolute structures. Before going into this, we will discuss some attempts to disable the trivialization strategies just outlined.

2.3 Strategies Against Triviality

Involving the Principle of Equivalence

As diffeomorphism *covariance* is a rather trivial requirement to satisfy, we will from now on only be concerned with diffeomorphism *invariance*. As we explained, it could be achieved by letting the Σ's 'change sides', i.e. become dynamical structures (γ's and Φ's), as schematically written down in (13). We seek sensible criteria that will limit the number of such renegades. A physical criterion that suggests itself is to allow only those Σ to change sides which are known to correspond to dynamical variables in a wider context. For example, we may allow the spacetime metric g to become formally dynamical, since we know that it describes the gravitational field, even if in the context at hand the self-dynamics of the gravitational field is not relevant and therefore, as a matter of approximation, fixed to some value (e.g. the Minkowski metric). Doing this would render the Maxwell equations (7) (plus the equations for g) diffeomorphism invariant. But this alone would not work for the diffusion equation, where n would still act as a non-dynamical structure.

Hence we see that the requirement to achieve diffeomorphism invariance by at most adjoining g to the dynamical variables is rather non-trivial and connects to Einstein's principle of equivalence. Let us quote Wolfgang Pauli in this context:

> Einen physikalischen Inhalt bekommt die allgemeine kovariante Formulierung der Naturgesetze erst durch das Äquivalenzprinzip, welches zur Folge hat, daß die Gravitation durch die g_{ik} *allein* beschrieben

wird, und daß diese nicht unabhängig von der Materie gegeben, sondern selbst durch die Feldgleichungen bestimmt sind. Erst deshalb können die g_{ik} als *physikalische Zustandsgrößen* bezeichnet werden.[7] ([17], p. 181; the emphases are Pauli's)

Absolute Structures

As already remarked, another strategy to render the requirement of diffeomorphism invariance non-trivial was suggested by Anderson [3] by means of his notion of 'absolute structures'. However, most commentators share the opinion that Anderson did not succeed to give a proper definition of this term. Even worse, some feel that so far nobody has, in fact, succeeded in giving a fully satisfying definition.

To see what is behind this somewhat unhappy state of affairs, let us start with a tentative definition that suggests itself from the discussion given above:

Definition 4 (tentative). *Any field which is either not dynamical, or whose solution space consists of a single* Diff(M)*-orbit, is called an* ***absolute structure****.*

In general terms, let $\mathcal{S}$ denote the space of solutions to a given theory. If the theory is Diff(M)-invariant $\mathcal{S}$ carries an action of Diff(M). The fields can be thought of as parameterising on $\mathcal{S}$. An absolute structure is a parameter which takes the same range of values in each Diff(M) orbit and therefore cannot separate any two of them. If we regard Diff(M) as a gauge group, i.e. that Diff(M)-related configurations are physically indistinguishable, then absolute structures carry no observable content.

Following our general strategy we could now attempt to give a definition of 'background independence':

Definition 5 (tentative). *A theory is called* ***background independent*** *iff its equations are* Diff(M)*-invariant in the sense of Definition 3 and its fields do not include absolute structures in the sense of Definition 4.*

Before discussing these proposal, let us look at some more examples.

2.4 More Examples

Scalar Gravity a la Einstein–Fokker

In 1913, just before the advent of general relativity, Gunnar Nordsröm invented a formally consistent Poincaré-invariant scalar theory of gravity; see,

[7] 'The generally covariant formulation of the physical laws acquires a physical content only through the principle of equivalence, in consequence of which gravitation is described *solely* by the g_{ik} and these latter are not given independently from matter, but are themselves determined by field equations. Only for this reason can the g_{ik} be described as *physical quantities*' ([16], p. 150).

e.g., the survey by von Laue [22]. Shortly after its publication it was pointed out by Einstein and Fokker that Nordström's (second) theory can be presented in a 'covariant' way. Explicitly they said,

> Im folgenden soll dargetan werden, daß man zu einer in formaler Hinsicht vollkommen geschlossenen und befriedigenden Darstellung der Theorie [Nordströms] gelangen kann, wenn man, wie dies bei der Einstein-Grossmannschen Theorie bereits geschehen ist, das invarianten-theoretische Hilfsmittel benutzt, welches uns in dem absoluten Differentialkalkül gegeben ist.[8] ([20], Vol. 4, Doc. 28, p. 321)

The essential observation is this: consider conformally flat metrics:

$$g_{\mu\nu} = \phi^2\, \eta_{\mu\nu}\ , \tag{16}$$

then the field equation is equivalent to

$$R[g] \;=\; 24\pi G\, g^{\mu\nu} T_{\mu\nu}\ , \tag{17a}$$

where $R[g]$ is the Ricci scalar for the metric g, whereas the equation of motion for the particle becomes the geodesic equation with respect to g:

$$\ddot{x}^\mu + \Gamma^\mu_{\alpha\beta}\dot{x}^\alpha\dot{x}^\beta = 0\ . \tag{17b}$$

Now, the system (17), considered as equations for the metric g and the trajectory x, is clearly Diff(M)-invariant. But Nordström's theory is equivalent to (17) *plus* (16). Here η is a non-dynamical field so that (16, 17) is only Diff(M)-covariant. According to the general scheme outlined above this could be remedied by letting the metric η be a new dynamical variable whose equation of motion just asserts its flatness:

$$\mathbf{Riem}[\eta] = 0\ . \tag{18}$$

But then η qualifies as an absolute structure according to Definition 4 and the theory (16, 17, 18) is not background independent. The subgroup $G \subset$ Diff(M) that stabilizes η is – by definition – the inhomogeneous Lorentz group, which had already been the invariance group of Nordström's theory. So no additional invariance has, in fact, been gained in the transition from Nordström's to the Einstein–Fokker formulation.

Sometimes the absolute structures are not so easy to find because the theory is formulated in such a way that they are not yet isolated as separate field. For example, in the case at hand, (16) and (18) together are clearly equivalent to the single condition that g be conformally flat, which in turn

[8] 'In the following we wish to show that one can arrive at a formally complete and satisfying presentation of the theory [Nordström's] if one uses the methods from the theory of invariants given by the absolute differential calculus, as it was already done in the Einstein–Grossman theory.'

is equivalent to the vanishing of the conformal curvature tensor for g (Weyl tensor):

$$\mathbf{Weyl}[g] = 0 \,. \tag{19}$$

The field $\eta_{\mu\nu}$ has now disappeared from the description and the theory does not explicitly display any absolute structure anymore. But, of course, it is still there; it is now part of the field g. To bring it back to light, make a field redefinition $g_{\mu\nu} \mapsto (\phi, h_{\mu\nu})$ which isolates the part determined by (19); for example,

$$\phi \;\; := [-\det\{g_{\mu\nu}\}]^{\frac{1}{8}} \,, \tag{20}$$

$$h_{\mu\nu} := g_{\mu\nu}\; [-\det\{g_{\mu\nu}\}]^{-\frac{1}{4}} \,. \tag{21}$$

Then any two solutions for the full set of equations are such that their component fields $h_{\mu\nu}$ and $h'_{\mu\nu}$ are related by a diffeomorphism. Hence $h_{\mu\nu}$ is an absolute structure.

Clearly there is a rather non-trivial mathematical theory behind the last statement of diffeomorphism equivalence of $h_{\mu\nu}$. We could not have made that statement had we not already been in possession of the full solution theory for (19) which, after all, is a complicated set of non-linear partial differential equations of second order.

A Massless Scalar Field from an Action Principle

Usually we require the equations of motion to be the Euler–Lagrange equations for some associated action principle. Would the somewhat bold strategy to render non-dynamical structures dynamical by adding *by hand* 'equations of motion' which fix them to their previous values also work if these added equations were required to be the Euler–Lagrange equations for some common action principle? The answer is by no means obvious, as the following simple example taken from [19] illustrates:

Consider a real massless[9] scalar field in Minkowski space:

$$\Box\phi := \eta^{\mu\nu}\nabla_\mu\nabla_\nu\phi = 0 \,. \tag{22}$$

According to standard strategy the non-dynamical Minkowski metric η is eliminated by introducing the dynamical variable g, replacing η in (22) by g, and adding the flatness condition

$$\mathbf{Riem}[g] = 0 \tag{23}$$

as new equation of motion. Is there an action principle whose Euler–Lagrange equations are (equivalent to) these equations? This seems impossible without

[9] This is just assumed for simplicity. The arguments work the same way if a mass term were included.

introducing yet another field λ (a Lagrange multiplier) whose variation just yields (23). The action would then be

$$S = \tfrac{1}{2}\int dV\ g^{\mu\nu}\nabla_\mu\phi\,\nabla_\nu\phi + \tfrac{1}{4}\int dV\ \lambda^{\alpha\beta\mu\nu}R_{\alpha\beta\mu\nu}\,, \tag{24}$$

where the symmetries of the tensor field λ are that of the Riemann tensor:

$$\lambda^{\alpha\beta\mu\nu} = \lambda^{[\alpha\beta][\mu\nu]} = \lambda^{\mu\nu\alpha\beta}\,. \tag{25}$$

Variation with respect to ϕ and λ yield (22) and (23) respectively, and variation with respect to g gives

$$\nabla_\mu\nabla_\nu\lambda^{\alpha\mu\beta\nu} = T^{\alpha\beta}\,, \tag{26}$$

where $T^{\alpha\beta}$ is the energy–momentum tensor for ϕ. These equations do not give a background independent theory for the fields (ϕ, g, λ) since g is an absolute structure. The solution manifold of the ϕ field is, in fact, the same as before. For this it is important to note that there is an integrability condition resulting from (23,26), namely $\nabla_\alpha T^{\alpha\beta} = 0$, which is however already implied by (22). Hence no extra constraints on ϕ result from (26).

However, the λ field seems to actually add more dimensions to the solution manifold and hence to the observable content of the theory. Indeed, using the Poincaré Lemma in flat space one shows that any divergenceless symmetric 2-tensor $T^{\mu\nu}$ can always be written as in (26), where λ has the symmetries (25). But this does not fix $\lambda^{\mu\alpha\nu\beta}$, so that the set of Diff(M)-equivalence classes of stationary points of (24) is strictly 'larger' than the set of solutions of (22). In other words, the (Diff(M) reduced) phase space for the theory described by (24) is 'larger' then that for (22).[10] A a result we conclude that the reformulation given here does *not* achieve an equivalent Diff(M)-invariant reformulation of (22) in terms of an action principle.

2.5 Problems with Absolute Structures

A first thing to realize from the examples above is that the notion of absolute structure should be slightly refined. More precisely, it should be made local in order to capture the idea that an absolute element in the theory does not represent local degrees of freedom. Rather than saying that a field corresponds to an absolute structure if its solution space consists of a single Diff(M)-orbit, we would like to make the latter condition local:

Definition 6. *Two fields T_1 and T_2 are said to be* ***locally diffeomorphism equivalent*** *iff for any point $p \in M$ there exits a neighbourhood U of p and a diffeomorphism $\phi_U : U \to U$ such that $\phi_U \cdot (T_1|_U) = T_2|_U$.*

[10] I am not aware of a reference where a Hamiltonian reduction of (24) is carried out.

Note that local diffeomorphism equivalence defines an equivalence relation on the set of fields. Accordingly, following a suggestion of Friedman [7], we should replace the tentative Definition 4 by the following:

Definition 7. *Any field which is either not dynamical or whose solutions are all locally diffeomorphism equivalent is called an* ***absolute structure****.*

In fact, this is what we implicitly used in the discussions above where we slightly oversimplified matters. For example, any two flat metrics g_1, g_2 (i.e. which satisfy $\mathbf{Riem}[g_{1,2}] = 0$) are generally only *locally* diffeomorphism equivalent. Likewise, a conformally flat metric g (i.e. which satisfy $\mathbf{Weyl[g]=0}$) is *locally* diffeomorphism equivalent to $f^2\eta$, where f is non-vanishing function and η is a fixed flat metric.

Having corrected this we should also adapt the tentative Definition 5:

Definition 8. *A theory is called* ***background independent*** *iff its equations are* Diff(M)*-invariant in the sense of Definition 3 and its fields do not include absolute structures in the sense of Definition 7.*

So far so good. Is this, then, the final answer? Unfortunately not! The standard argument against *this* notion of absolute structure is that it may render structures absolute that one would normally call dynamical. The canonical example, usually attributed to Robert Geroch [10], makes use of the well-known fact in differential geometry that nowhere vanishing vector fields are always locally diffeomorphism equivalent (see, e.g., Theorem 2.1.9 in [1]). Hence any diffeomorphism-invariant theory containing vector fields among their fundamental field variables cannot be background independent. For example, consider the coupled Einstein–Euler equations for a perfect fluid of density ρ and four-velocity u in spacetime with metric g. This system of equations is Diff(M)-invariant. By definition of a velocity field we have $g(u, u) = c^2$. This means that u cannot have zeros, even if for physical reasons we would usually assume the fluid to be present not everywhere in spacetime, i.e. the support of ρ is a proper subset of spacetime.[11] Then the four velocity of the fluid is an absolute structure, contrary to our physical intention.

I know of two suggestions how to avoid this conclusion in the present example. One is to use the 1-form $u_\mu\, dx^\mu$ rather than the vector field $u^\mu \partial_\mu$ as fundamental dynamical variable for the fluid. The point being that one-form fields are *not* locally diffeomorphism equivalent. For example, a closed (exact) one-form field will always be mapped into a closed (exact) one-form field, and hence cannot be locally diffeomorphism equivalent to a non-closed field. Another suggestion, in fact the only one that I have seen in the literature ([8] p. 59 footnote 9 and [21], p. 99, footnote 8) is to take the energy–momentum density Π rather than u as fundamental variable. To be sure, on

[11] It seems a little strange to be forced to consider velocity fields u in regions where $\rho = 0$, i.e. where there is no fluid matter. Velocity of what? one might ask. In a concrete application this means that we have to extend u beyond the support of ρ and that the physical prediction is independent of that extension.

the support of Π we can think of it as equal to ρu, but on the complement of its support there is no need to define a u. This avoids the unwanted conclusion whenever Π indeed has zeros; otherwise the argument given above for u just applies to Π.

An even simpler argument, which I have not seen in the physics literature, even applies to pure gravity. It rests on the following theorem from differential geometry, an elegant proof of which was given by Moser [12]: given two compact-oriented n-dimensional manifolds V_1 and V_2 with n-forms μ_1 and μ_2 respectively. There exists an orientation-preserving diffeomorphism $\phi : V_1 \rightarrow V_2$ such that $\phi^* \mu_2 = \mu_1$ iff the μ_1-volume of V_1 equals the μ_2-volume of V_2, i.e. iff

$$\int_{V_1} \mu_1 = \int_{V_2} \mu_2 \, . \tag{27}$$

If we take $V_1 = V_2$ to be the closure of an open neighbourhood U in the spacetime manifold M, this theorem implies that the metric volume forms, written in coordinates as

$$\mu = \sqrt{|\det[g(\partial_\mu, \partial_\nu)]|}\, dx^1 \wedge \cdots \wedge dx^n \, , \tag{28}$$

are locally diffeomorphism equivalent iff they assign the same volume to U. Hence it follows that the metric volume elements modulo constant factors are absolute elements in pure gravity. Note that this implies that for any metric g any point $p \in M$ there is always a local coordinate system $\{x^\mu\}$ in an open neighbourhood U of p such that $\sqrt{|\det[g(\partial_\mu, \partial_\nu)]|} = 1$.

3 Conclusion

Background independence is one of the central strategic issues in discussions on competing approaches to quantum gravity. This clearly emerges from the contributions of Kiefer, Thiemann, Nicolai and Peeters, Lauscher and Reuter, and Louis, Mohaupt, and Theisen to this book. Given the impressive amount of effort that is devoted to analyse the consequences of these different approaches, it seems a little strange to me that the very notion of background independence is tolerated to be in the state of relative elusiveness in which it appears to be. Clearly, in *specific* situations it is usually not difficult to associate a mathematically well-defined meaning to an 'intuitively obvious' interpretation of such a requirement of background independence. But when used as *general* strategic criterion one should, I think, come up with a generally valid and mathematically well-defined definition. I am not aware of such a definition. Attempts were made in the past, but they run into the problems outlined here. Hence the problem must be regarded as an outstanding one.

References

1. Ralph Abraham and Jerrold E. Marsden. *Foundations of Mechanics.* The Benjamin/Cummings Publishing Company, Reading, Massachusetts, second edition, 1978.
2. James L. Anderson. Relativity principles and the role of coordinates in physics. In Hong-Yee Chiu and William F. Hoffmann, editors, *Gravitation and Relativity*, pp. 175–194. W.A. Benjamin, Inc., New York and Amsterdam, 1964.
3. James L. Anderson. *Principles of Relativity Physics.* Academic Press, New York, 1967.
4. Julian Barbour. On general covariance and best matching. In Craig Callender and Nick Huggett, editors, *Physics Meets Philosophy at the Planck Scale – Contemporary Theories in Quantum Gravity*, pp. 199–212. Cambridge University Press, Cambridge (England), 2001.
5. Elie Cartan. Sur les varietes a connexion affine et la théorie de la relativité généralisée. *Annales Scientifiques de l'École Normale Supérieure*, 40:325–412, 1923.
6. Elie Cartan. Sur les varietes a connexion affine et la théorie de la relativité généralisée. *Annales Scientifiques de l'École Normale Supérieure*, 41:1–15, 1924.
7. Michael Friedman. Relativity principles, absolute objects and symmetry groups. In Patrick Suppes, editor, *Space, Time and Geometry*, pp. 296–320, Dordrecht, Holland, 1973. D. Reidel Publishing Company.
8. Michael Friedman. *Foundations of Space-Time Theories.* Princeton University Press, Princeton, New Jersey, 1983.
9. Kurt Friedrichs. Eine invariante Formulierung des Newtonschen Gravitationsgesetzes und des Grenzüberganges vom Einsteinschen zum Newtonschen Gesetz. *Mathematische Annalen*, 98:566–575, 1927.
10. Roger Jones. The special and general principles of relativity. In P. Barker and C.G. Shugart, editors, *After Einstein*, pp. 159–173, Memphis, USA, 1981. Memphis State University Press.
11. Erich Kretschmann. Über den physikalischen Sinn der Relativitätspostulate, A. Einsteins neue und seine ursprüngliche Relativitätstheorie. *Annalen der Physik*, 53:575–614, 1917.
12. Jürgen Moser. On the volume elements on a manifold. *Transactions of the American Mathematical Society*, 120(2):286–294, 1965.
13. John Norton. Einstein, Nordström and the early demise of scalar Lorentz-covariant theories of gravitation. *Archive for the History of Exact Sciences*, 45:17–94, 1992.
14. John Norton. General covariance and the foundations of general relativity: Eight decades of dispute. *Reports of Progress in Physics*, 56:791–858, 1993.
15. John Norton. General covariance, gauge theories and the Kretschmann objection. In Katherine Brading and Elena Castellani, editors, *Symmetries in Physics: Philosophical Reflections*, pp. 110–123. Cambridge University Press, Cambridge, UK, 2003.
16. Wolfgang Pauli. *Theory of Relativity.* Dover Publications, Inc., New York, 1981. Unabridged and unaltered republication of the english translation by G. Field that was first published in 1958 by Pergamon Press.

17. Wolfgang Pauli. *Relativitätstheorie.* Springer Verlag, Berlin, 2000. Reprint of the original 'Encyclopädie-Artikel' from 1921 with various additions, including Pauli's own supplementary notes from 1956. Edited and annotated by Domenico Giulini.
18. Lee Smolin. The case for background independence. *www.arxiv.org/abs/ hep-th/0507235.*
19. Rafael Sorkin. An example relevant to the Kretschmann-Einstein debate. *Modern Physics Letters A*, 17(11):695–700, 2002.
20. John Stachel et al., editors. *The Collected Papers of Albert Einstein, Vols. 1–9.* Princeton University Press, Princeton, New Jersey, 1987–2005.
21. Norbert Straumann. *General Relativity.* Springer Verlag, Berlin, 2004.
22. Max von Laue. Die Nordströmsche Gravitationstheorie. *Jahrbuch der Radioaktivität und Elektronik*, 14(3):263–313, 1917.

Part V

Quantum Gravity

Why Quantum Gravity?

C. Kiefer

Institut für Theoretische Physik, Universität zu Köln, Zülpicher Straße 77, 50937 Köln, Germany
kiefer@thp.uni-koeln.de

Quantum theory seems to be a universal framework for physical theories. In fact, most of the interactions found in Nature are already successfully accommodated into this framework. The only interaction for which this has not yet been achieved is gravity. All manifestations of the gravitational field known so far can be understood from a classical theory—Einstein's theory of general relativity (GR), also called geometrodynamics. It is defined by the Einstein–Hilbert action

$$S_{\mathrm{EH}} = \frac{c^4}{16\pi G} \int_{\mathcal{M}} \mathrm{d}^4x \, \sqrt{-g}\,(R - 2\Lambda) \; + \; \text{boundary term} + S_{\mathrm{m}} \, , \tag{1}$$

where R and Λ are the Ricci scalar and the cosmological constant, respectively, and where S_{m} denotes the action for non-gravitational fields from which one can derive the energy–momentum tensor according to

$$T_{\mu\nu}(x) = \frac{2}{\sqrt{-g}} \frac{\delta S_{\mathrm{m}}}{\delta g^{\mu\nu}(x)} \, . \tag{2}$$

There exist certain 'uniqueness theorems' which state that every reasonable theory of the gravitational field must contain GR (or its natural generalization, the Einstein–Cartan theory) in a certain limit. Details for this and the material discussed below can be found in [1–3] and the references therein.

In spite of the success of GR, there are many reasons to believe that the most fundamental theory of gravity is a *quantum* theory. Unfortunately, no experimental material is presently available, which would point in a definite direction. The reasons are therefore of a theoretical nature. The main motivations for quantum gravity are as follows:

- **Unification.** The history of science shows that a reductionist viewpoint has been very fruitful in physics. The standard model of particle physics is a *quantum* field theory which has partially united all non-gravitational

C. Kiefer: *Why Quantum Gravity?*, Lect. Notes Phys. **721**, 123–130 (2007)
DOI 10.1007/978-3-540-71117-9_7

interactions through the gauge group SU(3) × SU(2) × U(1), cf. the contribution by H.-G. Dosch to this book. The universal coupling of gravity to all forms of energy would make it plausible that gravity has to be implemented in a quantum framework, too. Moreover, attempts to construct an exact semiclassical theory, where gravity stays classical but all other fields are quantum, have failed up to now. This demonstrates in particular that classical and quantum *concepts* (phase space versus Hilbert space, etc.) are most likely incompatible.

- **Cosmology and black holes**. As the *singularity theorems* and the ensuing breakdown of GR demonstrate, a fundamental understanding of the early universe—in particular its initial conditions near the 'big bang'—and of the final stages of black-hole evolution requires an encompassing theory. From the historical analogue of quantum mechanics (which due to the existence of stationary states has rescued the atoms from collapse) the general expectation is that this encompassing theory is a *quantum* theory. It must be emphasized that *if* gravity is quantized, the kinematical non-separability of quantum theory demands that the whole Universe must be described in quantum terms. This leads necessarily to the concepts of quantum cosmology and the wave function of the universe.
- **Problem of time**. Quantum theory and GR (in fact, any generally covariant theory) contain drastically different concepts of time (and spacetime). Strictly speaking, they are incompatible. In quantum theory, time is an external (absolute) element, *not* described by an operator (in special relativistic quantum field theory, the role of absolute time is played by the external Minkowski spacetime). In contrast, spacetime is a dynamical object in GR. It is clear that a unification with quantum theory must lead to modifications of the concept of time. Related problems concern the role of background structures in quantum gravity, the role of the diffeomorphism group (Poincaré invariance, as used in ordinary quantum field theory, is no longer a symmetry group), and the notion of 'observables'.
- **Avoidance of divergences**. It has long been speculated that quantum gravity may lead to a theory devoid of the ubiquitous divergences arising in quantum field theory. This may happen, for example, through the emergence of a natural cutoff at small distances (large momenta). In fact, modern approaches such as string theory or loop quantum gravity provide indications for a discrete structure at small scales.

What are the relevant scales on which effects of quantum gravity should be unavoidable? As has already been shown by Max Planck in 1899, the fundamental constants speed of light (c), gravitational constant (G), and quantum of action ($\hbar$) can be combined in a unique way (up to a dimensionless factor) to yield units of length, time, and mass. In Planck's honour they are called

Planck length, $l_{\rm P}$, Planck time, $t_{\rm P}$, and Planck mass, $m_{\rm P}$, respectively. They are given by the expressions

$$l_{\rm P} = \sqrt{\frac{\hbar G}{c^3}} \approx 1.62 \times 10^{-33}\ \text{cm}\ , \tag{3}$$

$$t_{\rm P} = \frac{l_{\rm P}}{c} = \sqrt{\frac{\hbar G}{c^5}} \approx 5.40 \times 10^{-44}\ \text{s}\ , \tag{4}$$

$$m_{\rm P} = \frac{\hbar}{l_{\rm P} c} = \sqrt{\frac{\hbar c}{G}} \approx 2.17 \times 10^{-5}\ \text{g} \approx 1.22 \times 10^{19}\ \text{GeV}\ . \tag{5}$$

The Planck mass seems to be a rather large quantity by microscopic standards. One has to keep in mind, however, that this mass (energy) must be concentrated in a region of linear dimension $l_{\rm P}$ in order to see direct quantum-gravity effects. In fact, the Planck scales are attained for an elementary particle whose Compton wavelength is (apart from a factor of 2) equal to its Schwarzschild radius,

$$\frac{\hbar}{m_{\rm P} c} \approx R_{\rm S} \equiv \frac{2Gm_{\rm P}}{c^2}\ ,$$

which means that the spacetime curvature of such an elementary particle is non-negligible. A truly unified theory may, of course, contain other parameters. An example is string theory where the fundamental 'string length' $l_{\rm s}$ appears as the fundamental scale instead of the Planck length (which there is a derived quantity).

The ratio of atomic scales to the Planck scale is expressed by the 'fine structure constant of gravity',

$$\alpha_{\rm g} = \frac{Gm_{\rm pr}^2}{\hbar c} \equiv \left(\frac{m_{\rm pr}}{m_{\rm P}}\right)^2 \approx 5.91 \times 10^{-39}\ , \tag{6}$$

where $m_{\rm pr}$ denotes the proton mass. Its smallness is responsible for the unimportance of quantum-gravitational effects on laboratory and astrophysical scales, and for the separation between micro- and macrophysics. It is interesting that structures in the universe occur for masses which can be expressed as simple powers of $\alpha_{\rm g}$ in units of $m_{\rm pr}$, cf. [4]. For example, stellar masses are of the order $\alpha_{\rm g}^{-3/2} m_{\rm pr}$, while stellar lifetimes are of the order $\alpha_{\rm g}^{-3/2} t_{\rm P}$. It is also interesting to note that the size of human beings is roughly the geometric mean of Planck length and size of the observable universe. It is an open question whether a fundamental theory of quantum gravity can provide an explanation for such values, for example, for the ratio $m_{\rm pr}/m_{\rm P}$, or not. If not, only an anthropic principle could yield a—not very satisfying—'explanation'. The challenge is to find a non-anthropic solution.

Below the level of full quantum gravity one can distinguish from a conceptual point of view at least two other levels. The first, lowest, level deals with quantum *mechanics* in *external* gravitational fields (either described by GR or its Newtonian limit). No back reaction on the gravitational field is taken into account. This is the only level where experiments exist so far. Already in the 1970s, experiments of neutron interferometry were performed in the gravitational field of the Earth. It was possible, in particular, to show that the weak equivalence principle holds at the given level of precision. More recently, gravitational quantum bound states of neutrons in the field of the Earth have been measured. A detailed review of the interaction between gravity and mesoscopic quantum systems can be found in [5].

The second level concerns quantum *field* theory in *external* gravitational fields. One has attempted to include the back reaction of the quantum fields onto the gravitational field in various models, but without final result. Although experimental data are still lacking, there exist on this level at least precise predictions. The most important one concerns Hawking radiation for black holes [6]. A black hole radiates with temperature

$$T_{\mathrm{H}} = \frac{\hbar\kappa}{2\pi k_{\mathrm{B}} c} , \tag{7}$$

where κ is the surface gravity of a stationary black hole which by the no-hair theorem is uniquely characterized by its mass M, its angular momentum J, and its electric charge Q. In the particular case of the spherically symmetric Schwarzschild black hole one has $\kappa = c^4/4GM = GM/R_{\mathrm{S}}^2$ and therefore

$$T_{\mathrm{H}} = \frac{\hbar c^3}{8\pi k_{\mathrm{B}} GM} \approx 6.17 \times 10^{-8} \left(\frac{M_{\odot}}{M}\right) \mathrm{K} . \tag{8}$$

The black hole shrinks due to Hawking radiation and possesses a finite lifetime. The final phase, where γ-radiation is being emitted, could be observable. The temperature (8) is unobservably small for black holes that result from stellar collapse. One would need primordial black holes produced in the early universe because they could possess a sufficiently low mass, cf. [7]. For example, black holes with an initial mass of 5×10^{14} g would evaporate at the present age of the universe. In spite of several attempts, no experimental hint for black-hole evaporation has been found. Primordial black holes can result from density fluctuations produced during an inflationary epoch. However, they can only be produced in sufficient numbers if the scale invariance of the power spectrum is broken at some scale, cf. [8]. It should also be mentioned that small black holes may even be produced in future accelerators such as the Large Hadron Collider (LHC) if the idea is correct that spacetime possesses more than four dimensions with properties such that the higher-dimensional Planck energy is much smaller than m_{P}. Motivation for such extra dimensions comes from string theory, cf. the contribution by Louis et al. to this book.

It must be emphasized that the expression for T_{H} contains all fundamental constants of nature. One may speculate that this expression—relating the macroscopic parameters of a black hole with thermodynamic quantities—plays a similar role for quantum gravity as de Broglie's relations $E = \hbar\omega$ and $p = \hbar k$ once played for the development of quantum theory [9]. Hawking radiation was derived in the semiclassical limit in which the gravitational field can be treated classically. According to (8), the black hole loses mass through its radiation and becomes hotter. After it has reached a mass of the size of the Planck mass (5), the semiclassical approximation breaks down and the full theory of quantum gravity should be needed. Black-hole evaporation thus plays a crucial role in any approach to quantum gravity.

Since black holes radiate thermally, they also possess an *entropy*, the 'Bekenstein–Hawking entropy', which is given by the expression

$$S_{\mathrm{BH}} = \frac{k_{\mathrm{B}} c^3 A}{4G\hbar} = k_{\mathrm{B}} \frac{A}{4l_{\mathrm{P}}^2} \,, \tag{9}$$

where A is the surface area of the event horizon. For a Schwarzschild black hole with mass M, this reads

$$S_{\mathrm{BH}} \approx 1.07 \times 10^{77} k_{\mathrm{B}} \left(\frac{M}{M_{\odot}}\right)^2 . \tag{10}$$

Since the Sun has an entropy of about $10^{57} k_{\mathrm{B}}$, this means that a black hole resulting from the collapse of a star with a few solar masses would experience an increase in entropy by 20 orders of magnitude during its collapse. It is one of the challenges of any approach to provide a microscopic explanation for this entropy, i.e. to derive (9) from a counting of microscopic quantum gravitational states according to

$$S_{\mathrm{BH}} = -k_{\mathrm{B}} \mathrm{tr}(\rho \ln \rho) \,, \tag{11}$$

where ρ denotes the reduced density matrix of the system.

There exists a related effect to (7) in flat Minkowski space. An observer with uniform acceleration a experiences the standard Minkowski vacuum not as empty, but as filled with *thermal* radiation of temperature

$$T_{\mathrm{DU}} = \frac{\hbar a}{2\pi k_{\mathrm{B}} c} \approx 4.05 \times 10^{-23} \, a \left[\frac{\mathrm{cm}}{\mathrm{s}^2}\right] \, \mathrm{K} \,. \tag{12}$$

This temperature is often called the 'Davies-Unruh temperature'. Formally, it arises from (7) through the substitution of κ by a. This can be understood from the fact that *event horizons* are present in both the black-hole case and the acceleration case. Although (12) seems to be a small effect, it was suggested to search for it in accelerators or in experiments with ultra-intense lasers, without definite success up to now.

As we have already mentioned above, experimental clues for quantum gravity are elusive. A direct probe of the Planck scale (5) in high-energy experiments would be illusory. In fact, an accelerator using current laws would have to have the size of several thousand light years in order to probe the Planck energy $m_{\mathrm{P}} c^2 \approx 10^{19}$ GeV. However, it is imaginable that effects of quantum gravity can in principle occur at lower energy scales. Possibilities could be non-trivial applications of the superposition principle for the quantized gravitational field or the existence of discrete quantum states in black-hole physics or the early universe. But one might also be able to observe quantum-gravitational correction terms to established theories, such as correction terms to the functional Schrödinger equation in an external spacetime or effective terms violating the weak equivalence principle. Such effects could potentially be measured in the anisotropy spectrum of the cosmic microwave background radiation or in the forthcoming satellite tests of the equivalence principle, such as MICROSCOPE or STEP.

Various approaches to quantum gravity give a hint at the existence of a discrete structure of spacetime at the smallest scales. Such a microstructure could be recognizable, for example, from modified dispersion relations of electromagnetic radiation coming from far-away objects (e.g. γ-ray bursts).

A truly fundamental theory should have such a rigid structure that all phenomena in the low-energy regime, such as particle masses or coupling constants, could be predicted in an unambiguous way. As there is no direct experimental hint yet, most work in quantum gravity focuses on the attempt to construct a mathematically and conceptually consistent (and appealing) framework. But one should also keep in mind Einstein's dictum that only the final theory specifies what can be observed.

There is, of course, no a priori given starting point in the methodological sense. In this context Chris Isham makes a distinction between a 'primary theory of quantum gravity' and a 'secondary theory' [10]. In the primary approach, one starts with a given classical theory and applies heuristic quantization rules. This is the approach usually adopted, and it was successfully applied, for example, in quantum electrodynamics (QED). In most cases, the starting point is GR, leading to 'Quantum General Relativity', but one could also start from another classical theory such as the Brans–Dicke theory. One usually distinguishes between 'canonical' and 'covariant' approaches. Covariant approaches employ four-dimensional covariance at some stage of the formalism. Examples include perturbation theory (Feynman-diagrammatic expansion of the Einstein–Hilbert action), renormalization-group approaches, and path integral methods. Canonical approaches, on the other hand, employ a Hamiltonian formalism and thus need an identification of the canonical variables and their conjugate momenta. Examples are quantum geometrodynamics and loop quantum gravity, their difference lying in the choice of variables and momenta. Details of these approaches can be found in [1] and in the other contributions to the quantum-gravity part of this book. The main advantage of quantum GR is that the starting point is given—the classical theory. The main disadvantage is that one does not arrive immediately at a unified theory of all interactions.

The opposite holds for a 'secondary theory'. The ambition is to start with a fundamental quantum framework of all interactions and trying to derive (quantum) GR in certain limiting situations, for example, through an energy expansion. The most important example here is string theory (M-theory). The main advantage is that the fundamental quantum theory automatically yields a unification. The main disadvantage is that the starting point is entirely speculative. In practice, however, string theory is in its present state also defined by the detour of quantizing a classical theory (such as the Nambu–Goto string). Concerning *this* aspect, quantum GR and string theory use similar methods.

Even if quantum GR were superseded by a more fundamental theory such as string theory, it should nevertheless be valid at least as an *effective theory* in some appropriate limit. The reason is that far away from the Planck scale, classical GR is the appropriate theory, which in turn must be the classical limit of an underlying quantum theory. Except perhaps close to the Planck scale itself, quantum GR should be a viable framework (such as QED, which is also supposed to be only an effective theory). Nevertheless, quantum GR is a candidate for a theory of quantum gravity at all scales.

Quantum GR is perturbatively a non-renormalizable theory (the gravitational constant G has negative mass dimension). In spite of this, it may lead to definite results on the effective level. For example, Bjerrum-Bohr et al. [11] have calculated the quantum gravitational corrections to the Newtonian potential between two masses m_1 and m_2,

$$V(r) = -\frac{Gm_1m_2}{r}\left(1 + 3\frac{G(m_1+m_2)}{rc^2} + \frac{41}{10\pi}\frac{G\hbar}{r^2c^3}\right) . \tag{13}$$

This result is independent of the ambiguities that are present at higher energies. In fact, effective field theories are successfully applied elsewhere, for example 'chiral perturbation theory' in QCD (where one considers the limit of the pion mass $m_\pi \to 0$) is such an effective theory.

An important question in the formal quantization of a given classical theory is which of the structures in the classical theory should be quantized, i.e. should be subject to the superposition principle, and which structures should remain as classical (or absolute, non-dynamical) structures. Isham distinguishes the following hierarchy of structures [12]:

> Point set of events $\longrightarrow$ topological structure $\longrightarrow$ differentiable manifold $\longrightarrow$ causal structure $\longrightarrow$ Lorentzian structure.

Most approaches subject the Lorentzian and therefore the causal structure to quantization, but keep the manifold structure fixed. This is, however, not the only possibility. It might be that even the topological structure is fundamentally quantized. According to the Copenhagen interpretation of quantum theory, all these structures would probably have to stay classical, because they are thought to be necessary ingredients for the measurement process. For the purpose of quantum gravity, such a viewpoint is, however, insufficient and probably inconsistent. The modern attitude is to try to avoid in quantum gravity all absolute structures—this is referred to as *background independence*.

Quantum geometrodynamics and loop quantum gravity seem to implement this principle, whereas in the present state of string theory it is only partially implemented.

References

1. C. Kiefer, *Quantum Gravity*, 2nd edn (Oxford University Press, Oxford, 2007).
2. C. Kiefer, Quantum gravity: general introduction and recent developments. *Annalen der Physik*, **15**, 129–148 (2006).
3. D. Giulini, C. Kiefer, and C. Lämmerzahl, eds. *Quantum Gravity–From Theory to Experimental Search.* Lecture Notes in Physics **631** (Springer, Berlin, 2003).
4. M. Rees, *Perspectives in Astrophysical Cosmology* (Cambridge University Press, Cambridge, 1995).
5. C. Kiefer and C. Weber, On the interaction of mesoscopic quantum systems with gravity. *Annalen der Physik*, **14**, 253–278 (2005).
6. S. W. Hawking, Particle creation by black holes. *Communications in Mathematical Physics*, **43**, 199–220 (1975).
7. B. Carr, Primordial black holes: Do they exist and are they useful? `astro-ph/0511743` (2005).
8. D. Blais, T. Bringmann, C. Kiefer, and D. Polarski, Accurate results for primordial black holes from spectra with a distinguished scale. *Physical Review D*, **67**, 024024 [11 pages] (2003).
9. H. D. Zeh, *The physical basis of the direction of time*, 5th edn (Springer, Berlin, 2007). See also `http://www.time-direction.de`.
10. C. J. Isham, Quantum gravity. In: *General relativity and gravitation*, ed. by M. A. H. Mac Callum (Cambridge University Press, Cambridge), pp. 99–129 (1987).
11. N. E. J. Bjerrum-Bohr, J. F. Donoghue, and B. R. Holstein, Quantum gravitational corrections to the nonrelativistic scattering potential of two masses. *Phys. Rev. D*, **67**, 084033 [12 pages] (2003).
12. C. J. Isham, Prima facie questions in quantum gravity. In: *Canonical gravity: From classical to quantum*, ed. by J. Ehlers and H. Friedrich (Springer, Berlin), pp. 1–21 (1994).

The Canonical Approach to Quantum Gravity: General Ideas and Geometrodynamics

D. Giulini[1] and C. Kiefer[2]

[1] Physikalisches Institut, Universität Freiburg, Hermann-Herder-Straße 3, 79104 Freiburg, Germany
giulini@idefix.physik.uni-freiburg.de
[2] Institut für Theoretische Physik, Universität zu Köln, Zülpicher Straße 77, 50937 Köln, Germany
kiefer@thp.uni-koeln.de

1 Introduction

The really novel feature of general relativity (henceforth abbreviated GR), as compared to other field theories in physics, is that spacetime is not a fixed background arena that merely stages physical processes. Rather, spacetime is itself a dynamical entity, meaning that its properties depend in parts on its specific matter content. Hence, contrary to the Newtonian picture, in which spacetime acts (via its inertial structure) but is not acted upon by matter, the interaction between matter and spacetime now goes both ways.

Saying that the spacetime is 'dynamic' does not mean that it 'changes' with respect to any given external time. Time is clearly within, not external to spacetime. Accordingly, solutions to Einstein's equations, which are whole spacetimes, do not as such describe anything evolving. In order to take such an evolutionary form, which is, for example, necessary to formulate an initial value problem, we have to re-introduce a notion of 'time' with reference to which we may speak of 'evolution'. This is done by introducing a structure that somehow allows to split spacetime into space and time.

Let us explain this in more detail: suppose we are given a spacetime, that is, a four-dimensional differentiable manifold M with Lorentzian metric g. We assume that M can be foliated by a family $\{\Sigma_t \mid t \in \mathbb{R}\}$ of spacelike leaves. That is, for each number t there is an embedding of a fixed three-dimensional manifold Σ into M,

$$\mathcal{E}_t : \Sigma \to M \ , \tag{1}$$

whose image $\mathcal{E}_t(\Sigma) \subset M$ is just Σ_t, which is a spacelike submanifold of M; see Fig. 1. It receives a Riemannian metric by restricting the Lorentzian metric

D. Giulini and C. Kiefer: *The Canonical Approach to Quantum Gravity: General Ideas and Geometrodynamics*, Lect. Notes Phys. **721**, 131–150 (2007)
DOI 10.1007/978-3-540-71117-9_8

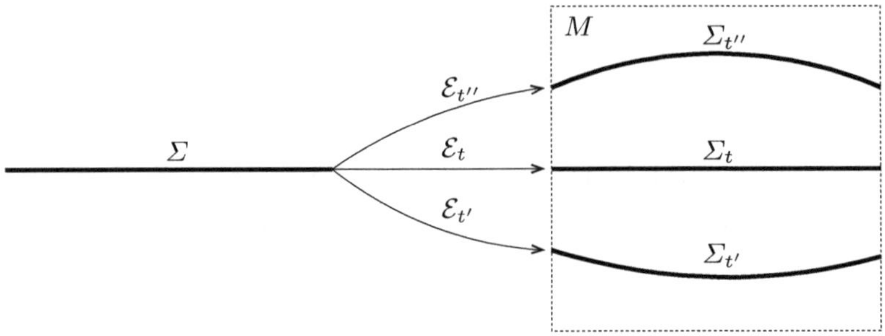

Fig. 1. Foliation of spacetime M by a one-parameter family of embeddings $\mathcal{E}_t$ of the 3-manifold Σ into M. Σ_t is the image in M of Σ under $\mathcal{E}_t$. Here the leaf $\Sigma_{t'}$ is drawn to lie to the past and $\Sigma_{t''}$ to the future of Σ_t

g of M to the tangent vectors of Σ_t. This can be expressed in terms of the 3-manifold Σ. If we endow Σ with the Riemannian metric

$$h_t := \mathcal{E}_t^* g \,, \tag{2}$$

then (Σ, h_t) is isometric to the submanifold Σ_t with the induced metric.

Each three-dimensional leaf Σ_t now corresponds to an instant of time t, where t is so far only a topological time: it faithfully labels instants in a continuous fashion, but no implication is made as to its relation to actual clock readings. The statement of such relations can eventually only be made on the basis of dynamical models for clocks coupled to the gravitational field.

By means of the foliation we now recover a notion of time: we view spacetime, (M, g), as the one-parameter family of spaces, $t \mapsto (\Sigma, h_t)$. Spacetime then becomes nothing but a 'trajectory of spaces'. In this way we obtain a dynamical system whose configuration variable is the Riemannian metric on a 3-manifold Σ. It is to make this point precise that we carefully distinguish between the manifold Σ and its images Σ_t in M. In the dynamical formulation given now, there simply is no spacetime to start with and hence no possibility to embed Σ into something. Only *after* solving the dynamical equations can we construct spacetime and interpret the time dependence of the metric of Σ as being brought about by 'wafting' Σ through M via a one-parameter family of embeddings $\mathcal{E}_t$. But initially there is only a 3-manifold Σ of some topological type[1] and the equations of motion together with some suitable initial data. For a fuller discussion we refer to the comprehensive work by Isham and Kuchař [13, 14].

[1] It can be shown that the Einstein equations do not pose any obstruction to the topology of Σ, that is, solutions exist for *any* topology. However, one often imposes additional requirements on the solution. For example, one may require that there exists a moment of time symmetry, which will make the corresponding instant Σ_t a totally geodesic submanifold of M, like e.g. in recollapsing cosmological models at the moment of maximal expansion. In this case the topology of Σ will be severely restricted. In fact, most topologies Σ will only support geometries that always expand or contract somewhere.

2 The Initial-Value Formulation of GR

Whereas a specified motion of Σ through M, characterized by the family of embeddings (1), gives rise to a one-parameter family of metrics h_t, the converse is not true. That is to say, it is not true that *any* one-parameter family of metrics h_t of Σ can be obtained from a given spacetime (M, g) and a one-parameter family of embeddings $\mathcal{E}_t$, such that (2) holds.

Moreover, there is clearly a huge redundancy in creating (M, g) from the family $\{(\Sigma, h_t) \mid t \in \mathbb{R}\}$, since there are obviously many different motions of Σ through the same M, which give rise to apparently different solution curves h_t. This redundancy can be locally parameterized by four functions, on Σ: a scalar field α and a vector field β. In the embedding picture, they describe the components of the velocity vector field

$$\frac{\partial}{\partial t} := \frac{d}{dt}\mathcal{E}_t \tag{3}$$

normal and tangential to the leaves Σ_t respectively. We write

$$\frac{\partial}{\partial t} = \alpha n + \beta \, , \tag{4}$$

where n is the normal to Σ_t. The tangential component, β, just generates intrinsic diffeomorphisms on each Σ_t, whereas the normal component, α, really advances one leaf Σ_t to the next one; see Fig. 2.

For the initial-value problem it is the derivative along the normal n of the 3-metric h, denoted by K, that gives the essential information. Hence we write

$$\frac{\partial h_t}{\partial t} = \alpha K_t + L_\beta h_t \, . \tag{5}$$

In the embedding picture, K_t is the extrinsic curvature of Σ_t in M.

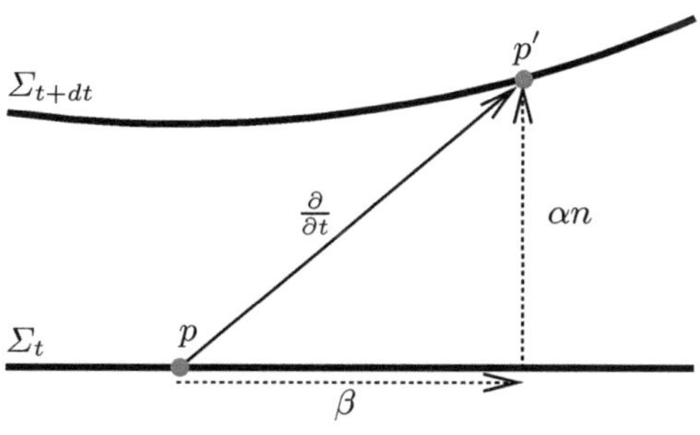

Fig. 2. Infinitesimally nearby leaves Σ_t and Σ_{t+dt}. For some point $q \in \Sigma$, the image points $p = \mathcal{E}_t(q)$ and $p' = \mathcal{E}_{t+dt}(q)$ are connected by the vector $\partial/\partial t|_p$, whose components tangential and normal to Σ_t are β and αn, respectively. n is the normal to Σ_t in M, β is called the 'shift vector-field' and α the 'lapse function' on Σ_t

The first-order evolution equations that result from Einstein's field equations are then of the general form

$$\frac{\partial h_t}{\partial t} = F_1(h_t, K_t; \alpha, \beta) , \tag{6}$$

$$\frac{\partial K_t}{\partial t} = F_2(h_t, K_t; \alpha, \beta; \text{matter}) , \tag{7}$$

where F_1 in (6) is given by the right-hand side of (5). F_2 is a more complicated function whose precise structure need not interest us now and which also depends on matter variables; see e.g. [8].

3 Why Constraints

As we have seen, the initial data for the gravitational variables consist of a differentiable 3-manifold Σ, a Riemannian metric h – the configuration variable, and another symmetric second rank tensor field K on Σ – the velocity variable. However, the pair (h, K) cannot be chosen arbitrarily. This is because there is a large redundancy in describing a fixed spacetime M by a foliation (1). On the infinitesimal level this gauge freedom is just the freedom of choosing α and β. The gauge transformations generated by β are just the spatial diffeomorphisms of Σ. β may be an arbitrary function of t, which corresponds to the fact that we may arbitrarily permute the points in each leaf Σ_t separately (only restricted by some differentiability conditions). The gauge transformations generated by α correspond to pointwise changes in the velocities with which the leaves Σ_t push through M. These too may vary arbitrarily within the leaves as well as with coordinate time t.

Whenever there is gauge freedom in a dynamical theory, there are so-called *constraints*, that is, conditions which restrict the initial data; see e.g. [10]. For each gauge freedom parameterized by an arbitrary function, there is one functional combination of the initial data which has to vanish. In our case there are four gauge functions, α, and the three components of β. Accordingly there are four constraints, which group into one *scalar or Hamiltonian constraint*, $H[h, K] = 0$, and three combined in the *vector or diffeomorphism constraint*, $D[h, K] = 0$. Their explicit expressions are[2]

$$H[h, K] = (2\kappa)^{-1} G^{ab\,cd} K_{ab} K_{cd} - (2\kappa)^{-1} \sqrt{h}\left({}^{(3)}R - 2\Lambda\right) + \sqrt{h}\rho , \tag{8}$$

$$D^a[h, K] = -\kappa^{-1} G^{ab\,cd} \nabla_b K_{cd} + \sqrt{h} j^a . \tag{9}$$

Here ρ and j^a are the energy and momentum densities of the matter, ∇ and ${}^{(3)}R$ are the Levi-Civita connection and its associated scalar curvature of

[2] Here and below we shall write $\sqrt{h} := \sqrt{\det\{h_{ab}\}}$ and use the abbreviation $\kappa = 8\pi G/c^4$. Hence κ has the physical dimension of $\mathrm{s}^2 \cdot \mathrm{m}^{-1} \cdot \mathrm{kg}^{-1}$. We shall set $c = 1$ throughout.

(Σ, h). Finally $G^{ab\,cd}$ is the so-called 'DeWitt metric', which at each point of Σ defines an h-dependent Lorentzian metric on the $1+5$-dimensional space of symmetric second-rank tensors at that point.[3] Its explicit form is given by

$$G^{ab\,cd} = \tfrac{\sqrt{h}}{2}\left(h^{ac}h^{bd} + h^{ad}h^{bc} - 2h^{ab}h^{cd}\right) . \tag{10}$$

Note that the linear space of symmetric second-rank tensors is viewed here as the tangent space ('velocity space') of the space Riem(Σ) of Riemannian metrics on Σ. From (10) one sees that it is the trace part of the 'velocities', corresponding to changes of the scale (conformal part) of the Riemannian metric, which span the negative-norm velocity directions.

4 Comparison with Conventional Form of Einstein's Equations

The presence of constraints and their relation to the evolution equations is the key structure in canonical GR. It is therefore instructive to point out how this structure arises from the conventional, four-dimensional form of Einstein's equations. Before doing this, it is useful to first remind ourselves on the analogous situation in electrodynamics.

So let us first consider electrodynamics in Minkowski space. As usual, we write the field tensor F as exterior differential of a vector potential A, that is, $F = dA$. In components this reads $F_{\mu\nu} = \partial_\mu A_\nu - \partial_\nu A_\mu$. Here $E_i = F_{0i}$ are the components of the electric, $B_i = -F_{jk}$ of the magnetic field, where ijk is a cyclic permutation of 123. The homogeneous Maxwell equations now simply read $dF = 0$, whereas the inhomogeneous Maxwell equations are given by (in components)

$$M^\mu := \partial_\nu F^{\mu\nu} + \tfrac{4\pi}{c} j^\mu = 0 , \tag{11}$$

where here j^μ is the electric four-current. Due to its antisymmetry, the field tensor obeys the identity

$$\partial_\mu \partial_\nu F^{\mu\nu} \equiv 0 . \tag{12}$$

Taking the divergence of (11) and using (12) leads to

$$\partial_\mu M^\mu \equiv \tfrac{4\pi}{c} \partial_\mu j^\mu = 0 , \tag{13}$$

showing the well-known fact that Maxwell's equations imply charge conservation as integrability condition.

Let us now interpret the role of charge conservation in the initial-value problem. Decomposing (12) into space and time derivatives yields

$$\partial_0 \partial_\nu F^{0\nu} \equiv -\partial_a \partial_\nu F^{a\nu} . \tag{14}$$

---

[3] The Lorentzian signature of the DeWitt metric has nothing to do with the Lorentzian signature of the spacetime metric: it persists in Euclidean gravity.

Even though the right-hand side contains third derivatives in the field A_μ, time derivatives appear at most in second order (since ∂_a is spatial). Hence, since it is an identity, $\partial_\nu F^{0\nu}$ contains time derivatives only up to first order. But the initial data for the second-order equation (11) consist of the field A_μ and its first-time derivative. Hence the time component M^0 of Maxwell's equations gives a relation amongst initial data; in other words, it is a *constraint.* Clearly this is just the Gauß constraint $\boldsymbol{\nabla} \cdot \boldsymbol{E} - 4\pi\rho = 0$ (here ρ is the electric charge density). Only the three spatial components of (11) contain second time derivatives and hence propagate the fields. They provide the evolutionary part of Maxwell's equations.

Now, assume we are given initial data satisfying the constraint $M^0 = 0$, which we evolve according to $M^a = 0$. How can we be sure that the evolved data again satisfy the constraint? To see when this is the case, we use the identity (13) and solve it for the time derivative of M^0:

$$\partial_0 M^0 \equiv -\partial_a M^a + \tfrac{4\pi}{c}\partial_\mu j^\mu \, . \tag{15}$$

This shows that if initially $M^a = 0$ (and hence $\partial_a M^a = 0$), then the constraint $M^0 = 0$ is preserved in time if and only if $\partial_\mu j^\mu = 0$. Charge conservation is thus recognized as the necessary and sufficient condition for the compatibility between the constraint part and the evolutionary part of Maxwell's equations.

Finally we wish to make another remark concerning the interplay between constraints and evolution equations. It is clear that a solution $F^{\mu\nu}$ to (11) satisfies the constraint on *any* simultaneity hypersurface of an inertial observer (i.e. spacelike plane). If the normal to the hypersurface is n_μ, this just states that $M^\mu = 0$ implies $M^\mu n_\mu = 0$. But the converse is obviously also true: if $M^\mu n_\mu = 0$ for all timelike n_μ, then $M^\mu = 0$. In words: given an electromagnetic field that satisfies the constraint (for given external current j^μ) on *any* spacelike plane in Minkowski space, then this field must necessarily satisfy Maxwell's equations. In this sense, Maxwell's equations are the *unique* propagation law that is compatible with Gauß constraint.

After this digression we return to GR, where we can perform an entirely analogous reasoning. We start with Einstein's equations, in which the spacetime metric $g_{\mu\nu}$ is the analog of A_μ and the Einstein tensor $G^{\mu\nu} := R^{\mu\nu} - \frac{1}{2}g^{\mu\nu}R$ is the analog of $\partial_\nu F^{\mu\nu}$. They read

$$E^{\mu\nu} := G^{\mu\nu} - \Lambda - \kappa T^{\mu\nu} = 0 \, . \tag{16}$$

Due to four-dimensional diffeomorphism invariance, we have the identity (twice contracted second Bianchi-Identity):

$$\nabla_\mu G^{\mu\nu} \equiv 0 \, , \tag{17}$$

which is the analog of (12). Taking the covariant divergence of (16) and using (17) yields

$$\nabla_\mu E^{\mu\nu} = -\kappa \nabla_\mu T^{\mu\nu} = 0 \, , \tag{18}$$

which is the analog of (13). Hence the vanishing covariant divergence of $T^{\mu\nu}$ is an integrability condition of Einstein's equations, just as the divergencelessness of the electric four-current was an integrability condition of Maxwell's equations.[4]

In order to talk about 'evolution', we consider the foliation (1) of M and locally use coordinates $\{x^0, x^a\}$ such that $\partial/\partial x^0$ is the normal n to the leaves and all $\partial/\partial x^a$ are tangential. Expanding (17) in terms of partial derivatives gives

$$\partial_0 G^{0\nu} = -\partial_a G^{a\nu} - \Gamma^{\mu}_{\mu\lambda} G^{\lambda\nu} - \Gamma^{\nu}_{\mu\lambda} G^{\mu\lambda} \,, \tag{19}$$

which is the analog of (14). Now, since the $G^{\mu\nu}$ contain at most second and the $\Gamma^{\lambda}_{\mu\nu}$ at most first derivatives of the metric $g_{\mu\nu}$, this identity immediately shows that the four components $G^{0\nu}$ ($\nu = 0,1,2,3$) contain at most first-time derivatives $\partial/\partial x^0$. But Einstein's equations are of second order, hence the four equations $E^{0\nu} = 0$ are relations amongst the initial data, rather than being evolution equations. In fact, up to a factor of -2 they are just the constraints (8–9):

$$H = -2E^{00} = -2(G^{00} - \Lambda - \kappa T^{00}) \,, \tag{20}$$

$$D^a = -2E^{0a} = -2(G^{0a} - \Lambda - \kappa T^{0a}) \,. \tag{21}$$

Moreover, the remaining purely spatial components of Einstein's equations are equivalent to the 12 first-order evolution equations (6–7).

The interplay between constraints and evolution equations can now be followed along the very same lines as for the electrodynamic analogy. Expanding the left equality of (18) in terms of partial derivatives gives

$$\partial_0 E^{0\nu} = -\partial_a E^{a\nu} - \Gamma^{\mu}_{\mu\lambda} E^{\lambda\nu} - \Gamma^{\nu}_{\mu\lambda} E^{\mu\lambda} - \kappa\nabla_\mu T^{\mu\nu} \,, \tag{22}$$

which is the analog of (15). It shows that the constraints are preserved by the evolution if and only if the energy–momentum tensor of the matter has vanishing covariant divergence.

Let us now turn to the last analogy: the uniqueness of the evolution that preserves constraints. Clearly Einstein's equations $E^{\mu\nu}$ imply $E^{\mu\nu} n_\mu = 0$ for any timelike vector field n_μ. Hence the constraints are satisfied on any spacelike slice through spacetime. Again the converse is also true: given a gravitational field such that $E^{\mu\nu} n_\mu = 0$ for any timelike n_μ (and given external $T^{\mu\nu}$), then this field must necessarily satisfy Einstein's equations. In this sense Einstein's equations follow uniquely from the condition of constraint preservation.

This property will be crucial for the interpretation of the quantum theory discussed below. We know from quantum mechanics that the classical trajectories have completely disappeared at the fundamental level. As we have

[4] There is, however, a notable difference in the physical interpretation of divergencelessness of a tensor field on one hand and a vector field on the other: $\nabla_\mu T^{\mu\nu} = 0$ does not as such imply a conservation law. Only in presence of a spacetime symmetry, that is, a Killing vector field K_ν, the current $J^\mu = T^{\mu\nu} K_\nu$ is conserved, $\nabla_\mu J^\mu = 0$, and hence gives rise to a conserved quantity.

discussed above, the analogue to a trajectory is in GR provided by a spacetime given as a set of three-dimensional geometries. In quantum gravity, the spacetime will therefore disappear like the classical trajectory in quantum mechanics. It is therefore not surprising that the evolution equations (6) and (7) will be absent in quantum gravity. All the information will be contained in the quantized form of the constraints (8) and (9).

5 Canonical Gravity

We have seen above that Einstein's equations can be written as a dynamical system (6–7) with constraints (8–9). Here we wish to give its canonical formulation. Basically this means to introduce momenta for the velocities and write the first-order equations of motions as Hamilton equations. For this we have to identify the Poisson structure and the Hamiltonian. The result is this: as before, the configuration variable is the Riemannian metric h_{ab} on Σ. Its canonical momentum is now given by

$$\pi^{ab} = (2\kappa)^{-1}\, G^{ab\,cd} K_{cd} = (2\kappa)^{-1}\, \sqrt{h}(K^{ab} - h^{ab} K^c_c) \,, \tag{23}$$

so that the Poisson brackets are

$$\{h_{ab}(x), \pi^{cd}(y)\} = \tfrac{1}{2}(\delta^c_a\delta^d_b + \delta^d_a\delta^c_b)\delta^{(3)}(x,y) \,, \tag{24}$$

where $\delta^{(3)}(x,y)$ is the Dirac distribution on Σ.

Elimination of K_{ab} in favour of π^{ab} in the constraints leads to their canonical form:

$$H[h,\pi] \;=\; 2\kappa\, G_{ab\,cd}\pi^{ab}\pi^{cd} \;-\; (2\kappa)^{-1}\sqrt{h}({}^{(3)}\!R - 2\Lambda) \;+\; \sqrt{h}\rho \,, \tag{25}$$

$$D^a[h,\pi] \;=\; -2\nabla_b\pi^{ab} \;+\; \sqrt{h}j^a \,, \tag{26}$$

where now[5]

$$G_{ab\,cd} = \frac{1}{2\sqrt{h}}(h_{ac}h_{bd} + h_{ad}h_{bc} - h_{ab}h_{cd}) \,. \tag{27}$$

Likewise, rewriting (6–7) in terms of the canonical variables shows that they are just the flow equations for the following Hamiltonian:

$$\mathcal{H}[h,\pi] = \int_\Sigma d^3x\{\alpha(x)H[h,\pi](x) + \beta^a(x)D_a[h,\pi](x)\} + \text{boundary terms.} \tag{28}$$

The crucial observation to be made here is that, up to boundary terms, the total Hamiltonian is a combination of pure constraints. The boundary terms

[5] Note the difference in the factor of two in the last term, as compared to (10). $G_{ab\,cd}$ is the inverse to $G^{ab\,cd}$, that is, $G^{ab\,nm}G_{nm\,cd} = \frac{1}{2}(\delta^a_c\delta^b_d + \delta^a_d\delta^b_c)$, and *not* obtained by simply lowering the indices using h_{ab}.

generally appear if Σ is non-compact, as it will be the case for the description of isolated systems, like stars or black holes. In this case the boundary terms are taken over closed surfaces at spatial infinity and represent conserved Poincaré charges, like energy, linear and angular momentum, and the quantity associated with asymptotic boost transformations. If, however, Σ is closed (i.e. compact without boundary) all of the evolution will be generated by constraints, that is, pure gauge transformations! In that case, evolution, as described here, is not an observable change. For that to be the case we would need an extrinsic clock, with respect to which 'change' can be defined. But a closed universe already contains – by definition – everything physical, so that no external clock exists. Accordingly, there is no external time parameter. Rather, all physical time parameters are to be constructed from within our system, that is, as functional of the canonical variables. A priori there is no preferred choice of such an intrinsic time parameter. The absence of an extrinsic time and the non-preference of an intrinsic one is commonly known as the *problem of time* in Hamiltonian (quantum-)cosmology.

Finally we turn to the commutation relation between the various constraints. For this it is convenient to integrate the local constraints (25–26) over lapse and shift functions. Hence we set (suppressing the phase-space argument $[h, \pi]$)

$$\mathcal{H}(\alpha) = \int_\Sigma d^3x\, H(x)\, \alpha(x) \;, \tag{29}$$

$$\mathcal{D}(\beta) = \int_\Sigma d^3x\, D^a(x)\, \beta_a(x) \;. \tag{30}$$

A straightforward but slightly tedious computation gives

$$\{\mathcal{D}(\beta), \mathcal{D}(\beta')\} = \mathcal{D}([\beta, \beta']) \;, \tag{31}$$

$$\{\mathcal{D}(\beta), \mathcal{H}(\alpha)\} = \mathcal{H}(\beta(\alpha)) \;, \tag{32}$$

$$\{\mathcal{H}(\alpha), \mathcal{H}(\alpha')\} = \mathcal{D}(\alpha\nabla\alpha' - \alpha'\nabla\alpha) \;. \tag{33}$$

There are three remarks we wish to make concerning these relations. First, (31) shows that the diffeomorphism generators form a Lie subalgebra. Second, (32) shows that this Lie subalgebra is not a Lie ideal. This means that the flow of the Hamiltonian constraint does not leave invariant the constraint hypersurface of the diffeomorphism constraint. Finally, the term $\alpha\nabla\alpha' - \alpha'\nabla\alpha$ in (33) contains the canonical variable h, which is used implicitly to raise the index in the differential in order to get the gradient ∇. This means that the relations above do not make the set of all $\mathcal{H}(\alpha)$ and all $\mathcal{D}(\beta)$ into a Lie algebra.[6]

[6] Sometimes this is expressed by saying that this is an 'algebra with structure functions'.

6 The General Kinematics of Hypersurface Deformations

In this section we wish to point out that the relations (31–33) follow a general pattern, namely to represent the 'algebra' of hypersurface deformations, or in other words, infinitesimal changes of embeddings $\mathcal{E} : \Sigma \to M$. To make this explicit, we introduce local coordinates x^a on Σ and y^μ on M. An embedding is then locally given by four functions $y^\mu(x)$, such that the 3×4 matrix $y^\mu_{,a}$ has its maximum rank 3 (we write $y^\mu_{,a} := \partial_a y^\mu$). The components of the normal to the image $\mathcal{E}(\Sigma) \subset M$ are denoted by n^μ, which should be considered as functional of $y^\mu(x)$. The generators of normal and tangential deformations of $\mathcal{E}$ with respect to the lapse function α and shift vector field β are then given by

$$N_\alpha = \int_\Sigma d^3x \; \alpha(x)\, n^\mu[y(x)]\, \frac{\delta}{\delta y^\mu(x)} \,, \tag{34}$$

$$T_\beta = \int_\Sigma d^3x \; \beta^a(x)\, y^\mu_{,a}(x)\, \frac{\delta}{\delta y^\mu(x)} \,, \tag{35}$$

which may be understood as tangent vectors to the space of embeddings of Σ into M. A calculation[7] then leads to the following commutation relations

$$[T_\beta, T_{\beta'}] = -T_{[\beta,\beta']} \,, \tag{36}$$

$$[T_\beta, N_\alpha] = -N_{\beta(\alpha)} \,, \tag{37}$$

$$[N_\alpha, N_{\alpha'}] = -T_{\alpha\nabla\alpha'-\alpha'\nabla\alpha} \,. \tag{38}$$

Up to the minus signs this is just (31–33). The minus signs are just the usual ones that one always picks up when going from the action of vector fields to the Poisson action of the corresponding phase-space functions. (In technical terms, the mapping from vector fields to phase-space functions is a Lie-*anti*-homomorphism.)

This shows that (31–33) just mean that we have a Hamiltonian realization of hypersurface deformations. In particular, (31–33) is neither characteristic of the action nor the field content: *Any* four dimensional diffeomorphism invariant theory will gives rise to this very same 'algebra'. It can be shown that under certain general locality assumptions the expressions (25) and (26) give the unique 2-parameter (here κ and Λ) family of realizations for N and T satisfying (36–38) on the phase space parameterized by (h_{ab}, π^{ab}); see [11] and also [18].

---

[7] Equation (36) is immediate. To verify (37–38) one needs to compute $\delta n^\mu[y(x)]/\delta y^\nu(x')$. This can be done in a straightforward way by varying

$$g(y(x))_{\mu\nu} n^\mu[y(x)] n^\nu[y(x)] = -1 \quad \text{and} \quad g_{\mu\nu}(y(x))\, y^\mu_{,a}(x)\, n^\nu[y(x)] = 0$$

with respect to $y(x)$.

7 Topological Issues

As we have just discussed, Einstein's equations take the form of a constrained Hamiltonian system if put into canonical form. The unconstrained configuration space is the space of all Riemannian metrics on some chosen 3-manifold Σ. This space is denoted by $\mathrm{Riem}(\Sigma)$. Any two Riemannian metrics that differ by an action of the diffeomorphism constraint are gauge equivalent and hence to be considered as physically indistinguishable. Let us briefly mention that the question of whether and when the diffeomorphism constraint actually generates all diffeomorphisms of Σ is rather subtle. Certainly, what is generated lies only in the identity component of the latter, but even on that it may not be onto. This occurs, for example, in the case where Σ contains asymptotically flat ends with non-vanishing Poincaré charges associated. Asymptotic Poincaré transformations are then not interpreted as gauge transformations (otherwise the Poincaré charges were necessarily zero), but as proper physical symmetries (i.e. changes of state that are observable in principle).

Leaving aside the possible difference between what is generated by the constraints and the full group $\mathrm{Diff}(\Sigma)$ of diffeomorphisms of Σ, we may consider the quotient space $\mathrm{Riem}(\Sigma)/\mathrm{Diff}(\Sigma)$ of Riemannian *geometries*. This space is called *superspace* in the relativity community (this has nothing to do with supersymmetry), which we denote by $\mathcal{S}(\Sigma)$. Now from a topological viewpoint $\mathrm{Riem}(\Sigma)$ is rather trivial. It is a cone[8] in the (infinite dimensional) vector space of all symmetric second-rank tensor fields. But upon factoring out $\mathrm{Diff}(\Sigma)$ the quotient space $\mathcal{S}(\Sigma)$ inherits some of the topological information concerning Σ, basically because $\mathrm{Diff}(\Sigma)$ contains that information [6]. This is schematically drawn in Fig. 3.

In a certain generalized sense, GR is a dynamical system on the phase space (i.e. cotangent bundle) built over superspace. The topology of superspace is characteristic for the topology of Σ, though in a rather involved way. Note that, by construction, the Hamiltonian evolution is that of a varying embedding of Σ into spacetime. Hence the images Σ_t are all of the same topological type. This is why canonical gravity in the formulation given here cannot describe transitions of topology.

Note, however, that this is not at all an implication by Einstein's equations. Rather, it is a consequence of our restriction to spacetimes that admit a global spacelike foliation. There are many solutions to Einstein's equations that do not admit such foliations globally. This means that these spacetimes cannot be constructed by integrating the equations of motions (6–7) successively from some initial data. Should we rule out all other solutions? The general feeling seems to be that, at least in quantum gravity, topology changing classical solutions should not be ruled out as possible contributors in the sum over histories (path integral). Figure 4 shows two such histories. Whereas in the left

[8] Any real positive multiple λh of $h \in \mathrm{Riem}(\Sigma)$ is again an element of $\mathrm{Riem}(\Sigma)$.

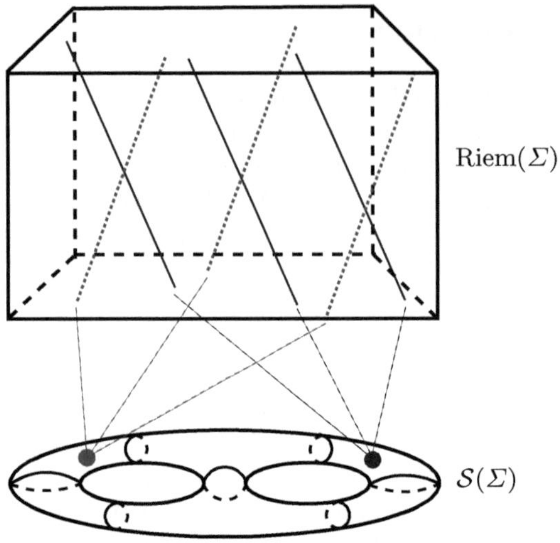

Fig. 3. The topologically trivial space Riem(Σ), here drawn as the box above, is fibered by the action of the diffeomorphism group. The fibers are the straight lines in the box, where the sets consisting of three dashed and three solid lines, respectively, form one fiber each. In the quotient space $\mathcal{S}(\Sigma)$ each fiber is represented by one point only. By taking the quotient, $\mathcal{S}(\Sigma)$ receives the non-trivial topology from Diff(Σ). To indicate this, $\mathcal{S}(\Sigma)$ is represented as a double torus

picture the universe simply 'grows a nose', it bifurcates in the right example to become disconnected.

One may ask whether there are topological restrictions to such transitions. First of all, it is true (though not at all obvious) that for any given two 3-manifolds $\Sigma_{\rm i}$, $\Sigma_{\rm f}$ (neither needs to be connected) there is a 4-manifold M whose boundary is just $\Sigma_{\rm i} \cup \Sigma_{\rm f}$. In fact, there are infinitely many such M. Amongst them, one can always find some which can be endowed with a globally regular Lorentz metric g, such that $\Sigma_{\rm i}$ and $\Sigma_{\rm f}$ are spacelike. However, if topology changes, (M, g) necessarily contains closed timelike curves [2]. This fact has sometimes been taken as rationale for ruling out topology change in (classical) GR. But it should be stressed that closed timelike curves do not necessarily ruin conventional concepts of predictability. In any case, let us accept this slight pathology and ask what other structures we wish to define on M. For example, in order to define fermionic matter fields on M we certainly wish to endow M with a $SL(2, \mathbb{C})$ spin structure. This is where now the first real obstructions for topological transitions appear [3].[9] It is then possible to translate them into selection rules for transitions between all known 3-manifolds [4].

9 Their result is the following: let $\Sigma = \Sigma_{\rm i} \cup \Sigma_{\rm f}$ be the spacelike boundary of the Lorentz manifold M, then $\dim\big(H^0(\Sigma, \mathbf{Z}_2)\big) + \dim\big(H^1(\Sigma, \mathbf{Z}_2)\big)$ has to be even for M to admit an $SL(2, \mathbb{C})$ spin structure.

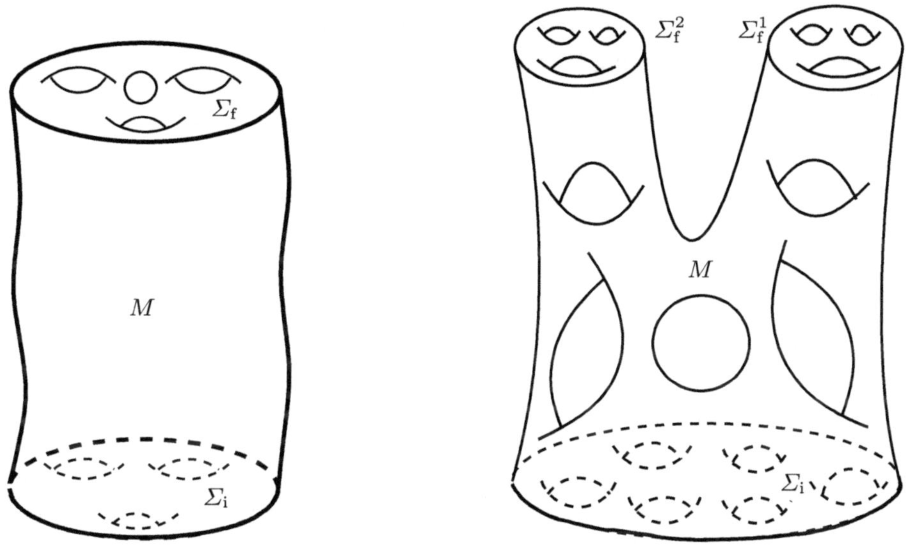

Fig. 4. Spacetimes in which spatial sections change topology. In the (**left**) picture the initial universe Σ_i has three, the final Σ_f four topological features (*'holes'*) – it 'grows a nose' while staying connected. In the (**right**) picture the initial universe Σ_i splits into two copies $\Sigma_\mathrm{f}^{1,2}$, so that $\Sigma_f = \Sigma_\mathrm{f}^1 \cup \Sigma_\mathrm{f}^2$. In both cases, the interpolating spacetime M can be chosen to carry a Lorentzian metric with respect to which initial and final hypersurfaces are spacelike, possibly at the price of making M topologically complicated, like indicated in the right picture

So far the considerations were purely kinematical. What additional obstructions arise if the spacetime (M, g) is required to satisfy the field equations? Here the situation becomes worse. It is, for example, known that any topology-changing spacetime that satisfies Einstein's equations with matter that satisfies the weak-energy condition $T_{\mu\nu}l^\mu l^\nu \geq 0$ for all lightlike l^μ must necessarily be singular.[10] Hence it seems that we need to consider degenerate metrics already on the classical level if topology change is to occur. Can we relax the notion of 'solution to Einstein's equations' so as to contain these degenerate cases as well? The answer is 'yes' if instead of taking the metric as basic variable we rewrite the equations in terms of vierbeine and connections (first-oder formalism). It turns out that the kind of singularities one has to cope with are very mild indeed: the vierbeine become degenerate on sets of measure zero but, somewhat surprisingly, the curvature stays bounded everywhere. In fact, there is a very general method to generate an abundance of such solutions [12].

It is a much-debated question whether topology-changing amplitudes are suppressed or, to the contrary, needed in quantum gravity. On one hand, it has

[10] In fact, this result can be considerably strengthened: instead of invoking Einstein's equations we only need to require $R_{\mu\nu}l^\mu l^\nu \geq 0$ for all lightlike l^μ.

been shown in the context of specific lower-dimensional models that matter fields on topology-changing backgrounds may give rise to singularities corresponding to infinite densities of particle production [1]. On the other hand, leaving out topology-changing amplitudes in the sum-over-histories approach is heuristically argued to be in conflict with expected properties of localized pseudo-particle-like excitations in gravity (so-called 'geons'), like, for example, the usual spin-statistic relation [19]. Here there still seems to be much room for speculations.

8 Geometric Issues

Just in the same way as any Lagrangian theory endows the configuration space with the kinetic-energy metric, Riem(Σ) inherits a metric structure from the 'kinetic-energy' part of (8). Tangent vectors at $h \in$ Riem(Σ) are symmetric second-rank tensor fields on Σ and their inner product is given by the so-called *Wheeler–DeWitt metric*:

$$\mathcal{G}_h(V, V') = \int_\Sigma d^3x \, G^{ab\,cd} V_{ab} V'_{cd} \,. \tag{39}$$

Due to the pointwise Lorentzian signature (1+5) of $G^{ab\,cd}$ it is of a hyper-Lorentzian structure with infinitely many negative, null, and positive directions each. However, not all directions in the tangent space T_h(Riem(Σ)) correspond to physical changes. Those generated by diffeomorphism, which are of the form $V_{ab} = \nabla_a \beta_b + \nabla_b \beta_a$ for some vector field β on Σ, are pure gauge. We call them *vertical*. The diffeomorphism constraint (26) for $j^a = 0$ – a case to which we now restrict for simplicity – now simply says that V must be $\mathcal{G}$–orthogonal to such vertical directions. We call such orthogonal directions *horizontal*. Moreover, it is easily seen that the inner product (39) is invariant under Diff(Σ). All this suggests how to endow superspace, $\mathcal{S}(\Sigma)$, with a natural metric: take two tangent vectors at a point $[h]$ in $\mathcal{S}(\Sigma)$, lift them to horizontal vectors at h in Riem(Σ), and there take the inner product according to (39).

However, this procedure only works if the horizontal subspace of T_h(Riem(Σ)) is truly complementary to the vertical space of gauge directions. But this is not guaranteed due to $\mathcal{G}$ not being positive definite: whenever there are vertical directions of zero $\mathcal{G}$-norm, there will be non-trivial intersections of horizontal and vertical spaces. Sufficient conditions on h for this *not* to happen can be derived: for example, a strictly negative Ricci tensor [7]. The emerging picture is that there are open sets in $\mathcal{S}(\Sigma)$ in which well-defined hyper-Lorentzian geometries exist, which are separated by closed transition regions in which the signature of these metrics change. The transition regions precisely consist of those geometries $[h]$ which possess vertical directions of zero $\mathcal{G}$-norm; see Fig. 5.

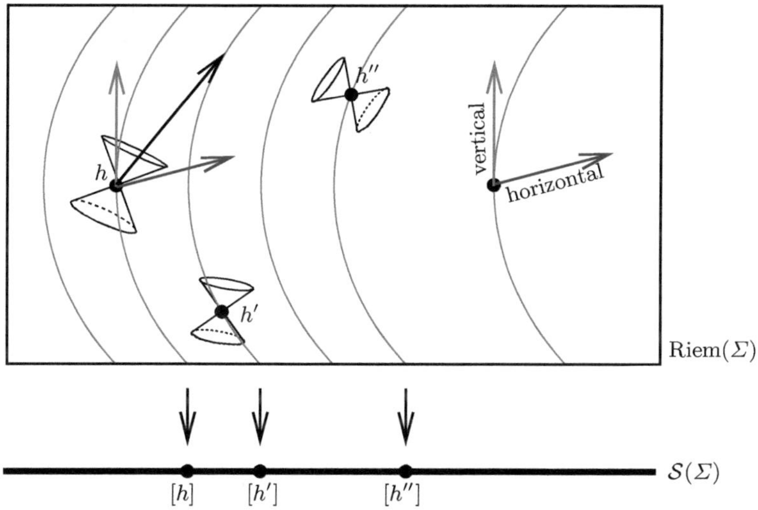

Fig. 5. The space Riem(Σ), fibred by the orbits of Diff(Σ) (*curved vertical lines*). Tangent directions to these orbits are called 'vertical', the $\mathcal{G}$-orthogonal directions 'horizontal'. Horizontal and vertical directions intersect whenever the 'hyper-light-cone' touches the vertical directions, as in point h'. At h, h', and h'' the vertical direction is depicted as time-, light-, and spacelike respectively. Hence $[h']$ corresponds to a transition point where the signature of the metric in superspace changes

9 Quantum Geometrodynamics

Einstein's theory of GR has now been brought into a form where it can be subject to the procedure of canonical quantization. As we have argued above, all the information that is needed is encoded in the constraints (25) and (26). However, quantizing them is far from trivial [16]. One might first attempt to solve the constraints on the classical level and then to quantize only the reduced, physical, degrees of freedom. This has not even been achieved in quantum electrodynamics (except for the case of freely propagating fields), and it is illusory to achieve in GR. One therefore usually follows the procedure proposed by Dirac and tries to implement the constraints as conditions on physically allowed wave functionals. The constraints (25) and (26) then become the quantum conditions

$$\hat{H}\Psi = 0 , \tag{40}$$

$$\hat{D}^a\Psi = 0 , \tag{41}$$

where the 'hat' is a symbolic indication for the replacements of the classical expressions by operators. This procedure also applies if other variables instead of the three-metric and its momentum are used; for example, such quantum constraints also play the role in loop quantum gravity, cf. the contributions of Nicolai and Peeters as well as Thiemann to this book. In the present case the resulting formalism is called quantum geometrodynamics.

Quantum geometrodynamics is defined by the transformation of $h_{ab}(x)$ into a multiplication operator and π^{cd} into a functional derivative operator, $\pi^{cd} \to -\mathrm{i}\hbar\delta/\delta h_{cd}(x)$. The constraints (25) and (26) then assume the form, restricting here to the vacuum case for simplicity,

$$\hat{H}\Psi \equiv \left(-2\kappa\hbar^2 G_{abcd}\frac{\delta^2}{\delta h_{ab}\delta h_{cd}} - (2\kappa)^{-1}\sqrt{h}\left({}^{(3)}R - 2\Lambda\right)\right)\Psi = 0\,, \quad (42)$$

$$\hat{D}^a\Psi \equiv -2\nabla_b\frac{\hbar}{\mathrm{i}}\frac{\delta\Psi}{\delta h_{ab}} = 0\,. \quad (43)$$

Equation (42) is called the *Wheeler–DeWitt equation* in honour of the work by Bryce DeWitt and John Wheeler; see e.g. [16] for details and references. In fact, these are again infinitely many equations (one equation per space point). The constraints (43) are called the *quantum diffeomorphism (or momentum) constraints*. Occasionally, both (42) and (43) are referred to as Wheeler–DeWitt equations. In the presence of non-gravitational fields, these equations are augmented by the corresponding terms.

The argument of the wave functional Ψ is the three-metric $h_{ab}(x)$ (plus non-gravitational fields). However, because of (43), Ψ is invariant under coordinate transformations on three-dimensional space (it may acquire a phase with respect to 'large diffeomorphisms' that are not connected with the identity). A most remarkable feature of the quantum constraint equations is their 'timeless' nature – the external parameter t has completely disappeared.[11] Instead of an external time one may consider an 'intrinsic time' that is distinguished by the kinetic term of (42). As can be recognized from the signature of the DeWitt metric (10), the Wheeler–DeWitt equation is locally hyperbolic, that is, it assumes the form of a local wave equation. The intrinsic timelike direction is related to the conformal part of the three-metric. With respect to the discussion in the last section one may ask whether there are regions in superspace where the Wheeler–DeWitt metric exists and has precisely one negative direction. In that case the Wheeler–DeWitt equation would be strictly hyperbolic (rather than ultrahyperbolic) in a neighbourhood of that point. It has been shown that such regions indeed exist and that they include neighbourhoods of the standard round three-sphere geometry [7]. This implies that the full Wheeler–DeWitt equation that describes fluctuations around the positive curvature Friedmann universe is strictly hyperbolic. In this case the scale factor of the Friedmann universe could serve as an intrinsic time. The indefinite nature of the kinetic term reflects the fact that gravity is attractive [5].

[11] In the case of asymptotic spaces such a parameter may be present in connection with Poincaré transformations at spatial infinity. We do not consider this case here.

There are many problems associated with the quantum constraints (42) and (43). An obvious problem is the 'factor-ordering problem': the precise form of the kinetic term is open – there could be additional terms proportional to $\hbar$ containing at most first derivatives in the metric. Since second functional derivatives at the same space point usually lead to undefined expressions such as $\delta(0)$, a regularization (and perhaps renormalization) scheme has to be employed. Connected with this is the potential presence of anomalies, cf. the contribution by Nicolai and Peeters. Another central problem is what choice of Hilbert space one has to make, if any, for an interpretation of the wave functionals. No final answer to this problem is available in this approach [16].

What about the semiclassical approximation and the recovery of an appropriate external time parameter in some limit? For the full quantum constraints this can at least be achieved in a formal sense (i.e. treating functional derivatives as if they were ordinary derivatives and neglecting the problem of anomalies); see [16, 17]. The discussion is also connected to the question: where does the imaginary unit i in the (functional) Schrödinger equation come from? The full Wheeler–DeWitt equation is real, and one would thus also expect real solutions for Ψ. An approximate solution is found through a Born–Oppenheimer type of scheme, in analogy to molecular physics. The state then assumes the form

$$\Psi \approx \exp(\mathrm{i}S_0[h]/\hbar)\,\psi[h,\phi]\;, \tag{44}$$

where h is an abbreviation for the three-metric and ϕ stands for non-gravitational fields. In short, one finds that

- S_0 obeys the Hamilton–Jacobi equation for the gravitational field and thereby defines a classical spacetime which is a solution to Einstein's equations (this order is formally similar to the recovery of geometrical optics from wave optics via the eikonal equation).
- ψ obeys an approximate (functional) Schrödinger equation,

$$\mathrm{i}\hbar\underbrace{\nabla S_0\,\nabla\psi}_{\frac{\partial\psi}{\partial t}} \approx H_{\mathrm{m}}\,\psi\;, \tag{45}$$

  where H_{m} denotes the Hamiltonian for the non-gravitational fields ϕ. Note that the expression on the left-hand side of (45) is a shorthand notation for an integral over space, in which ∇ stands for functional derivatives with respect to the three-metric. Semiclassical time t is thus defined in this limit from the dynamical variables.
- The next order of the Born–Oppenheimer scheme yields quantum gravitational correction terms proportional to the inverse Planck mass squared, $1/m_{\mathrm{P}}^2$. The presence of such terms may in principle lead to observable effects, for example, in the anisotropy spectrum of the cosmic microwave background radiation.

The Born–Oppenheimer expansion scheme distinguishes a state of the form (44) from its complex conjugate. In fact, in a generic situation both states will decohere from each other, that is, they will become dynamically independent [15]. This is a type of symmetry breaking, in analogy to the occurrence of parity violating states in chiral molecules. It is through this mechanism that the i and the t in the Schrödinger equation emerge.

The recovery of the Schrödinger equation (45) raises an interesting issue. It is well known that the notion of Hilbert space is connected with the conservation of probability (unitarity) and thus with the presence of an external time (with respect to which the probability is conserved). The question then arises whether the concept of a Hilbert space is still required in the *full* theory where no external time is present. It could be that this concept makes sense only on the semiclassical level where (45) holds.

10 Applications

The major physical applications of quantum gravity concern cosmology and black holes. Although the above-presented formalism exists, as yet, only on a formal level, one can study models that present no mathematical obstacles. Typically, such models are obtained by imposing symmetries on the solutions of the equations [16]. Examples are spherical symmetry (useful for black holes) and homogeneity and isotropy (useful for cosmology).

Quantum cosmology is the application of quantum theory to the universe as a whole. Let us consider a simple example: a Friedmann universe with scale factor $a \equiv \mathrm{e}^{\alpha}$ containing a massive scalar field ϕ. In this case, the diffeomorphism constraints (43) are identically fulfilled, and the Wheeler–DeWitt equation (42) reads

$$\hat{H}\psi \equiv \left(G\hbar^2 \frac{\partial^2}{\partial\alpha^2} - \hbar^2 \frac{\partial^2}{\partial\phi^2} + m^2\phi^2 \mathrm{e}^{6\alpha} - \frac{\mathrm{e}^{4\alpha}}{G} \right) \psi(\alpha, \phi) = 0 \, . \tag{46}$$

This equation is simple enough to find solutions (at least numerically) and to study physical aspects such as the dynamics of wave packets and the semiclassical limit [16].

There is one interesting aspect in quantum cosmology that possesses far-reaching physical consequences. Because (42) does not contain an external time parameter t, the quantum theory exhibits a kind of determinism drastically different from the classical theory [16, 20]. Consider a model with a two-dimensional configuration space spanned by the scale factor, a, and a homogeneous scalar field, ϕ, see Fig. 6. (Such a model is described, for example, by (46) with $m = 0$.) The classical model be such that there are solutions where the universe expands from an initial singularity, reaches a maximum, and recollapses to a final singularity. Classically, one would impose, in a Lagrangian formulation, $a, \dot{a}, \phi, \dot{\phi}$ (satisfying the constraint) at some t_0 (for

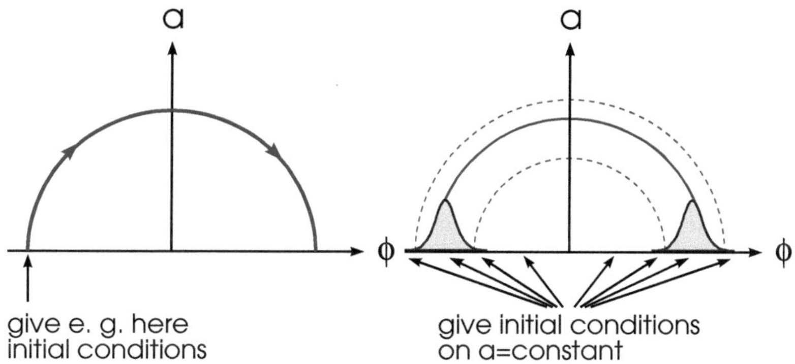

Fig. 6. The classical and the quantum theory of gravity exhibit drastically different notions of determinism

example, at the left leg of the trajectory), and then the trajectory would be determined. This is indicated on the left-hand side of Fig. 6. In the quantum theory, on the other hand, there is no t. The hyperbolic nature of a minisuperspace equation such as (46) suggests to impose boundary conditions at $a =$ constant. In order to represent the classical trajectory by narrow wave packets, the 'returning part' of the packet must be present 'initially' (with respect to a). The determinism of the quantum theory then proceeds from small a to large a, not along a classical trajectory (which does not exist). This behaviour has consequences for the validity of the semiclassical approximation and the arrow of time. In fact, it may in principle be possible to understand the origin of irreversibility from quantum cosmology, by the very fact that the Wheeler–DeWitt equation is asymmetric with respect to the intrinsic time given by a. The framework of canonical quantum cosmology is also suitable to address the quantum-to-classical transition for cosmological variables such as the volume of the universe [15, 16]. Using the approach of loop quantum gravity (see Thiemann's contribution) one arrives at a Wheeler–DeWitt equation in cosmology which is fundamentally a difference equation instead of a differential equation of the type (46). In the ensuing framework of loop quantum gravity it seems that the classical singularities of GR can be avoided.

Singularity avoidance for collapse situations can also be found from spherically symmetric models of quantum geometrodynamics. For example, in a model with a collapsing null dust cloud, an initially collapsing wave packet evolves into a superposition of collapsing and expanding packet [9]. This leads to destructive interference at the place where the singularity in the classical theory occurs. Other issues, such as the attempt to give a microscopic derivation of the Bekenstein–Hawking entropy (see the contribution by C. Kiefer to this book), have been mainly addressed in loop quantum gravity. A final, clear-cut, derivation remains, however, elusive.

References

1. Anderson, A. and DeWitt, B. (1986). Does the topology of space fluctuate. *Foundations of Physics*, **16**, 91.
2. Geroch, R. (1967). Topology in General Relativity. *Journal of Mathematical Physics*, **8**, 782–786.
3. Gibbons, G.W. and Hawking, S.W. (1992). Selection rules for topology change. *Communications in Mathematical Physics*, **148**, 345–352.
4. Giulini, D. (1992). On the selection rules for spin-Lorentz cobordisms. *Communications in Mathematical Physics*, **148**, 353–357.
5. Giulini, D. and Kiefer, C. (1994). Wheeler–DeWitt metric and the attractivity of gravity. *Physics Letters A*, **193**, 21–24.
6. Giulini, D. (1995a). On the Configuration-Space Topology in General Relativity *Helvetica Physica Acta*, **68**, 87–111.
7. Giulini, D. (1995b). What is the geometry of superspace. *Physical Review D*, **51**, 5630–5635.
8. Giulini, D. (1998). On the construction of time-symmetric black-hole initial data. In *Black Holes: Theory and Observation* (eds. F. Hehl, C. Kiefer, and R. Metzler), pp. 224–243. Lecture Notes in Physics 514. Springer, Berlin.
9. Hájíček, P. (2003). Quantum theory of gravitational collapse (lecture notes on quantum conchology). In *Quantum Gravity: From Theory to Experimental Search* (eds. D. Giulini, C. Kiefer, and C. Lämmerzahl), pp. 255–299. Lecture Notes in Physics 631. Springer, Berlin.
10. Henneaux, M. and Teitelboim, C. (1992). *Quantization of Gauge Systems* (Princeton University Press, Princeton).
11. Hojman, S.A., Kuchař, K., and Teitelboim, C. (1976). Geometrodynamics regained. *Annals of Physics*, **96**, 88–135.
12. Horowitz, G. (1991). Topology change in classical and quantum gravity. *Classical and Quantum Gravity*, **8**, 587–602.
13. Isham, C. and Kuchař, K. (1985a). Representations of spacetime diffeomorphisms I: canonical parametrised spacetime theories. *Annals of Physics* **164**, 288–315.
14. Isham, C. and Kuchař, K. (1985b). Representations of spacetime diffeomorphisms II: canonical geometrodynamics. *Annals of Physics* **164**, 316–333.
15. Joos, E., Zeh, H.D., Kiefer, C., Giulini, D., Kupsch, J., and Stamatescu, I.-O. (2003). *Decoherence and the Appearance of a Classical World in Quantum Theory*, 2nd edn (Springer, Berlin).
16. Kiefer, C. (2007). *Quantum Gravity, 2nd edn* (Oxford University Press, Oxford).
17. Kiefer, C. (2006). Quantum gravity: general introduction and recent developments. *Annalen der Physik*, **15**, 129–148.
18. Kuchař, K. (1974). Geometrodynamics regained: a Lagrangian approach. *Journal of Mathematical Physics*, **15**, 708–715.
19. Sorkin, R. (1997). Forks in the road, on the way to quantum gravity. *International Journal of Theoretical Physics*, **36**, 2759–2781.
20. Zeh, H.D. (2007). *The Physical Basis of the Direction of Time*, 5th edn (Springer, Berlin). See also `http://www.time-direction.de`.

Loop and Spin Foam Quantum Gravity: A Brief Guide for Beginners

H. Nicolai[1] and K. Peeters[2]

[1] Max-Planck-Institut für Gravitationsphysik, Albert-Einstein-Institut, Am Mühlenberg 1, 14476 Golm, Germany
Hermann.Nicolai@aei.mpg.de

[2] Max-Planck-Institut für Gravitationsphysik, Albert-Einstein-Institut, Am Mühlenberg 1, 14476 Golm, Germany
Kasper.Peeters@aei.mpg.de

1 Quantum Einstein Gravity

The assumption that Einstein's classical theory of gravity can be quantised non-perturbatively is at the root of a wide variety of approaches to quantum gravity. The assumption constitutes the basis of several discrete methods [1], such as dynamical triangulations and Regge calculus, but it also implicitly underlies the older Euclidean path integral approach [2, 3] and the somewhat more indirect arguments which suggest that there may exist a non-trivial fixed point of the renormalisation group [4–6]. Finally, it is the key assumption which underlies loop and spin foam quantum gravity. Although the assumption is certainly far-reaching, there is to date no proof that Einstein gravity cannot be quantised non-perturbatively, either along the lines of one of the programs listed above or perhaps in an entirely different way.

In contrast to string theory, which posits that the Einstein–Hilbert action is only an effective low-energy approximation to some other, more fundamental, underlying theory, loop and spin foam gravity takes Einstein's theory in four space-time dimensions as the basic starting point, either with the conventional or with a (constrained) 'BF-type' formulation.[1] These approaches are background independent in the sense that they do not presuppose the existence of a given background metric. In comparison to the older geometrodynamics approach (which is also formally background independent) they make use of many new conceptual and technical ingredients. A key role is played by the reformulation of gravity in terms of connections and holonomies. A related feature is the use of spin networks in three (for canonical formulations)

[1] In the remainder, we will often follow established (though perhaps misleading) custom and summarily refer to this framework of ideas simply as 'Loop Quantum Gravity', or LQG for short.

H. Nicolai and K. Peeters: *Loop and Spin Foam Quantum Gravity: A Brief Guide for Beginners*, Lect. Notes Phys. **721**, 151–184 (2007)
DOI 10.1007/978-3-540-71117-9_9

and four (for spin foams) dimensions. These, in turn, require other mathematical ingredients, such as non-separable ('polymer') Hilbert spaces and representations of operators which are not weakly continuous. Undoubtedly, novel concepts and ingredients such as these will be necessary in order to circumvent the problems of perturbatively quantised gravity (that novel ingredients are necessary is, in any case, not just the point of view of LQG but also of most other approaches to quantum gravity). Nevertheless, it is important not to lose track of the physical questions that one is trying to answer.

Evidently, in view of our continuing ignorance about the 'true theory' of quantum gravity, the best strategy is surely to explore all possible avenues. LQG, just like the older geometrodynamics approach [7], addresses several aspects of the problem that are currently outside the main focus of string theory, in particular the question of background independence and the quantisation of geometry. Whereas there is a rather direct link between (perturbative) string theory and classical space-time concepts, and string theory can therefore rely on familiar notions and concepts, such as the notion of a particle and the S-matrix, the task is harder for LQG, as it must face up right away to the question of what an observable quantity is in the absence of a proper semi-classical space-time with fixed asymptotics.

The present text, which is based in part on the companion review [8], is intended as a brief introductory and critical survey of loop and spin foam quantum gravity,[2] with special attention to some of the questions that are frequently asked by non-experts, but not always adequately emphasised (for our taste, at least) in the pertinent literature. For the canonical formulation of LQG, these concern in particular the definition and implementation of the Hamiltonian (scalar) constraint and its lack of uniqueness. An important question (which we will not even touch on here) concerns the consistent incorporation of matter couplings, and especially the question as to whether the consistent quantisation of gravity imposes any kind of restrictions on them. Establishing the existence of a semi-classical limit, in which classical space-time and the Einstein field equations are supposed to emerge, is widely regarded as the main open problem of the LQG approach. This is also a prerequisite for understanding the ultimate fate of the non-renormalisable UV divergences that arise in the conventional perturbative treatment. Finally, in any canonical approach there is the question whether one has succeeded in achieving (a quantum version of) full space-time covariance, rather than merely covariance under diffeomorphisms of the three-dimensional slices. In [8] we have argued (against a widely held view in the LQG community) that for this, it is not enough to check the closure of two Hamiltonian constraints on diffeomorphism invariant states, but that it is rather the *off-shell closure* of the constraint algebra that should be made the crucial requirement in establishing quantum space-time covariance.

[2] Whereas [8] is focused on the 'orthodox' approach to loop quantum gravity, to wit the Hamiltonian framework.

Many of these questions have counterparts in the spin foam approach, which can be viewed as a 'space-time covariant version' of LQG, and at the same time as a modern variant of earlier attempts to define a discretised path integral in quantum gravity. For instance, the existence of a semi-classical limit is related to the question whether the Einstein–Hilbert action can be shown to emerge in the infrared (long distance) limit, as is the case in (2+1) gravity in the Ponzano–Regge formulation, cf. (38). Regarding the non-renormalisable UV divergences of perturbative quantum gravity, many spin foam practitioners seem to hold the view that there is no need to worry about short distance singularities and the like because the divergences are simply 'not there' in spin foam models, due to the existence of an intrinsic cutoff at the Planck scale. However, the same statement applies to any regulated quantum field theory (such as lattice gauge theory) before the regulator is removed, and on the basis of this more traditional understanding, one would therefore expect the 'correct' theory to require some kind of refinement (continuum) limit,[3] or a sum 'over all spin foams' (corresponding to the 'sum over all metrics' in a formal path integral). If one accepts this point of view, a key question is whether it is possible to obtain results which do not depend on the specific way in which the discretisation and the continuum limit are performed (this is also a main question in other discrete approaches which work with reparametrisation invariant quantities, such as in Regge calculus). On the other hand, the very need to take such a limit is often called into question by LQG proponents, who claim that the discrete (regulated) model correctly describes physics at the Planck scale. However, it is then difficult to see (and, for gravity in (3+1) dimensions, has not been demonstrated all the way in a single example) how a classical theory with all the requisite properties, and in particular full space-time covariance, can emerge at large distances. Furthermore, without considering such limits, and in the absence of some other unifying principle, one may well remain stuck with a multitude of possible models, whose lack of uniqueness simply mirrors the lack of uniqueness that comes with the need to fix infinitely many coupling parameters in the conventional perturbative approach to quantum gravity.

Obviously, a brief introductory text such as this cannot do justice to the numerous recent developments in a very active field of current research. For this reason, we would like to conclude this introduction by referring readers to several 'inside' reviews for recent advances and alternative points of view, namely [9–11] for the canonical formulation, [12–14] for spin foams, and [15] for both. A very similar point of view to ours has been put forward in [16, 17].[4] Readers are also invited to have a look at [18] for an update on the very latest developments in the subject.

[3] Unless quantum gravity is ultimately a *topological* theory, in which case the sequence of refinements becomes stationary. Such speculations have also been entertained in the context of string and M theory.

[4] However, [16, 17] only addresses the so-called 'm-ambiguity', whereas we will argue that there are infinitely many other parameters which a microscopic theory of quantum gravity must fix.

2 The Kinematical Hilbert Space of LQG

There is a general expectation (not only in the LQG community) that at the very shortest distances, the smooth geometry of Einstein's theory will be replaced by some quantum space or space-time, and hence the continuum will be replaced by some 'discretuum'. Canonical LQG does not do away with conventional spacetime concepts entirely, in that it still relies on a spatial continuum Σ as its 'substrate', on which holonomies and spin networks live (or 'float') – of course, with the idea of eventually 'forgetting about it' by considering abstract spin networks and only the combinatorial relations between them. On this substrate, it takes as the classical phase space variables the holonomies of the Ashtekar connection,

$$h_e[A] = \mathcal{P} \exp \int_e A_m^a \tau_a \mathrm{d}x^m \,, \quad \text{with} \quad A_m^a := -\tfrac{1}{2}\epsilon^{abc}\omega_{mbc} + \gamma\, K_m^a \,. \tag{1}$$

Here, τ_a are the standard generators of SU(2) (Pauli matrices), but one can also replace the basic representation by a representation of arbitrary spin, denoted by $\rho_j(h_e[A])$. The Ashtekar connection A is thus a particular linear combination of the spin connection ω_{mbc} and the extrinsic curvature K_m^a which appear in a standard (3+1) decomposition. The parameter γ is the so-called 'Barbero-Immirzi parameter'. The variable conjugate to the Ashtekar connection turns out to be the inverse densitised dreibein $\tilde{E}_a{}^m := e\, e_a{}^m$. Using this conjugate variable, one can find the objects which are conjugate to the holonomies. These are given by integrals of the associated two-form over two-dimensional surfaces S embedded in Σ,

$$F_S[\tilde{E}, f] := \int_S \epsilon_{mnp} \tilde{E}_a^m\, f^a\, \mathrm{d}x^n \wedge \mathrm{d}x^p \,, \tag{2}$$

where $f^a(x)$ is a test function. This flux vector is indeed conjugate to the holonomy in the sense described in Fig. 1: if the edge associated to the holonomy intersects the surface associated to the flux, the Poisson bracket between the two is non-zero,

$$\Big\{ (h_e[A])_{\alpha\beta},\, F_S[\tilde{E}, f] \Big\} = \pm\, \gamma\, f_a(P) \big(h_{e_1}[A]\, \tau^a\, h_{e_2}[A] \big)_{\alpha\beta} \,, \tag{3}$$

where $e = e_1 \cup e_2$ and the sign depends on the relative orientation of the edge and the two-surface. This Poisson structure is the one which gets promoted to a commutator algebra in the quantum theory.

Instead of building a Hilbert space as the space of functions over configurations of the Ashtekar connection, i.e. instead of constructing wave-functionals $\Psi[A_m(\mathbf{x})]$, LQG uses a Hilbert space of wave functionals which "probe" the geometry only on one-dimensional submanifolds, the so-called *spin networks*. The latter are (not necessarily connected) graphs Γ embedded in Σ consisting of finitely many edges (links). The wave functionals are functionals over the

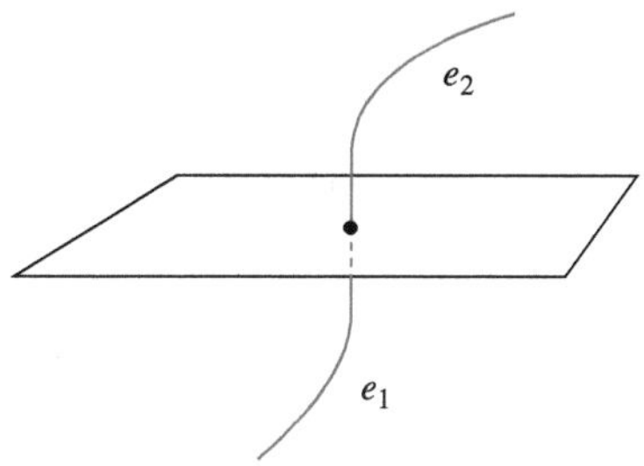

Fig. 1. LQG employs holonomies and fluxes as elementary conjugate variables. When the edge of the holonomy and the two-surface element of the flux intersect, the canonical Poisson bracket of the associated operators is non-vanishing, and inserts a τ-matrix at the point of intersection, cf. (3)

space of holonomies. In order to make them $\mathbb{C}$-valued, the SU(2) indices of the holonomies have to be contracted using invariant tensors (i.e. Clebsch–Gordan coefficients). The wave function associated to the spin network in Fig. 2 is, for instance, given by

$$\Psi[\text{fig.2}] = \Big(\rho_{j_1}(h_{e_1}[A])\Big)_{\alpha_1\beta_1}\Big(\rho_{j_2}(h_{e_2}[A])\Big)_{\alpha_2\beta_2} \times \Big(\rho_{j_3}(h_{e_3}[A])\Big)_{\alpha_3\beta_3} C^{j_1 j_2 j_3}_{\beta_1\beta_2\beta_3}\cdots\,, \tag{4}$$

where dots represent the remainder of the graph. The spin labels $j_1,\dots$ must obey the standard rules for the vector addition of angular momenta, but otherwise can be chosen arbitrarily. The spin network wave functions Ψ are thus labelled by Γ (the spin network graph), by the spins $\{j\}$ attached to the edges, and the intertwiners $\{C\}$ associated to the vertices.

At this point, we have merely defined a space of wave functions in terms of rather unusual variables, and it now remains to define a proper Hilbert space structure on them. The discrete kinematical structure which LQG imposes does, accordingly, *not* come from the description in terms of holonomies and fluxes. After all, this very language can also be used to describe ordinary Yang–Mills theory. The discrete structure which LQG imposes is also entirely different from the discreteness of a lattice or naive discretisation of space (i.e. of a finite or countable set). Namely, it arises by 'polymerising' the continuum via an unusual *scalar product*. For any two spin network states, one defines this scalar product to be

$$\big\langle \Psi_{\Gamma,\{j\},\{C\}} \,\big|\, \Psi'_{\Gamma',\{j'\},\{C'\}}\big\rangle = \begin{cases} 0 & \text{if } \Gamma \neq \Gamma'\,, \\ \displaystyle\int \prod_{e_i\in\Gamma} \mathrm{d}h_{e_i}\; \bar{\psi}_{\Gamma,\{j\},\{C\}}\, \psi'_{\Gamma',\{j'\},\{C'\}} & \text{if } \Gamma = \Gamma'\,, \end{cases} \tag{5}$$

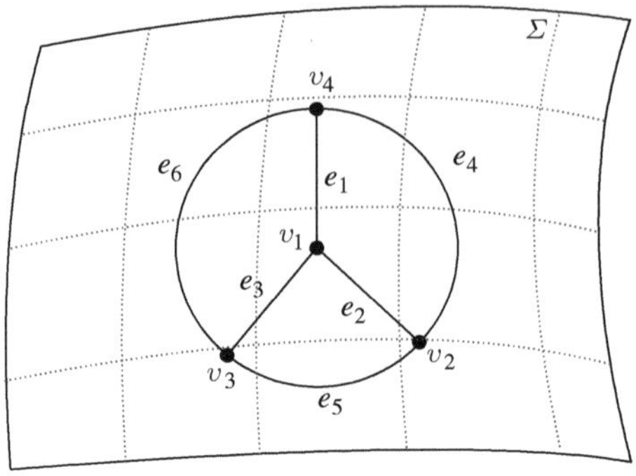

Fig. 2. A simple spin network, embedded in the spatial hypersurface Σ. The hypersurface is only present in order to provide coordinates which label the positions of the vertices and edges. Spin network wave functions only probe the geometry along the one-dimensional edges and are insensitive to the geometry elsewhere on Σ

where the integrals $\int dh_e$ are to be performed with the SU(2) Haar measure. The spin network wave functions ψ depend on the Ashtekar connection only through the holonomies. The *kinematical Hilbert space* $\mathcal{H}_{\text{kin}}$ is then defined as the completion of the space of spin network wave functions w.r.t. this scalar product (5). The topology induced by the latter is similar to the discrete topology ('pulverisation') of the real line with countable unions of points as the open sets. Because the only notion of 'closeness' between two points in this topology is whether or not they are coincident, whence *any* function is continuous in this topology, this raises the question as to how one can recover conventional notions of continuity in this scheme.

The very special choice of the scalar product (5) leads to representations of operators which need not be weakly continuous: this means that expectation values of operators depending on some parameter do not vary continuously as these parameters are varied. Consequently, the Hilbert space does not admit a countable basis, hence is *non-separable*, because the set of all spin network graphs in Σ is uncountable, and non-coincident spin networks are orthogonal w.r.t. (5). Therefore, any operation (such as a diffeomorphism) which moves around graphs continuously corresponds to an uncountable sequence of mutually orthogonal states in $\mathcal{H}_{\text{kin}}$. That is, no matter how 'small' the deformation of the graph in Σ, the associated elements of $\mathcal{H}_{\text{kin}}$ always remain a finite distance apart, and consequently, the continuous motion in 'real space' gets mapped to a highly discontinuous one in $\mathcal{H}_{\text{kin}}$. Although unusual, and perhaps counter-intuitive, as they are, these properties constitute a cornerstone for the hopes that LQG can overcome the seemingly unsurmountable problems of conventional geometrodynamics: if the representations used in LQG were equivalent to the ones of geometrodynamics, there would be no reason to expect LQG not to end up in the same quandary.

Because the space of quantum states used in LQG is very different from the one used in Fock space quantisation, it becomes non-trivial to see how semi-classical 'coherent' states can be constructed, and how a smooth classical space-time might emerge. In simple toy examples, such as the harmonic oscillator, it has been shown that the LQG Hilbert space indeed admits states (complicated linear superpositions) whose properties are close to those of the usual Fock space coherent states [19]. In full (3+1)-dimensional LQG, the classical limit is, however, far from understood (so far only kinematical coherent states are known [20–25], i.e. states which do not satisfy the quantum constraints). In particular, it is not known how to describe or approximate classical space-times in this framework that 'look' like, say, Minkowski space, or how to properly derive the classical Einstein equations and their quantum corrections. A proper understanding of the semi-classical limit is also indispensable to clarify the connection (or lack thereof) between conventional perturbation theory in terms of Feynman diagrams and the non-perturbative quantisation proposed by LQG.

However, the truly relevant question here concerns the structure (and definition!) of *physical* space and time. This, and not the kinematical 'discretuum' on which holonomies and spin networks 'float', is the arena where one should try to recover familiar and well-established concepts like the Wilsonian renormalisation group, with its *continuous* 'flows'. Because the measurement of lengths and distances ultimately requires an operational definition in terms of appropriate matter fields and states obeying the physical state constraints, 'dynamical' discreteness is expected to manifest itself in the spectra of the relevant physical observables. Therefore, let us now turn to a discussion of the spectra of three important operators and to the discussion of physical states.

3 Area, Volume, and the Hamiltonian

In the current setup of LQG, an important role is played by two relatively simple operators: the 'area operator' measuring the area of a two-dimensional surface $S \subset \Sigma$ and the 'volume operator' measuring the volume of a three-dimensional subset $V \subset \Sigma$. The latter enters the definition of the Hamiltonian constraint in an essential way. Nevertheless, it must be emphasised that the area and volume operators are *not* observables in the Dirac sense, as they do not commute with the Hamiltonian. To construct *physical* operators corresponding to area and volume is more difficult and would require the inclusion of matter (in the form of 'measuring rod fields').

The area operator is most easily expressed as

$$A_S[g] = \int_S \sqrt{\mathrm{d}F^a \cdot \mathrm{d}F^a}\,, \quad \text{with} \quad \mathrm{d}F_a := \epsilon_{mnp}\tilde{E}_a^m \mathrm{d}x^n \wedge \mathrm{d}x^p \tag{6}$$

(the area element is here expressed in terms of the new 'flux variables' $\tilde{E}_a^m$, but is equal to the standard expression $\mathrm{d}F_a := \epsilon_{abc} e_m{}^b e_n{}^c \mathrm{d}x^m \wedge \mathrm{d}x^n$). The next step is to rewrite this area element in terms of the spin network variables, in particular the momentum $\tilde{E}_a{}^m$ conjugate to the Ashtekar connection. In order to do so, we subdivide the surface into infinitesimally small surfaces S_I as in Fig. 3. Next, one approximates the area by a Riemann sum (which, of course, converges for well-behaved surfaces S), using

$$\int_{S_I} \sqrt{\mathrm{d}F^a \cdot \mathrm{d}F^a} \approx \sqrt{F^a_{S_I}[\tilde{E}]\, F^a_{S_I}[\tilde{E}]}\,. \tag{7}$$

This turns the operator into the expression

$$A_S[\tilde{E}^a_m] = \lim_{N\to\infty} \sum_{I=1}^{N} \sqrt{F^a_{S_I}[\tilde{E}]\, F^a_{S_I}[\tilde{E}]}\,. \tag{8}$$

If one applies the operator (8) to a wave function associated with a fixed graph $\varGamma$ and refines it in such a way that each elementary surface S_I is pierced by only *one* edge of the network, one obtains, making use of (3) twice,

$$\hat{A}_S \varPsi = 8\pi l_p^2 \gamma \sum_{p=1}^{\#\text{edges}} \sqrt{j_p(j_p+1)}\, \varPsi\,. \tag{9}$$

These spin network states are thus eigenstates of the area operator. The situation becomes considerably more complicated for wave functions which contain a spin network vertex which lies in the surface S; in this case the area operator does not necessarily act diagonally anymore (see Fig. 4). Expression (9) lies at the core of the statement that areas are quantised in LQG.

The construction of the volume operator follows similar logic, although it is substantially more involved. One starts with the classical expression for the volume of a three-dimensional region $\varOmega \subset \Sigma$,

$$V(\varOmega) = \int_\varOmega \mathrm{d}^3x \sqrt{\left|\frac{1}{3!}\epsilon_{abc}\epsilon^{mnp}\tilde{E}^a_m \tilde{E}^b_n \tilde{E}^c_p\right|}\,. \tag{10}$$

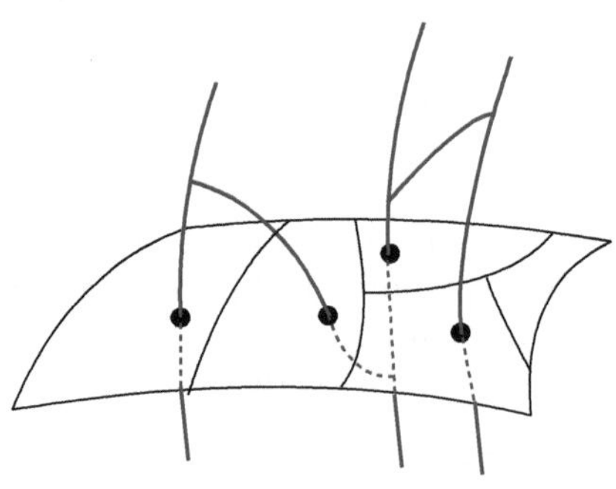

Fig. 3. The computation of the spectrum of the area operator involves the division of the surface into cells, such that at most one edge of the spin network intersects each given cell

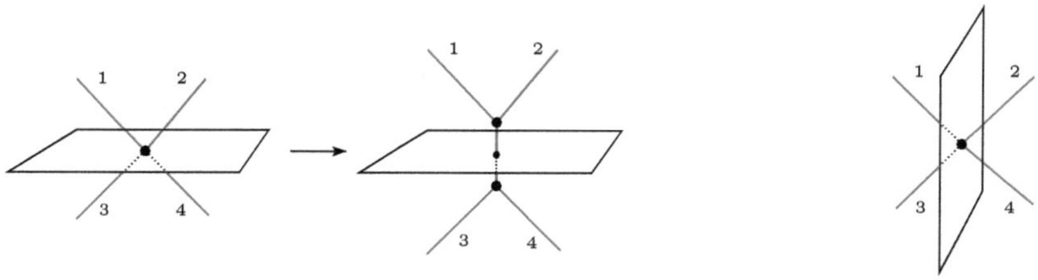

Fig. 4. The action of the area operator on a node with intertwiner $C^{j_1 j_2 k}_{\alpha_1 \alpha_2 \beta} C^{j_3 j_4 k}_{\alpha_3 \alpha_4 \beta}$. Whether or not this action is diagonal depends on the orientation of the surface associated to the area operator. In the figure on the (**left**), the location of the edges with respect to the surface is such that the invariance of the Clebsch–Gordan coefficients can be used to evaluate the action of the area operator. The result can be written in terms of a 'virtual' edge. In the figure on the (**right**), however, this is not the case, a recoupling relation is needed, and the spin network state is not an eigenstate of the corresponding area operator

Just as with the area operator, one partitions Ω into small cells $\Omega = \cup_I \Omega_I$, so that the integral can be replaced with a Riemann sum. In order to express the volume element in terms of the canonical quantities introduced before, one then again approximates the area elements $\mathrm{d}F^a$ by the small but finite area operators $F^a_S[\tilde{E}]$, such that the volume is obtained as the limit of a Riemann sum

$$V(\Omega) = \lim_{N \to \infty} \sum_{I=1}^{N} \sqrt{\left| \frac{1}{3!} \epsilon_{abc} F^a_{S^1_I}[\tilde{E}]\, F^b_{S^2_I}[\tilde{E}]\, F^c_{S^3_I}[\tilde{E}] \right|} \,. \tag{11}$$

The main problem is now to choose appropriate surfaces $S_{1,2,3}$ in each cell. This should be done in such a way that the r.h.s. of (11) reproduces the correct classical value. For instance, one can choose a point inside each cube Ω_I, then connect these points by lines and 'fill in' the faces. In each cell Ω_I one then has three lines labelled by $a = 1, 2, 3$; the surface S^a_I is then the one that is traversed by the a-th line. With this choice it can be shown that the result is insensitive to small 'wigglings' of the surfaces, hence independent of the shape of S^a_I, and the above expression converges to the desired result. See [26, 27] for some recent results on the spectrum of the volume operator.

The key problem in canonical gravity is the definition and implementation of the Hamiltonian (scalar) constraint operator, and the verification that this operator possesses all the requisite properties. The latter include (quantum) space-time covariance as well as the existence of a proper semi-classical limit, in which the classical Einstein equations are supposed to emerge. It is this operator which replaces the Hamiltonian evolution operator of ordinary quantum mechanics, and encodes all the important dynamical information of the theory (whereas the Gauss and diffeomorphism constraints are merely 'kinematical'). More specifically, together with the kinematical constraints, it defines the *physical states* of the theory, and thereby the physical Hilbert space $\mathcal{H}_{\rm phys}$ (which may be separable [28], even if $\mathcal{H}_{\rm kin}$ is not).

To motivate the form of the *quantum Hamiltonian* one starts with the classical expression, written in loop variables. To this aim one rewrites the Hamiltonian in terms of Ashtekar variables, with the result

$$H[N] = \int_\Sigma \mathrm{d}^3x\, N \frac{\tilde{E}_a^m \tilde{E}_b^n}{\sqrt{\det \tilde{E}}} \Big(\epsilon^{abc} F_{mnc} - \frac{1}{2}(1+\gamma^2) K_{[m}{}^a K_{n]}{}^b \Big) \,. \tag{12}$$

For the special values $\gamma = \pm i$, the last term drops out, and the Hamiltonian simplifies considerably. This was indeed the value originally proposed by Ashtekar, and it would also appear to be the natural one required by local Lorentz invariance (as the Ashtekar variable is, in this case, just the pullback of the four-dimensional spin connection). However, imaginary γ obviously implies that the phase space of general relativity in terms of these variables would have to be complexified, such that the original phase space could be recovered only after imposing a reality constraint. In order to avoid the difficulties related to quantising this reality constraint, γ is now usually taken to be real. With this choice, it becomes much more involved to rewrite (12) in terms of loop and flux variables.

4 Implementation of the Constraints

In canonical gravity, the simplest constraint is the Gauss constraint. In the setting of LQG, it simply requires that the SU(2) representation indices entering a given vertex of a spin network enter in an SU(2) invariant manner. More complicated are the diffeomorphism and Hamiltonian constraint. In LQG these are implemented in two entirely different ways. Moreover, the implementation of the Hamiltonian constraint is not completely independent, as its very definition relies on the existence of a subspace of diffeomorphism invariant states.

Let us start with the diffeomorphism constraint. Unlike in geometrodynamics, one cannot immediately write down formal states which are manifestly diffeomorphism invariant, because the spin network functions are not supported on all of Σ, but only on one-dimensional links, which 'move around' under the action of a diffeomorphism. A formally diffeomorphism invariant state is obtained by 'averaging' over the diffeomorphism group, and more specifically by considering the formal sum

$$\eta(\Psi)[A] := \sum_{\phi \in \mathrm{Diff}(\Sigma|\Gamma)} \Psi_\Gamma[A \circ \phi] \,. \tag{13}$$

Here $\mathrm{Diff}(\Sigma|\Gamma)$ is obtained by dividing out the diffeomorphisms leaving invariant the graph Γ. Although this is a continuous sum which might seem to be ill-defined, it can be given a mathematically precise meaning because the unusual scalar product (5) ensures that the inner product between a state and a diffeomorphism-averaged state,

$$\langle \eta(\Psi_{\Gamma'}) \,|\, \Psi_\Gamma \rangle = \sum_{\phi \in \mathrm{Diff}(\Sigma|\Gamma')} \langle \phi^* \circ \Psi_{\Gamma'} \,|\, \Psi_\Gamma \rangle \,, \tag{14}$$

consists at most of a *finite* number of terms. It is this fact which ensures that $\langle \eta(\Psi_\Gamma)|$ is indeed well defined as an element of the space dual to the space of spin networks (which is dense in $\mathcal{H}_{\mathrm{kin}}$). In other words, although $\eta(\Psi)$ is certainly outside of $\mathcal{H}_{\mathrm{kin}}$, it does make sense *as a distribution*. On the space of diffeomorphism averaged spin network states (regarded as a subspace of a distribution space) one can now again introduce a Hilbert space structure 'by dividing out' spatial diffeomorphisms, namely

$$\langle\!\langle \eta(\Psi)|\eta(\Psi')\rangle\!\rangle := \langle \eta(\Psi)|\Psi' \rangle \,. \tag{15}$$

The completion by means of this scalar product defines the space $\mathcal{H}_{\mathrm{diff}}$; but note that $\mathcal{H}_{\mathrm{diff}}$ is *not* a subspace of $\mathcal{H}_{\mathrm{kin}}$!

As mentioned above, however, it is the Hamiltonian constraint which plays the key role in canonical gravity, as it this operator which encodes the dynamics. Implementing this constraint on $\mathcal{H}_{\mathrm{diff}}$ or some other space is fraught with numerous choices and ambiguities, inherent in the construction of the quantum Hamiltonian as well as the extraordinary complexity of the resulting expression for the constraint operator [29]. The number of ambiguities can be reduced by invoking independence of the spatial background [10], and indeed, without making such choices, one would not even obtain sensible expressions. In other words, the formalism is partly 'on-shell' in that the very existence of the (unregulated) Hamiltonian constraint operator depends very delicately on its 'diffeomorphism covariance', and the choice of a proper 'habitat', on which it is supposed to act in a well-defined manner. A further source of ambiguities, which, for all we know, has not been considered in the literature so far, consists in possible $\hbar$-dependent 'higher order' modifications of the Hamiltonian, which might still be compatible with all consistency requirements of LQG.

In order to write the constraint in terms of only holonomies and fluxes, one has to eliminate the inverse square root $\tilde{E}^{-1/2}$ in (12) as well as the extrinsic curvature factors. This can be done through a number of tricks found by Thiemann [30]. The vielbein determinant is eliminated using

$$\epsilon_{mnp}\epsilon^{abc}\tilde{E}^{-1/2}\tilde{E}_b{}^n\tilde{E}_c{}^p = \frac{1}{4\gamma}\Big\{A_m{}^a(\mathbf{x}), V\Big\} \,. \tag{16}$$

where $V \equiv V(\Sigma)$ is the total volume, cf. (10). The extrinsic curvature is eliminated by writing it as

$$K_m{}^a(\mathbf{x}) = \frac{1}{\gamma}\{A_m{}^a(\mathbf{x}) \,,\, \bar{K}\} \qquad \text{where} \qquad \bar{K} := \int_\Sigma \mathrm{d}^3x\, K_m{}^a\tilde{E}_a{}^m \,, \tag{17}$$

and then eliminating the integrand of $\bar{K}$ using

$$\begin{aligned} \bar{K}(\mathbf{x}) &= \frac{1}{\gamma^{3/2}}\Big\{\frac{\tilde{E}_a{}^m\tilde{E}_b{}^n}{\sqrt{\tilde{E}}}\epsilon^{abc}F_{mnc}(\mathbf{x}),\, V\Big\} \\ &= \frac{1}{4\gamma^{5/2}}\epsilon^{mnp}\Big\{\{A_m{}^a\,,\,V\}F_{npa}\,,\,V\Big\}\,, \end{aligned} \tag{18}$$

i.e. writing it as a nested Poisson bracket. Inserting these tricks into the Hamiltonian constraint, one replaces (12) with the expression

$$\begin{aligned} H[N] = \int_\Sigma \mathrm{d}^3x\, N\epsilon^{mnp}\mathrm{Tr}\Big(&F_{mn}\{A_p, V\} \\ &- \frac{1}{2}(1+\gamma^2)\{A_m,\bar{K}\}\{A_n,\bar{K}\}\{A_p,V\}\Big)\,, \end{aligned} \tag{19}$$

with $\bar{K}$ understood to be eliminated using (18). This expression, which now contains only the connection A and the volume V, is the starting point for the construction of the quantum constraint operator.

In order to quantise the classical Hamiltonian (19), one next elevates all classical objects to quantum operators as described in the foregoing sections, and replaces the Poisson brackets in (19) by quantum commutators. The resulting *regulated Hamiltonian* then reduces to a sum over the vertices v_α of the spin network with lapses $N(v_\alpha)$

$$\begin{aligned} \hat{H}[N,\epsilon] = \sum_\alpha N(v_\alpha)\,\epsilon^{mnp} \\ \times\,\mathrm{Tr}\Big\{&\big(h_{\partial P_{mn}(\epsilon)} - h^{-1}_{\partial P_{mn}(\epsilon)}\big)\,h_p^{-1}\,\big[h_p,\hat{V}\big] \\ &- \tfrac{1}{2}(1+\gamma^2)h_m^{-1}\,\big[h_m,\bar{K}\big]\,h_n^{-1}\,\big[h_n,\bar{K}\big]\,h_p^{-1}\big[h_p,\hat{V}\big]\Big\}\,, \end{aligned} \tag{20}$$

where $\partial P_{mn}(\epsilon)$ is a small loop attached to the vertex v_α that must eventually be shrunk to zero. In writing the above expression, we have furthermore assumed a specific (but, at this point, not specially preferred) ordering of the operators.

Working out the action of (20) on a given spin network wave function is rather non-trivial, and we are not aware of any concrete calculations in this regard, other than for very simple special configurations (see, e.g., [31]); to get an idea of the complications, readers may have a look at a recent analysis of the volume operator and its spectrum in [32]. In particular, the available calculations focus almost exclusively on the action of the first term in (20), whereas the second term (consisting of multiply nested commutators, cf. (18)) is usually not discussed in any detail. At any rate, this calculation involves a number of choices in order to fix various ambiguities, such as the ordering ambiguities in both terms in (20). An essential ingredient is the action of the operator $h_{\partial P_{mn}(\epsilon)} - h^{-1}_{\partial P_{mn}(\epsilon)}$, which is responsible for the addition of a plaquette to the spin network. The way in which this works is depicted (schematically) in Fig. 5. The plaquette is added in a certain SU(2) representation, corresponding to the representation of the trace in (20). This representation label j is arbitrary, and constitutes a quantisation ambiguity (often called 'm-ambiguity').

Having defined the action of the regulated Hamiltonian, the task is not finished, however, because one must still take the limit $\epsilon \to 0$, in which the attached loops are shrunk to zero. As it turns out, this limit cannot be taken straightforwardly: due to the scalar product (5) and the non-separability of $\mathcal{H}_{\text{kin}}$ the limiting procedure runs through a sequence of mutually orthogonal states, and therefore does not converge in $\mathcal{H}_{\text{kin}}$. For this reason, LQG must resort to a weaker notion of limit, either by defining the limit as a weak limit on a (subspace of the) algebraic dual of a dense subspace of $\mathcal{H}_{\text{kin}}$ [11, 33] or by taking the limit in the weak * operator topology [10]. In the first case, the relevant space (sometimes referred to as the 'habitat') is a distribution space which contains the space $\mathcal{H}_{\text{diff}}$ of formally diffeomorphism invariant states as a subspace, but its precise nature and definition is still a matter of debate. In the second case, the limit is implemented (in a very weak sense) on the original kinematical Hilbert space $\mathcal{H}_{\text{kin}}$, but that space will not contain any diffeomorphism invariant states other than the 'vacuum' $\Psi = \mathbf{1}$. The question of the proper 'habitat' on which to implement the action of the Hamiltonian constraint is thus by no means conclusively settled.

From a more general point of view, it should be noted that the action of the Hamiltonian constraint is always 'ultralocal': all changes to the spin network are made in an $\epsilon \to 0$ neighbourhood of a given vertex, while the spin network graph is kept fixed [34–36]. Pictorially speaking, the only action

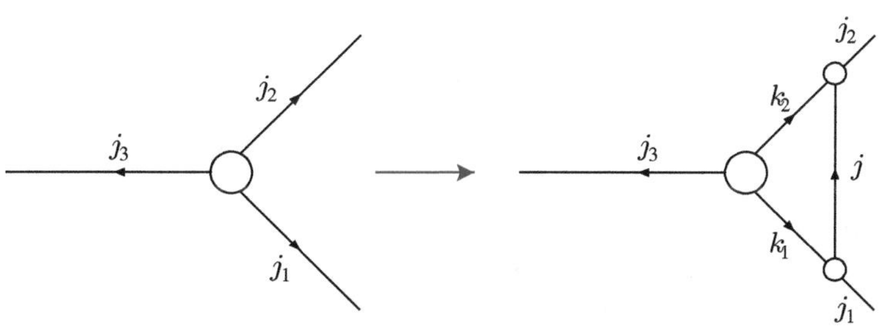

Fig. 5. Schematic depiction of the action of the Hamiltonian constraint on a vertex of a spin network wave function. Two new vertices are introduced, and the original vertex is modified. Note that in order for this to be true, particular choices have been made in the quantisation prescription

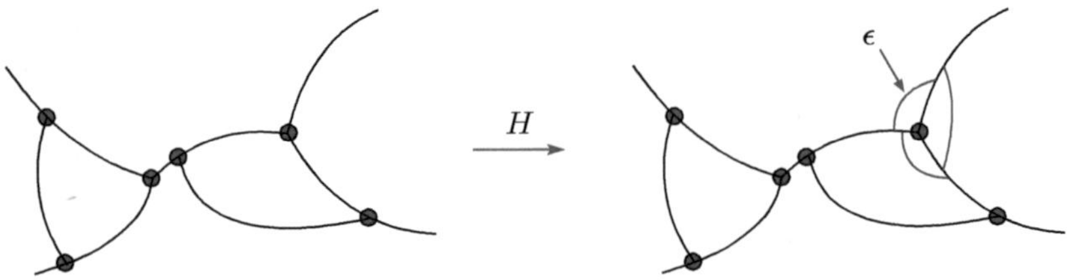

Fig. 6. The action of the Hamiltonian constraint is 'ultra-local', in the sense that it acts only in a neighbourhood of 'size' ϵ around a spin network vertex

of the (regulated) Hamiltonian is to dress up the vertices with 'spiderwebs', see Fig. 6. More specifically, it has been argued [33] that the Hamiltonian acts at a particular vertex only by changing the intertwiners at that vertex. This is in stark contrast to what happens in lattice field theories. There the action of the Hamiltonian always links two different existing nodes, the plaquettes are by construction always spanned between existing nodes, and the continuum limit involves the lattice as a whole, not only certain sub-plaquettes that shrink to a vertex. This is also what one would expect on physical grounds for a theory with non-trivial dynamics.

The attitude often expressed with regard to the ambiguities in the construction of the Hamiltonian is that they correspond to *different* physics, and therefore the choice of the correct Hamiltonian is ultimately a matter of physics (experiment?), and not mathematics. However, it appears unlikely to us that Nature will allow such a great degree of arbitrariness at its most fundamental level: in fact, our main point here is that the infinitely many ambiguities which killed perturbative quantum gravity are also a problem that other (to wit, non-perturbative) approaches must address and solve.[5]

5 Quantum Space-Time Covariance?

Space-time covariance is a central property of Einstein's theory. Although the Hamiltonian formulation is not manifestly covariant, full covariance is still present in the classical theory, albeit in a hidden form, via the classical (Poisson or Dirac) algebra of constraints acting on phase space. However, this is not necessarily so for the quantised theory. As we explained, LQG treats the diffeomorphism constraint and the Hamiltonian constraint in a very different manner. Why and how then should one expect such a theory to recover full space-time (as opposed to purely spatial) covariance? The crucial issue here

[5] The abundance of 'consistent' Hamiltonians and spin foam models (see below) is sometimes compared to the vacuum degeneracy problem of string theory, but the latter concerns *different solutions* of the *same* theory, as there is no dispute as to what (perturbative) string theory *is*. However, the concomitant lack of predictivity is obviously a problem for both approaches.

is clearly what LQG has to say about the quantum algebra of constraints. Unfortunately, to the best of our knowledge, the 'off-shell' calculation of the commutator of two Hamiltonian constraints in LQG – with an explicit operatorial expression as the final result – has never been fully carried out. Instead, a survey of the possible terms arising in this computation has led to the conclusion that the commutator vanishes on a certain restricted 'habitat' of states [33, 37, 38], and that therefore the LQG constraint algebra closes without anomalies. By contrast, we have argued in [8] that this 'on-shell closure' is not sufficient for a full proof of *quantum space-time covariance*, but that a proper theory of quantum gravity requires a constraint algebra that closes 'off shell', i.e. without prior imposition of a subset of the constraints. The fallacies that may ensue if one does not insist on off-shell closure can be illustrated with simple examples. In our opinion, this requirement may well provide the acid test on which any proposed theory of canonical quantum gravity will stand or fail.

While there is general agreement as to what one means when one speaks of 'closure of the constraint algebra' in classical gravity (or any other classical constrained system [39]), this notion is more subtle in the quantised theory.[6] Let us therefore clarify first the various notions of closure that can arise: we see at least three different possibilities. The strongest notion is 'off-shell closure' (or 'strong closure'), where one seeks to calculate the commutator of two Hamiltonians

$$\left[\hat{H}[N_1]\, ,\, \hat{H}[N_2]\right] = \hat{O}(N_1;N_2)\ . \tag{21}$$

Here we assume that the quantum Hamiltonian constraint operator,

$$\hat{H}[N] := \lim_{\epsilon\to 0} \hat{H}[N,\epsilon]\ , \tag{22}$$

has been rigorously defined as a suitably weak limit, and without further restrictions on the states on which (21) is supposed to hold. In writing the above equations, we have thus been (and will be) cavalier about habitat questions and the precise definition of the Hamiltonian; see, however, [8, 33, 38] for further details and critical comments.

Unfortunately, it appears that the goal of determining $\hat{O}(N_1;N_2)$ as a *bona fide* 'off-shell' operator on a suitable 'habitat' of states, and prior to the imposition of any constraints, is unattainable within the current framework of LQG. For this reason, LQG must resort to weaker notions of closure, by making partial use of the constraints. More specifically, (21) can be relaxed substantially by demanding only

$$\left[\hat{H}[N_1],\, \hat{H}[N_2]\right] |\mathcal{X}\rangle = 0\ , \tag{23}$$

but still with the unregulated Hamiltonian constraint $\hat{H}[N]$. This 'weak closure' should hold for all states $|\mathcal{X}\rangle$ in a restricted habitat of states that are

[6] For reasons of space, we here restrict attention to the bracket between two Hamiltonian constraints, because the remainder of the algebra involving the kinematical constraints is relatively straightforward to implement.

'naturally' expected to be annihilated by the r.h.s. of (21), and that are subject to the further requirement that the Hamiltonian can be applied twice without leaving the 'habitat'. The latter condition is, for instance, met by the 'vertex smooth' states of [33]. As shown in [33, 38], the commutator of two Hamiltonians indeed vanishes on this 'habitat', and one is therefore led to conclude that the full constraint algebra closes 'without anomalies'.

The same conclusion was already arrived at in an earlier computation of the constraint algebra in [30, 37], which was done from a different perspective (no 'habitats'), and makes essential use of the space of diffeomorphism invariant states $\mathcal{H}_{\text{diff}}$, the 'natural' kernel of the r.h.s. of (21). Here the idea is to verify that [30, 37]

$$\lim_{\substack{\epsilon_1 \to 0 \\ \epsilon_2 \to 0}} \langle \mathcal{X} \mid \big[\hat{H}[N_1, \epsilon_1],\, \hat{H}[N_2, \epsilon_2] \big]\, \Psi \rangle = 0 \;, \tag{24}$$

for all $|\mathcal{X}\rangle \in \mathcal{H}_{\text{diff}}$, and for all $|\Psi\rangle$ in the space of finite linear combinations of spin network states. As for the Hamiltonian itself, letting $\epsilon_{1,2} \to 0$ in this expression produces an uncountable sequence of mutually orthogonal states w.r.t. the scalar product (5). Consequently, the limit again does not exist in the usual sense, but only as a weak * limit. The 'diffeomorphism covariance' of the Hamiltonian is essential for this result. Let us stress that (23) and (24) are by no means the same: in (23) one uses the unregulated Hamiltonian (where the limit $\epsilon \to 0$ has already been taken), whereas the calculation of the commutator in (24) takes place inside $\mathcal{H}_{\text{kin}}$, and the limit $\epsilon \to 0$ is taken only *after* computing the commutator of two regulated Hamiltonians. These two operations (taking the limit $\epsilon \to 0$, and calculating the commutator) need not commute. Because with both (23) and (24) one forgoes the aim of finding an operatorial expression for the commutator $\big[\hat{H}[N_1], \hat{H}[N_2]\big]$, making partial use of the constraints, we say (in a partly supergravity inspired terminology) that the algebra closes 'on-shell'.

Although on-shell closure may perhaps look like a sufficient condition on the quantum Hamiltonian constraint, it is easy to see, at the level of simple examples, that this is not true. Consider, for instance, the Hamiltonian constraint of bosonic string theory, and consider modifying it by multiplying it with an operator which commutes with all Virasoro generators. There are many such operators in string theory, for instance the mass-squared operator (minus an arbitrary integer). In this way, we arrive at a realisation of the constraint operators which is very similar to the one used in LQG: the algebra of spatial diffeomorphisms is realised via a (projective) unitary representation, and the Hamiltonian constraint transforms covariantly (the extra factor does not matter, because it commutes with all constraints). In a first step, one can restrict attention to the subspace of states annihilated by the diffeomorphism constraint, the analogue of the space $\mathcal{H}_{\text{diff}}$. Imposing now the new Hamiltonian constraint (the one multiplied with the Casimir) on this subspace would produce a 'non-standard' spectrum by allowing extra diffeomorphism invariant states of a certain prescribed mass. The algebra would also still close on-shell, i.e. on the 'habitat' of states annihilated by the diffeomorphism constraint. The point here is not so much whether this new spectrum is 'right' or 'wrong', but rather that in allowing such modifications which are compatible with on-shell closure of the constraint algebra, we introduce an infinite ambiguity and arbitrariness into the definition of the physical states. In other words, if we only demand on-shell closure as in LQG, there is no way of telling whether or not the vanishing of a commutator is merely accidental, i.e. not really due to the diffeomorphism invariance of the state, but caused by some other circumstance.

By weakening the requirements on the constraint algebra and by no longer insisting on off-shell closure, crucial information gets lost. This loss of information is reflected in the ambiguities inherent in the construction of the LQG Hamiltonian. It is quite possible that the LQG Hamiltonian admits many further modifications on top of the ones we have already discussed, for which the commutator continues to vanish on a suitably restricted habitat of states – in which case neither (23) nor (24) would amount to much of a consistency test.

6 Canonical Gravity and Spin Foams

Attempts to overcome the difficulties with the Hamiltonian constraint have led to another development, *spin foam models* [40–42]. These were originally proposed as space-time versions of spin networks, to wit, evolutions of spin networks in 'time', but have since developed into a class of models of their own, disconnected from the canonical formalism. Mathematically, spin foam models represent a generalisation of spin networks, in the sense that group theoretical objects (holonomies, representations, intertwiners, etc.) are attached not only to vertices and edges (links), but also to higher-dimensional faces in a simplicial decomposition of space-time.

The relation between spin foam models and the canonical formalism is based on a few general features of the action of the Hamiltonian constraint operator on a spin network (for a review on the connection, see [43]). As we have discussed above, the Hamiltonian constraint acts, schematically, by adding a small plaquette close to an existing vertex of the spin network (as in Fig. 5). In terms of a space-time picture, we see that the edges of the spin network sweep out surfaces, and the Hamiltonian constraint generates new surfaces, as in Fig. 7; but note that this graphical representation does not capture the details of how the action of the Hamiltonian affects the intertwiners at the vertices. Instead of associating spin labels to the edges of the spin network, one now associates the spin labels to the surfaces, in such a way that the label of the surface is determined by the label of the edge which lies in either the initial or final surface.

In analogy with proper-time transition amplitudes for a relativistic particle, it is tempting to define the transition amplitude between an initial spin

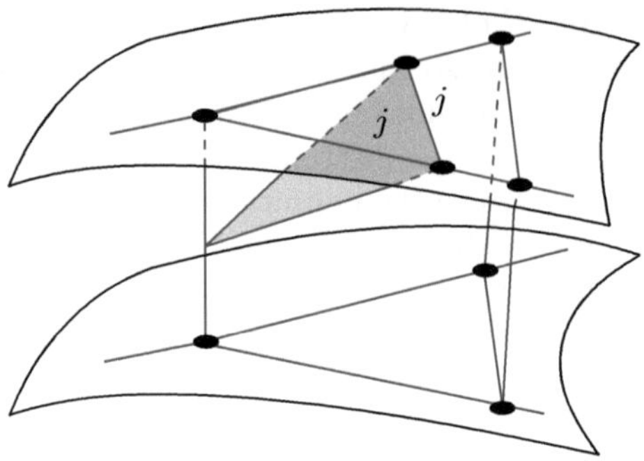

Fig. 7. From spin networks to spin foams, in (2+1) dimensions. The Hamiltonian constraint has created one new edge and two new vertices. The associated surface inherits the label j of the edge which is located on the initial or (in this case) final space-like surface

network state and a final one as

$$Z_T := \langle\psi_f|\exp\left(i\int_0^T \mathrm{d}t\, H\right)|\psi_i\rangle$$
$$= \sum_{n=0}^{\infty}\frac{(i\,T)^n}{n!}\int \mathrm{d}\psi_1\ldots\mathrm{d}\psi_n\,\langle\psi_f|H|\psi_1\rangle$$
$$\times\langle\psi_1|H|\psi_2\rangle\cdots\langle\psi_n|H|\psi_i\rangle\,, \quad (25)$$

where we have repeatedly inserted resolutions of unity. A (somewhat heuristic) derivation of the above formula can be given by starting from a formal path integral [41], which, after gauge fixing and choice of a global time coordinate T, and with appropriate boundary conditions, can be argued to reduce to the above expression. There are many questions one could ask about the physical meaning of this expression, but one important property is that (just as with the relativistic particle) the transition amplitude will project onto physical states (formally, this projection is effected in the original path integral by integrating over the lapse function multiplying the Hamiltonian density). One might thus consider (25) as a way of defining a physical inner product.

Because path integrals with oscillatory measures are notoriously difficult to handle, one might wonder at this point whether to apply a formal Wick rotation to (25), replacing the Feynman weight with a Boltzmann weight, as is usually done in Euclidean quantum field theory. This is also what is suggested by the explicit formulae in [41], where i in (25) is replaced by (-1). However, this issue is much more subtle here than in ordinary (flat space) quantum field theory. First of all, the distinction between a Euclidean (Riemannian) and a Lorentzian (pseudo-Riemannian) manifold obviously requires the introduction of a metric of appropriate signature. However, spin foam models, having their roots in (background independent) LQG, do not come with a metric, and thus the terminology is to some extent up to the beholder. To avoid confusion, let us state clearly that our use of the words 'Euclidean' and 'Lorentzian' here always refers to the use of oscillatory weights $e^{iS_{\mathrm{E}}}$ and $e^{iS_{\mathrm{L}}}$, respectively, where the actions S_{E} and S_{L} are the respective actions for Riemannian resp. pseudo-Riemannian metrics. The term 'Wick rotated', on the other hand, refers to the replacement of the oscillatory weight e^{iS} by the exponential weight e^{-S}, with either $S = S_{\mathrm{E}}$ or $S = S_{\mathrm{L}}$. However, in making use of this terminology, one should always remember that there is no Osterwalder–Schrader type reconstruction theorem in quantum gravity, and therefore any procedure (or 'derivation') remains formal. Unlike the standard Euclidean path integral [2, 3], the spin foam models to be discussed below are generally interpreted to correspond to path integrals *with oscillatory weights* e^{iS}, but come in both Euclidean and Lorentzian variants (corresponding to the groups SO(4) and SO(1,3), respectively). This is true even if the state sums involve only *real* quantities (nj-symbols, edge amplitudes, etc.), cf. the discussion after (38).

The building blocks $\langle\psi_k|H|\psi_l\rangle$ in the transition amplitude (25) correspond to elementary spin network transition amplitudes, as in Fig. 7. For a given value of n, i.e. a given number of time slices, we should thus consider objects of the type

$$Z_{\psi_1,\ldots,\psi_n} = \langle\psi_f|H|\psi_1\rangle\langle\psi_1|H|\psi_2\rangle\cdots\langle\psi_n|H|\psi_i\rangle\ . \tag{26}$$

Each of the building blocks depends only on the values of the spins at the spin network edges and the various intertwiners in the spin network state. The points where the Hamiltonian constraint acts non-trivially get associated to spin foam vertices; see Fig. 8. Instead of working out (26) directly from the action of the Hamiltonian constraint, one could therefore also define the amplitude directly in terms of sums over expressions which depend on the various spins meeting at the spin foam nodes. In this way, one arrives at the so-called *state sum* models, which we will describe in the following section.

A problematic issue in the relation between spin foams and the canonical formalism comes from covariance requirements. While tetrahedral symmetry (or the generalisation thereof in four dimensions) is natural in the spin foam picture, the action of the Hamiltonian constraint, depicted in Fig. 7, does not reflect this symmetry. The Hamiltonian constraint only leads to the so-called '$1 \to 3$ moves', in which a single vertex in the initial spin network is mapped to three vertices in the final spin network. In the spin foam picture, the restriction to only these moves seems to be in conflict with the idea that the slicing of space-time into a space+time decomposition can be chosen arbitrarily. For space-time covariance, one expects $2 \to 2$ and $0 \to 4$ moves (and their time-reversed partners) as well, see Fig. 9. These considerations show that there is no unique path from canonical gravity to spin foam models, and thus no unique model either (even if there was a unique canonical Hamiltonian).

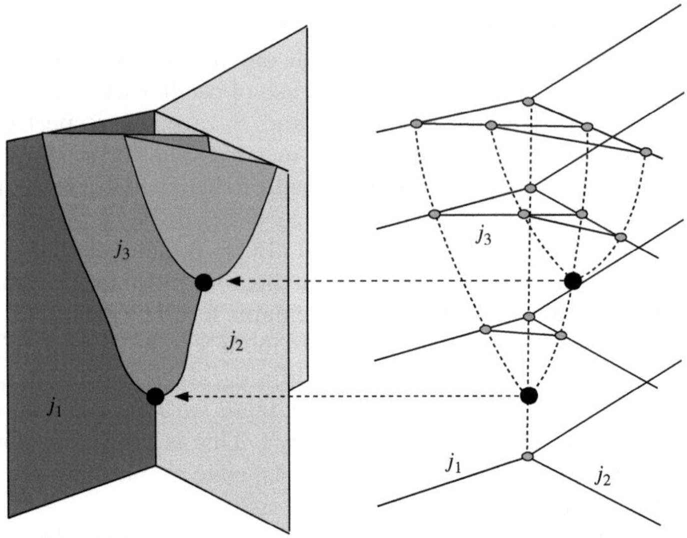

Fig. 8. A spin foam (**left**) together with its spin network evolution (**right**) in (2+1) dimensions. Spin foam nodes correspond to the places where the Hamiltonian constraint in the spin network acts non-trivially (*black dots*). Spin foam edges correspond to evolved spin network nodes (*grey dots*), and spin foam faces correspond to spin network edges. The spin labels of the faces are inherited from the spin labels of spin network edges. If all spin network nodes are three-valent, the spin foam nodes sit at the intersection of six faces, and the dual triangulation consists of tetrahedrons

It has been argued [41] that these missing moves can be obtained from the Hamiltonian formalism by a suitable choice of operator ordering. In Sect. 4 we have used an ordering, symbolically denoted by FEE, in which the Hamiltonian first opens up a spin network and subsequently glues in a plaquette.

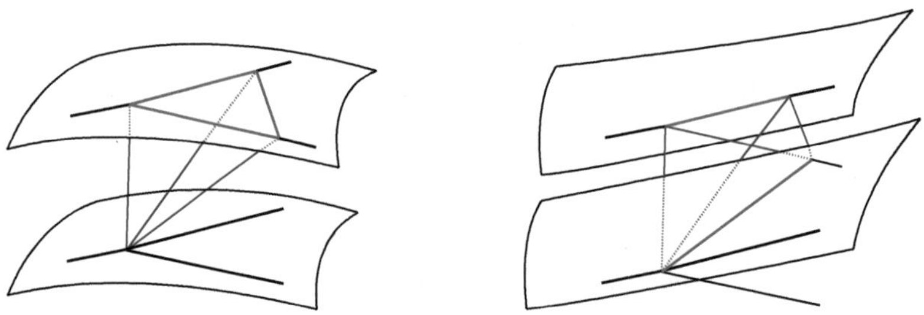

Fig. 9. The Hamiltonian constraint induces a $1 \to 3$ move in the spin foam formalism (**figure on the left**). However, by slicing space-time in a different way, one can equivalently interpret this part of the spin foam as containing a $2 \to 2$ move (**figure on the right**). This argument suggests that the ultra-local Hamiltonian may not be sufficient to achieve space-time covariance. For clarity, the network edges which lie in one of the spatial slices have been drawn as thick lines

If one chooses the ordering to be EEF, then the inverse densitised vielbeine can open the plaquette, thereby potentially inducing a $2 \to 2$ or $0 \to 4$ move. However, [30] has argued strongly against this operator ordering, claiming that in such a form the Hamiltonian operator cannot even be densely defined. In addition, the derivation sketched here is rather symbolic and hampered by the complexity of the Hamiltonian constraint [44]. Hence, to summarise, for (3+1) gravity a decisive proof of the connection between spin foam models and the full Einstein theory and its canonical formulation appears to be lacking, and it is by no means excluded that such a link does not even exist.

7 Spin Foam Models: Some Basic Features

In view of the discussion above, it is thus perhaps best to view spin foam models as models in their own right, and, in fact, as a novel way of defining a (regularised) path integral in quantum gravity. Even without a clear-cut link to the canonical spin network quantisation programme, it is conceivable that spin foam models can be constructed which possess a proper semi-classical limit in which the relation to classical gravitational physics becomes clear. For this reason, it has even been suggested that spin foam models may provide a possible 'way out' if the difficulties with the conventional Hamiltonian approach should really prove insurmountable.

The simplest context in which to study state sum models is (2+1) gravity, because it is a topological ('BF'-type') theory, i.e. without local degrees of freedom, which can be solved exactly (see e.g. [45–47] and [48] for a more recent analysis of the model within the spin foam picture). The most general expression for a state sum in (2+1) dimensions takes, for a given spin foam ϕ, the form

$$Z_\phi = \sum_{\text{spins } \{j\}} \prod_{f,e,v} A_f(\{j\})\, A_e(\{j\})\, A_v(\{j\}) \,, \tag{27}$$

where f, e, v denote the faces, edges, and vertices respectively. The amplitudes depend on all these sub-simplices, and are denoted by A_f, A_e, and A_v respectively. There are many choices which one can make for these amplitudes. In three Euclidean dimensions, space-time covariance demands that the contribution to the partition sum has tetrahedral symmetry in the six spins of the faces which meet at a node (here we assume a 'minimal' spin foam; models with more faces intersecting at an edge are of course possible).

Now, a model of this type has been known for a long time: it is the Ponzano–Regge model for three-dimensional gravity, which implements the above principles by defining the partition sum

$$Z_\phi^{\mathrm{PR}} = \sum_{\text{spins } \{j_i\}} \prod_{\text{faces } f} (2j_f + 1) \prod_{\text{vertices } v}$$

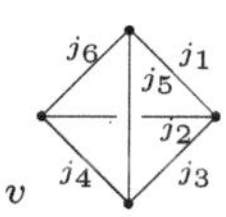

(28)

The graphical notation denotes the Wigner $6j$ symbol, defined in terms of four contracted Clebsch–Gordan coefficients as

$$\{6j\} \sim \sum_{m_1,\ldots,m_6} C^{j_1\ j_2\ j_3}_{m_1 m_2 m_3} C^{j_5\ j_6\ j_1}_{m_5 m_6 m_1} C^{j_6\ j_4\ j_2}_{m_6 m_4 m_2} C^{j_4\ j_5\ j_3}_{m_4 m_5 m_3} . \tag{29}$$

For SU(2) representations, the sum over spins in the Ponzano–Regge state sum (28) requires that one divides by an infinite factor in order to ensure convergence (more on finiteness properties below) and independence of the triangulation. The tetrahedron appearing in (28) in fact has a direct geometrical interpretation as the object dual to the spin foam vertex. The dual tetrahedron can then also be seen as an elementary simplex in the triangulation of the manifold. Three-dimensional state sums with boundaries, appropriate for the calculation of transition amplitudes between two-dimensional spin networks, have been studied in [49].

When one tries to formulate spin foam models in four dimensions, the first issue one has to deal with is the choice of the representation labels on the spin foam faces. From the point of view of the canonical formalism it would seem natural to again use SU(2) representations, as these are used to label the edges of a spin network in three spatial dimensions, whose evolution produces the faces (2-simplices) of the spin foam. However, this is not what is usually done. Instead, the faces of the spin foam are supposed to carry representations of SO(4) $\approx$ SO(3) $\times$ SO(3) [or SO(1,3) $\approx$ SL(2,$\mathbb{C}$) for Lorentzian space-times]. The corresponding models in four dimensions are purely topological theories, the so-called "BF models", where $F(A)$ is a field strength and B the Lagrange multiplier two-form field whose variation enforces $F(A) = 0$. Up to this point, the model is analogous to gravity in (2+1) dimensions, except that the relevant gauge group is now SO(4) [or SO(1,3)]. However, in order to recover general relativity and to re-introduce local (propagating) degrees of freedom into the theory, one must impose a constraint on B.

Classically, this constraint says that B is a 'bi-vector', i.e. $B^{ab} = e^a \wedge e^b$. The quantum mechanical analogue of this constraint amounts to a restriction to a particular set of representations of SO(4) = SU(2) $\otimes$ SU(2), namely those where the spins of the two factors are equal, the so-called *balanced representations*, denoted by (j, j) (for $j = \frac{1}{2}, 1, \frac{3}{2}, \ldots$). Imposing this restriction on the state sum leads to a class of models first proposed by Barrett and Crane [50, 51]. In these models the vertex amplitudes are given by combining the 10 spins of the faces which meet at a vertex, as well as possibly further 'virtual' spins associated to the vertices themselves, using an expression built from contracted Clebsch–Gordan coefficients. For instance, by introducing an extra 'virtual' spin i_k associated to each edge where four faces meet, one can construct an intertwiner between the four spins by means of the following expression:

$$I^{j_1\ \cdots j_4\ ;i_k}_{m_1 \cdots m_4} = \sum_{m_k} C^{j_1\ j_2\ i_k}_{m_1 m_2 m_k} C^{j_3\ j_4\ i_k}_{m_3 m_4 m_k} . \tag{30}$$

However, this prescription is not unique as we can choose between three different 'channels' (here taken to be $12 \leftrightarrow 34$); this ambiguity can be fixed by imposing symmetry, see below. Evidently, the number of channels and virtual spins increases rapidly with the valence of the vertex. For the above four-vertex, this prescription results in a state sum[7]

$$Z_\phi^{\{i_k\}} = \sum_{\text{spins } \{j_i\}} \prod_{\text{faces } f} \prod_{\text{edges } e} A_f(\{j\})\, A_e(\{j\})$$

$$\times \prod_{\text{vertices } v} \quad (31)$$

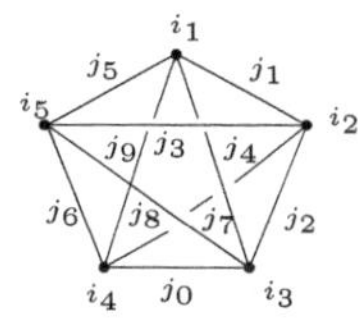

where the spins j denote spin labels of balanced representations (j, j) (as we already mentioned, without this restriction, the model above corresponds to the topological BF model [52–54]). The precise factor corresponding to the pentagon (or "$15j$" symbol) in this formula is explicitly obtained by multiplying the factors (30) (actually, one for each SO(3) factor in SO(4)), and contracting (summing) over the labels m_i,

$$\{15j\} = \sum_{m_i} I^{j_1\, j_4\, j_9\, j_5\, ;i_1}_{m_1 m_4 m_9 m_5} I^{j_1\, j_2\, j_7\, j_3\, ;i_2}_{m_1 m_2 m_7 m_3}$$

$$\times I^{j_4\, j_2\, j_8\, j_0\, ;i_3}_{m_4 m_2 m_8 m_0} I^{j_9\, j_7\, j_0\, j_6\, ;i_4}_{m_9 m_7 m_0 m_6} I^{j_5\, j_3\, j_8\, j_6\, ;i_5}_{m_5 m_3 m_8 m_6} . \quad (32)$$

There are various ways in which one can make (31) independent of the spins i_k associated to the edges. One way is to simply sum over these spins. This leads to the so-called '$15j$ BC model',

$$Z_\phi^{15j} = \sum_{\text{spins } \{j_i,\, i_k\}} \prod_{\text{faces } f} \prod_{\text{edges } e} A_f(\{j\})\, A_e(\{j\}) \times \prod_{\text{vertices } v} \{15j\} . \quad (33)$$

An alternative way to achieve independence of the edge intertwiner spins is to include a sum over the i_k in the definition of the vertex amplitude. These models are known as '$10j$ BC models',

$$Z_\phi^{10j} = \sum_{\text{spins } \{j_i\}} \prod_{\text{faces } f} \prod_{\text{edges } e} A_f(\{j\})\, A_e(\{j\})$$

$$\times \prod_{\text{vertices } v} \sum_{\text{spins } \{i_k\}} f(\{i_k\})\, \{15j\} , \quad (34)$$

[7] There is now no longer such a clear relation of the graphical object in (31) to the dual of the spin foam vertex: faces and edges of the spin foam map to faces and tetrahedrons of the dual in four dimensions, respectively, but these are nevertheless represented with edges and vertices in the figure in (31).

labelled by an arbitrary function $f(\{i_k\})$ of the intertwiner spins. Only for the special choice [50]

$$f(\{i_k\}) = \prod_{k=1}^{5} (2i_k + 1) \tag{35}$$

does the vertex amplitude have simplicial symmetry [55], i.e. is invariant under the symmetries of the pentagon (31) (where the pentagon really represents a 4-simplex).[8]

While the choice (34, 35) for the vertex amplitude $A_v(\{j\})$ is thus preferred from the point of view of covariance, there are still potentially many different choices for the face and edge amplitudes $A_f(\{j\})$ and $A_e(\{j\})$. Different choices may lead to state sums with widely varying properties. The dependence of state sums on the face and edge amplitudes is clearly illustrated by, e.g., the numerical comparison of various models presented in [57]. A natural and obvious restriction on the possible amplitudes is that the models should yield the correct classical limit – to wit, Einstein's equations – in the large j limit, corresponding to the infrared (see also the discussion in the following section). Therefore, any function of the face spins which satisfies the pentagon symmetries and is such that the state sum has appropriate behaviour in the large j limit is a priori allowed. Furthermore, the number of possible amplitudes, and thus of possible models, grows rapidly if one allows for more general valences of the vertices. In the literature, the neglect of higher-valence vertices is often justified by invoking the fact that the valence ≤ 4 spin network wave functions in the Hamiltonian formulation constitute a superselection sector in $\mathcal{H}_{\text{kin}}$ (because the 'spiderwebs' in Fig. 6 do not introduce higher valences). However, we find this argument unconvincing because (i) the precise relation between the Hamiltonian and the spin foam formulation remains unclear, and (ii) physical arguments based on ultralocality (cf. our discussion at the end of Sect. 6) suggest that more general moves (hence, valences) should be allowed.

Let us also mention that, as an alternative to the Euclidean spin foam models, one can try to set up *Lorentzian spin foam models*, as has been done in [58, 59]. In this case, the (compact) group SO(4) is replaced by the *non-compact* Lorentz group SO(1,3) [or SL(2, $\mathbb{C}$)]. Recall that in both cases we deal with oscillatory weights, not with a weight appropriate for a Wick-rotated model. It appears unlikely that there is any relation between the Lorentzian models and the Euclidean ones. Furthermore, the analysis of the corresponding

[8] There is an interesting way to express combinatorial objects such as the $10j$ symbol in terms of integrals over group manifolds, which goes under the name of 'group field theory' (see, e.g., [56]), and which also allows an interpretation in terms of 'Feynman diagrams'. The relation between spin foams and group field theory is potentially useful to evaluate state sums because the corresponding integrals can be evaluated using stationary phase methods. We will, however, not comment on this development any further since there is (under certain assumptions) a one-to-one map between spin foam models and group field theory models.

Lorentzian state sums is much more complicated due to the fact that the relevant (i.e. unitary) representations are now infinite-dimensional.

8 Spin Foams and Discrete Gravity

To clarify the relation between spin foam models and earlier attempts to define a discretised path integral in quantum gravity, we recall that the latter can be roughly divided into two classes, namely:

- *Quantum Regge Calculus* (see, e.g., [60]), where one approximates space-time by a triangulation consisting of a fixed number of simplices, and integrates over all edge lengths, keeping the 'shape' of the triangulation fixed;
- *Dynamical Triangulations* (see, e.g., [61–63]), where the simplices are assigned fixed edge lengths, and one sums instead over different triangulations, but keeping the number of simplices fixed (thus changing only the 'shape', but not the 'volume' of the triangulation).

Both approaches are usually based on a positive signature (Euclidean) metric, where the Boltzmann factor is derived from, or at least motivated by, some discrete approximation to the Einstein–Hilbert action, possibly with a cosmological constant (but see [64, 65] for some recent progress with a Wick-rotated 'Lorentzian' dynamical triangulation approach which introduces and exploits a notion of causality on the space-time lattice). In both approaches, the ultimate aim is then to recover continuum space-time via a refinement limit in which the number of simplices is sent to infinity. Establishing the existence of such a limit is a notoriously difficult problem that has not been solved for four-dimensional gravity. In fact, for quantum Regge models in two dimensions such a continuum limit does not seem to agree with known continuum results [66–69] (see, however, [70]).

From the point of view of the above classification, spin foam models belong to the first, 'quantum Regge', type, as one sums over all spins for a given spin foam, but does not add, remove, or replace edges, faces, or vertices, at least not in the first step. Indeed, for the spin foams discussed in the foregoing section, we have so far focused on the partition sum for a *single* given spin foam. An obvious question then concerns the next step, or more specifically the question how spin foam models can recover (or even only define) a continuum limit. The canonical setup, where one sums over all spin network states in expressions like (25), would suggest that one should sum over all foams,

$$Z^{\text{total}} = \sum_{\text{foams } \phi} w_\phi \, Z_\phi \,, \tag{36}$$

where Z_ϕ denotes the partition function for a given spin foam ϕ, and where we have allowed for the possibility of a non-trivial weight w_ϕ depending only on the topological structure ('shape') of the foam. The reason for this sum would be to achieve formal independence of the triangulations. In a certain sense this would mimic the dynamical triangulation approach (except that one now would also sum over foams with a different number of simplices and different edge lengths), and thus turn the model into a hybrid version of the above approaches. However, this prescription is far from universally accepted, and several other ideas on how to extract classical, continuum physics from the partition sum Z_ϕ have been proposed.

One obvious alternative is to *not* sum over all foams, but instead look for a refinement with an increasing number of cells,[9]

$$Z^\infty = \lim_{\#\ \text{cells}\to\infty} Z_\phi \,. \tag{37}$$

The key issue is then to ensure that the final result does not depend on the way in which the triangulations are performed and refined (this is a crucial step which one should understand in order to interpret results on single-simplex spin foams like those of [71, 72]). The refinement limit is motivated by the fact that it does appear to work in three space-time dimensions, where (allowing for some 'renormalisation') one can establish triangulation independence [73]. Furthermore, for large spins, the $6j$ symbol which appears in the Ponzano–Regge model approximates the Feynman weight for Regge gravity [74, 75]. More precisely, when all six spins are simultaneously taken large,

$$\{6j\} \sim \left(e^{iS_{\text{Regge}}(\{j\})+\frac{i\pi}{4}} + e^{-iS_{\text{Regge}}(\{j\})-\frac{i\pi}{4}}\right) \,. \tag{38}$$

Here $S_{\text{Regge}}(\{j\})$ is the Regge action of a tetrahedron, given by

$$S_{\text{Regge}}(\{j\}) = \sum_{i=1}^{6} j_i\,\theta_i \,, \tag{39}$$

where θ_i is the dihedral angle between the two surfaces meeting at the ith edge. Related results in four dimensions are discussed in [76] and, using group field theory methods, in [77]. We emphasise once more that this by no means singles out the $6j$ symbol as the unique vertex amplitude: we can still multiply it by any function of the six spins which asymptotes to one for large spins.

The $6j$ symbol is of course real, which explains the presence of a cosine instead of a complex oscillatory weight on the right-hand side of (38). Indeed, it seems rather curious that, while the left-hand side of (38) arises from an expression resembling a Boltzmann sum, the right-hand side contains oscillatory factors which suggest a path integral with oscillatory weights. In view of our remarks in Sect. 6, and in order to make the relation to Regge gravity

[9] But note that, formally, the sum over all foams can also be thought of as a refinement limit if one includes zero spin representations (meaning no edge) in the refinement limit.

somewhat more precise, one must therefore argue either that a proper path integral in gravity produces both terms, or otherwise that one can get rid of one of the terms by some other mechanisms. The first possibility appears to be realised in (2+1) gravity, because one can cast the gravitational action into Chern–Simons form $S = \int R \wedge e$, in which case a sum over orientations of the dreibein would lead to terms with both signs in the exponent. Unfortunately, this argument does not extend to four dimensions, where the gravitational action $S = \int R \wedge e \wedge e$ depends quadratically on the vierbein. For this reason, it has instead been suggested that one of the two oscillatory terms disappears for all physical correlation functions [71].

The vertex amplitudes represented by the $6j$ or $10j$ symbols only form part of the state sum (27). The known four-dimensional models depend rather strongly on the choice of the face and edge amplitudes: while some versions of the Barrett–Crane $10j$ model have diverging partition sums, others are dominated by configurations in which almost all spins are zero, i.e. configurations which correspond to zero-area faces [57]. Once more, it is important to remember that even in 'old' Regge models in two dimensions, where a comparison with exact computations in the continuum is possible [78–80], the continuum limit does not seem to agree with these exact results [66–69] (the expectation values of edge lengths do not scale as a power of the volume when a diffeomorphism invariant measure is used, in contrast to the exact results). Therefore, it is far from clear that (37) will lead to a proper continuum limit.

A third proposal is to take a fixed spin foam and to simply define the model as the sum over all spins [56, 81, 82]; this proposal differs considerably from both the Regge and dynamical triangulation approaches. Considering a fixed foam clearly only makes sense provided the partition sum is actually independent of the triangulation of the manifold (or more correctly, one would require that physical correlators are independent of the triangulation). Such a situation arises in the three-dimensional Ponzano–Regge model, but three-dimensional gravity does not contain any local degrees of freedom. For higher dimensions, the only triangulation-independent models known so far are topological theories, i.e. theories for which the local degrees of freedom of the metric do not matter. If one insists on triangulation independence also for gravity, then one is forced to add new degrees of freedom to the spin foam models (presumably living on the edges). In this picture, a change from a fine triangulation to a coarse one is then compensated by more information stored at the edges of the coarse triangulation. This then also requires (presumably complicated) rules which say how these new degrees of freedom behave under a move from one triangulation to another. Note that even when the partition sum is independent of the refinement of the triangulation, one would probably still want to deal with complicated cross sections of foams to describe 'in' and 'out' coherent states. At present, there is little evidence that triangulation independence can be realised in non-topological theories, or that the problems related to the continuum limit will not reappear in a different guise.

9 Predictive (Finite) Quantum Gravity?

Let us now return to the question as to what can be said about finiteness properties of spin foam models, and how they relate to finiteness properties (or rather, lack thereof!) of the standard perturbative approach – after all, one of the main claims of this approach is that it altogether avoids the difficulties of the standard approach. So far, investigations of finiteness have focused on the partition sum itself. Namely, it has been shown that for a variety of spin foam models, the partition sum for a *fixed* spin foam is finite,

$$\sum_{\text{spins } \{j\}} Z_\phi(\{j\}) = \text{finite} \,. \tag{40}$$

Even though a given spin foam consists of a finite number of links, faces, ..., divergences could arise in principle because the range of each spin j is infinite. One way to circumvent infinite sums is to replace the group SU(2) by the quantum group $\mathrm{SU}(2)_q$ (which has a finite number of irreps), or equivalently, by introducing an infinite positive cosmological constant [73]; in all these cases the state sum becomes finite.[10] A similar logic holds true in four dimensions and for Lorentzian models, although in the latter case the analysis becomes more complicated due to the non-compactness of the Lorentz group, and the fact that the unitary representations are all infinite dimensional [84]. Perhaps unsurprisingly, there exist choices for edge and surface amplitudes in four dimensions which look perfectly reasonable from the point of view of covariance, but which are nevertheless not finite [57].

It should, however, be emphasised that the finiteness of (40) is a statement about *infrared* finiteness. Roughly speaking, this is because the spin j corresponds to the 'length' of the link, whence the limit of large j should be associated with the *infinite volume limit*. In statistical mechanics, partition functions generically diverge in this limit, but in such a way that physical correlators possess a well-defined limit (as quotients of two quantities which diverge). From this point of view, the finiteness properties established so far say nothing about the UV properties of quantum gravity, which should instead follow from some kind of refinement limit, or from an averaging procedure where one sums over all foams, as discussed above. The question of convergence or non-convergence of such limits has so far not received a great deal of attention in the literature.

This then, in a sense, brings us back to square one, namely the true problem of quantum gravity, which lies in the ambiguities associated with an infinite number of non-renormalisable UV divergences. As is well known this

[10] The division by the infinite factor which is required to make the Ponzano–Regge state sum finite can be understood as dividing out the volume of the group of residual invariances of Regge models [83]. These invariances correspond to changes of the triangulation which leave the curvature fixed. However, dividing out by the volume of this group does not eliminate the formation of 'spikes' in Regge gravity.

problem was originally revealed in a perturbative expansion of Einstein gravity around a fixed background, which requires an infinite series of counterterms, starting with the famous two-loop result [85–87]

$$\Gamma^{(2)}_{\rm div} = \frac{1}{\epsilon}\frac{209}{2880}\frac{1}{(16\pi^2)^2}\int \mathrm{d}^4x\,\sqrt{g}\,C_{\mu\nu\rho\sigma}C^{\rho\sigma\lambda\tau}C_{\lambda\tau}{}^{\mu\nu}\ . \tag{41}$$

The need to fix an infinite number of couplings in order to make the theory predictive renders perturbatively quantised Einstein gravity useless as a physical theory. What we would like to emphasise here is that *any* approach to quantum gravity must confront this problem, and that the need to fix infinitely many couplings in the perturbative approach, and the appearance of infinitely many ambiguities in non-perturbative approaches are really just different sides of the same coin.

At least in its present incarnation, the canonical formulation of LQG does not encounter any UV divergences, but the problem reappears through the lack of uniqueness of the canonical Hamiltonian. For spin foams (or, more generally, discrete quantum gravity) the problem is no less virulent. The known finiteness proofs all deal with the behaviour of a single foam, but, as we argued, these proofs concern the infrared rather than the ultraviolet. Just like canonical LQG, spin foams thus show no signs of ultraviolet divergences so far, but, as we saw, there is an *embarras de richesse* of physically distinct models, again reflecting the non-uniqueness that manifests itself in the infinite number of couplings associated with the perturbative counterterms. Indeed, fixing the ambiguities of the non-perturbative models by ad hoc, albeit well-motivated, assumptions is not much different from defining the perturbatively quantised theory by fixing infinitely many coupling constants 'by hand' (and thereby remove all divergences). Furthermore, even if they do not 'see' any UV divergences, non-perturbative approaches cannot be relieved of the duty to explain *in all detail* how the 2-loop divergence (41) and its higher loop analogues 'disappear', be it through cancellations or some other mechanism.

Finally, let us remark that in lattice gauge theories, the classical limit and the UV limit can be considered and treated as separate issues. As for quantum gravity, this also appears to be the prevailing view in the LQG community. However, the continuing failure to construct viable *physical* semi-classical states, solving the constraints even in only an approximate fashion, seems to suggest (at least to us) that in gravity the two problems cannot be solved separately, but are inextricably linked – also in view of the fact that the question as to the precise fate of the two-loop divergence (41) can then no longer be avoided.

Acknowledgements

The first part of this paper is based on [8], written in collaboration with Marija Zamaklar. We thank Jan Ambjørn, Herbert Hamber, Claus Kiefer, Kirill Krasnov, Hendryk Pfeiffer, Martin Reuter, and Marija Zamaklar for discussions and correspondence. We are also grateful to Laurent Freidel for a transatlantic debate that helped clarify some points in the original version of this review.

References

1. R. Loll, 'Discrete approaches to quantum gravity in four dimensions', *Living Rev. Rel.* **1** (1998) 13, http://xxx.lanl.gov/abs/gr-qc/9805049 gr-qc/9805049.
2. G. W. Gibbons and S. W. Hawking, 'Action integrals and partition functions in quantum gravity', *Phys. Rev.* **D15** (1977) 2752–2756.
3. S. W. Hawking, 'The path-integral approach to quantum gravity', in '*An Einstein Centenary Survey*', S. Hawking and W. Israel, eds., pp. 746–789. Cambridge University Press, 1979.
4. S. Weinberg, 'Ultraviolet divergences in quantum gravity', in '*An Einstein Centenary Survey*', S. Hawking and W. Israel, eds., pp. 790–832. Cambridge University Press, 1979.
5. S. Weinberg, 'What is quantum field theory, and what did we think it was?", http://xxx.lanl.gov/abs/hep-th/9702027hep-th/9702027.
6. O. Lauscher and M. Reuter, "Is quantum Einstein gravity nonperturbatively renormalizable?', *Class. Quant. Grav.* **19** (2002) 483–492, http://xxx.lanl.gov/abs/hep-th/0110021hep-th/0110021.
7. C. Kiefer, 'Quantum Gravity', Clarendon Press, 2004.
8. H. Nicolai, K. Peeters, and M. Zamaklar, 'Loop quantum gravity: an outside view', *Class. Quant. Grav.* **22** (2005) R193–R247, http://xxx.lanl.gov/abs/hep-th/0501114hep-th/0501114.
9. R. Gambini and J. Pullin, 'Loops, knots, gauge theories and quantum gravity', Cambridge University Press, 1996.
10. T. Thiemann, 'Introduction to modern canonical quantum general relativity', http://xxx.lanl.gov/abs/gr-qc/0110034gr-qc/0110034.
11. A. Ashtekar and J. Lewandowski, "Background independent quantum gravity: A status report", *Class. Quant. Grav.* **21** (2004) R53, http://xxx.lanl.gov/abs/gr-qc/0404018gr-qc/0404018.
12. A. Perez, 'Introduction to loop quantum gravity and spin foams', http://xxx.lanl.gov/abs/gr-qc/0409061gr-qc/0409061.
13. J. C. Baez, "An introduction to spin foam models of BF theory and quantum gravity", *Lect. Notes Phys.* **543** (2000) 25–94, http://xxx.lanl.gov/abs/gr-qc/9905087gr-qc/9905087.
14. A. Perez, "Spin foam models for quantum gravity", *Class. Quant. Grav.* **20** (2003) R43, http://xxx.lanl.gov/abs/gr-qc/0301113gr-qc/0301113.
15. C. Rovelli, 'Quantum gravity', Cambridge University Press, 2004.
16. A. Perez, "On the regularization ambiguities in loop quantum gravity", http://xxx.lanl.gov/abs/gr-qc/0509118gr-qc/0509118.
17. A. Perez, "The spin-foam-representation of loop quantum gravity", http://xxx.lanl.gov/abs/gr-qc/0601095gr-qc/0601095.
18. "Loops '05", http://loops05.aei.mpg.de/.
19. A. Ashtekar, S. Fairhurst, and J. L. Willis, "Quantum gravity, shadow states, and quantum mechanics", *Class. Quant. Grav.* **20** (2003) 1031–1062, http://xxx.lanl.gov/abs/gr-qc/0207106gr-qc/0207106.

20. T. Thiemann, "Gauge field theory coherent states (GCS). I: General properties", *Class. Quant. Grav.* **18** (2001) 2025–2064, http://xxx.lanl.gov/abs/hep-th/0005233 hep-th /0005233.
21. T. Thiemann and O. Winkler, "Gauge field theory coherent states (GCS). II: Peakedness properties", *Class. Quant. Grav.* **18** (2001) 2561–2636, http://xxx.lanl.gov/abs/hep-th/0005237hep-th/0005237.
22. T. Thiemann and O. Winkler, "Gauge field theory coherent states (GCS) III: Ehrenfest theorems", *Class. Quant. Grav.* **18** (2001) 4629–4682, http://xxx.lanl.gov/abs/hep-th/0005234hep-th/0005234.
23. T. Thiemann and O. Winkler, "Gauge field theory coherent states (GCS). IV: Infinite tensor product and thermodynamical limit", *Class. Quant. Grav.* **18** (2001) 4997–5054, http://xxx.lanl.gov/abs/hep-th/0005235hep-th/0005235.
24. T. Thiemann, "Complexifier coherent states for quantum general relativity", http://xxx.lanl.gov/abs/gr-qc/0206037gr-qc/0206037.
25. H. Sahlmann, T. Thiemann, and O. Winkler, "Coherent states for canonical quantum general relativity and the infinite tensor product extension", *Nucl. Phys.* **B606** (2001) 401–440, http://xxx. lanl.gov/abs/gr-qc/0102038gr-qc/0102038.
26. J. Brunnemann and T. Thiemann, "On (cosmological) singularity avoidance in loop quantum gravity", http://xxx.lanl.gov/abs/gr-qc/0505032gr-qc/0505032.
27. K. A. Meissner, "Eigenvalues of the volume operator in loop quantum gravity", http://xxx.lanl.gov/abs/gr-qc/0509049gr-qc/0509049.
28. W. Fairbairn and C. Rovelli, "Separable Hilbert space in loop quantum gravity", *J. Math. Phys.* **45** (2004) 2802–2814, http://xxx.lanl.gov/abs/gr-qc/0403047gr-qc/0403047.
29. R. Borissov, R. De Pietri, and C. Rovelli, "Matrix elements of Thiemann's Hamiltonian constraint in loop quantum gravity", *Class. Quant. Grav.* **14** (1997) 2793–2823, http://xxx.lanl.gov/abs/gr-qc/9703090gr-qc/9703090.
30. T. Thiemann, "Quantum spin dynamics (QSD)", *Class. Quant. Grav.* **15** (1998) 839–873, http://xxx.lanl.gov/abs/gr-qc/9606089gr-qc/9606089.
31. R. De Pietri and C. Rovelli, "Geometry eigenvalues and scalar product from recoupling theory in loop quantum gravity", *Phys. Rev.* **D54** (1996) 2664–2690, http://xxx.lanl.gov/abs/gr-qc/9602023gr-qc/9602023.
32. J. Brunnemann and T. Thiemann, "Simplification of the spectral analysis of the volume operator in loop quantum gravity", http://xxx.lanl.gov/abs/gr-qc/0405060gr-qc/0405060.
33. J. Lewandowski and D. Marolf, "Loop constraints: A habitat and their algebra", *Int. J. Mod. Phys.* **D7** (1998) 299–330, http://xxx.lanl.gov/abs/gr-qc/9710016gr-qc/9710016.
34. L. Smolin, "The classical limit and the form of the Hamiltonian constraint in non-perturbative quantum general relativity", http://xxx.lanl.gov/abs/gr-qc/9609034gr-qc/9609034.
35. D. E. Neville, "Long range correlations in quantum gravity", *Phys. Rev.* **D59** (1999) 044032, http://xxx.lanl.gov/abs/gr-qc/9803066gr-qc/9803066.
36. R. Loll, "On the diffeomorphism-commutators of lattice quantum gravity", *Class. Quant. Grav.* **15** (1998) 799–809, http://xxx.lanl.gov/abs/gr-qc/9708025gr-qc/9708025.

37. T. Thiemann, "Anomaly-free formulation of non-perturbative, four-dimensional Lorentzian quantum gravity", *Phys. Lett.* **B380** (1996) 257–264, http://xxx.lanl.gov/abs/gr-qc/9606088gr-qc/9606088.
38. R. Gambini, J. Lewandowski, D. Marolf, and J. Pullin, "On the consistency of the constraint algebra in spin network quantum gravity", *Int. J. Mod. Phys.* **D7** (1998) 97–109, http://xxx.lanl.gov/abs/gr-qc/9710018gr-qc/9710018.
39. M. Henneaux and C. Teitelboim, "Quantization of gauge systems", Princeton University Press, 1992.
40. M. P. Reisenberger, "World sheet formulations of gauge theories and gravity", http://xxx.lanl.gov/abs/gr-qc/9412035gr-qc/9412035.
41. M. P. Reisenberger and C. Rovelli, "'Sum over surfaces' form of loop quantum gravity", *Phys. Rev.* **D56** (1997) 3490–3508, http://xxx.lanl.gov/abs/gr-qc/9612035gr-qc/9612035.
42. J. C. Baez, "Spin foam models", *Class. Quant. Grav.* **15** (1998) 1827–1858, http://xxx.lanl.gov/abs/gr-qc/9709052gr-qc/9709052.
43. R. De Pietri, "Canonical "loop" quantum gravity and spin foam models", http://xxx.lanl.gov/abs/gr-qc/9903076gr-qc/9903076.
44. C. Rovelli, "The projector on physical states in loop quantum gravity", *Phys. Rev.* **D59** (1999) 104015, http://xxx.lanl.gov/abs/gr-qc/9806121gr-qc/9806121.
45. S. Deser, R. Jackiw, and G. 't Hooft, "Three-dimensional Einstein gravity: dynamics of flat space", *Ann. Phys.* **152** (1984) 220.
46. E. Witten, "(2+1)-Dimensional gravity as an exactly soluble system", *Nucl. Phys.* **B311** (1988) 46.
47. A. Ashtekar, V. Husain, C. Rovelli, J. Samuel, and L. Smolin, "(2+1)-Quantum gravity as a toy model for the (3+1) theory", *Class. Quant. Grav.* **6** (1989) L185.
48. K. Noui and A. Perez, "Three dimensional loop quantum gravity: Physical scalar product and spin foam models", *Class. Quant. Grav.* **22** (2005) 1739–1762, http://xxx.lanl.gov/abs/gr-qc/0402110gr-qc/0402110.
49. M. Karowski, W. Muller, and R. Schrader, "State sum invariants of compact three manifolds with boundary and $6j$ symbols", *J. Phys.* **A25** (1992) 4847–4860.
50. J. W. Barrett and L. Crane, "Relativistic spin networks and quantum gravity", *J. Math. Phys.* **39** (1998) 3296–3302, http://xxx.lanl.gov/abs/gr-qc/9709028gr-qc/9709028.
51. L. Freidel and K. Krasnov, "Spin foam models and the classical action principle", *Adv. Theor. Math. Phys.* **2** (1999) 1183–1247, http://xxx.lanl.gov/abs/hep-th/9807092hep-th/9807092.
52. H. Ooguri, "Topological lattice models in four dimensions", *Mod. Phys. Lett.* **A7** (1992) 2799–2810, http://xxx.lanl.gov/abs/hep-th/9205090hep-th/9205090.
53. L. Crane and D. Yetter, "A categorical construction of 4-D topological quantum field theories", in 'Quantum Topology', L. Kauffman and R. Baadhio, eds., pp. 120–130. World Scientific, Singapore, 1993. http://xxx.lanl.gov/abs/hep-th/9301062hep-th/9301062.
54. L. Crane, L. H. Kauffman, and D. N. Yetter, "State sum invariants of four-manifolds I", *J. Knot Theor Ramifications* **6** (1997) 177–234, http://xxx.lanl.gov/abs/hep-th/9409167hep-th/9409167.

55. M. P. Reisenberger, "On relativistic spin network vertices", *J. Math. Phys.* **40** (1999) 2046–2054, http://xxx.lanl.gov/abs/gr-qc/9809067 gr-qc/9809067.
56. L. Freidel, "Group field theory: An overview", *Int. J. Theor. Phys.* **44** (2005) 1769–1783, http://xxx.lanl.gov/abs/hep-th/0505016hep-th/ 0505016.
57. J. C. Baez, J. D. Christensen, T. R. Halford, and D. C. Tsang, "Spin foam models of Riemannian quantum gravity", *Class. Quant. Grav.* **19** (2002) 4627–4648, http://xxx.lanl.gov/abs/gr-qc/0202017gr-qc/0202017.
58. J. W. Barrett and L. Crane, "A Lorentzian signature model for quantum general relativity", *Class. Quant. Grav.* **17** (2000) 3101–3118, http://xxx.lanl.gov/abs/gr-qc/9904025gr-qc/9904025.
59. A. Perez and C. Rovelli, "Spin foam model for Lorentzian general relativity", *Phys. Rev.* **D63** (2001) 041501, http://xxx.lanl.gov/abs/gr-qc/0009021gr-qc/0009021.
60. R. M. Williams and P. A. Tuckey, "Regge calculus: A bibliography and brief review", *Class. Quant. Grav.* **9** (1992) 1409–1422.
61. D. V. Boulatov, V. A. Kazakov, I. K. Kostov, and A. A. Migdal, "Analytical and numerical study of the model of dynamically triangulated random surfaces", *Nucl. Phys.* **B275** (1986) 641.
62. A. Billoire and F. David, "Scaling properties of randomly triangulated planar random surfaces: a numerical study", *Nucl. Phys.* **B275** (1986) 617.
63. J. Ambjørn, B. Durhuus, J. Frohlich, and P. Orland, "The appearance of critical dimensions in regulated string theories", *Nucl. Phys.* **B270** (1986) 457.
64. J. Ambjørn, J. Jurkiewicz, and R. Loll, "Dynamically triangulating Lorentzian quantum gravity", *Nucl. Phys.* **B610** (2001) 347–382, http://xxx.lanl.gov/abs/hep-th/0105267hep-th/0105267.
65. J. Ambjørn, J. Jurkiewicz, and R. Loll, "Emergence of a 4D world from causal quantum gravity", *Phys. Rev. Lett.* **93** (2004) 131301, http://xxx.lanl.gov/abs/hep-th/0404156hep-th/0404156.
66. W. Bock and J. C. Vink, "Failure of the Regge approach in two-dimensional quantum gravity", *Nucl. Phys.* **B438** (1995) 320–346, http://xxx.lanl.gov/abs/hep-lat/9406018hep-lat/9406018.
67. C. Holm and W. Janke, "Measure dependence of 2D simplicial quantum gravity", *Nucl. Phys. Proc. Suppl.* **42** (1995) 722–724, http://xxx.lanl.gov/abs/hep-lat/9501005hep-lat/9501005.
68. J. Ambjørn, J. L. Nielsen, J. Rolf, and G. K. Savvidy, "Spikes in quantum Regge calculus", *Class. Quant. Grav.* **14** (1997) 3225–3241, http://xxx.lanl.gov/abs/gr-qc/9704079gr-qc/9704079.
69. J. Rolf, "Two-dimensional quantum gravity", PhD thesis, University of Copenhagen, 1998. http://xxx.lanl.gov/abs/hep-th/9810027 hep-th/9810027.
70. H. W. Hamber and R. M. Williams, "On the measure in simplicial gravity", *Phys. Rev.* **D59** (1999) 064014, http://xxx.lanl.gov/abs/hep-th/9708019hep-th/9708019.
71. C. Rovelli, "Graviton propagator from background-independent quantum gravity", http://xxx.lanl.gov/abs/gr-qc/0508124gr-qc/0508124.
72. S. Speziale, "Towards the graviton from spinfoams: The 3d toy model", http://xxx.lanl.gov/abs/gr-qc/0512102gr-qc/0512102.
73. H. Ooguri, "Partition functions and topology changing amplitudes in the 3D lattice gravity of Ponzano and Regge", *Nucl. Phys.* **B382** (1992) 276–304, http://xxx.lanl.gov/abs/hep-th/9112072hep-th/9112072.

74. G. Ponzano and T. Regge, "Semiclassical limit of Racah coefficients", in "Spectroscopic and group theoretical methods in physics". North-Holland, 1968.
75. J. Roberts, "Classical $6j$-symbols and the tetrahedron", *Geom. Topol.* **3** (1999) 21–66, http://xxx.lanl.gov/abs/math-ph/9812013math-ph/9812013.
76. J. W. Barrett and R. M. Williams, "The asymptotics of an amplitude for the 4-simplex", *Adv. Theor. Math. Phys.* **3** (1999) 209–215, http://xxx.lanl.gov/abs/gr-qc/9809032gr-qc/9809032.
77. L. Freidel and K. Krasnov, "Simple spin networks as Feynman graphs", *J. Math. Phys.* **41** (2000) 1681–1690, http://xxx.lanl.gov/abs/hep-th/9903192hep-th/9903192.
78. V. G. Knizhnik, A. M. Polyakov, and A. B. Zamolodchikov, "Fractal structure of 2d-quantum gravity", *Mod. Phys. Lett.* **A3** (1988) 819.
79. F. David, "Conformal field theories coupled to 2-d gravity in the conformal gauge", *Mod. Phys. Lett.* **A3** (1988) 1651.
80. J. Distler and H. Kawai, "Conformal field theory and 2-d quantum gravity or who's afraid of Joseph Liouville?", *Nucl. Phys.* **B321** (1989) 509.
81. J. W. Barrett, "State sum models for quantum gravity", http://xxx.lanl.gov/abs/gr-qc/0010050gr-qc/0010050.
82. H. Pfeiffer, "Diffeomorphisms from finite triangulations and absence of 'local' degrees of freedom", *Phys. Lett.* **B591** (2004) 197–201, http://xxx.lanl.gov/abs/gr-qc/0312060gr-qc/0312060.
83. L. Freidel and D. Louapre, "Diffeomorphisms and spin foam models", *Nucl. Phys.* **B662** (2003) 279–298, http://xxx.lanl.gov/abs/gr-qc/0212001gr-qc/0212001.
84. L. Crane, A. Perez, and C. Rovelli, "A finiteness proof for the Lorentzian state sum spinfoam model for quantum general relativity", http://xxx.lanl.gov/abs/gr-qc/0104057gr-qc/0104057.
85. M. H. Goroff and A. Sagnotti, "Quantum gravity at two loops", *Phys. Lett.* **B160** (1985) 81.
86. M. H. Goroff and A. Sagnotti, "The ultraviolet behavior of Einstein gravity", *Nucl. Phys.* **B266** (1986) 709.
87. A. E. M. van de Ven, "Two loop quantum gravity", *Nucl. Phys.* **B378** (1992) 309–366.

Loop Quantum Gravity: An Inside View

T. Thiemann[1,2]

[1] Max-Planck-Institut für Gravitationsphysik, Albert-Einstein-Institut, Am Mühlenberg 1, 14476 Golm, Germany
tthiemann@aei.mpg.de
[2] Perimeter Institute for Theoretical Physics, Waterloo, Canada
tthiemann@perimeterinstitute.ca

1 Introduction

Loop quantum gravity (LQG) [1–3] has become a serious competitor to string theory as a candidate theory of quantum gravity. Its popularity is steadily growing because it has transpired that the major obstacle to be solved in combing the principles of general relativity and quantum mechanics is to preserve a key feature of Einstein's theory, namely *background independence*. LQG, in contrast to the present formulation of string theory, has background independence built in by construction.

Loosely speaking, background independence means that the spacetime metric is not an external structure on which matter fields and gravitational perturbations propagate. Rather, the metric is a dynamical entity which becomes a fluctuating quantum operator. These fluctuations will be huge in extreme astrophysical (centre of black holes) and cosmological (big bang singularity) situations and the notion of a (smooth) background metric disappears, the framework of quantum field theory on (curved) background metrics [4, 5] becomes meaningless. Since quantum gravity is supposed to take over as a more complete theory precisely in those situations when there is no meaningful concept of a (smooth) metric *at all* available, background independence is *a necessary feature* of a successful quantum gravity theory.

Indeed, the modern formulation of ordinary QFT on background spacetimes uses the algebraic approach [6] and fundamental for this framework is the locality axiom: Two (scalar) field operators $\widehat{\phi(f)}$, $\widehat{\phi(f')}$ which are smeared with test functions f, f' whose supports are *spacelike separated* are axiomatically required to commute. In other words, the causality structure of the external background metric *defines the algebra of field operators* $\mathfrak{A}$. One then looks for Hilbert space representations of $\mathfrak{A}$. It follows that without a background metric one *cannot even define* quantum fields in the usual setting.

Notice that background independence implies that a non-perturbative formulation must be found. For, if we split the metric as $g = g_0 + h$ where g_0 is

T. Thiemann: *Loop Quantum Gravity: An Inside View*, Lect. Notes Phys. **721**, 185–263 (2007)
DOI 10.1007/978-3-540-71117-9_10

a given background metric and treat the deviation h (graviton) as a quantum field propagating on g_0, then we break background independence because we single out g_0. We also break the symmetry group of Einstein's theory which is the group of diffeomorphisms of the given spacetime manifold M. Moreover, it is well known that this perturbative formulation leads to a non-renormalisable theory, with [7] or without [8] supersymmetry. String theory [9] in its current formulation is also background dependent because one has to fix a target space background metric (mostly Minkowski space or maybe AdS) and let strings propagate on it. Some excitations of the string are interpreted as gravitons and one often hears that string theory is perturbatively finite to all orders, in contrast to perturbative quantum gravity. However, this has been established only to second order and only for the heterotic string [10] which is better but still far from a perturbatively finite theory. In fact, even perturbative finiteness is not the real issue because one can formulate perturbation theory in such a way that UV divergences never arise [11]. The issue is (1) whether only a finite number of free renormalisation constants need to be fixed (predictability) and (2) whether the perturbation series converges. Namely, in a fundamental theory as string theory claims to be, there is no room for singularities such as a a divergent perturbation series. This is different from QED which is believed to be only an effective theory. Hence, before one does not prove convergence of the perturbation series, string theory has not been shown to be a fundemantal theory. All these issues are obviously avoided in a manifestly non-perturbative formulation.

One of the reasons why LQG is gaining in its degree of popularity as compared to string theory is that LQG has "put its cards on the table". LQG has a clear conceptual setup which follows from physical considerations and is based on a rigorous mathematical formulation. The "rules of the game" have been written and are not tempered with. This makes it possible even for outsiders of the field [12, 13] to get a relatively good understanding of the physical and mathematical details. There is no room for extra, unobserved structures, the approach is intendedly minimalistic. In LQG one just tries to make quantum gravity and general relativity work together harmonically. However, in order to do so one must be ready to go beyond some of the mathematical structures that we got used to from ordinary QFT as we have explained. Much of the criticism against LQG of which some can be found in [12] has to do with the fact that physicists equipped with a particle physics background feel uneasy when one explains to them that in LQG we cannot use perturbation theory, Fock spaces, background metrics etc. This is not the fault of LQG. It will be a common feature of all quantum gravity theories which preserve background independence. In such theories, the task is to construct a new type of QFT, namely a QFT on a *differential manifold M* rather than a QFT on a background spacetime (M, g_0). Since such a theory "quantises all backgrounds at once" in a coherent fashion, the additional task is then to show that for any background metric g_0 the theory contains a semiclassical sector which looks like ordinary QFT on (M, g_0). This is what LQG is designed to do, not more and not less.

Another criticism which is raised against LQG is that it is not a unified theory of all interactions in the sense that string theory claims to be. Indeed, in LQG one quantises geometry together with the fields of the (supersymmetric extension of the) standard model. At present there seem to be no constraints on the particle content or the dimensionality of the world. In fact, this is not quite true because the size of the physical Hilbert space of the theory may very well depend on the particle content, and moreover the concrete algebra $\mathfrak{A}$ which one quantises in LQG is available only in 3+1 dimensions. But apart from that it is certainly true that LQG cannot give a prediction of the matter content. The fact that all matter can be treated may however be an advantage because given the fact that in the past 100 years we continuously discovered substructures of particles up to the subnuclear scale makes it likely that we find even more structure until the Planck scale which is some 16 orders of magnitude smaller than what the LHC can resolve. Hence, LQG is supposed to deliver a universal framework for combining geometry and matter, however, it does not uniquely predict which matter; and does not want to. Notice that while theorists would find a "unique" theory aesthetical, there is no logical reason why a fundamental theory should be mathematically unique.

In this context we would like to point out the following: One often hears that string theory is mathematically unique, predicting supersymmetry, the dimensionality (ten) of the world and the particle content. What one means by this is that a consistent quantum string theory based on the Polyakov action on the Minkowski target spacetime exists only if one is in ten dimensions and only if the theory is supersymmetric and there only five such theories. However, this is not enough in order to have a unique theory because string theory must be decompactified from ten to four dimensions with supersymmetry being broken at sufficiently high energies in order to be phenomenologically acceptable. Recent findings [14] show that for Minkowski space there are an at least countably infinite number of physically different, phenomenologically acceptable ways to compactify string theory from ten to four dimensions. These possibilities are labelled by flux vacua and the resulting collection of quantum string theories is called *the landscape*. In this sense, *string theory is far from being mathematically unique.* The presence of an infinite number of possibilities could be interpreted as the loss of predictability of string theory and the use of the anthropic principle was proposed [15]. The question, whether a physical theory that needs the anthropic principle still can be called a fundamental theory, was discussed in [16].

Our interpretation of the landscape is the following which is in agreement with [17]:
The anthropic principle should be avoided by all means in a fundamental theory, hence a new idea is needed and it is here where background independence could help. We notice that each landscape vacuum is based on a different background structure (flux charges, moduli). In addition, a landscape will

exist for each of the uncountably infinite number of target space background spacetimes.[1] Thus, the full string theory landscape is presumably labelled not only by flux vacua on a given spacetime but also by the background spacetimes. Suppose one could rigorously quantise string theory on all of these background structures. Then, if one knew how to combine all of these distinct quantum theories into a single one, thus achieving *background independence*, then the landscape would have disappeared. The understanding of the present author is that this is what M-theory is supposed to achieve but currently, to the best knowledge of the author, there seems to be no convincing idea for how to do that.

In this chapter we summarise the status of the quantum dynamics of LQG which has been the focal point of criticism in [12]. Our intention is to give a self-contained inside point of view of the subject which is complementary to [12] in the sense that we explain in some detail why the constructions used in LQG are physically well motivated and sometimes even mathematically unique, hence less ambiguous than described in [12]. We exactly define what is meant by canonical quantisation of general relativity, indicating explicitly the freedom that one has at various stages of the quantisation programme. We will see that the theory has much less freedom than [12] makes it look like. In particular, we evaluate "what has been gained in LQG as compared to the old geometrodynamics approach" and we will see that the amount of progress is non-trivial. We also include more recent results such as the master constraint programme [19] which tightens the implementations of the quantum dynamics and enables to systematically construct the physical inner product, which was not possible as of 3 years ago. Furthermore, in order to show that there is not only mathematical progress in LQG, we also fill the gaps that [12] did not report on such as the semiclassical sector of the theory, matter coupling, quantum black hole physics, quantum cosmology and LQG phenomenology. Finally we also show that the key technique that was used to make the Wheeler–DeWitt operator well defined [20] is also the underlying reason for the success of loop quantum cosmology (LQC) which is the usual cosmological minisuperspace toy model quantised with the methods of LQG.

Here we only sketch these results since we want to reach a rather general audience. All the technical details can be found in the comprehensive and self-contained monograph [2].

Remark

Since no theoretical Ansatz concerning quantum gravity has been yet brought to completion each of these Ansätze is understandably subject to criticisms.

[1] Currently string theory can be quantised only on a handful of target space background spacetimes, mostly only on those on which the theory becomes a free field theory. For instance, on $\mathrm{AdS}_5 \times \mathrm{S}^5$, which is much discussed in the context of the famous AdS/CFT correspondence [18], string theory is interacting and to the best knowledge of the author currently no quantum string theory on this spacetime was constructed.

This is clearly the case also for LQG, and in particular such criticisms can be found, e.g., in [12], or in the paper by Nicolai and Peeters [13] in this book. We want to stress that the purpose of this chapter, while in part a response to [12], is *not* to criticise [12]. In fact, the considerable effort of the non-expert authors of [12] to present LQG in as much technical detail as they did from a particle physicist's perspective is highly appreciated. Rather, what we have in mind is to draw a more optimistic picture than [12] did, to hopefully resolve confusions that may have arisen from gaps in [12] and to give a more complete picture of all the research being done in LQG than [12] did. The discussion will be kept objectively, problems with the present formulation of LQG will not be swept under the rug but rather discussed in great detail together with their possible solutions.

2 Classical Preliminaries

The starting point is a Lagrangean formulation of the classical field theory, say the Einstein–Hilbert–Lagrangean for general relativity. Hence one has an action

$$S = \int_M d^{n+1}X\ L(\Phi, \partial\Phi) \tag{1}$$

where L is the Lagrangean density, that is, a scalar density of weight one constructed in a covariant fashion from the fields Φ and their first partial derivatives[2] which is sufficient for gravity and all known matter. Here Φ stands for a collection of fields including the metric and all known matter. M is an $(n+1)$-dimensional, smooth differential manifold.

If one wants to have a well-posed initial value formulation, then the metric fields g that live on M are such that (M, g) is globally hyperbolic which implies [23] that M is diffeomorphic to the direct product $\mathbb{R} \times \sigma$ where σ is an n-dimensional smooth manifold.[3] Since the action (1) is invariant under Diff(M), the diffeomorphisms $Y:\ \mathbb{R} \times \sigma \to M;\ (t, x) \mapsto Y_t(x)$ are a symmetry of the action. For each Y we obtain a foliation of M into a one-parameter family of spacelike hypersurfaces $\Sigma_t = Y_t(\sigma)$. One now pulls all fields back by Y and obtains an action on $\mathbb{R} \times \sigma$. Due to the arbitrariness of Y, this action contains $n+1$ fields, usually called lapse and shift fields, which appear without time derivatives, they are Lagrange multipliers. The Legendre transformation is therefore singular and leads to constraints on the resulting phase space [25, 26]. They can be obtained by extremisation of the action with respect to the Lagrange multipliers.

Hence, after the Legendre transformation we obtain a phase space $\mathcal{M}$ of canonical fields ϕ which are the pull-backs to σ of the spacetime fields Φ

[2] Higher derivative theories can also be treated canonically [21]; however, they are generically pathological, that is, unstable [22].

[3] Unless otherwise stated we take σ to be compact without boundary in order to avoid a lengthy discussion of boundary terms.

together with their canonically conjugate momenta π. The symplectic manifold $\mathcal{M}$ with coordinates $(\phi(x), \pi(x))_{x\in\sigma}$ equipped with the corresponding canonical bracket[4] $\{\phi(x), \pi(x')\} = \delta^{(n)}(x, x')$ is a time t independent object.

As one can show by purely geometric arguments [27], these constraints are automatically of first class in the terminology of Dirac, that is, they close under their mutual Poisson brackets, irrespective of the matter content of the theory. As we will need them in some detail later on, let us display this so called Dirac algebra $\mathfrak{D}$ in more detail

$$\begin{aligned} \{D(\boldsymbol{N}), D(\boldsymbol{N}')\} &= 8\pi G_{\text{Newton}}\ D(\mathcal{L}_{\boldsymbol{N}}\boldsymbol{N}') \\ \{D(\boldsymbol{N}), H(N')\} &= 8\pi G_{\text{Newton}}\ H(\mathcal{L}_{\boldsymbol{N}}N') \\ \{H(N), H(N')\} &= 8\pi G_{\text{Newton}}\ D(q^{-1}(NdN' - N'dN) \end{aligned} \tag{2}$$

The notation is as follows: We distinguish between the so-called spatial diffeomorphism constraints $D_a(x),\ a = 1,..,n;\ x \in \sigma$ and the Hamiltonian constraints $H(x),\ x \in \sigma$. Notice that these are infinitely many constraints, $n+1$ per $x \in \sigma$. We smear them with test functions N^a, N, specifically $D(\boldsymbol{N}) = \int_\sigma d^3x\ N^a D_a$ and $H(N) = \int_\sigma d^3x\ NH$. Finally, q_{ab} is the pull-back to σ of the spacetime metric with inverse q^{ab} and $\mathcal{L}$ denotes the Lie derivative.

The algebra $\mathfrak{D}$ is *universal* and underlies the canonical formulation of any field theory based on an action which is Diff(M) invariant and contains general relativity in $n+1$ dimensions as, for instance, the closed bosonic string[5] [28].

As we can read off from (2), it has the following structure: The first line in (2) says that elements of the form $D(\boldsymbol{N})$ generate a subalgebra which can be identified with the Lie algebra diff(σ) of the spatial diffeomorphism group Diff(σ) of σ. This is why the $D(\boldsymbol{N})$, where the $\boldsymbol{N}$ are arbitrary smooth vector fields on σ of rapid decrease, are called spatial diffeomorphism constraints. The second line in (2) says that diff(σ) is not an ideal of $\mathfrak{D}$ because the Hamiltonian constraints $H(N)$, where the N are arbitrary smooth functions on σ of rapid decrease, are not diff(σ) invariant. The name Hamiltonian constraint stems from the fact that the Hamiltonian flow of this constraint on the phase space generates gauge motions which, when the equations of motion hold, can be identified with spacetime diffeomorphisms generated by vector fields orthogonal to the hypersurfaces Σ_t. Finally the third line in (2) says that (2) is not a Lie algebra in the strict sense of the word because, while the right-hand side of the Poisson bracket between two Hamiltonian constraints is a linear combination of spatial diffeomorphism constraints, the coefficients

[4] We suppress spatial tensor and internal Lie algebra indices.

[5] A closed bosonic string is an embedding of a circle $\sigma := S^1$ into a $D{+}1$ dimensional target space background manifold, mostly $D+1$ dimensional Minkowski space. The spacetime manifold of the string is therefore $M = \mathbb{R} \times S^1$, that is, a 2D cylinder. In 2D gravity is topological, hence gravity is also trivially contained in string theory.

in that linear combination have non-trivial phase space dependence through the tensor $q^{ab}(x)$.

A pecularity happens in the case $n = 1$, such as the closed bosonic string: In $n = 1$ dimensions, p-times contravariant and q-times covariant tensors are the same thing as scalar densities of weight $q - p$. For this reason, in contrast to $n > 1$ dimensions, in $n = 1$ dimensions the constraints come with a natural density weight of two rather than one while the smearing functions acquire density weight -1 rather than 0. One can think of this as if the actual constraints had been multiplied by a factor of $\sqrt{q}$ while the smearing functions had been multiplied by a factor of $\sqrt{q}^{-1}$ which, however, does not change the Poisson bracket because gravity is not dynamical in 2D. For this reason the factor q^{-1} must be absent in the third relation in (2) in order to match the density weights on both sides of the equations. This is why *only in 2D* the Dirac algebra $\mathfrak{D}$ is a true Lie algebra. In fact, using the combinations $V_\pm := H \pm D$ one realises that the Dirac algebra acquires the structure a direct sum of loop algebras (Lie algebra of the diffeomorphism group of the circle) $\mathfrak{D} \cong \text{diff}(S^1) \oplus \text{diff}(S^1)$, see [28] for all the details. Thus, *in 2D the Dirac algebra trivialises*. It does not even faintly display the complications that come with the non-Lie algebra structure of $\mathfrak{D}$ in realistic field theories, that is, $D = 4$. Therefore, any comparisons made between structures in 2D and 4D which hide this important difference are void of any lesson.

Proceeding with the general classical theory, what we are given is a phase space $\mathcal{M}$ subject to a collection of constraints $C_I,\ I \in \mathcal{I}$ where in our case the labelling set comprises the $N, \boldsymbol{N}$. These constraints force us to consider the constraint hypersurface $\overline{\mathcal{M}} := \{m \in \mathcal{M};\ C_I(m) = 0\ \forall\ I \in \mathcal{I}\}$. The closure of $\mathfrak{D}$ means that the Hamiltonian flow of the C_I preserves $\overline{\mathcal{M}}$. Since the C_I generate gauge transformations (namely spacetime diffeomorphisms), all the points contained in the gauge orbit $[m]$ through $m \in \overline{\mathcal{M}}$ must be identified as physically equivalent. As one can show in general [29], the set of orbits $\widehat{\mathcal{M}} := \{[m];\ m \in \overline{\mathcal{M}}\}$ is again a symplectic manifold and known as the reduced phase space.

It is mathematically more convenient to consider functions on all of $\mathcal{M}$ which are invariant under gauge transformations, called Dirac observables. Their restrictions to $m \in \overline{\mathcal{M}}$ are completely determined by $[m]$. The physical idea to construct such functions is due to Rovelli [30] and its mathematical implementation has been much improved recently in [31] (see also [32]). For a particularly simple realisation of this so-called "relational Ansatz" in terms of suitable matter, see [33]. We consider functions T_I on $\mathcal{M}$ which have the property that the matrix with entries $A_{IJ} := \{C_I, T_J\}$ is invertible (at least locally). Let X_I be the Hamiltonian vector field of the constraint $C'_I := \sum_J (A^{-1})_{IJ} C_J$. The set of constraints C'_I is equivalent to the set of the C_I but the C'_I have the advantage that the vector fields X_I are weakly (i.e. on $\overline{\mathcal{M}}$) mutually commuting. Now, given any smooth function f on $\mathcal{M}$ and any real numbers τ_I, in the range of the T_I, consider

$$O_f(\tau) := [\alpha_t(f)]_{t=\tau-T} \ , \quad \alpha_t(f) := [\exp(\sum_I t_I X_I) \cdot f] \tag{3}$$

Notice that one is supposed to first evaluate $\alpha_t(f)$ with t_I considered as real numbers and then evaluate the result at the phase space dependent point $t_I = \tau_I - T_I$. It is not difficult to show that (3) is a weak Dirac observable, that is $\{C_I, O_f(\tau)\}_{|\overline{\mathcal{M}}} = 0$. It has the physical interpretation of displaying the value of f in the gauge $T_I = \tau_I$. Equivalently, it is the gauge-invariant extension of f off the gauge cut $T = \tau$ and in fact can be expanded in a power series in $\tau - T$ by expanding the exponential function in (3).

The relational Ansatz solves the problem of time of canonical quantum gravity: By this one means that in generally covariant systems there is no Hamiltonian, there are only Hamiltonian constraints. Since the observables of the theory are the gauge-invariant functions on phase space, that is, the Dirac observables, "nothing moves in canonical quantum gravity" because the Poisson brackets between the Hamiltonian constraints and the Observables vanishes (weakly) by construction. The missing evolution of the Dirac observables $O_f(\tau)$ is now supplied as follows: Using the fact that the map α_t in (3) is actually a Poisson automorphism (i.e. a canonical transformation) one can show that (1) if the phase space coordinates can be subdivided into canonical pairs (T_I, π_I) and (q^a, p_a) and (2) if f is a function of only[6] q^a, p_a then the evolution in τ_I has a Hamiltonian generator [32]. That is, there exist Dirac observables $H_I(\tau)$ such that $\partial O_f(\tau)/\partial \tau_I = \{O_f(\tau), H_I(\tau)\}$.

The task left is then to single out a one-parameter family $s \mapsto \tau_I(s)$ such that the corresponding Hamiltonian

$$H(s) = \sum_I \dot{\tau}_I(s) H_I(\tau(s)) \tag{4}$$

is positive, s-independent, and reduces to the Hamiltonian of the standard model on flat space. This has been achieved recently in [33] using suitable matter which supplies the clocks T_I with the required properties. It follows that the gauge-invariant functions $O_f(s)$ then evolve according to the physical Hamiltonian H. Moreover, they satisfy the algebra $\{O_f(s), O_{f'}(s)\} = O_{\{f,f'\}}(s)$ because the s evolution has the canonical generator H.

2.1 Summary

Classical canonical gravity has a clear conceptual and technical formulation with no mysteries or unsolved conceptual problems. Certainly classical general relativity is not an integrable system and thus not everything is technically solvable (for instance, not all solutions to the field equations are known)

6 Nothing is lost by this assumption because T_I is pure gauge and the constraints can be solved for π_I in terms of q^a, p_a, T_I.

but one exactly knows what to do in order to try to solve a given problem. The canonical formulation that we have used here for a generally covariant field theory is widely used in numerical general relativity with great success. General covariance is manifestly built into the framework and is faithfully represented in terms of the Dirac algebra $\mathfrak{D}$ (2) which is the key object to construct the invariants (3) of the theory and the physical Hamiltonian H (4) according to which they evolve. At no point in those constructions did one use a background metric or did one violate spacetime diffeomorphism invariance. This is because, while one did use a split of spacetime into space and time, one did consider all splits simultaneously which is reflected in the constraints that in turn enforce spacetime diffeomorphism invariance.

For clarity we mention that diffeomorphism invariance should not be confused with Poincaré invariance. Poincaré invariance is an invariance of a special solution to Einstein's vacuum equations. It is not a symmetry or a gauge invariance of the theory. The gauge group is Diff(M) which is a background metric independent object because it only refers to the differential manifold M but to no metric. In fact, if σ is compact as appropriate for certain cosmological models, then the Poincaré group $\mathcal{P}$ has no place in the theory. If M is equipped with asymptotically flat boundary conditions then in fact one can in addition define Poincaré generators of $\mathcal{P}$ as functions on phase space, called ADM charges [24]. These are particular Dirac observables. Notice that $\mathcal{P}$ is not contained in Diff(M) because diffeomorphisms are of rapid decrease at spatial infinity (at least they vanish there). This must be because $\mathcal{P}$ is a symmetry and not a local gauge invariance like Diff(M).

3 Canonical Quantisation Programme

The programme of canonical quantisation is a mathematical formalism which seeks to provide a quantum field theory from a given classical field theory. There are several choices to be made within the formalism and the outcome depends on it. This applies to ordinary field theories such as free scalar fields on Minkowski space as well as to more complicated situations. In the presence of constraints such as in general relativity one would ideally solve the constraints classically before quantising the theory. That is, one studies the representation theory of the algebra of invariants such as (3). Unfortunately, this is generically too difficult because the algebra of invariants is complictated and thus usually prevents one from using standard representations for simple algebras such as a Fock representation for usual CCR (canonical commutation relation) or CAR (canonical anticommuttion relation) algebras.

Thus, in order to start the quantisation process one follows Dirac [25] and starts with a redundant set of functions on phase space which generate a sufficiently simple Poisson algebra so that suitable representations thereof can be found. These functions are not gauge invariant but provide a system of coordinates for $\mathcal{M}$. Then, in a second step, provided that the constraints themselves can be represented on the chosen, so-called "kinematical", Hilbert space as closable[7] and densely defined operators, one looks for the generalised joint kernel of the constraint operators. Here generalised refers to the fact that the joint kernel typically has trivial intersection with the Hilbert space, that is, the non-zero solutions of the constraints are not normalisable. Rather, they are elements of the physical Hilbert space which is not a subspace of the kinematical Hilbert space. The physical Hilbert space is induced from the kinematical Hilbert space by applying standard spectral theory to the constraint operators. Once the physical Hilbert space is known, at least in principle, it automatically carries a self-adjoint representation of the algebra of strong observables, that is, those operators that commute with all quantum constraints and for which (3) and (4) are the classical counterparts.

All of this is of course difficult, if not impossible, to carry out exactly and in full completeness for general relativity because, after all, one is dealing with a rather non-linear and highly interacting QFT. Hence, in praxis one will have to develop and rely on approximation schemes. However, these are only technical difficulties coming from the complexity of the theory. There are no in principle obstacles, the programme of canonical quantisation follows a clear sequence of steps at each of which one knows exactly what one has to do and sometimes one has a certain freedom which one will exploit using physical intuition.

After the above sketch of the programme, we will now become somewhat more detailed and pin down explicitly the freedom that one has and the choices that one has to make.

The starting point is then a symplectic manifold $\mathcal{M}$ subject to real valued, first-class constraints $C_I,\ I \in \mathcal{I}$. That is, we have $\{C_I, C_J\} = f_{IJ}\ ^K\ C_K$ for some, possibly phase space dependent functions $f_{IJ}\ ^K$, called structure functions. We will assume for simplicity, as it is the case in general relativity, that we are dealing with a completely parametrised system, that is, there is no a priori gauge-invariant Hamiltonian. In order to simplify the discussion for the purposes of this short review, we display here for concreteness a recently proposed strategy [19, 34] to deal with those constraints: Consider instead the individual constraints C_I the single master constraint $M := \sum_I C_I K_{IJ} C_J$. Here $K = (K_{IJ})$ is a positive definite matrix valued function on phase space. The master constraint contains the same information about the gauge redundancy of the system as the individual C_I since $M = 0$ is equivalent with $C_I = 0$ for all I and the equation $\{O, \{O, M\}\}_{M=0}$ is equivalent with $\{O, C_I\}_{M=0}$ for all I. Hence the master constraint selects the same reduced phase space as the original set of constraints. The reason for using the matrix K is that we can and often must use the associated freedom to regularise the square of the constraints: namely, typically the C_I become operator-valued distributions and their square is therefore ill-defined. By a judicious choice of K (which also becomes an operator) one can remove the corresponding UV singularity. See, e.g., [35] for examples.

[7] That is, the adjoint is also densely defined.

Given this set-up, the programme of canonical quantisation consists of the following:[8]

I. *Algebra of elementary functions* $\mathfrak{E}$
Select a Poisson sub$^*-$algebra $\mathfrak{E}$ of $C^\infty(\mathcal{M})$, called elementary functions, which separates the points of $\mathcal{M}$. That is, $\mathfrak{E}$ should be closed under taking Poisson brackets and complex conjugation and for any $m \neq m'$ there exists $e \in \mathfrak{E}$ such that $e(m) \neq e(m')$. The latter property implies that any $f \in C^\infty(\mathcal{M})$ can be thought of as a function of the elements of $\mathfrak{E}$ so that $\mathfrak{E}$ is a sytstem of coordinates for $\mathcal{M}$ (which maybe redundant). The choice of $\mathfrak{E}$ will be guided by mathematical convenience and physical intuition: One will try to use bounded functions, such as the Weyl elements used in free field theories, in order to deal with bounded operators later on which avoids domain questions. Of course, the algebra $\mathfrak{E}$ should be sufficiently simple in order that one can manage to find representations of the corresponding quantum algebra at all. Furthermore, one will choose $\mathfrak{E}$ in such a way that its elements transform in a simple way under the gauge group of the system in question.

II. *Quantum* $^*-$*algebra* $\mathfrak{A}$
One now constructs a $^*-$algebra $\mathfrak{A}$ using the following well-known procedure: We consider the free algebra $\mathfrak{F}$ of finite linear combinations of formal words. A word is a formal finite sequence of elements $w = (e_1..e_N)$. Multiplication of words consists in combining sequences, e.g. $w \cdot w' = (e_1..e_N) \cdot (e'_1..e'_{N'}) := (e_1..e_N e'_1..e'_{N'})$. The involutive structure is defined by $w^* := (e^*_N..e^*_1)$. We now consider the two-sided ideal $\mathfrak{I}$ generated by elements of the form (1). $ee' - e'e - i\hbar\{e,e'\}$ and (2) $e^* - \overline{e}$. The quantum algebra is the quotient $\mathfrak{A} := \mathfrak{F}/\mathfrak{I}$.

III. *Kinematical Hilbert space*
Next we study the representation theory of $\mathfrak{A}$. As is well known, for field theories such as general relativity the number of unitarily inequivalent representations is usually uncountably infinite. For instance, all Fock representations of a free massive scalar field with different masses are unitarily inequivalent. This follows by a simple application of Haag's theorem [6]. Hence, in order to select from this multitude of possibilities one must use dynamical input, such as the mass in the scalar field example. In the case of the presence of the constraints, dynamical input into the representation problem is provided for instance by asking that (parts of) the gauge group be represented unitarily on the corresponding Hilbert space or that the constraints be represented at all as closable and densely defined operators, possibly subject to some choice of factor ordering and maybe after some sort of regularisation and renormalisation. For the purpose of this discussion it will be sufficient to insist that the kinematical Hilbert space $\mathcal{H}$ carries the master constraint operator $\widehat{M}$ as a positive and self-adjoint operator.

---

[8] What follows is still a simplified version. See [2] for a complete discussion.

IV. *Physical Hilbert space*

The idea to solve the master constraint is to apply spectral theory to it [34]. Suppose that the Hilbert $\mathcal{H}$ decomposes into separable $\widehat{M}$-invariant subspaces $\mathcal{H}_\theta$ where θ labels the corresponding sectors. Then it is well known that $\mathcal{H}_\theta$ is unitarily equivalent to a direct integral Hilbert space

$$\mathcal{H}_\theta \cong \mathcal{H}_\theta^\oplus := \int_{\mathrm{spec}(\widehat{M})}^\oplus d\mu(\lambda)\, \mathcal{H}_\lambda^\theta \tag{1}$$

Here the measure class μ on the spectrum $\mathrm{spec}(\widehat{M})$ of the master constraint is unique and the multiplicities $\dim(\mathcal{H}_\lambda^\theta)$ are unique up to μ-measure zero sets. The unitary map $U : \mathcal{H}_\theta \to \mathcal{H}_\theta^\oplus;\ \psi \mapsto (\tilde{\psi}(\lambda))_\lambda$ is a generalisation of the Fourier transform and is such that $U\widehat{M}\psi = (\lambda\tilde{\psi}(\lambda))_\lambda$, that is, $U\widehat{M}U^{-1}$ is represented as multiplication by λ on $\mathcal{H}_\lambda^\theta$. We have

$$<\psi,\psi'>_{\mathcal{H}_\theta}=<U\psi,U\psi'>_{\mathcal{H}_\theta^\oplus}=\int_{\mathrm{spec}(\widehat{M})} d\mu(\lambda)\ <\tilde{\psi}(\lambda),\tilde{\psi}'(\lambda)>_{\mathcal{H}_\lambda^\theta} \tag{2}$$

The physical Hilbert space is now defined as

$$\mathcal{H}_{\mathrm{phys}} := \oplus_\theta\, \mathcal{H}_{\lambda=0}^\theta \tag{3}$$

There are several remarks in order about (3):

1. In order that this works one must Lebesgue decompose every space $\mathcal{H}_\theta$ into the $\widehat{M}$-invariant pure point, absolutely continious and continuous singular pieces and then decompose them as a direct integral.
2. The spaces $\mathcal{H}_{\lambda=0}^\theta$ are uniquely determined in the pure point case but in the absolutely continuous case (and continuous singular case, which usually is absent in practice) further input is required because here the set $\{\lambda = 0\}$ is of μ-measure zero. Roughly speaking, one requires that the space $\mathcal{H}_0^\theta$ carries a non-trivial, irreducible representation of the algebra of (strong) observables. See [34] for details.
3. Due to a bad choice of factor ordering involved in the construction of $\widehat{M}$ it may happen that $0 \notin \mathrm{spec}(\widehat{M})$. This typically happens when the quantum constraints $\widehat{C}_I$ that enter the definition of $\widehat{M}$ are anomalous, that is, if they do not close as a quantum algebra. This can easily happen especially in the case that the classical constraint algebra involves non-trivial structure functions rather than structure constants. Hence, although the master constraint always trivially forms a non anomalous algebra, possible anomalies in the original algebra are detected by it, so nothing is swept under the rug. In this case, following [36], we propose to replace $\widehat{M}$ by $\widehat{M}' = \widehat{M} - \lambda_0$ where $\lambda_0 = \min(\mathrm{spec}(\widehat{M}))$. Here λ_0 should be finite and $\lim_{\hbar\to 0}\lambda_0 = 0$ in order that both $\widehat{M}$, $\widehat{M}'$ have the same classcial limit. This has worked so far in all studied cases [35] where λ_0 is related to a reordering or normal ordering of the constraints into a non-anomalous form.

4. In case that the constraints can be exponentiated to a Lie group $\mathfrak{G}$ one can avoid the construction of the master constraint and apply a more heuristic technique called group averaging [37]. This is at most possible if the constraints form an honest Lie algebra with structure constants rather than structure functions. Since we may assume without loss of generality that the constraints and the structure constants are real valued, we assume that we are given a unitary representation U of $\mathfrak{G}$ on $\mathcal{H}$. Assume also that there is a Haar measure ν on $\mathfrak{G}$, that is, a not necessarily normalised but bi-invariant (with respect to group translations) positive measure on $\mathfrak{G}$. Fix a dense domain $\mathcal{D}$ and let $\mathcal{D}^*$ be the algebraic dual of $\mathcal{D}$, that is, linear functionals on $\mathcal{D}$ with the topology of pointwise convergence of nets. We define the rigging map

$$\eta : \mathcal{D} \to \mathcal{D}^*;\ f \mapsto \int_{\mathfrak{G}} d\nu(\mathfrak{g}) \ < U(\mathfrak{g})f, . > \tag{4}$$

The reason for restricting the domain of η to a dense subset $\mathcal{D}$ of $\mathcal{H}$ is that in general only then (4) defines an element of $\mathcal{D}^*$.
The image of η are solutions to the constraints in the sense that[9]

$$[\eta(f)](U(\mathfrak{g})f') = [\eta(f)](f') \tag{5}$$

for all $\mathfrak{g} \in \mathfrak{G}$ and all $f' \in \mathcal{D}$. Notice that if we would identify the distribution $\eta(f)$ with the formal vector

$$\eta'(f) := \int_{\mathfrak{G}} d\nu(\mathfrak{g})\ U(\mathfrak{g})f \tag{6}$$

then its norm diverges unless ν is normalisable, that is, unless $\mathfrak{G}$ is compact so that $\eta'(f)$ is not an element of $\mathcal{H}$ in general. However, formally we have $< \eta'(f), f' >= [\eta(f)](f')$ and thus

$$< U(\mathfrak{g})\eta'(f), f' >=< \eta'(f), U(\mathfrak{g}^{-1})f' >= \eta(f)[U(\mathfrak{g}^{-1})f'] =< \eta'(f), f' > \tag{7}$$

for all $\mathfrak{g},\ f'$. Thus formally $U(\mathfrak{g})\eta'(f) = \eta'(f)$ which shows that $\eta'(f)$ is a (generalised, since not normalisable) eigenvector of all the $U(\mathfrak{g})$ with eigenvalue equal to one as appropriate for a solution to the constraints. Hence (5) is the rigorous statement of the formal computation (7).
We define the physical inner product on the image of η by

$$< \eta(f), \eta(f') >_{\text{phys}} := \eta(f')[f] \tag{8}$$

and the physical Hilbert space is the completion of $\eta(\mathcal{D})$ in the corresponding norm.[10]

[9] In general, given an operator O which together with its adjoint $O^\dagger$ is densely defined on $\mathcal{D} \subset \mathcal{H}$ and preserves $\mathcal{D}$ we define the dual O' on the algebraic dual $\mathcal{D}^*$ by $[O'l](f) := l(O^\dagger f)$ for all $f \in \mathcal{D}$.

[10] Provided that (8) is positive semidefinite and with removal of zero norm vectors understood.

5. The spectral decomposition solution of the constraint can be seen as a special case of group averaging in the sense that in case of a single self-adjoint constraint $\widehat{M}$ we can indeed exponentiate it to obtain a one-parameter unitary, Abelean group $U(t) = \exp(it\widehat{M})$. The Haar measure in this case would seem to be the Lebesgue measure $d\nu(t) = dt/(2\pi)$. We then formally have (we drop the label θ)

$$\begin{aligned} < \eta(f), \eta(f') >_{\text{phys}} &= \int_{\mathbb{R}} d\nu(t) \ < U(t)f', f > \\ &= \int_{\mathbb{R}} d\nu(t) \int_{\text{spec}(\widehat{M})} d\mu(\lambda)\ e^{-it\lambda} < \tilde{f}'(\lambda), \tilde{f}(\lambda) >_{\mathcal{H}_\lambda} \\ &= \int_{\text{spec}(\widehat{M})} d\mu(\lambda) \ < \tilde{f}'(\lambda), \tilde{f}(\lambda) >_{\mathcal{H}_\lambda} \int_{\mathbb{R}} d\nu(t)\ e^{-it\lambda} \\ &= c < \tilde{f}'(0), \tilde{f}(0) >_{\mathcal{H}_0} \end{aligned} \tag{9}$$

where $c = [\int_{\text{spec}(\widehat{M})} d\mu(\lambda)\ \delta(\lambda)]$. This calculation is formal in the sense that we have interchanged the sequence of the integrations. Also the constant c can be vanishing or divergent which is one of the reasons why group averaging is only formal. For instance in the case of a pure point spectrum the appropriate measure is not the Lebesgue measure but rather the Haar measure on the Bohr compactification of the real line. See [34] for the details. However, at least heuristically one sees how these methods are related.

This ends the outline of the quantisation programme. We now apply it to general relativity.

4 Status of the Quantisation Programme for Loop Quantum Gravity (LQG)

In this section we describe to what extent the canonical quantisation programme has been implemented for general relativity, that is, we give the status of loop quantum gravity. As an aside we sketch the historical development of the subject. We will mostly consider pure gravity; matter coupling works completely similar [38]. Also, in order to avoid technicalities about boundary terms (which can be dealt with [38]) consider compact σ without boundary unless stated otherwise.

4.1 Canonical Quantum Gravity before LQG

The canonical quantisation of general relativity in terms of the ADM variables [4] culminated in the seminal work by DeWitt [39] which *formally* carried out many of the steps outlined in the previous section. These crucial papers laid the foundations for a substantial amount of work on the canonical quantisation of general relativity that followed. However, the stress is here on the word *formally*. We mention just some problems with these pioneering papers.

1. *Kinematics:*
   The Hilbert space representation used there was given in terms of a formal path integral which must be called ill-defined by the standards of a mathematical physicist. For instance, the "measure" was defined to be an infinite Lebesgue measure $[Dq]$ over a space of three metrics, an object that does not exist mathematically; the integration space, which should be given the appropriate structure of a measurable space was not specified etc.
   Nonetheless, if one defines the three metric operator as a multiplication operator and the conjugate momentum operator as a functional differential operator times $i\ell_P^2$ then one arrives at a formal representation of the canonical commutation relations such that the canonical coordinates are represented as formally symmetric operators. Moreover, the formal Lebesgue measure is formally invariant under infinitesimal spatial diffeomorphisms.
2. *Dynamics*
   2a. *Spatial Diffeomorphism Invariance:*
   In order to solve the spatial diffeomorphism constraint one can assume that wave functions are normalisable functionals of spatially diffeomorphism invariant functions of the tree metric such as integrals over σ of scalar densities of weight one constructed from the metric, the curvature tensor and all its covariant derivatives. In order that those derivatives make sense one must assume that the functional integral is over smooth three metrics. However, even if the wave function is, say, of the form $\exp(-\int_\sigma d^3x\sqrt{\det(q)})$ which is damped for large q then the functional integral is ill-defined: Due to spatial diffeomorphism invariance of the wave function and measure, the infinite volume of $\mathrm{Diff}(\sigma)$ must be factored out. But even after that, the space of smooth metrics is typically of measure zero with respect to the Gaussian type measure $[Dq]\ \exp(-2\int_\sigma d^3x\sqrt{\det(q)})$. Finally the function $\det(q)$ can stay small while components of q_{ab} can become large, hence the exponent has flat directions so that the integral also has divergent modes. Hence the norm of these type of states are dangerously close to being either plain infinite or plain zero.
   2b. *Hamiltonian Constraints*
   The infinite number of Hamiltonian constraints were formally given as a functional differential equation of second order, which goes by the famous name Wheeler–DeWitt equations. However, these "operators", which are really products of operator-valued distributions multiplied at the same point in σ, are hopelessly divergent on the space of wave functions just specified where the divergence really orginates from

the product of operator-valued distributions. There was no "normal ordering" or renormalisation possible because no exact vaccum state could be found with respect to which one should normally order.

It is therefore fair to say that canonical quantum gravity got stuck at the level of [39] in the mid-1960s.

4.2 The New Phase Space

In a sense, in terms of the ADM variables one could never even find a proper, background-independent representation of the canonical commutation relations. Thus, even leaving the dynamics aside, one could never even finish the kinematical part of the programme.

With the advent of the new variables [40] there was new hope. Initially the new variables consisted, instead of a three metric and (essentially) the extrinsic curvature as a canonical pair, of an $SL(2,\mathbb{C})$ connection $A^{\mathbb{C}}$ and an imaginary $sl(2,\mathbb{C})$ valued, pseudo-two-form[11] $E^{\mathbb{C}}$. This was attractive because the Hamiltonian constraint, after multiplying it with the non-polynomial factor[12] $\sqrt{\det(q)}$, becomes a *fourth order polynomial* $\tilde{H} = \sqrt{\det(q)}H$ in terms of $A^{\mathbb{C}}$, $E^{\mathbb{C}}$ which is no worse than in Yang–Mills theory. Hence the dynamics seemed to be drastically simplified as compared to the ADM formulation with its non-polynomial Hamiltonian constraint.

The catch, however, was in the reality conditions: Namely, in order to deal with real rather than complex general relativity one had to impose the reality conditions

$$\overline{A^{\mathbb{C}}} + A^{\mathbb{C}} = 2\Gamma, \ \overline{E^{\mathbb{C}}} + E^{\mathbb{C}} = 0 \tag{1}$$

where Γ is spin connection of the triad e determined by the three metric. Since essentially $E^{\mathbb{C}} = -i\sqrt{\det(q)}e$ it follows that Γ and thus (1) take a highly non-polynomial form. In fact, Γ is a fraction whose numerator and denominator are homogeneous polynomials of degree three in $E^{\mathbb{C}}$ and its first partial derivatives. It is clear that to find a representation of the formal $^*-$algebra $\mathfrak{A}$ with (1) as $^*-$relations is hopeless and to date nobody was successful.

Despite this, in [41, 42] an honest representation for a canonical theory based on an $SU(2)$ connection A and a real $su(2)$ valued pseudo-two-form E was constructed.[13] More precisely, [41] constructs a measurable space of generalised (distributional) connections $\overline{\mathcal{A}}$ which turns out to be the Gel'fand spectrum of an Abelean C^*–subalgebra of the corresponding kinematical algebra $\mathfrak{A}$. In [42] a (regular, Borel, probability) measure μ_0 on $\overline{\mathcal{A}}$ was constructed.

---

[11] A pseudo-two-form is dual, via the totally antisymmetric, metric-independent symbol, to a vector density.

[12] The determinant of the three metric is required to be everywhere non-vanishing classically, hence the modified constraint captures the same information about the reduced phase space as the original one.

[13] That in this representation the pseudo-two-form is indeed an essentially self-adjoint operator-valued distribution was only shown later in [43].

Thus, the corresponding Hilbert space $\mathcal{H} := L_2(\overline{\mathcal{A}}, d\mu_0)$ is a space of square integrable functions on $\overline{\mathcal{A}}$. As expected [44], the space of classical (smooth) connections $\mathcal{A}$ is contained in a measurable subset of $\overline{\mathcal{A}}$ of measure zero. Hence, any (formal) state which requires to be restricted to smooth connections in order that, say the Hamiltonian constraint be defined on it, has zero norm and thus can be discarded from $\mathcal{H}$. For the first time, these and related questions could be answered with absolute precision.

However, what does this Hilbert space have to do with general relativity if the true phase space is in terms of $SL(2,\mathbb{C})$ plus complicated reality conditions rather than $SU(2)$ with simple reality conditions? In [45] it was pointed out that the Hilbert space $\mathcal{H}$ can still be considered as a representation space for the quantum kinematics of general relativity. In fact, the connection A and pseudo-two-form E are related to triad e and extrinsic curvature K by[14]

$$A_a^j = \Gamma_a^j + \beta K_{ab} e_j^b \ , \ E_j^a = \sqrt{\det(q)} e_j^a / \beta \tag{2}$$

where $a, b, c, .. = 1, 2, 3$ are spatial tensor indices, where $j, k, l, .. = 1, 2, 3$ are $su(2)$ Lie algebra indices and the real number β is called the Immirzi parameter [46]. For any (non-vanishing) value of β, the variables (A, E) are canonically conjugate and thus can be used as a kinematical starting point for the quantisation programme.

The price to pay is that, in order to keep it polynomial, one has to multiply the Hamiltonian constraint (which of course depends explicitly on β) by a sufficiently large power of $\det(q) = |\det(E)|$. This was considered to be rather unattractive because these very high degree polynomials would intuitively drastically worsen the UV singularity structure of the Hamiltonian constraint as compared to the ADM formulation. In fact, this UV problem was already noticed at a rather formal level with the quantum version of $\tilde{H}$ in terms of the complex variables [47]: All the formal solutions to the Hamiltonian constraint were solutions at the regularised level only (in some ordering). When taking the (point splitting) regulator away, the result would be of the type zero times infinity. These problems were expected to even worsen when increasing the polynomial degree of the Hamiltonian constraint. Hence, the initial excitement that formally Wilson loop functions of smooth and non-intersecting loops were formally annihilated by the Hamiltonian constraint dropped significantly.

Hence, a critic could have said at this point,

> *You have made the theory more complicated and you have not gained anything: You may have a rigorous kinematical framework but that framework does not support the quantum dynamics of the theory.*

In [48] it was demonstrated how these obstacles can be overcome:
One can show that general relativity or any other background-independent quantum field theory is *UV self-regulating* provided one equips the Hamiltonian constraint with its natural density weight equal to one as it automatically

[14] If the $SU(2)$ Gauss law holds.

appears in the classical analysis. Notice that the Hamiltonian of the standard model on Minkowski space has density weight two rather than one. This is the reason why in background-dependent quantum field theories UV singularities appear. One can intuitively understand this as follows: In background-dependent theories, Hamiltonians are spatial integrals over sums of products of operator-valued distributions evaluated at the same point. Such products are therefore divergent. In background-independent theories such polynomials P also appear; however, they appear as numerators in a fraction P/Q. The denominator Q of that fraction is an appropriate power of $\sqrt{\det(q)}$ such that P/Q is a scalar density of weight one. As one can show, if the numerator has the singularity structure of the $(n+1)$th power of the δ-distribution[15] (and its spatial derivatives) then the denominator has the singularity structure of the nth power. This must be the case in order that the oparator valued distribution has the correct density weight. Hence, in a proper (point splitting) regularisation of P/Q one can, intuitively speaking, "factor out" out n of those δ-distributions and one is left with a well-defined integral after removing the regulator.

In other words, it was wrong to assume that the Hamiltonian constraint should be polynomial. Rather, *it must be non-polynomial* in order that it is well defined. Of course, the details are not as simple as that and we will explain the open issues in the next section. However, even at this stage one can say,

> *What has been gained is that not only a rigorous kinematical framework has been erected, that framework also supports the quantum dynamics. In particular, the original problem of the reality conditions is completely resolved.*

One of the most important issues is whether that dynamics defined by the final, regulator free, Hamiltonian constraint operator, which underwent rather non-trivial regularisation steps until one removed the regulator, reduces to the classical one in an appropriate classical limit. We will have much to say about this point in subsequent sections.

Before we close this section, let us comment on some criticism that one might have encountered [49, 50]: The complex connection is actually the pull-back to σ of the (anti) self-dual part of the 4D spin connection. Hence it has a covariant interpretation. The real valued connection is not related to a covariant action as simply as that. The relation is as follows: Additional to the Palatini action one considers a term which is topological on shell which amounts to the total action

$$S = \int_M F_{IJ} \wedge *(e^I \wedge e^J) + \frac{1}{\beta} \int_M F_{IJ} \wedge (e^I \wedge e^J) \tag{3}$$

Here $I, J, K, .. = 0, 1, 2, 3$ are Lorentz indices, e^I is the cotetrad one form and F_{IJ} is the curvature for a Lorentz connection A_{IJ}. The first term in (3)

[15] Notice that the δ-distribution $\delta(x, y)$ transforms as a density of weight one in say x and as a scalar in say y.

is the Palatini action. The second term is a total derivative when substituting the equation of motion for the connection A^{IJ}. Now when performing the Legendre transform of (3) one encounters second-class constraints [26]. These must be eliminated by using the Dirac bracket or by partially fixing the Lorentz gauge symmetry $SO(1,3)$ (or its universal cover $SL(2,\mathbb{C})$) to $SO(3)$ (or $SU(2)$) respectively, called the time gauge.

Using the Dirac bracket leads to a Poisson structure with respect to which connections are not Poisson commuting. Hence, while manifestly originating from a covariant action, the Lorentz connections cannot be used as a configuration space in the quantisation programme, that is, they cannot be represented as (commuting) multiplication operators [50]. In fact, to date there is no honest representation based on Lorentz connections available. On the other hand, the time gauge immediately leads to the phase space description sketched above with Immirzi parameter β. The manifest covariant origin of the phase space is lost due to the gauge fixing of the Lorentz group [49], however, one can show easily [2, 3] that symplectic reduction with respect to the $SU(2)$ Gauss constraint results in the manifestly covariant ADM phase space. Hence, both criticisms are of purely aesthetical nature and do not give rise to either an obstacle or an insight concerning the quantisation.

4.3 Quantum Kinematics

Elementary Functions

Having convinced ourselves that the cotangent bundle $\mathcal{M} := T^*(\mathcal{A})$ over the space of smooth $SU(2)$ connections is an appropriate kinematical phase space of general relativity we are supposed to choose an appropriate Poisson $^*-$subalgebra $\mathfrak{E}$ of elementary functions. Experience from lattice gauge theory [51] shows that it is convenient to work with $SU(2)$ valued magnetic holonomies

$$A(e) := \mathcal{P}\exp(\int_e A) \tag{4}$$

and real valued electric fluxes

$$E_f(S) := \int_S \mathrm{Tr}(\mathrm{n} * \mathrm{E}) \tag{5}$$

Here e is a path in σ, S is a two surface in σ and n is a Lie algebra valued scalar.[16] These functions separate the points of $\mathcal{M}$ since $G = SU(2)$ is compact [52]. Moreover, they satisfy the reality conditions

$$\overline{A(e)} = [A(e^{-1})]^T \ , \ \ \overline{E_n(S)} = E_n(S) \tag{6}$$

[16] For simplicity we assume that the $SU(2)$ principal bundle is trivial which is always possible. The final quantum theory turns out not to be affected by this assumption. The paths and surfaces are piecewise analytic for technical reasons.

as well as the Poisson brackets

$$\{A(e), A(e')\} = 0 \ , \ \ \{E_f(S), A(e)\} = 8\pi G_{\text{Newton}} \beta A(e_1) f(S \cap e) A(e_2) \quad (7)$$

Here we have assumed that e and S intersect transversally in an interior point of both S, e thus splitting the path e at $S \cap e$ as $e = e_1 \circ e_2$ and G is Newton's constant. Similar formulae can be derived if S, e intersect in a more complicated way.

The algebra $\mathfrak{E}$ can now be described as follows: Consider the algebra Cyl of cylindrical functions, that is, those which depend on a finite number of holonomies only. Hence, a cylindircal function is of the form $f(A) = f_\gamma(\{A(e)\}_{e\in E(\gamma)})$ where γ is an oriented graph (a collection of paths, called edges, which intersect in their end points only), $E(\gamma)$ denotes the set of edges of γ and f_γ is a complex valued function on $SU(2)^N$ where N is the number of edges of γ. Next, consider the vector field $u_{S,n}$, considered as a derivation on Cyl, defined by

$$u_{S,n}[f] := \{E_n(S), f\} = \sum_{e\in E(\gamma)} \{E_n(S), [A(e)]_{mn}\} \frac{\partial f_\gamma}{\partial [A(e)]_{mn}} \quad (8)$$

The algebra $\mathfrak{E}$ is now defined as the Lie algebra generated by the pairs (f, u) where $f\in$Cyl and u is a derivation on Cyl which is either of the form of a finite linear combinations of the $u_{n,S}$ or which is generated from those by the Lie bracket $\{(f,u),(f',u')\} = (u[f'] - u'[f], [u,u'])$ where $[u, u']$ denotes the commutator of vector fields.

Notice that, in this sense, the Poisson bracket between the $E_n(S)$ is generically non-vanishing. Its result is such that, formally, the Jacobi identity holds in $\mathfrak{E}$. The reason for this is that we do not smear the fields in three but in less dimensions. If we would smear in three dimensions as usual, then the smeared electric fields would Poisson commute. See [55] for more details on this point. The reason for why we do not smear the fields in three dimensions is due to the fact that it is natural in a background-independent theory to smear one form in one dimension and two forms in two dimensions. This way we do not need a background metric in order raise or lower indices. Moreover, our holonomies and fluxes transform in a simple way under the kinematical part of the gauge group, that is, $SU(2)$ gauge transformations and spatial diffeomorphisms. In fact, consider the smeared Gauss constraint and spatial diffeomorphism constraint respectively given by

$$G(\Lambda) := \int \sigma \, d^3x \, \Lambda^j \, G_j \ , \ \ D(v) := \int \sigma \, d^3x \, v^a \, C_a \quad (9)$$

(where Λ, v are test functions) where[17]

$$G_j = \mathrm{Tr}(\tau_j \mathcal{D}_a E^a) \ , \ \ D_a = \mathrm{Tr}(F_{ab} E^b) \quad (10)$$

[17] $\mathcal{D}$ and F are respectively the covariant differential and curvature determined by A and τ_j, $j = 1, 2, 3$ denotes a basis of $su(2)$.

and the one-parameter families of canonical transformations generated by them. These are explicitly given by, say on $f \in \mathrm{Cyl}$,

$$\begin{aligned} \alpha_{\exp(t\Lambda)}(f) &:= \sum_{n=0}^{\infty} \frac{t^n}{n!}\{C(\Lambda), f\}_{(n)} \\ \alpha_{\varphi_v^t}(f) &:= \sum_{n=0}^{\infty} \frac{t^n}{n!}\{C(v), f\}_{(n)} \end{aligned} \tag{11}$$

where $b(e)$, $f(e)$ denote respectively the beginning and final point of e. Here $t \mapsto \varphi_v^t$ is the one-parameter family of diffeomorphisms generated by v. It is not difficult to see that (11) is the restriction to local gauge transformations of the form $g = \exp(t\Lambda)$ and spatial diffeomorphisms of the form $\varphi = \varphi_v^t$ of the following action of the semidirect product $\mathfrak{G} = \mathcal{G} \rtimes \mathrm{Diff}(\sigma)$ on Cyl given by

$$\begin{aligned} [\alpha_g(f)](A) &= f_\gamma(\{g(b(e))A(e)g(f(e))^{-1}\}_{e\in E(\gamma)}) \\ [\alpha_\varphi(f)](A) &= f_\gamma(\{A(\varphi(e))\}_{e\in E(\gamma)}) \end{aligned} \tag{12}$$

There is a similar action on the vector fields $u_{n,S}$. As the notation suggests, the maps α_g, α_φ are automorphisms of $\mathfrak{E}$, that is, $\alpha_.(\{a,b\}) = \{\alpha_.(a), \alpha_.(b)\}$ for any $a, b \in \mathfrak{E}$ as one can easily verify. Hence we have a representation of $\mathfrak{G}$ as automorphisms on $\mathfrak{E}$.

Quantum *−Algebra

We follow the standard construction of Sect. 3:
Consider the free *−algebra $\mathfrak{F}$ generated by $\mathfrak{E}$. That is, we consider finite linear combinations of "words" w constructed from $\mathfrak{E}$. A word is simply a formal finite sequence $w = (a_1..a_N)$ of elements a_k of $\mathfrak{E}$. Multiplication of words is defined as $w \cdot w' = (a_1..a_N a'_1..a'_{N'})$ where $w = (a_1..a_N)$, $w' = (a'_1..a'_{N'})$. The *operation on $\mathfrak{F}$ is $w^* = (\bar{a}_N..\bar{a}_1)$.

Consider the two-sided ideal $\mathfrak{I}$ in $\mathfrak{F}$ generated by elements of the form

$$(a) \cdot (b) - (b) \cdot (a) - i\hbar(\{a,b\}) \tag{13}$$

for all $a, b \in \mathfrak{E}$. Then the quantum *−algebra is defined as the quotient

$$\mathfrak{A} := \mathfrak{F}/\mathfrak{I} \tag{14}$$

We can now simply lift the automorphisms labelled by $\mathfrak{G}$ from $\mathfrak{E}$ to $\mathfrak{A}$ by $\alpha_.(w) = (\alpha_.(a_1)..\alpha_.(a_N))$.

Representations of $\mathfrak{A}$

In quantum field theory, representations of $\mathfrak{A}$ are never unique in contrast to the situation in quantum mechanics where the Stone–von Neumann theorem guarantees that irreducible and weakly continuous representations of the

Weyl algebra generated by the unitary operators $U(x) = \exp(ixq)$, $V(y) = \exp(iyp)$, $x, y \in \mathbb{R}$ are automatically unitarily equivalent to the Schrödinger representation. For example, an appeal to Haag's theorem [6] reveals that Fock representations for free massive scalar fields with different masses are unitarily inequivalent representations of the corresponding Weyl algebra. Hence already in this simplest case we have an uncountably infinite number of unitarily inequivalent representations of the canonical commutation relations and all of them satisfy the Wightman axioms, e.g. Poincaré invariance. In order to single out preferred representations one must use additional criteria from physics. In the case of the scalar field, the representation is fixed if we insist on the Wightman axioms plus specifying the mass of the scalar field. Hence we need dynamical input as pointed out in [56].

In the case of our algebra $\mathfrak{A}$ the idea is to use dynamical input from the kinematical gauge algebra $\mathfrak{A}$. Namely, we want a unitary representation of $\mathfrak{G}$ on the Hilbert space. To do this, recall that for any $^*-$algebra such as our $\mathfrak{A}$ it is true that any representation is a (possibly uncountably infinite) direct sum of cyclic representations. Hence it is sufficient to consider cyclic representations. Next, any cyclic representation is in one-to-one correspondence with a state ω on $\mathfrak{A}$ via the GNS construction [6]. Here a *state* is defined as a positive linear functional on $\mathfrak{A}$, that is, $\omega(w^* w) \geq 0$ for all $w \in \mathfrak{A}$. It is not to be confused with *vectors*, that is, elements of some Hilbert space. Hence, it suffices to consider states on $\mathfrak{A}$.

The physical input to have a unitary representation of $\mathfrak{A}$ on the GNS Hilbert space $\mathcal{H}_\omega$ determined by ω now amounts to asking that the state ω be $\mathfrak{G}$-invariant. To see this we have to recall some elements of the GNS construction.

The GNS construction means that there is a one-to-one correspondence between states ω on a (unital) $^*-$algebra $\mathfrak{A}$ and GNS data $(\mathcal{H}_\omega,\ \pi(\omega),\ \Omega_\omega)$. Here $\mathcal{H}_\omega$ is a Hilbert space, π_ω is a representation of $\mathfrak{A}$ by densely defined and closable operators on $\mathcal{H}_\omega$ and Ω_ω is a cyclic vector in $\mathcal{H}_\omega$. Cyclic means that $\pi_\omega(\mathfrak{A})\Omega_\omega$ is dense in $\mathcal{H}_\omega$. This is done as follows: Consider the subspace of $\mathfrak{A}$ (considered as vector space) defined by $\mathfrak{J} := \{w \in \mathfrak{A};\ \omega(w^* w) = 0\}$. It is not difficult to show that this is a left ideal. Consider the equivalence classes $[w] := \{w + w'; w' \in \mathfrak{J}\}$. Then $\mathcal{H}_\omega$ is the closure of the vector space $\mathfrak{A}/\mathfrak{J}$ of equivalence classes, $\Omega_\omega := [1]$ and $\pi_\omega(w)[w'] := [ww']$. The scalar product is defined as $< [w], [w'] >_{\mathcal{H}_\omega} := \omega(w^* w')$. Now if ω is in addition $\mathfrak{G}$ invariant then by using the automorphism property it is easy to see that $U_\omega(\mathfrak{g})[w] := [\alpha_{\mathfrak{g}}(w)]$ is a unitary representation of $\mathfrak{G}$ with $\mathfrak{G}$-invariant cyclic vector Ω_ω.

The surprising result is now the following structural theorem [57].

Theorem 4.1. *The only $\mathfrak{G}$-invariant state on the holonomy–flux algebra $\mathfrak{A}$ is the Ashtekar–Isham–Lewandowski state ω_{AIL} whose GNS data coincide with the Ashtekar–Isham–Lewandowski representation.*

The surprising aspect to this theorem is that not the full gauge symmetry of the theory associated with the Hamiltonian constraint had to be used. In fact it is actually sufficient to just use the spatial diffeomorphism invariance in order to prove the theorem.[18]

The assumptions of the theorem are fairly weak as one can see. A possible generalisation is as follows: We have implicitly assumed that the flux operators themselves exist as self-adjoint operators on the Hilbert space. This is equivalent to asking that the state is regular, that is, weakly continuous with respect to the one parameter unitary groups they generate. This need not to be the case. In [58] it was shown that including non-regular states into the analysis does not change the uniqueness result modulo a slight additional assumption that one has to make. This is to say that the uniqueness result is fairly robust. It is rather important in the following sense: Suppose we had found a multitude of representations which satisfy the physical criterion of $\mathfrak{G}$-invariance. Then each of them would be a bona fide kinematical starting point for the Dirac quantisation programme which would amount to a large amount of ambiguity. The uniqueness result excludes this possibility and we can thus be confident to use the Hilbert space $\mathcal{H}_{AIL}$.

The Kinematical Hilbert Space and Its Properties

There are several complementary characterisations of the kinematical Hilbert space $\mathcal{H} := \mathcal{H}_\omega$ which are useful in different contexts. This section is for the mathematically inclined reader and can be skipped by readers interested only in the conceptual framework.

1. *Positive linear functional characterisation*
   We notice first of all that every word w can be written, using the commutation relations (7), (13) as a finite linear combination of reduced words. A reduced word is of the form $f u_{n_1 S_1} .. u_{n_N S_N}$ with $f \in$Cyl and arbitrary n_k, S_k and $N = 0, 1, ...$ Due to linearity it suffices to specify ω on reduced words. The definition is

$$\omega(w) = \begin{cases} 0 & \text{if } N > 0 \\ \omega_0(f) & \text{if } N = 0 \end{cases} \tag{15}$$

   Here ω_0 is the so-called "Ashtekar–Lewandowski positive linear functional" on the C^*-algebra completion $\overline{\text{Cyl}}$ of Cyl with respect to the sup norm. It can be explicitly written as

$$\omega_0(f) = \int_{SU(2)^n} d\mu_H(h_1)..d\mu_H(h_n)\; f_\gamma(h_1, .., h_n) \tag{16}$$

[18] The careful statement of the theorem uses semianalytic rather than smooth structures on σ. For every smooth structure there is always a semianalytic structure and semianalytic charts are equivalent up to smooth diffeomorphisms. Semianalyticity is the rigorous formulation of the more intuitive notion of piecewise analyticity. See [57] for details.

for $f(A) = f_\gamma(A(e_1), .., A(e_n)$, that is, f cylindrical over a graph with n edges. Here μ_H is the Haar measure on $SU(2)$. The Hilbert space $\mathcal{H}$ is the GNS Hilbert space derived from (15).

2. *C^*–algebraic characterisation*
The completion $\overline{\text{Cyl}}$ of Cyl with respect to the sup norm $||f|| := \sup_{A\in\mathcal{A}} |f(A)|$ defines an Abelean C^*–algebra [59]. Define the space of generalised connections $\overline{\mathcal{A}}$ as its Gel'fand spectrum[19] $\Delta(\overline{\text{Cyl}})$ [59], also called the Ashtekar–Isham space of generalised connections. By the Gel'fand isomorphism we can think of $\overline{\text{Cyl}}$ as the space $C(\overline{\mathcal{A}})$ of continuous functions on the spectrum. The spectrum of an Abelean C^*–algebra is a compact Hausdorff space if equipped with the Gel'fand topology of pointwise convergence of nets. Hence, by the Riesz–Markov theorem [60] the positive linear functional ω_0 in (16) is in one-to-one correspondence with a (regular, Borel) measure μ_0 on $\overline{\mathcal{A}}$ also called the Ashtekar–Lewandowski measure. The Hilbert space $\mathcal{H} := L_2(\overline{\mathcal{A}}, d\mu_0)$ is the space of square integrable functions on $\overline{\mathcal{A}}$ with respect to that measure.

3. *Projective limit characterisation*
The spectrum of $\overline{\text{Cyl}}$ abstractly defined above can be given a concrete geometric interpretation. It can be identified set theoretically and topologically as the set of homomorphisms from the groupoid $\mathcal{P}$ of paths into $SU(2)$, that is, there is a homeomorphism $\overline{\mathcal{A}} \to \text{Hom}(\mathcal{P}, SU(2))$ [61]. Here the groupoid of paths is defined, roughly speaking, as the set of (piecewise analytic) paths modulo retracings and reparametrisations together with the operations of (1) connecting paths with common beginning or end point and (2) inversion of orientation. Now recall that an element $A \in \overline{\mathcal{A}}$ is a homomorphism from $\overline{\text{Cyl}}$ into the complex numbers. Consider a function $f \in$Cyl cylindrical over some graph γ. Since A is a homomorphism we have $A(f) = f_\gamma(\{A(h_e)\}_{e\in E(\gamma)})$ where for $A \in \mathcal{A}$, $h_e(A) = A(e)$ is the holonomy map. Hence it suffices to consider the action of $A \in \overline{\mathcal{A}}$ on holonomy maps. Now since $h_e h_{e'} = h_{e\circ e'}$, $h_e^{-1} = (h_e)^{-1}$ and A is a homomorphism it follows that every point in the spectrum defines an element of $\text{Hom}(\mathcal{P}, SU(2))$. That also the converse is true is shown, e.g., in [62], hence there is a bijection.

To see that this bijection is a homeomorphism we must specify a topology on $\text{Hom}(\mathcal{P}, SU(2))$. To do this, we describe the space $\text{Hom}(\mathcal{P}, SU(2))$ as a projective limit: For every graph γ we consider the space $\text{Hom}(\gamma, SU(2))$ of homomorphisms from the subgroupoid of paths within γ (also denoted γ) into the gauge group. Since such homomorphisms are completely specified by their action on the edges of the graph, the sets $\text{Hom}(\gamma, SU(2))$ are identified topologically with $SU(2)^n$ where n is the number of edges of the graph. As such, $\text{Hom}(\gamma, SU(2))$ is a compact Hausdorff space. The set of subgroupoids is partially ordered and directed with respect to the inclusion relation, that is, for any two γ, γ'

[19] That is, the set of all homomorphisms from the algebra into the complex numbers.

there is $\tilde{\gamma}$ (e.g. $\gamma \cup \gamma'$) such that $\gamma, \gamma' \subset \tilde{\gamma}$. Given $\gamma \subset \gamma'$ we say that $A_{\gamma'} \in \text{Hom}(\gamma', SU(2))$ is compatible with $A_\gamma \in \text{Hom}(\gamma, SU(2))$ provided that the restriction of $A_{\gamma'}$ to γ coincides with A_γ, that is, $A_{\gamma'|\gamma} = A_\gamma$. The projective limit $\text{Hom}(\mathcal{P}, SU(2))$ (of the spaces $\text{Hom}(\gamma, SU(2))$) is the (automatically closed) subset of the infinite direct product $\overline{X}$ of the $\text{Hom}(\gamma, SU(2))$ restricted to the compatible points. The space $\overline{X}$ carries the natural Tychonov topology [63] with respect to which it is compact and Hausdorff. This property is inherited by the closed subset $\text{Hom}(\mathcal{P}, SU(2))$ in the subspace topology. As one can show, the compact Hausdorff topologies on $\overline{\mathcal{A}}$ and $\text{Hom}(\mathcal{P}, SU(2))$ are identified by the above mentioned bijection $A \mapsto (A_{|\gamma})_\gamma$ where $A_{|\gamma}$ is the restriction of A to γ.

Also the measure μ_0 abstractly defined via the Riesz–Markov theorem can be given a nice projective description: On each subgroupoid γ we consider the product Haar measure $\mu_{0,\gamma}$ as in (16). Let p_γ: $\text{Hom}(\mathcal{P}, SU(2)) \to \text{Hom}(\gamma, SU(2))$; $A \mapsto A_{|\gamma}$ be the restriction map. The system of measures $\mu_{0,\gamma}$ satisfies the following compatibility condition: For every $\gamma \subset \gamma'$ we have $\int d\mu_{0,\gamma'} f_\gamma = \int d\mu_{0,\gamma} f_\gamma$ for every $f = f_\gamma \circ p_\gamma$ cylindrical over γ. This property qualifies the $\mu_{0,\gamma}$ as the cylindrical projections [64] $\mu_{0,\gamma} = \mu_0 \circ p_\gamma^{-1}$ of a measure on the projective limit. Here the translation invariance and normalisation of the Haar measure are absolutely crucial to establish this property.

4. *Inductive limit characterisation*
   We consider the Hilbert spaces $\mathcal{H}_\gamma(\text{Hom}(\gamma, SU(2)), d\mu_{0,\gamma})$. For every $\gamma \subset \gamma'$ there is an isometric embedding $U_{\gamma\gamma'} : \mathcal{H}_\gamma \to \mathcal{H}_{\gamma'}$. These isometries satisfy $U_{\gamma\tilde{\gamma}} = U_{\gamma'\tilde{\gamma}} U_{\gamma\gamma'}$ for all $\gamma \subset \gamma' \subset \tilde{\gamma}$. This qualifies the $\mathcal{H}_\gamma$ as an inductive system of Hilbert spaces. The Hilbert space $\mathcal{H}$ is the corresponding inductive limit.

It is not difficult to show that this representation of $\mathfrak{A}$ is irreducible [65]. Moreover, it turns out that the Hilbert space $\mathcal{H}$ has an orthonormal basis over which one has complete control, the spin network basis [66]. These provide an indispensible tool in all analytical calculations in LQG. A *spin network* (SNW) is a quadruple[20] $s = (\gamma, j, m, n)$ consisting of a graph γ, a collection of spin quantum numbers $j = \{j_e\}_{e \in E(\gamma)}$ and two collections of magnetic quantum numbers $m = \{m_e\}_{e \in E(\gamma)}$, $n = \{m_e\}_{e \in E(\gamma)}$ subject to the conditions $j_e = 1/2, 1, 3/2, ..$ and $m_e,\ n_e \in \{-j_e, -j_e+1, .., j_e\}$. The analytical expression for a *spin network function* (SNWF) is given by (we write $A(e) := A(h_e)$ for $A \in \overline{\mathcal{A}}$)

$$T_s(A) := \prod_{e \in E(\gamma)} \sqrt{2j_e + 1}\, [\pi_{j_e}(A(e))]_{m_e n_e} \tag{17}$$

[20] It is understood that at bivalent vertices such that the incident edges are at least $C^{(1)}$ continuations of each other, then in the intertwiner decomposition of the state (see below) no trivial representation occurs. Otherwise this leads to an overcounting problem. Hence, if the intertwiner is trivial then such points are not counted as vertices.

Here π_j is the spin j irreducible representation of $SU(2)$. Its dimension is $2j+1$ and we label the entries of the corresponding matrices by $[\pi(h)]_{\rm mn}$.

Three important properties of $\mathcal{H}$ follow from the existence of the SNW basis:

1. Since the set of finite graphs is an uncountably infinite set, the kinematical Hilbert space is therefore non-separable since it does not have a countable basis.
2. Consider a vector field v on σ and let $t \mapsto \varphi^v_t$ be the one-parameter family of spatial diffeomorphisms generated by it.[21] Then the one-parameter unitary group $t \mapsto U(\varphi^v_t)$ is not weakly continuous, that is, it does not hold that $\lim_{t\to 0} < \psi, U(\varphi^u_t)\psi' >=< \psi, \psi' >$ for all $\psi, \psi' \in \mathcal{H}$. To see this choose $\psi = \psi' = T_s$ such that the graph γ underlying s has support in the support of v. Then $< T_s, U(\varphi^v_t)T_s >= 0$ for all $\epsilon > |t| > 0$ for some ϵ because $U(\varphi)T_s = T_{\varphi\cdot s}$ where $\varphi \cdot s = (\varphi(\gamma), j, m, n)$ if $s = (\gamma, j, m, n)$. By Stone's theorem [67] this means that the infinitesimal generators of spatial diffeomorphisms do not exist as (self-adjoint) operators on $\mathcal{H}$.
3. On SNWFs the operators $A(e)$ act by multiplication while $E_{\rm n,S} := u_{\rm n,S}$ becomes a linear combination of the right invariant vector fields $X^j_e = {\rm Tr}([\tau_j A(e)]^T \partial/\partial A(e))$ on a copy of $SU(2)$ coordinatised by $A(e)$.

4.4 Quantum Dynamics

The quantum dynamics consists of two steps: (1) Reduction of the system with respect to the gauge transformations generated by the constraints and (2) Introduction of a notion of time with respect to which observables (gauge invariant operators) evolve. It is convenient to subdivide the discussion of the reduction step into the gauge transformations corresponding to $\mathfrak{G}$ and those generated by the Hamiltonian constraint. We will also mention spin foam models which are the path integral formultion of LQG. Spin foam models were completely neglected in [12] although half of the current activity in LQG is devoted to them. This was partly corrected in [13]. The presentation will be brief since our main focus is on the criticisms of [12] towards the canonical formulation.

Reduction of Gauss– and Spatial Diffeomorphism Constraint

Gauss Constraint

The SNWF are not invariant under $\mathcal{G}$. It is easy to make them gauge invariant as follows: Pick a vertex $v \in V(\gamma)$ in the vertex set of γ and consider the edges $e_1, .., e_N$ incident at it. Let us assume for simplicity that the edges are all outgoing from v, the general case is similar but requires more book keeping. It is

[21] These are obtained by computing the integral curves $c^v_x(t)$ defined by $\dot{c}^v_x(t) = v(c^v_x(t))$, $c^v_x(0) = x$ and setting $\varphi^v_t(x) := c^v_x(t)$.

easy to see that at v the state transforms in the tensor product representation $j_1 \otimes .. \otimes j_n$ where $j_k := j_{e_k}$. Hence in order to make the state gauge invariant, all we need to do is to couple the N spins $j_1, .., j_N$ to resulting spin zero. This is familiar from the quantum mechanics of the angular momentum: We begin with

$$|j_1 m_1 > \otimes |j_2 m_2 >= \sum_{j_1 2} < j_{12} m_1 + m_2 | j_1 m_1; j_2 m_2 > \ |j_{12} m_1 + m_2 > \quad (18)$$

The recoupling quantum numbers take range in $j_{12} \in \{|j_1 - j_2|, .., j_1 + j_2\}$ and $< j_{12} m_1 + m_2 | j_1 m_1; j_2 m_2 >$ is the familiar Clebsch–Gordan coefficient (CGC). Next we repeat (18) with the substitutions $(j_1, m_1; j_2, m_2) \to (j_{12}, m_1 + m_2; j_3, m_3)$.

The procedure is now iterated until all spins have been recoupled to total angular momentum $J = 0$ and total magnetic quantum number $M = m_1 + .. + m_N = 0$. Consider the corresponding coefficients $< j_1 m_1; ..; j_N m_N | JM >$ in the decomposition of $|j_1 m_1 > \otimes .. \otimes |j_N m_N >$ into the $|JM >$. As we just showed, these can be written explicitly as polynomials of CGCs. We are interested only in those coefficients with $J = 0$, called intertwiners I_v. This imposes some restriction on the range of the j_k in order that this is possible at all. The number of those intertwiners does not depend on the sequence in which we couple those spins which is called a recoupling scheme. Different recoupling schemes are related by a unitary transformation. We now take one of those intertwiners and sum the SNWF times the intertwiner over all $m_k \in \{-j_k, .., j_k\}$. The result is a state which is gauge invariant at v. Now repeat this for all vertices.

The resulting states are gauge invariant and orthonormal with respect to the kinematical inner product by the properties of the CGCs and they define an orthonormal basis of the $\mathcal{G}$ invariant Hilbert space. We will also denote them by T_s where now $s = (\gamma, j, I)$ and $I = \{I_v\}_{v \in V(\gamma)}$.

Spatial Diffeomorphism Constraint

While the solutions to the Gauss constraint were normalisable with respect to the kinematical inner product, this turns out to be no longer the case with respect to the spatial diffeomorphism constraint. Let $\mathcal{D}$ be the finite linear span of SNWFs which by construction is dense in $\mathcal{H}$. We will look for solutions to the spatial diffeomorphism constraint in the algebraic dual $\mathcal{D}^*$ of $\mathcal{D}$. The algebraic dual of $\mathcal{D}$ are simply linear functionals on $\mathcal{D}$ equipped with the topology of pointwise convergence of nets (weak $*-$topology). It is clear that an element $l \in \mathcal{D}^*$ is completely specified by the numbers $l_s := l(T_s)$. Hence we can write any element of $\mathcal{D}^*$ formally as the uncountable direct sum

$$l = \sum_s l_s \ < T_s, . > \quad (19)$$

where the sum is over all gauge-invariant spin network labels.

Following the group-averaging technique described earlier, we say that an element $l \in \mathcal{D}^*$ is spatially diffeomorphism invariant provided that

$$l(U(\varphi)f) = l(f) \tag{20}$$

for all $\varphi \in$Diff(σ) and all $f \in \mathcal{D}$. As we have seen, this definition is a direct generalisation from vectors $\psi \in \mathcal{H}$ to distributions of the equation $U(\varphi)\psi = \psi$ for all $\varphi \in$Diff(σ). The latter equation has only one solution (up to a constant) in $\mathcal{H}$, namely $\psi = \Omega_\omega = 1$, the trivial spin network state.

In order to see what this requirement amounts to we notice that it is sufficient to restrict attention to the $f = T_s$. Let $[s] = \{\varphi \cdot s;\ \varphi \in \text{Diff}(\sigma)\}$ be the orbit of s. Then it is not difficult to see that (20) amounts to asking that $l_s = l_{s'}$ whenever $[s] = [s']$. Thus $l_s = l_{[s]}$ just depends on the orbit and not on the representative. It is therefore clear that no non-zero solution except for the vector 1 is normalisable with respect to the kinematical inner product $< l, l' >:= \sum_s \overline{l_s} l'_s$. Interestingly, the solutions are labelled by *generalised knot classes* where generalised refers to the fact that we allow for knots with intersections. Any solution obviously is a linear combination of the elementary solutions $T_{[s]} := \sum_{s' \in [s]} < T_{s'}, . >$.

We therefore have to define a new inner product on the solution space $\mathcal{D}^*_{\text{Diff}}$. This can be systematically done using the group-averaging technique described in Sect. 3. The only known Haar measure on Diff(σ) is the counting measure. Indeed, it is almost true that $T_{[s]}$ coincides with the image of the rigging map

$$\eta(T_s) := \sum_{\varphi \in \text{Diff}(\sigma)} < U(\varphi)T_s, . > \tag{21}$$

if it was not for fact that Diff(σ) contains an uncountably infinite number of elements which have trivial action on any given s. These trivial action diffeomorphisms form a subgroup (but not an invariant one) but that subgroup evidently depends on s. Hence one cannot take a universal factor group (rather: coset) for the averaging in order to get rid of the associated infinity. However, we notice that formally $\eta(T_s)[T'_s] = 0$ whenever $[s] \neq [s']$. Hence it is justified to decompose the kinematical Hilbert space into the direct sum of Diff(σ) invariant subspaces $\mathcal{H}_{[\gamma]}$ consisting of the finite linear span of SNWFs over the graphs γ' in the orbit $[\gamma]$ of γ. The group averaging can now be done on these subspaces seperately because in any case their images under (21) would be orthogonal. This is done by identifying a subset Diff$_{[\gamma]}(\sigma)$ which is in one-to-one correspondence[22] with the points in $[\gamma]$. When restricting (21) only to those diffeomorphisms and a discrete set of additional graph symmetries[23] then we indeed get $\eta(T_s) = k_{[s]} T_{[s]}$ where $k_{[s]}$ is a positive constant

[22] That is, fix a representative γ_0 in every orbit and select diffeomorphisms which map γ_0 to every point in the orbit.

[23] These are diffeomorphisms which leave the range of the representative γ_0 invariant but permute the edges among each other.

which is of the form of a positive number $k_{[\gamma(s)]}$ times an integer which is the ratio of the orbit sizes of the least symmetric $[s']$ with $[\gamma(s')] = [\gamma(s)]$ and the orbit size of $[s]$. See [43] for the details. It follows that the spatially diffeomorphism invariant inner product is determined by

$$< T_{[s]}, T_{[s']} >_{\text{Diff}} = \frac{1}{k_{[s]} k_{[s']}} < \eta(T_s), \eta(T_{s'}) >_{\text{Diff}} = \frac{1}{k_{[s]} k_{[s']}} \eta(T_{s'})[T_s] = \frac{\delta_{[s],[s']}}{k_{[s]}} \tag{22}$$

Notice, however, that the relative normalisation of the $T_{[s]}$ is only fixed for those s which have diffeomorphic underlying graphs because we applied the averaging to all those "sectors" separately. In order to fix the normalisations between the sectors one needs to consider diffeomorphism invariant operators which are classically real valued and map between these sectors and require that they be self-adjoint (or at least symmetric).

Finally we mention that $\mathcal{H}_{\text{Diff}}$ just like $\mathcal{H}$ is still not separable because the set of singular knot classes $[\gamma]$ has uncountably infinite cardinality [68]. This is easy to understand from the fact that the group of semianalytic diffeomorphism reduces to $GL(3,\mathbb{R})$ at each vertex. Hence, for vertices of valence higher than nine one cannot arbitrarily change, in a coordinate chart, all the angles between the tangents of the adjacent edges. It turns out that valence five is already sufficient, that is, there are diffeomorphism invariant "angles", called moduli θ in all vertices of valence five or higher. There are several proposals for an enlargement of the group of diffeomorphisms [69–71], however, these groups do not interact well with certain crucial operators in the theory such as the volume operator which depend on at least $C^{(1)}(\sigma)$ structures while those extensions basically replace diffeomorphisms by homeomorphisms or even mor general bijective maps on σ. We will see, however, that the non-separability of $\mathcal{H}_{\text{Diff}}$ is immaterial when we pass to the physical Hilbert space $\mathcal{H}_{\text{phys}}$.

Reduction of the Hamiltonian Constraint

The informed reader knows that the implementation of the Hamiltonian constraint is the most important technical problem in canonical quantum gravity ever since. The source of these technical problems within LQG can be appreciated when recalling the Dirac algebra $\mathfrak{D}$ (2):

1. The first relation in (2) means that diff(σ) is a subalgebra. However, the second relation says that this subalgebra is not an ideal. In other words, the Hamiltonian constraints are not spatially diffeomorphism invariant. In particular, if there is a quantum operator $\widehat{H}(N)$ associated with $H(N)$ then it *cannot be defined* on $\mathcal{H}_{\text{Diff}}$. We stress this simple observation here because one often hears statements saying the contrary in the literature. Spatially diffeomorphism invariant states do play a role but a quite different one as we will see shortly. The constraint operators $\widehat{H}(N)$ must be defined on the kinematical Hilbert space $\mathcal{H}$ and nowhere else. One could

try, as suggested in [72, 73] to define the dual $\widehat{H}'(N)$ of the constraint operator on some subspace $\mathcal{D}^*_\star$ of $\mathcal{D}^*$ invariant under the $\widehat{H}'(N)$ via

$$[\widehat{H}'(N)l](f) := l(\widehat{H}(N)^\dagger f) \tag{23}$$

for all $f \in \mathcal{D}$. However, in order to solve all constraints, eventually one wants to restrict $\mathcal{D}^*_\star$ to the space of spatially diffeomorphism invariant distributions on which $\widehat{H}'(N)$ is ill-defined. Hence the definition on $\mathcal{H}$ is the only option.

2. We have seen that the kinematical Hilbert space is, under rather mild assumptions, uniquely selected. In other words, there is no other choice. Unfortunately, as we have seen, in this representation the diffeomorphisms are not represented weakly continuously and there is no way out of this fact. This poses a problem in representing (2) on $\mathcal{H}$ because evidently (2) involves the infinitesimal generators $D(\boldsymbol{N})$ of spatial diffeomorphisms which are obstructed to exist as quantum operators as we just have said. As far as the first two relations in (2) are concerned, there is a substitute involving finite (exponentiated) diffeomorphisms only. It is given by

$$\begin{aligned} U(\varphi)U(\varphi')U(\varphi)^{-1} &= U(\varphi \circ \varphi' \circ \varphi^{-1}) \\ U(\varphi)\widehat{H}(N)U(\varphi)^{-1} &= \widehat{H}(N \circ \varphi) \end{aligned} \tag{24}$$

Indeed, if $\widehat{D}(\boldsymbol{N})$ would exist then one parameter subgroups of spatial diffeomorphisms would be given by $U(\varphi_t^{\boldsymbol{N}}) = \exp(it\widehat{D}(\boldsymbol{N})/(\hbar 8\pi G_{\text{Newton}}))$ and then (24) would be equivalent to

$$\begin{aligned} [\widehat{D}(\boldsymbol{N}), \widehat{D}(\boldsymbol{N}')] &= i8\pi G_{\text{Newton}}\hbar \;\; \widehat{D}(\mathcal{L}_{\boldsymbol{N}}\boldsymbol{N}') \\ [\widehat{D}(\boldsymbol{N}), \widehat{H}(N')] &= i8\pi G_{\text{Newton}}\hbar \;\; \widehat{H}(\mathcal{L}_{\boldsymbol{N}}N') \end{aligned} \tag{25}$$

upon taking the derivative at $t = 0$.

3. In other words there is a finite diffeomorphism reformulation of the first two relations in (2). However, this is no longer possible for the third relation in (2). The problem is the structure function involved on the right-hand side of this relation which prevents us from exponentiating the Hamiltonian constraints. The commutator algebra of the Hamiltonian constraints is simply so complicated that the Dirac algebra $\mathfrak{D}$ is no longer a Lie group. Therefore we cannot exponentiate the third relation in (2) and there seems to be no chance to find a substitute involving finite diffeomorphisms only.
4. Even if the problem just mentioned could be solved, we would still have no idea for how to find the physical inner product because group averaging only works for Lie algebra valued constraints.

These remarks sound like an obstruction to implement the operator version of the Hamiltonian constraint in LQG at all. In what follows we describe the progress that has been made over the past 10 years with regard to this task. There are two constructions: The first, surprisingly, indeed proposes a quantisation of the Hamiltonian constraints as operators on the kinematical Hilbert space. The algebra of these operators *is non-Abelean and closes* in a precise sense as we will see. We again stress this because one sometimes reads that the Hamiltonian constraint algebra is Abelean [72, 73] which is simply wrong. However, no physical scalar product using these operators has so far been constructed due to the non-Lie algebra structure mentioned above. Also, so far the correctness of the semiclassical limit of these constraint operators has not been established, in particular it is unsettled in which sense the third relation in (2) is implemented in the quantum theory.

To make progress on these two open problems, the construction of the physical scalar product and the establishment of the correct classical limit which are interlinked in a complicated way as we will see, the master constraint programme was launched [19, 34, 35, 75]. We have outlined it already in Sect. 3 for a general theory and will apply it to the Hamiltonian constraints below.

This section is organised as follows:

The master constraint programme overcomes many of the shortcomings of the Hamiltonian constraint and is the modern version of the implementation of the Hamiltonian constraint in LQG. We will still describe first the old Hamiltonian constraint programme [20, 38, 75] in order to address the criticisms spelled out in [12] and because the quantisation technique in [19, 75] is still heavily based on the *key techniques* developed in [20]. Indeed, without the techniques developed in [20] the recent intriguing results of loop quantum cosmology (LQC)[24] [76] such as avoidance of the big bang singularity could never have been achieved. A large amount of the success of LQC is a direct consequence of [20].

Then we describe the master constraint programme which was not mentioned at all in [12] although it removes many of the criticisms stated there. In particular we describe recent progress made in a particular version of the master constraint programme called algebraic quantum gravity (AQG) [77] which establishes that the master constraint operator has the correct semiclassical limit. The work [77] also removes the criticism of [12] that no calculations involving the book operator can be carried out in LQG. This, together with the general direct integral construction of the physical inner product already described make the master constraint programme a promising step forward in LQG.

[24] LQC is the usual homogeneous (and isotropic) cosmological model quantised by LQG methods. It is not the cosmological sector of LQG because LQG is a quantum field theory (infinite number of degrees of freedom) while LQC is a quantum mechanical toy model (finite number of degrees of freedom) in which the inhomogeneous excitations are switched off by hand.

Unfortunalely the subsequent discussion is rather complictaed because the problems of anomaly freeness, semiclassical limit, dense definition, representation of the Dirac algebra etc. for the Hamiltonian constraints are interlinked in a complex way. In order to appreciate these interdependencies we have to go into some detail about the actual constructions. We will try our best at keeping the discussion as non-technical as possible.

Hamiltonian Constraint

The task is to quantise the Hamiltonian constraints which on the new phase space are given by[25]

$$H(N)=\int_\sigma d^3x\, N(x)\,\frac{\mathrm{Tr}(F_{\mathrm{ab}}E^aE^b)}{\sqrt{|\det(E)|}}(x) \tag{26}$$

where N is a test function. The non-polynomial character of (26) is evident and it is hard to imagine that there is any way to tame (26) when replacing A, E by their operator equivalents.

We now sketch the *key tools* developed in [20]. Let R_x be any region in σ containing x as an interior point, then

$$e_a^j(x)=-\{A_a^j(x),V(R_x)\}/\kappa \tag{27}$$

where $\kappa=8\pi G_{\mathrm{Newton}}$ and

$$V(R_x):=\int_{\mathrm{R_x}} d^3y\,\sqrt{|\det(E)|}(y) \tag{28}$$

is the volume of the region R_x. Using the relation $e_j^a=E_j^a/\sqrt{|\det(E)|}$ we can rewrite (26) as

$$H(N)=-\frac{1}{\kappa}\int_\sigma N(x)\,\mathrm{Tr}(F(x)\wedge\{A(x),V(R_x)\}) \tag{29}$$

where now all the dependence on E resides in the volume function $V(R_x)$. The point of doing this is that $V(R_x)$ admits a well-defined quantisation as a positive essentially self-adjoint operator[26] $\widehat{V}(R_x)$ on $\mathcal{H}$. Following the rules of canonical quantisation one would then like to replace the Poisson bracket between the functions appearing in (29) by the commutator between the corresponding operators divided by $i\hbar$.

---

[25] This is only the simplest piece of the geometry part of the constraint, the remaining piece as well as matter contributions can be treated analogously [38] and will be neglected here for pedagogical reasons.

[26] Actually there are two inequivalent volume operators [78, 79] which result from using two different background-independent regularisation techniques. However, in a recent mathematical self-consistency analysis [80] the operator [78] could be ruled out.

The problem is that the connection operator $\widehat{A}(x)$ does not exist on the Hilbert space $\mathcal{H}$. To see this, note the classical identity $A_a(x)\dot{p}^a(0) = (d/dt)_{t=0}A(p_t)$ where $p: [0,1] \to \sigma;\ s \mapsto p(s)$ is a path, $p_t(s) = p(ts)$ for $t \in [0,1]$, $p(0) = x$ and $A(p_t)$ is the holonomy along p_t. By varying the path we can recover the connection from the holonomy. Hence we would like to define the connection operator by this formula from the holonomy operator. However, this does not work since the family of operators $t \mapsto A(p_t)$ is not weakly continuous on $\mathcal{H}$. Hence the derivative at $t = 0$ is ill-defined. It follows that the UV singularity structure of the Hamiltonian constraints is not at all determined by the E dependence but rather by the A dependence. In particular, the ambiguities discussed below coming from the loop attachment purely stem from the A dependence.

It is at this point where we must regularise (29). We consider a triangulation τ of σ by tetrahedra Δ. For each Δ, let us single out a corner $p(\Delta)$ and denote the edges of Δ outgoing from $p(\Delta)$ by $s_I(\Delta),\ I = 1, 2, 3$. Likewise, denote by $s_{IJ}(\Delta)$ the edges of Δ connecting the end points of $s_I(\Delta),\ s_J(\Delta)$ such that the loop $\beta_{IJ}(\Delta) = s_I(\Delta) \circ s_{IJ}(\Delta) s_J(\Delta)^{-1}$ is the boundary of a face of Δ. In particular $s_{JI}(\Delta) = s_{IJ}(\Delta)^{-1}$. It is then not difficult to see that

$$
\begin{aligned}
&H_\tau(N) : \qquad (30)\\
&= \frac{1}{\kappa}\sum_{\Delta\in\tau} N(p_\Delta) \sum_{IJK} \epsilon^{IJK}\, \mathrm{Tr}(A(\beta_{IJ}(\Delta))A(s_K(\Delta))\{A(s_K(\Delta))^{-1}, V(R_{p(\Delta)})\})
\end{aligned}
$$

is a Riemann sum approximation to (29), that is, it converges to (29) as we refine the triangulation to the continuum. We will denote the refinement limit by $\tau \to \sigma$.

Since (30) is now written in terms of quantities of which the quantisation is known we immediately get a regularised Hamiltonian constraint operator on $\mathcal{H}$ given by

$$
\begin{aligned}
&\widehat{H}^\dagger_\tau(N) : \qquad (31)\\
&= \frac{1}{i\ell_P^2}\sum_{\Delta\in\tau} N(p_\Delta) \sum_{IJK} \epsilon^{IJK}\mathrm{Tr}(A(\beta_{IJ}(\Delta))A(s_K(\Delta))[A(s_K(\Delta))^{-1}, \widehat{V}(R_{p(\Delta)})])
\end{aligned}
$$

The reason for the adjoint operation in (31) is due to the definition of the dual action in the footnote before (5) on elements in $\mathcal{D}^*$ which in turn would coincide with the action of $\widehat{H}(N)$ if elements of $\mathcal{D}^*$ would be normalisable. Notice that (30) is real valued so that classically $H_\tau(N) = \overline{H_\tau(N)}$ so we may denote its operator equivalent with or without adjoint operation. It is not difficult to see that in this ordering the operator is densely defined[27] on $\mathcal{D}$ and closable (its adjoint is also densely defined on $\mathcal{D}$). However, it is

[27] This is basically due to the properties of the volume operator $\widehat{V}(R)$: If the graph of a spin network state does not contain a vertex inside the region R then it is annihilated.

not even symmetric in this ordering. This may seem strange at first; however, it is not logically required because we are only interested in the zero point of its spectrum. It is not even possible to have a symmetric ordering as pointed out in [81] where it is shown that for reasons of anomaly freeness in constraint algebras with structure functions symmetric orderings are ruled out.

What we are interested in is in which operator topology (if any) the limit $\tau \to \sigma$ exists. Since $\widehat{H}_\tau(N)$ is not bounded, convergence in the uniform topology is ruled out. For the same reason that connection operators are not defined, convergence in the weak (and thus also strong) operator topology is ruled out. Hence we are looking for a weaker topology. There is only one *natural* candidate available: The weak* topology with respect to the algebraic dual $\mathcal{D}^*$ or a suitable subspace thereof. The only *natural* subspace is the space of spatially diffeomorphism invariant distributions $\mathcal{D}^*_{\rm Diff}$ (finite linear combinations of the $T_{[s]}$ defined in Sect. 4.4).

Before we do this, we must tame the limit $\tau \to \sigma$ somewhat: Notice that classically $\lim_{\tau\to\sigma} H_\tau(N) = H(N)$ no matter how we refine the triangulation. This observation suggests the following strategy: Given a graph γ we consider a family $\epsilon \mapsto \tau^\epsilon_\gamma$ of triangulations adapted to γ where ϵ denotes the fineness of the triangulation and $\epsilon \to 0$ corresponds to $\tau \to \sigma$. This family is equipped with the following properties: For each vertex $v \in V(\gamma)$ and each triple of edges $e_1, e_2, e_3 \in E(\gamma)$ incident at v consider a tetrahedron $\Delta^\epsilon_v(e_1, e_2, e_3)$ such that $p(\Delta^\epsilon_v(e_1, e_2, e_3)) = v$, such that $s_I(\Delta^\epsilon_v(e_1, e_2, e_3))$ is a proper segment of e_I, such that the $s_{\rm IJ}(\Delta^\epsilon_v(e_1, e_2, e_3))$ do not intersect γ except in their endpoints and such that the $\Delta^\epsilon_v(e_1, e_2, e_3)$ are diffeomorphic for different values of ϵ. That such tetrahedra always exist is proved in [20].
Consider seven additional tetrahedra $\Delta^\epsilon_{\rm v,1}(e_1, e_2, e_3), .., \Delta^\epsilon_{\rm v,7}(e_1, e_2, e_3)$ which are obtained by analytically continuing[28] the segments $s_I(\Delta^\epsilon_v(e_1, e_2, e_3))$ through the vertex so that we obtain altogether eight tetrahedra of equal coordinate volume which are like the eight octants of a Cartesian coordinate system. Denote by $W^\epsilon_v(e_1, e_2, e_3)$ the neighbourhood of v they fill. Let W^ϵ_v be the region occupied by the union of the $W^\epsilon_v(e_1, e_2, e_3)$ as we vary the unordered triples of edges incident at v. For sufficiently fine triangulation, the W^ϵ_v are mutually disjoint. Finally let W^ϵ_γ be the union of the W^ϵ_v. We have the following identity for any classical integral

$$\int_\sigma = [\int_{\sigma - \mathrm{W}^\epsilon_\gamma}] + \sum_{\mathrm{v}\in \mathrm{V}(\gamma)} \frac{1}{\binom{n_v}{3}} \sum_{\mathrm{e_1 \cap e_2 \cap e_3 = v}} \{[\int_{\mathrm{W}^\epsilon_\mathrm{v} - \mathrm{W}^\epsilon_\mathrm{v}(\mathrm{e_1,e_2,e_3})}] + \int_{\mathrm{W}^\epsilon_\mathrm{v}(\mathrm{e_1,e_2,e_3})}]\} \tag{32}$$

where n_v is the valence of v. We now triangulate the regions $\sigma - W^\epsilon_\gamma$, $W^\epsilon_v - W^\epsilon_v(e_1, e_2, e_3)$ arbitrarily and use the classical approximation $\int_{\mathrm{W}^\epsilon_\mathrm{v}(\mathrm{e_1,e_2,e_3})} \approx 8 \int_{\Delta^\epsilon_\mathrm{v}(\mathrm{e_1,e_2,e_3})}$. Then, the tetrahedra within $\sigma - W^\epsilon_\gamma$, $W^\epsilon_v - W^\epsilon_v(e_1, e_2, e_3)$ can be shown not to contribute to the action of the operator $\widehat{H}_{\tau^\epsilon_\gamma}(N)$ on any SNWF

[28] For sufficiently fine triangulation the segments can be taken to be analytic.

T_s over γ so that we obtain

$$\widehat{H}^{\dagger}_{\tau^{\epsilon}_{\gamma}}(N)T_s = \frac{1}{i\ell_P^2} \sum_{\mathrm{v} \in \mathrm{V}(\gamma)} N(v) \frac{8}{\binom{n_v}{3}} \sum_{\mathrm{e}_1 \cap \mathrm{e}_2 \cap \mathrm{e}_3} \sum_{\mathrm{IJK}} \epsilon^{IJK} \times \tag{33}$$
$$\times \mathrm{Tr}(A(\beta_{\mathrm{IJ}}(\Delta^{\epsilon}_v(e_1,e_2,e_3)))$$
$$A(s_K(\Delta^{\epsilon}_v(e_1,e_2,e_3)))[A(s_K(\Delta^{\epsilon}_v(e_1,e_2,e_3)))^{-1}, \widehat{V}(R_{\mathrm{p}(\Delta)})]) \; T_s$$

For each γ choose[29] some ϵ_γ once and for all such that $\tau_\gamma := \tau_\gamma^{\epsilon_\gamma}$ satisfies the required properties and define $\widehat{H}^{\dagger}(N)T_s := \widehat{H}^{\dagger}_{\tau_\gamma}T_s$ and $\widehat{H}^{\dagger}_{\epsilon}(N)T_s := \widehat{H}^{\dagger}_{\tau^{\epsilon}_{\gamma}}T_s$ whenever $s = (\gamma, j, I)$. Then, *due to spatial diffeomorphism invariance* we have the following notion of convergence

$$\lim_{\epsilon \to 0} |l(\widehat{H}^{\dagger}(N)f) - l(\widehat{H}^{\dagger}_{\epsilon}(N)f)| = 0 \tag{34}$$

for all $l \in \mathcal{D}^*_{\mathrm{Diff}}$ and all $f \in \mathcal{D}$. It is quite remarkable that precisely the space of diffeomorphism invariant distribtions which is selected by one of the gauge symmetries of the theory naturally allows us to define an appropriate operator topology with respect to which it is possible to remove the regulator of the Hamiltonian constraints. Notice that despite the fact that we have worked with triangulations adapted to a graph, the operator is a linear operator on $\mathcal{H}$ where it is, together with its adjoint, densely defined on $\mathcal{D}$.

One of the most striking features is that the Hamiltonian constraint operators *do not suffer from UV singularities* as we anticipated in a background-independent theory.

Several remarks are in order:

1. *Quantum Spin Dynamics (QSD)*
Intuitively, the action of the Hamiltonian constraint operator on spin network functions over a graph γ is by creating the new edges $s_{\mathrm{IJ}}(\Delta_v^{\epsilon_\gamma}(e_1,e_2,e_3))$ coloured with the spin $1/2$ representation and by changing the spin on the segements $s_I(\Delta_v^{\epsilon_\gamma}(e_1,e_2,e_3))$ from j to $j \pm 1/2$. Hence in analogy to QCD one could LQG call QSD.
2. *Locality*
The action of the Hamiltonian constraint operator has been criticised to be too local [82] in the following sense: The modifications that the Hamiltonian constraint operator performs at a given vertex do not propagate over the whole graph but are confined to a neighbourhood of the vertex. In fact, repeated action of the Hamiltonian generates more and more new edges ever closer to the vertex never intersecting each other thus producing a fractal structure. In particular there is no action at the new vertices created. This is not what happens in lattice gauge theory where no new edges are created.

[29] Use the axiom of choice for each diffeomorphism equivalence class of loop assignments.

Notice, however, that there is a large conceptual difference between lattice gauge theory which is a background-dependent and regulator-dependent (discretised) theory while LQG is a background-independent and regulator-independent (continuum) theory. Even the role of the *single* QCD Hamiltonian (generator of physical time evolution) and the *infinite number* of Hamiltonian constraints (generator of unphysical time reparametrisations) is totally different. Hence there is no logical reason why one should compare the lattice QCD Hamiltonian with the QSD (or LQG) Hamiltonian constraints. In particular, by inspection the infinite number of constraints $H(x) = 0$ have a more local structure than a Hamiltonian $H = \int_\sigma d^3x H(x)$.

Next, it is actually technically incorrect that the actions of the Hamiltonian constraints $\widehat{H}_v$, $\widehat{H}_{v'}$ at different vertices v, v' do not influence each other: In fact, these two operators do not commute, for instance if v, v' are next neighbour, because for any choice function $\gamma \mapsto \epsilon_\gamma$ what is required is that the loop attachments at v, v' do not intersect which requires that the action at v' after the action at v attaches the loop at v' closer to v' than it would before the action at v and vice versa.

Finally, the action of the Hamiltonian constraints on spin network states did not fall from the sky but was derived from a proper regularisation. In particular it is not difficult to see that the operator would become anomalous (see below) if it would act at the vertices that it creates. This would indeed happen if one used the volume operator [78] rather than [79]. Fortunately, the volume operator [78] was shown to be inconsistent [80] for totally independent reasons.

In summary, there is no conclusive reason for why this locality property of the constraints is a bad feature. In fact, in 3D [38] the solution space of those constraints selects precisely the physical Hilbert space of [83].

3. *Ambiguities*

   The final Hamiltonian constraint operators seem to be highly ambiguous. There are several qualitatively different sources of ambiguities:

   3a. *Factor ordering ambiguities*

   We decided to order the E-dependent terms in (30) to the right of the A-dependent terms. Could we have reversed the order? The answer is negative [20]: Any other ordering results in an expression which is no longer densely defined because the operator would map any spin network state to a state which is a linear combination of SNWFs whose underlying graphs are all tetrahedra of the triangulation. The resulting "state" is not normalisable in the infinite refinement limit. Thus the factor ordering chosen is in fact *unique*.

   3b. *Representation ambiguities*

   When replacing connections by holonomies we have used the holonomy in the defining representation of $SU(2)$. However, as pointed out in [84] we could also work with higher spin representations without affecting the limit of the Riemann sum approximation. Recently it was shown [85] that higher spin leads to spurious solutions to the Hamiltonian constraints in 3D (where all solutions are known) and therefore very likey also in 4D. Hence this representation ambiguity is very likely to be *absent*.

Notice also that such kind of ambiguities are also present in ordinary QFT: Consider a $\lambda\phi^4$ QFT. Classically we could replace $\pi(x)$ by $\pi_f(x) := e^{i\phi(f)}\pi(x)e^{-i\phi(f)}$ in the Hamiltonian where $\phi(f) = \int_{\mathbb{R}^3} d^3x f(x)\phi(x)$ with some suitable test function f and ϕ, π are canonically conjugate. One could even replace $\exp(i\phi(f))$ by some other invertible functional F of ϕ and consider $[F\pi F^{-1} + \bar{F}^{-1}\pi\bar{F}]/2$. In quantum theory the Hamiltonian does change when performing this substitution leading to a different spectrum. Of course, in QFT one would never do that because this factor-ordering ambiguity generically spoils polynomiality of the Hamiltonian, so one is guided by some simplicity or naturalness principle. In general relativity the Hamiltonian constraint is non-polynomial from the outset; however, still $j = 1/2$ is the simplest choice.

3c. *Loop assignment ambiguities*

The largest source of ambiguities is in the choice of the family of triangulations $\epsilon \mapsto \tau_\gamma^\epsilon$ adapted to a graph. In particular, while it is natural to align the edges of the tetrahedra of the triangulations with the beginning segments of the edges of the graph[30] because there are no other natural terahedra available in the problem, it is not the only logically possible choice. For instance, one could slightly detach the loops $\beta_{\mathrm{IJ}}(\Delta_v^\epsilon(e_1, e_2, e_3))$ from the beginning segments of e_1, e_2, e_3 as mentioned in the review by Ashtekar and Lewandowski in [3] which found its way into [12]. Our statement here is as follows: First of all there is an additional, heuristic argument in favour of the alignment. Secondly, even if one does not accept that argument, all of these uncountably infinite number of ambiguities at the level of $\mathcal{H}$ are reduced to a countable number at the level of $\mathcal{H}_{\mathrm{phys}}$ of which all but a few are rather pathological in the sense that one could also use them in lattice gauge theory but does not due to reasons of naturalness.

Concerning the first claim, we want to point out that one of the reasons for why we have decided to work with the expression $\{A_a^j(x), V(R_x)\}$ rather than $(\epsilon_{\mathrm{abc}} E_k^b E_l^c \epsilon_{\mathrm{jkl}}/\sqrt{|\det(E)|})(x)$ is that direct quantisation of the latter would formally result on a spin network state over a graph γ in an expression of the form (before introducing the point splitting regulator)

30 There is no ambiguity in the fact that the only contributions of the operator result from the vertices of the graph. This is a direct consequence of the properties of the volume operator [79] and the unique factor ordering mentioned above.

$$\int_\sigma d^3x \frac{1}{\sum_{\mathrm{v}'\in \mathrm{V}(\gamma)} \delta(x,v')\widehat{V}_v} \sum_{\mathrm{v}\in \mathrm{V}(\gamma)} \sum_{\mathrm{e}_1\cap \mathrm{e}_2=\mathrm{v}} \int_0^1 dt\; \dot{e}_1^a(t)$$
$$\delta(x,e_1(t)) \int_0^1 ds\; \dot{e}_2^b(s)\; \delta(x,e_2(s)) \times$$
$$\times F^j_{\mathrm{ab}}(\frac{e_1(t)+e_2(s)}{2})\epsilon_{\mathrm{jkl}} X^k_{\mathrm{e}_1} X^l_{\mathrm{e}_2} \tag{35}$$

where X^j_e is a right invariant vector field on the copy of $SU(2)$ corresponding to $A(e)$. Likewise $\widehat{V}_v$ is a well-defined operator (not an operator-valued distribution) built from those vector fields. Clearly (35) involves the holonomy of an infinitesimal loop whose tangents at the v are pairs of edges incident at v. This motivates the alignment mentioned above.[31] The only reason why (35) is not used in place of (30) is that the the operator $\widehat{V}_v$ has zero modes so that its inverse is not even densely defined.

Concerning the second claim we notice that solutions to all constraints will be elements l of $\mathcal{D}^*_{\mathrm{Diff}}$ which satisfy $l(\widehat{H}(N)^\dagger f) = 0$ for all N and all $f \in \mathcal{D}$. Now since l is spatially diffeomorphism invariant, the space of solutions to all constraints *only depends on the spatially, piecewise analytic diffeomorphism invariant characteristics* of the loops $\beta_{\mathrm{IJ}}(\Delta_v^{\epsilon_\gamma}(e_1,e_2,e_3))$. Hence it matters whether or not the tetrahedra Δ are just continuous at their corners or of higher differentiability class, how the additional edges are routed or braided through the edges of the graph and whether they are aligned or not. Concerning the braiding, a natural choice is the one displayed in [20] which makes use of Puisseaux' theorem.[32] It follows that the seemingly uncountably infinite set of possible loop assignments is reduced

---

[31] A careful. point splitting regularsation removes the δ-distribution in both numerator and the denominator as well as the the integral over σ leaving only an integral over s,t with support in infinitesimal neighbourhoods of the vertices of the graph in question.

[32] Basically one wants that the arcs intersect the graph only in their end points which for sufficiently fine triangulations can only happen for edges e incident at the vertex in question. One first shows that there always exists an adapted frame, that is, a frame such that s_I, s_J lie in the x,y plane for sufficiently short s_I, s_J. Now one shows that for any other edge e of the graph whose beginning segment is not aligned with either s_I or s_J there are only two possibilities: A. Either for all adapted frames the beginning segment of e lies above or below the x,y plane and whether it is above or below is independent of the adapted frame. B. Or there exists an adapted frame such that e lies above the x,y plane. This can be achieved simultaneously for all edges incident at the vertex in question. The natural prescription is then to let the edge s_{IJ} be the straight line in the selected frame connecting the end points of s_I, s_J at which it intersects transversally.

to a *discrete number of choices*, of which all but a finite number is unnatural,[33] once we construct solutions of all constraints.

In summary, the most natural proposal is such that the edges $s_{IJ}(\Delta_v^{\epsilon_\gamma}(e_1, e_2, e_3))$ intersect the graph γ transversally with the braiding described in [20]. This defines a concrete and non-ambiguous operator in the sense that it uniquely selects a subspace of $\mathcal{D}^*_{\text{Diff}}$ as the space of solutions to all constraints.

3d. *Habitat ambiguities*

In [12] we find an extensive discussion about "habitats" $\mathcal{D}^*_\star$. A habitat is a subspace of $\mathcal{D}^*$ containing $\mathcal{D}^*_{\text{Diff}}$ with the minimal requirement that it is preserved by the dual action of the Hamiltonian constraints. Habitats were introduced in [72, 73]. The idea was to take the limit $\epsilon \to 0$ for the duals of the Hamiltonian constraints on such a habitat in the sense of pointwise convergence. The habitat ambiguity is that there maybe zillions of habitats on which a limit of this kind can be performed. As was shown in those papers, there exists at least one such habitat and it has the property that the limit dual operators are Abelian.

We now show that this habitat ambiguity is actually absent: Namely, the habitat spaces must be genuine extensions of $\mathcal{D}^*_{\text{Diff}}$. Hence these spaces are not in the kernel of the spatial diffeomorphism constraint and are therefore unphysical. Hence the only domain where to define the Hamiltonian constraints (rather than their duals) is on $\mathcal{D}$, that is, on a dense subspace of the kinematical Hilbert space $\mathcal{H}$. This is the same domain as for the spatial diffeomorphism constraints which thus treats both types of constraints democratically. This fact is widely appreciated in the LQG community and not a matter of debate; the habitat construction presented in [12] is outdated. *Habitats are unphysical and completely irrelevant in LQG.*

On the kinematical Hilbert space the Hamiltonian constraints are non-commuting, see below. The apparent contradiction with the Abelean nature of the limits of the duals on the aforementioned habitat is resolved by the fact that effectively the commutator of the limiting duals on the habitat is the dual of the commutator on $\mathcal{D}$. While the commutator on $\mathcal{D}$ is non-vanishing, its dual annihilates $\mathcal{D}^*_{\text{Diff}}$ and also the habitat $\mathcal{D}^*_\star$ chosen which is a sufficiently small extension of $\mathcal{D}^*_{\text{Diff}}$.

Hence we see that the amount of ambiguity is far less severe than [12] perhaps make it sound once we pass to $\mathcal{H}_{\text{phys}}$. In fact, there are only a handful of natural proposals available.

---

[33] Like winding the segments s_I of the tetrahedra of the triangulation an arbitrary number of times around the edges e_I of the graph. Such a ridiculous choice could also be made in lattice gauge theory but is not considerded there due to reasons of naturalness.

4. *Anomalies*
As already mentioned, the constraint algebra can only be checked in the form that only involves finite diffeomorphisms. Indeed it is not difficult to see that the first two relations of the Dirac algebra (2) really hold in the form (24) on $\mathcal{H}$ up to a spatial diffeomorphism [2]. Likewise one can check that
$$l([\widehat{H}^\dagger(N), \widehat{H}^\dagger(N')]f) = 0 \tag{36}$$
for all test functions N, N', all $f \in \mathcal{D}$ and all $l \in \mathcal{D}^*_{\mathrm{Diff}}$. This can be read as an implementation of the third relation in (2) because that relation involves an infinitesimal spatial diffeomorphism constraint whose dual action should annihilate $\mathcal{D}^*_{\mathrm{Diff}}$. Of course, the commutator does not involve an infinitesimal diffeomorphism which does not exist in our theory. Rather what happens is the following: The commutator $[\widehat{H}^\dagger(N), \widehat{H}^\dagger(N')]$ is non-vanishing on $\mathcal{H}$. However, it can be shown that on SNWFs it is a finite linear combination of terms of the form $[U(\varphi) - U(\varphi')]\widehat{O}$ where $\widehat{O}$ is some operator on $\mathcal{H}$.
5. *On shell closure versus off shell closure*
As we just just saw, the quantum constraint algebra is consistent, that is, non-anomalous. More precisely, the first relation in (2) holds, in exponentiated form, on $\mathcal{H}$ exactly, it is non-anomalous in every sense. The second relation in (2) also holds in exponentiated form on $\mathcal{H}$ but only modulo a spatial diffeomorphism. How about the third relation in (2)? In [38] it is shown that an independent quantisation of the classical function $D(q^{-1}(N'dN - NdN'))$ appearing on the right-hand side of (2) can be given. There is no contradiction to the non-existence of the operator corresponding to $D(\boldsymbol{N})$ because $D(q^{-1}(N'dN - NdN'))$ is not of the form $D(\boldsymbol{N})$ due to the structure function q^{-1} which is responsible for the existence of the composite operator corresponding to $D(q^{-1}(N'dN - NdN'))$. That operator is constructed in a way analogous to the Hamiltonian constraint operator and is formulated in terms of the operators $U(\varphi)$. The duals of both operators $[\widehat{H}^\dagger(N), \widehat{H}^\dagger(N')]$ and $D(q^{-1}(\widehat{N'dN - NdN'}))$ annihilate $\mathcal{D}^*_{\mathrm{Diff}}$.
Hence what one can say is the following: We define two operators on $\mathcal{H}$ as equivalent $\widehat{O}_1 \sim \widehat{O}_2$ provided that the dual of $\widehat{O}_1 - \widehat{O}_2$ annihilates $\mathcal{D}^*_{\mathrm{Diff}}$. Then the classical identities (2) holds on $\mathcal{H}$ in the sense of equivalence classes (the first relation even identically).
One could call this partly on-shell closure (partly because we did not use the full $\mathcal{H}_{\mathrm{phys}}$ but only $\mathcal{H}_{\mathrm{Diff}}$ in the equivalence relation). While it would be more satisfactory to have full off-shell closure, it is not logically required: At the end we are only interested in physical states and these are in particular spatially diffeomorphism invariant. Those states cannot distinguish between different representatives of the equivalence classes.
6. *Semiclassical limit*
The problem with demonstrating off-shell closure is that, in contrast to the first two, the third relation in (2) does not hold by inspection, not even

modulo a diffeomorphism. This is not surprising because even classically one needs a full page of calculation in order to bring the Poisson bracket between two Hamiltonian constraints into the form of the right-hand side of the third relation in (2). This calculation involves reordering of terms, differential geometric identities and integrations by parts etc. which are difficult to perform at the operator level. In order to make progress on this issue one would therefore like to probe the Dirac algebra with semiclassical states, the idea being that in expectation values with respect to semiclassical states the operators can be replaced by their corresponding classical functions and commutators by Poisson brackets, up to $\hbar$ corrections.
There are two immediate obstacles with this idea:

The first is that the volume operator involved is not analytically diagonisable. Recently, however, it was shown that analytical calculations involving the volume operator can be performed precisely using coherent states on $\mathcal{H}$ [77, 86], so this problem has been removed.[34] The second is that the existing semiclassical tools are only appropriate for *graph non-changing operators* such as the volume operator. Namely, as we will see, in order to be normalisable, coherent states are (superpositions of) states defined on specific graphs. The Hamiltonian constraint operator, however, is graph changing. This means that it creates new modes on which the coherent state does not depend and whose fluctuations are therefore not suppressed. Therefore the existing semiclassical tools are insufficient for graph-changing operators such as the Hamiltonian constraint. The development of improved tools is extremely difficult and currently out of reach.

7. *Solutions and physical inner product*
Solutions to all constraints can be constructed algorithmically [38]. These are the full LQG analogues of the LQC solutions of the difference equation that results from the single Hamiltonian constraint of LQC. They are the first rigorous solutions ever constructed in canonical quantum gravity, have non-zero volume and are labelled by *fractal knot classes* because the iterated action of the Hamiltonian constraint creates a self-similar structure (spiderweb) around each vertex. However, as in LQC these solutions are not systematically derived from a rigging map which is why a physical inner product is currently missing for those solutions.

This finishes the discussion of the properties of the Hamiltonian constraint operators. We want to stress that while evidently several issues need to be resolved, this is the first time in history that canonical quantum gravity was brought to a level such that
1. these and related questions could meaningfully be asked and analysed with mathematical precision;
2. a concrete, natural proposal for the Hamiltonian constraint operators can be derived which is consistent (anomaly free), namely the one where the segments

[34] Notice that it is not possible to probe $\mathfrak{D}$ with semiclassical spatially diffeomorphism invariant states because none of the operators involved preserves $\mathcal{H}_{\mathrm{Diff}}$.

s_I and $s_{\rm IJ}$ respectively are aligned and transversal to the graph respectively and where the resulting loops $\beta_{\rm IJ}$ are in the $j = 1/2$ representation.

Nobody in the LQG community believes that this concrete model is the "right" or "final" one, but it provides a concrete proposal which can be studied and further improved.

As discussed, the most important open issues are the semiclassical limit and the physical inner product. These issues are overcome to a large extent by the master constraint programme.

Master Constraint Programme

The idea of the master constraint is to sidestep the complications of the Hamiltonian constraints that have their origin in the non-Lie algebra structure of the Dirac algebra $\mathfrak{D}$. Consider the master constraint

$$M := \int_\sigma d^3x \, \frac{H(x)^2}{\sqrt{\det(q)}(x)} \tag{37}$$

It is not difficult to see that (37) has the following properties:
1. $M = 0$ is equivalent with $H(N) = 0$ for all test functions N.
2. $\{F, \{F, M\}\}_{\rm M=0} = 0$ is equivalent with $\{F, H(N)\} = 0$ when $H(N') = 0$ for all test functions $N,\ N'$.
3. M is spatially diffeomorphism invariant.
The first property says that the single constraint $M = 0$ encodes the same constraint surface as the infinite number of Hamiltonian constraints while the second property says that the single double Poisson bracket with M selects the same weak Dirac observables as the infinite number of single Poisson brackets with Hamiltonian constraints. In other words the master constraints defines the same reduced phase space as the infinite number of Hamiltonian constraints.

The third property means that the complicated Dirac algebra $\mathfrak{D}$ can be replaced by the comparatively trivial Master Algebra $\mathfrak{M}$

$$\begin{aligned}
\{D(\boldsymbol{N}), D(\boldsymbol{N}')\} &= \kappa D(\mathcal{L}_{\boldsymbol{N}}\boldsymbol{N}') \\
\{D(\boldsymbol{N}), M\} &= 0 \\
\{M, M\} &= 0
\end{aligned} \tag{38}$$

which now is a *true Lie algebra.* This removes almost all obstacles that we encountered with the Hamiltonian constraints:

1. *Role of* $\mathcal{H}_{\rm Diff}$
   Since M is spatially diffeomorphism invariant, its operator version $\widehat{M}$ can be defined directly on the spatially diffeomorphism invariant Hilbert space $\mathcal{H}_{\rm Diff}$. In fact, if $\widehat{M}$ is a knot class changing spatially diffeomorphism invariant operator, then it *must* be defined on $\mathcal{H}_{\rm Diff}$, it cannot be defined on $\mathcal{H}$ [43]. In retrospect, this justifies the construction of $\mathcal{H}_{\rm Diff}$ because

for the solution of the Hamiltonian constraints the Hilbert space $\mathcal{H}_{\rm Diff}$ is unsuitable as an intermediate step towards the physical Hilbert space as it is not left invariant by the Hamiltonian constraints.

2. *Regulator removal*
Remember the awkward role of spatially diffeomorphism invariant states in the removal of the regulator of the Hamiltonian constraint operators using a special type of weak* operator topology. It turns out [74] that a knot class changing operator can indeed be constructed on $\mathcal{H}_{\rm Diff}$ by directly implementing the techniques of [20] sketched above. In the construction of this operator, the removal of the regulator is now in the standard weak operator topology of $\mathcal{H}_{\rm Diff}$.
3. *Physical Hilbert space*
Since the infinite number of Hamiltonian constraints was replaced by a single constraint, provided $\widehat{M}$ can be defined as a positive self-adjoint operator on $\mathcal{H}_{\rm Diff}$ and provided that $\mathcal{H}_{\rm Diff}$ decomposes into a direct sum of $\widehat{M}$-invariant, separable Hilbert spaces, we know that direct integral decomposition guarantees the existence of a physical Hilbert space with a positive definite inner product induced from $\mathcal{H}_{\rm Diff}$. In order to construct it explicitly one needs to know the projection valued measure associated with $\widehat{M}$, that is, only standard spectral theory is required. While this is a difficult task to carry out explicitly due to the complexity of the operator $\widehat{M}$, we therefore have *an existence proof* for $\mathcal{H}_{\rm phys}$.
4. *Anomalies: Ambiguities, locality and the semiclassical limit*
Since there is only one master constraint operator, it is trivially anomaly free. This enables one to consider a wider class of loop attachments, in particular those that would lead to an anomaly in the algebra of the Hamiltonian constraints. As examples show [35], such quantisations of the master constraint based on anomalous individual constraints lead to spectrum which does not include zero. However, the prescription to subtract the spectrum gap from the master constraint as mentioned in Sect. 3 works in all examples studied. One might worry that this spectrum gap, which in free field theories is related to a normal ordering constant, is infinite. However, this is not the case: The master constraint is not just a plain sum of squares of the individual constraints, it is a *weighted* sum. The weight in the case of gravity is the factor $1/\sqrt{\det(q)}$ in (37) which is the natural object to consider in order to make (37) spatially diffeomorphism invariant. For the case of the Maxwell field Gauss constraint studied in [35] the associated weight had to be a certain trace class operator on the one particle subspace of the Fock space. The weight function (operator) thus makes the normal ordering constant finite. Hence the master constraint programme can handle anomalous constraints.
For gravity of particular interest are constraints which are not graph changing, although the corresponding Hamiltonian constraints would be anomalous, for three reasons:

4a. *Ambiguities*
Since the loop to be attached is already part of the graph, the situation becomes closer to the situation in lattice gauge theory. This tremendously reduces the number of choices for the loop attachment and makes it no worse than the choice of a fundamental Hamiltonian in Wilson's approach to the renormalisation group.

4b. *Locality*
With this option we are free to consider for instance "next neighbour" loop attachments. This leads to a spreading of the influence of the action of the Hamiltonian constraint from one vertex to all others thus removing the criticism of [82].

4c. *Semiclassical limit*
A non-graph-changing master constraint can be defined on the kinematical Hilbert space. This has the advantage that the semiclassical states which so far in LQG are elements of $\mathcal{H}$ can be directly used to analyse the semiclassical properties of $\widehat{M}$. This has been done recently in [77] with the expected result that the infinitesimal generators (the Hamiltonian constraints) do have the correct semiclassical limit. Since these constraints determine the physical Hilbert space, this is an important step towards showing that gauge invariant operators commuting with $\widehat{M}$ have the correct semiclassical limit on the physical Hilbert space. The semiclassical limit is of course only reached on graphs which are sufficiently fine. Graphs with huge holes would correspond to spacetimes with degenerate metrics in macroscopic regions which is not allowed in classical general relativity.

Notice that semiclassical states have so far not been constructed on $\mathcal{H}_{\mathrm{Diff}}$. This means that the semiclassical limit of the graph-changing master constraint is currently out of reach, thus favouring the graph non-changing version.

In what follows we will sketch both the graph-changing master constraint operator on $\mathcal{H}_{\mathrm{Diff}}$ and the graph non-changing operator on $\mathcal{H}$.

Graph-Changing Master Constraint. We follow closely [74]. We notice that classically (τ is again a triangulation)

$$M = \lim_{\tau \to \sigma} \sum_{\Delta \in \tau} [\tilde{C}(\Delta)]^2 \tag{39}$$

where $\tilde{C}(\Delta)$ coincides with (30) for the smearing function $N = \chi_\Delta$ (the characteristic function for a tetrahedron) and with $V(\Delta)$ replaced by $\sqrt{V(\Delta)}$. Thus, the heuristic idea is to define the quadratic form on $\mathcal{D}^*_{\mathrm{Diff}}$ by

$$Q_M(l, l') := \lim_{\tau \to \sigma} \sum_{\Delta \in \tau} < l, [\widehat{\tilde{C}}'(\Delta)]^* [\widehat{\tilde{C}}'(\Delta)] l' >_{\mathrm{Diff}} \tag{40}$$

where the prime denotes the operator dual as usual and * denotes the adjoint on $\mathcal{H}_{\text{Diff}}$. Unfortunately (40) is ill-defined as it stands because the operators $\hat{\tilde{C}}'(\Delta)$ do not preserve $\mathcal{D}^*_{\text{Diff}}$. The cure is to extend $< .,. >_{\text{Diff}}$ to an inner product $< .,. >_*$ on all of $\mathcal{D}^*$. The final result turns out to be insensitive to the details of the extension because in the limit $\tau \to \sigma$ the Riemann sum becomes well defined on $\mathcal{D}^*_{\text{Diff}}$.

Rather than going through the rigorous argument which can be found in [74] we will present here the shortcut already sketched in [19]: Pretending that (40) is well defined we can insert a resolution of unity (we assume that the normalisation constants $k_{[s]}$ have been absorbed into the $T_{[s]}$ so that the $T_{[s]}$ form an orthonormal basis)

$$Q_M(T_{[s_1]}, T_{[s_2]}) := \lim_{\tau \to \sigma} \sum_{\Delta \in \tau} \sum_{[s]} < T_{[s_1]}, [\hat{\tilde{C}}'(\Delta)]^* T_{[s]} >_{\text{Diff}} < T_{[s]}, [\hat{\tilde{C}}'(\Delta)] l' >_{\text{Diff}} \tag{41}$$

Using the definition of the rigging map $\eta(T_s) = T_{[s]}$, the definition of the scalar product $< \eta(T_s), \eta(T_s') >_{\text{Diff}} = \eta(T_{s'})[T_s]$ and the definition of the dual operator $[O'l](f) = l(O^\dagger f)$ we obtain the now well-defined equation

$$Q_M(T_{[s_1]}, T_{[s_2]}) = \lim_{\tau \to \sigma} \sum_{\Delta \in \tau} \sum_{[s]} \overline{T_{[s_1]}(\hat{\tilde{C}}^\dagger(\Delta) T_{s_0([s])})}\; T_{[s_2]}(\hat{\tilde{C}}^\dagger(\Delta) T_{s_0([s])}) \tag{42}$$

where $s_0([s])$ is some representative of $[s]$. Now for fixed $[s_1], [s_2]$ the number of $[s]$ contributing to (41) is easily seen to be finite. Hence we can interchange the two sums in (41). Furthermore, for sufficiently fine τ we just need to consider those terahedra Δ_v containing a vertx of the graph underlying $s_0([s])$. One can then show that the limit $\tau \to \sigma$ becomes trivial due to the diffeomorphism invariance of $T_{[s_1]}$, $T_{[s_2]}$. Denoting $\hat{\tilde{C}}^\dagger(\Delta)$ for $v \in \Delta$ by $\hat{\tilde{C}}^\dagger_v$, which is the same as the coefficient of $N(v)$ in (33) just that $\hat{V}_v := V(R_v)$ is replaced by $\sqrt{\widehat{V_v}}$, we therefore obtain the final formula

$$Q_M(T_{[s_1]}, T_{[s_2]}) = \sum_{[s]} \sum_{v \in V(\gamma(s_0([s])))} \overline{T_{[s_1]}(\hat{\tilde{C}}^\dagger_v T_{s_0([s])})}\; T_{[s_2]}(\hat{\tilde{C}}^\dagger_v T_{s_0([s])}) \tag{43}$$

It is easy to show that (43) is independent of the representative $s_0([s])$ again due to spatial diffeomorphism invariance.

Expression (43) defines a positive quadratic form. However, it is not obvious that it presents the matrix elements of a positive operator. In [74] it is shown that (43) is closable thus presenting the matrix elements of a positive, self-adjoint operator $\widehat{M}$ on $\mathcal{H}_{\text{Diff}}$. Moreover, the non-separable Hilbert space $\mathcal{H}_{\text{Diff}}$ decomposes into an uncountable direct[35] sum $\mathcal{H}_{\text{Diff}} = \oplus_\theta \mathcal{H}^\theta_{\text{Diff}}$. Here the

[35] Modulo some subtleties which can be found in [74] and that can be dealt with.

sectors $\mathcal{H}^{\theta}_{\text{Diff}}$ are separable and are labelled by the angle moduli mentioned earlier. They are left invariant by $\widehat{M}$ basically because it only creates three valent vertices which do not have moduli. It follows that the direct integral method is applicable to $\widehat{M}$ thus resulting in the physical Hilbert space $\mathcal{H}_{\text{phys}}$ induced from $\mathcal{H}_{\text{Diff}}$.

It is easy to show that $\widehat{M}$ allows for an infinite number of zero eigenvectors (elements of $\mathcal{H}_{\text{Diff}}$). This follows immediately from the properties of the Hamiltonian constraints. One just has to choose $\gamma(s)$ to be out of the range of the graphs underlying the SNWs generated by the Hmiltonian constraints. Hence zero is contained in the point spectrum of this operator which constructed using non-anomalous constraints. However, due to the present lack of graph-changing and even spatially diffeomorphism-invariant coherent states, a verification of the correct semiclassical limit of the graph-changing $\widehat{M}$ is currently out of reach.

Non-Graph-Changing (Extended) Master Constraint. In order to have control on the semiclassical limit one must currently use a non-graph-changing operator and an operator which can be defined on $\mathcal{H}$. This can only be done by using underlying Hamiltonian constraints which are anomalous in the naive discretisation displayed below. However, there are techniques known from lattice gauge theory [87] which make use of the renormalisation group flow and which might enable one to work with non-anomalous constraints. This amounts to considering more sophisticated discretisations. We see here that the issues of the semiclassical limit and the anomaly freeness are interlinked in a complicated way. Fortunately, anomalies do not pose any obstacles to the master constraint programme.

In order to define such an operator we need the notion of a minimal loop: Given a vertex v of a graph γ and two edges e, e' outgoing from v, a loop $\beta(\gamma, v, e, e')$ within γ based at v, outgoing along e and incoming along e' is said to be *minimal* if there is no other loop within γ with the same properties and fewer edges traversed. Let $L(\gamma, v, e, e')$ be the set of minimal loops with the data indicated. Notice that this set is always non-empty but may consist of more than one element. We now define $\widehat{M}T_s := \widehat{M}_\gamma T_s$ on spin network states T_s over γ where

$$
\begin{aligned}
\widehat{M}_\gamma &:= \sum_{v \in V(\gamma)} \widehat{\tilde{C}}_v^\dagger \widehat{\tilde{C}}_v \qquad (44)\\
\widehat{\tilde{C}}_v &:= \frac{1}{|T(\gamma,v)|} \sum_{e_1,e_2,e_3 \in T(\gamma,v)} \frac{\epsilon_v(e_1,e_2,e_3)}{|L(\gamma,v,e_1,e_2)|} \sum_{\beta \in L(\gamma,v,e_1,e_2)} \\
&\quad \mathrm{Tr}([A(\beta) - A(\beta)^{-1}] A(e_3) [A(e_3)^{-1}, \sqrt{\widehat{V}_v}])
\end{aligned}
$$

Here $T(\gamma, v)$ is the number of ordered triples of edges incident at v (taken with outgoing orientation) whose tangents are linearly independent[36] and

[36] We set $\widehat{\tilde{C}}_v = 0$ if $T(\gamma, v) = \emptyset$.

$\epsilon_v(e_1, e_2, e_3) = \text{sgn}(\det(\dot{e}_1(0), \dot{e}_2(0), \dot{e}_3(0))$. The volume operator is given explicitly by

$$\widehat{V}_v = \sqrt{|\frac{i}{48} \sum_{\mathrm{e_1,e_2,e_3 \in T(\gamma,v)}} \epsilon_v(e_1, e_2, e_3)\epsilon_{\mathrm{jkl}} X^j_{\mathrm{e}_1} X^k_{\mathrm{e}_2} X^l_{\mathrm{e}_3}|} \tag{45}$$

where $X^j_e T_s = \text{Tr}([\tau_j A(e)]^T \partial/\partial A(e))$ is the right invariant vector field on the copy of $SU(2)$ determined by the holonomy $A(e)$ as introduced earlier.

It is easy to see that the definition (44) is spatially diffeomorphism invariant. Moreover, the results of [77] imply that expectation values with respect to the coherent states constructed in [86] which were barely mentioned in [12], defined on graphs which are sufficiently fine, the zeroth order in $\hbar$ of $\widehat{M}_\gamma$ coincides with the classical expression. In other words, the correctness of the classical limit of $\widehat{M}$ has been established recently. The results of [77] also imply that the commutator between the $\sum_v N_v \widehat{C}_v$ reproduces the third relation in (2) in the sense of expectation values with respect to coherent states where $\widehat{C}_v$ is the same as $\widehat{\widehat{C}}_v$ in (44) just that $\sqrt{\widehat{V}_v}$ is replaced by $\hat{V}_v$. This removes a further criticism spelled out in [12], namely we have off-shell closure of the Hamiltonian constraints to zeroth order in $\hbar$. Possible higher order corrections (anomalies) are no obstacle for the master constraint programme as already said.

Brief Note on the Volume Operator

In order to show this, one has to calculate the matrix elements of (45) which is non-trivial because the spectrum of that operator is not accessible exactly.[37] However, one can perform an error-controlled $\hbar$ expansion within coherent state matrix elements and compute the matrix elements of every term in that expansion analytically [77]. The idea is extremely simple and it will surprise nobody that this works: In applications we are interested in expressions of the form Q^r where Q is a positive operator, $0 < r \leq 1/4$ is a rational number and its relation to the volume operator is $V = \sqrt[4]{Q}$. The matrix elements of Q in coherent states can be computed in closed form. Now use the Taylor expansion of the function $f(x) = (1+x)^r$ up to some order N including the remainder with $x = Q/ < Q > -1$ where $< Q >$ is the expectation value of Q with respect to the coherent state of interest. The operators x^n in that expansion can be explicitly evaluated in the coherent state basis while the remainder can be estimated from above and provides a higher $\hbar$ correction than any of the x^n, $0 \leq n \leq N$. This completely removes the criticism of [12] that "nothing can be computed".

In [12] we also find a lengthy discussion about the regularisation of the volume operator in terms of flux operators. Actually the discussion in [12]

[37] The matrix elements of the argument of the square root are known in closed form [88].

follows closely [79]; however, the additional avaraging step performed in [79] is left out in [12] for reasons unclear to the present author. However, even if one considers that averaging procedure unconvincing or unmotivated, there are completely independent abstract reasons for why (45) is the only possibility to define the volume operator which were spelled out in [79]: Namley, the argument of the volume operator, which classically is given as the integral of $\sqrt{|\det(E)|}$ must be a completely skew expression in the right invariant vector fields because the only way to regularise it is in terms of flux operators. Now the relative coefficients between the terms for each triple are fixed, up to an overall constant, by spatial diffeomorphism covariance and cylindrical consistency.[38] The task to do was to show that a regularisation indeed exists which produces (45) which was done in [79] and to fix the constant which was done in [80]. In addition, there is an alternative point splitting regularisation [89] which does not use the averaging which also results in [79]. Hence there can be absolutely no debate [12] about the correctness of (45) in particular that now we know from [77] that its classical limit is correct.

Finally, [12] stresses that the final operator that one gets should be independent of the regularisation scheme and it is criticised that the regularisation scheme that one uses for the volume operator seems to depend on ad hoc choices so that different choices could give a different operator. Again we state that [79, 80] fix the volume operator uniquely. Apart from that we would like to stress that in ordinary QFT there are only a handful of regularisation schemes that one tests: Pauli–Villars, minimal subtraction, dimensional regularisation, point splitting. Here two different schemes were used [79, 89] which resulted in the same operator and thus the test is of the same order of "generality".

Algebraic Quantum Gravity (AQG)

Notice that the framework of algebraic quantum gravity (AQG) proposed in [77] in many ways supersedes LQG: In contrast to LQG, AQG is a purely combinatorial theory, that is, topology and differential structure of σ are semiclassical notions and not elements of the combinatorial formulation. Next, there are not an uncountably infinite number of finite, embedded graphs, there is only one countably infinite algebraic (or abstract) graph [90]. In particular, the theory loses its graph dependence; only in the semiclassical sectors (corresponding to different σ) do embedded graphs play a role. Hence AQG can possibly deal with topology change.

The Hilbert space of AQG is still not separable, but for an entirely different reason than in LQG: Since the graph is infinite we have to deal with an infinite tensor product of Hilbert spaces [86]. However, as von Neumann showed, these Hilbert spaces naturally decompose into separable Hilbert spaces which in our case turn out to be invariant under the algebraic version of $\widehat{M}$ so that on

[38] That is, the expression (45) for a given graph reduces to the one on any smaller graph when applied to spin network functions over the smaller graphs.

each sector the physical inner product exists by direct integral decomposition. Hence, non-separability poses absolutely no obstacle. Some of these sectors can presumably be identified as approximations to quantum field theories on curved backgrounds (namely when the geometry fluctuations around that background are small). In some sense, all QFTs on curved spacetimes are included which must be the case in order to have a background-independent theory. The Hilbert space therefore *has to be non-separable* for we do not expect QFTs on different backgrounds to be unitarily equivalent and there are certainly uncountably many non-diffeomorphic backgrounds.

Finally, since the natural representation $U(\varphi)$ of Diff(σ) is not available in the combinatorial theory (there is no σ), spatial diffeomorphism invariance has to be dealt with in an algebraic way. This is possible by using the *extended* master constraint whose classical expression for given σ is given by

$$M_E = M + \int_\sigma d^3x \; \frac{q^{ab} D_a D_b}{\sqrt{\det(q)}} \tag{46}$$

It turns out that the additional piece in (46) just like M itself can be lifted to the algebraic level thus abstracting from the given σ. Actually, the additional piece could also be defined in LQG[39] and the results of [77] also imply that M_E has the correct classical limit in both LQG and AQG. However, within LQG M_E is somewhat unmotivated because one already has the representation $U(\varphi)$ of spatial diffeomorphisms. In AQG, on the other hand, there is no choice and the advantage of M_E is that it treats the Hamiltonian constraint and the spatial diffeomorphism constraint on equal footing (rather than defining the infinitesimal generator for the Hamiltonian constraints but only exponentiated diffeomorphisms).

We refrain from displaying more details about AQG here as this is a rather recent proposal and because this is a review about LQG. The interested reader is referred to [77].

Dirac Observables and Physical Hamiltonian

As mentioned in Sect. 2, general relativity is an already parametrised system and in order to extract gauge invariant information and a notion of physical time evolution among observables one must deparamterise it, e.g. using the relational framework sketched in Sect. 2. There are many ways to do this but a minimal requirement is that the physical Hamiltonian is close to the Hamiltonian of the standard model at least when spacetime is close to being flat. In [33] a particularly simple way of deparametrisation which fulfils this requirement has been recently proposed using scalar phantom matter. In fact one can write the Hamiltonian constraints in the equivalent form $H(x) = \pi(x) + C(x)$ where π is the momentum conjugate to the phantom field ϕ and

[39] Despite the fact that D_a does not exist as an operator-valued distribution in LQG. The too singular D_a are tamed by the additional operator $q^{ab}/\sqrt{\det(q)}$.

C is a positive function on phase space which depends on all remaining matter and geometry only. Let now for any real number τ

$$h_\tau := \int_\sigma d^3x \, (\tau - \phi)(x) \, C(x), \quad h := \int_\sigma d^3x \, C(x) \tag{47}$$

Given a spatially diffeomorphism invariant function F we set

$$F(\tau) := \sum_{\mathrm{n}=0}^{\infty} \frac{1}{n!} \{h_\tau, F\}_{(\mathrm{n})} \tag{48}$$

Then $F(\tau)$ is a one parameter family of Dirac observables and $dF(\tau)/d\tau = \{h, F(\tau)\}$. In particular, h is itself a Dirac observable, namely the physical Hamiltonian that drives the physical time evolution of the Dirac observables.

This holds for the classical theory. In quantum theory (48) should be replaced by

$$\widehat{F}(\tau) := \exp(\hat{h}_\tau/(i\hbar)) \, \widehat{F} \exp(-\hat{h}_\tau/(i\hbar)) \tag{49}$$

provided we can make sense out of $\hat{h}_\tau$ as a self-adjoint operator. This is work in progress.

Brief Note on Spin Foam Models

Spin foam models [91] are an attempt at a path integral definition of LQG. They were heuristically defined in the seminal work [92] which attempted at the construction of the physical inner product via the formal exponentiation of the Hamiltonian constraints of [20]. The reason that this approach was formal is that the Hamiltonian constraints do not form a Lie algebra and they are not even self-adjoint. Thus, there are mathematical (exponentiation of non-normal operators) and physical (non-Lie group structure of the constraints) prohibiting the possibility that functional integration over N of $\exp(i\widehat{H}(N))$ leads to a (generalised) projector) issues with this proposal.

This is why spin foam models nowadays take a different starting point. Namely, one starts from the Palatini action and writes it as a topological BF theory $S_{\mathrm{BF}} = \int_M \mathrm{Tr}(B \wedge F)$ together with additional simplicity constraint action $S(\Lambda, B) = \int_M \mathrm{Tr}(\Lambda \otimes B \wedge B)$ where Λ is a Lagrange multiplier tensor field with certain symmetry properties. Extremisation with respect to Λ imposes the condition that the B field two form comes from the wedge product of two tetrads. The advantage of this formulation is that a lot is known about the topological BF theory and one can regard the additional simplicity constraint as a kind of "interaction" term in addition to the "free" BF term. In order to define the spin foam model one has to regularise it as in the canonical theory by introducing a finite 4D trinagulation τ and a corresponding discretisation of the action like Wilson's action for Yang–Mills theory. The connection A underlying the curvature F is located as a holonomy on the edges of the dual triangulation τ^* while the B field is located on the faces of τ. One integrates

$\exp(iS_{\mathrm{BF}} + iS(\Lambda, B))$ over A with respect to the Haar measure and over B and the Lagrange multiplier Λ with respect to Lebesgue measure. The integral over Λ results in a δ-distribution in B. This can be heuristically replaced by a δ distribution in the right invariant vector fields corresponding to the holonomies of the connection. One can then perform the B integral resulting in an additional δ distribution in the holonomies which then are written as a sum over representations using the Peter&Weyl theorem. The simplicity constraints in terms of the right invariant vectoir fields then impose restrictions on the occurring representations on the edges and intertwiners at the vertices.

These steps are simplest illustrated by modelling the situation by a one-dimensional system $S_{\mathrm{BF}} = BF$, $S(\Lambda, B) = \Lambda B^2$. Then formally

$$\int dF \int dB \int d\Lambda \exp(i[BF + \Lambda B^2]) = \int dF \int dB \int \delta(B^2) \exp(iBF)$$
$$= \int dF \int dB \int \delta(-(d/dF)^2) \exp(iBF) = \int dF \int \delta(-(d/dF)^2) \delta(F) \quad (50)$$

This brief paragraph does not reflect at all the huge body of research performed on spin foam models; we have barely touched only those aspects directly connected with the canonical formulation. Please refer to [91] and references therein for a more complete picture describing the beautiful connection with state sum models, TQFTs, categorification, 4D manifold invariants (Donaldson theory), non-commutative geometry, emergence of Feynman graph language and renormalisation groups etc.

From the canonical perspective, spin foam models are very important as they provide a manifestly spacetime diffeomorphism covariant formulation of LQG. In order to reach this goal, the following issues have to be overcome:

1. The relation with the canonical formulation is somehow lost. In fact, it is well known how to obtain a path integral formulation of a given canonical constrained theory [26]. The integration measure cannot be the naive one as used above if there are second-class constraints. That this is indeed the case has been shown in the important work [93] which is, in the mind of the author, not sufficiently appreciated.
2. While the simplicity constraints expressed in terms of B are mutually commuting as operators, their replacement in terms of right invariant vector fields do not and in fact they do not form a closed algebra. Hence, considered as quantum constraints they are anomalous and it is remarkable that there exists a unique non-trivial solution to the simplicity constraints [94] at all. For the corresponding model one can show that the path integral is dominated by degenerate metrics [95], hence it seems not to have the correct semiclassical behaviour which is then maybe not too surprising. There should be a way to implement the simplicity constraints in their non-anomalous form.
3. In contrast to pure BF theory these constrained BF theories are no longer topological and thus not independent of the triangulation. Thus, in order

to get rid of the triangulation dependence one could sum over triangulations, and the weights with which this should be done are motivated by group field theory [96]. The result is supposed to give a formula defining a rigging map. While there are attractive features such as an emergent Feynman graph language, it is presently unclear whether the sum converges (as it should in a fundamental theory) or whether it is maybe not more appropriate to perform a refinement limit as in the theory of dynamical triangulations [97].

5 Physical Applications

We have so far mostly reported about the status of the quantisation programme. Since LQG is a non-perturbative approach, preferrably one would complete the quantisation programme before one studies physical applications. Since the programme reached its current degree of maturity only relatively recently, physical applications could so far not attract much attention. Certainly what is needed in the future is an approximation scheme with respect to which physical states, the physical inner product, Dirac observables, and the physical Hamiltonian can be computed with sufficient detail. The semiclassical states [86] provide a possible avenue especially with respect to applications for which the quantum geometry can be regarded as almost classical. Namely we can consider kinematical semiclassical states which are peaked on the constraint surface and on the gauge cut defined by the clock variables. These states are then approximately annihilated by the master constraint and the power series defining the Dirac Observables can be terminated after a few terms just like in perturbation theory. This procedure could be called *quantum gauge fixing* because we do not fix a gauge classically but rather suppress the fluctuations off the constraint surface and off the gauge cut.

Despite the fact that such an approximation scheme has so far not been worked out in sufficient detail[40] there are already some physical applications of LQG which are insensitive to the details of such an approximation scheme. These are (1) matter coupling, (2) kinematical geometrical operators, (3) Quantum black hole physics, (4) semiclassical states, (5) loop quantum cosmology and (6) LQG phenomenology. We will say only very little about these topics here because our main focus is on the mathematical structure of LQG. Hence we will restrict ourseleves to the salient results and ideas.

5.1 Matter Coupling

We have so far hardly mentioned matter. However, in LQG all (supersymmetric) standard matter can be straightforwardly coupled as well [38, 75]. As far

[40] In particular one would like to know how close the approximate kinematical calculations are to the actual calculations on the physical Hilbert space.

as the kinematics is concerned, the background-independent representation for the gauge fields of the standard matter is the same as for the gravitational sector because all the constructions work for an arbitrary compact gauge group. For Higgs fields, which are located at the vertices of the graph and other scalar matter, one has a similar construction just that now states are labelled by points rather than edges. Finally for fermionic matter one uses a standard Berezin integral kind of construction. As far as the dynamics is concerned, the key technique of section "Hamiltonian Constraint" applies. All negative powers of $\det(q)$ which appear in the matter terms and which are potentially singular can be replaced by commutators between fractional powers of the volume operator and gravitational holonomy operators.

The corresponding contributions to the Hamiltonian constraint have to be added up and are then squared in the master constraint again without picking up an UV divergence. It is often criticised that LQG therefore does not impose any restriction on the allowed matter coupling. While that may turn out to be phenomenologically attractive for the reasons mentioned in the introduction, it may actually be technically incorrect: For it could be that the answer to the question, whether the spectrum of the master constraint contains zero, critically depends on which type of matter we couple. This is due to the fact that the shift of the minimum of the spectrum of the master constraint away from zero is typically due to a kind of normal ordering correction. Now intuition from ordinary QFT suggests that there must be a critical balance between bosonic and fermionic matter in order that positive bosonic corrections cancel negative fermionic ones. Hence, maybe after all the spectrum only contains zero if we allow for supersymmetric matter. In order to decide this a more detailed knowledge of the spectrum of the master constraint is required.

5.2 Kinematical Geometric Operators

One of the most cited results of LQG is the discreteness of the spectrum of kinematical geometric operators such as the volume operator, the area operator [78, 98] or the length operator [99]. The origin of this pure point spectrum is that these operators are functions of right invariant vector fields on various copies of $SU(2)$ and thus they are diagonalised by linear combinations spin network states with fixed graph and edge spin but varying intertwiners. Since for fixed edge spin the space of intertwiners is finite dimensional, it follows that these operators reduce to finite dimensional Hermitean matrices on these fixed graph and spin subspaces.

However, one should stress that the discreteness of the spectrum is a kinematical feature: None of these operators commutes with the spatial diffeomorphism or the master constraint. Whether or not these operators retain this property after having them made true Dirac observables via the relational machinery depends on the clock matter that is used to deparametrise the theory. See [100] for a discussion.

However, if the discreteness of the spectrum is retained then this could be interpreted as saying that in LQG the geometry is discontinous or distributive at Planck scale. At macroscopic scales there is a correspondence principle at work, that is, the difference between subsequent eigenvalues rapidly decays for large eigenvalues.

5.3 Semiclassical States

As already mentioned, the development of semiclassical tools represent an important area of research in the development of LQG because they allow to test whether LQG is really a quantum theory of General Theory and not of some pathological phase thereof. These developments were hardly mentioned in [12]. Semiclassical states for LQG [86, 101, 102] have so far been constructed only for the kinematical Hilbert space because the primary goal was so far to test the semiclassical limit of the constraint operators which by definition annihilate physical semiclassical states and thus cannot be tested by them. However, we will present some ideas of how spatially diffeomorphism invariant or even physical semiclassical states might be constructed.

The kinematical semiclassical states are actually coherent states and are all based on the *complexifier technique* [86] which we will briefly sketch below.

Suppose that we are given a phase space of cotangent bundle structure $\mathcal{M} = T^*\mathcal{A}$ where $\mathcal{A}$ is the configuration space. A *complexifier* is, roughly speaking, a positive function C on $\mathcal{M}$ with the dimension of an action which grows stronger than linearly as $E^I \to \infty$ where E^I denotes the momentum coordinates on $\mathcal{M}$ and $I \in \mathcal{I}$ is a labelling set. Denoting the points in $\mathcal{A}$ by A_I we define complex configuration coordinates

$$Z_I = \sum_{\mathrm{n}=0}^{\infty} \frac{i^n}{n!}\{A_I, C\} \tag{1}$$

explaining the name complexifier. Suppose that the theory can be canonically quantised such that $\hat{C}$ becomes a positive, self-adjoint operator on a Hilbert space $\mathcal{H} = L_2(\overline{\mathcal{A}}, d\mu)$ of square integrable functions on some distributional extension $\overline{\mathcal{A}}$ of $\mathcal{A}$ with respect to some measure μ. The quantum analogue of (1) becomes, upon replacing Poisson brackets by commutators divided by $i\hbar$, the *annihilation operator*

$$\hat{Z}_I = e^{-\hat{C}/\hbar}\, \hat{A}_I\, e^{\hat{C}/\hbar} \tag{2}$$

which explains the dimensionality of C. The operators $\hat{Z}_I$ are mutually commuting. The exponentials are defined via the spectral theorem. Let δ_A be the δ-distribution with respect to μ with support at A and consider the distribution

$$\Psi_A := e^{-\hat{C}/\hbar}\delta_A \tag{3}$$

Due to the positivity of $\hat{C}$ the operator $\exp(-\hat{C}/\hbar)$ is a smoothening operator, explaining the required positivity of C. In fact, if $\mathcal{H}$ is separable, then (3) will be an element of $\mathcal{H}$ (normalisable) if C is suitably chosen. Now the growth condition in the definition of C typically ensures that Ψ_A is *analytic* in A and hence can be analytically continued, as an L_2 function, to Z. Denoting the analytically continued object by Ψ_Z we obtain immediately the defining property of a *coherent state* to be a simultaneous eigenstate of the annihilation operators

$$\hat{Z}_I \, \Psi_Z = Z_I \, \Psi_Z \tag{4}$$

Notice, however, that if $\mathcal{H}$ is not separable then Ψ_Z is only a coherent distribution even if C has all the required properties.

This construction in fact covers all coherent states that have been considered for finite or infinite systems of uncoupled harmonic oscillators, in particular the "classical" coherent states for the Maxwell field (QED). For the Maxwell field the complexifier turns out to be

$$C = \frac{1}{2e^2} \int_{\mathbb{R}^3} d^3x \; \delta_{\mathrm{ab}} \; E^a (-\Delta)^{-1/2} E^b \tag{5}$$

where e is the electric charge, E^a the electric field and Δ the flat space Laplacian.

In fact, quadratic expressions in the momentum operators always are good choices for C. However, for LQG we may not use background-dependent objects such as Δ. In [86] quadratic expressions in the area operator (see below) were used and semiclassical properties such as peakedness in phase space, infinitesimal Ehrenfest property, overcompleteness, semiclassical limit and small fluctuations were established. Of course, since the kinematical LQG Hilbert space $\mathcal{H}$ is not separable, one must restrict the complexifier construction to separable subspaces. Natural candidates are the Hilbert spaces $\mathcal{H}_\gamma$ (closure of the span of SNWFs over γ) and $\mathcal{H}'_\gamma$ (closure of the span of SNWFs over all subgraphs of γ). The resulting states $\Psi_{\mathrm{Z},\gamma} = \sum_{\gamma(\mathrm{s})=\gamma} \Psi_{\mathrm{Z,s}} T_s$ and $\Psi'_{\mathrm{Z},\gamma} = \sum_{\gamma(\mathrm{s})\subset\gamma} \Psi_{\mathrm{Z,s}} T_s$ are respectively the spin network or cylindrical projections of the distributions $\Psi_Z = \sum_s \Psi_{\mathrm{Z,s}} T_s$ (the sum is over all SNWs) and are called shadows [102] or cut-offs [86] of Ψ_Z respectively.[41]

This graph dependence of the present semiclassical framework of LQG is an unpleasant feature which so far has prevented one from establishing the semiclassical limit of graph-changing operators such as the Hamiltonian constraint. This is because the Hamiltonian constraint creates new edges whose fluctuations are not controlled by these graph-dependent states. Hence the above-mentioned semiclassical properties only hold for graph non-changing operators and this is why the graph non-changing master constraint is under much better control than the Hamiltonian constraints. In AQG [77] even the graph dependence is lost because there is only one fundamental graph.

---

41 In order to avoid confusion which may arise form corresponding remarks in [12]: These states are functions of distributional connections $A \in \overline{\mathcal{A}}$ labelled by smooth fields Z. This is even the case for Maxwell coherent states. Hence one can surely get back the smooth fields of the classical theory in the classical limit.

Finally, let us address the question of spatially diffeomorphism invariant or physical states. These Hilbert spaces do not obviously have a representation as L_2 spaces, and moreover it is not easy to find complexifiers with the required properties which are either spatially diffeomorphism invariant or Dirac observables. Hence the complexifier idea is not immediately applicable. However, we have shown that there are (anti-linear) rigging maps $\eta_{\text{Diff}} : \mathcal{D} \to \mathcal{H}_{\text{Diff}}$ and $\eta_{\text{phys}} : \mathcal{D}^*_{\text{Diff}} \to \mathcal{H}_{\text{phys}}$ respectively. Now, given a, say, cut-off state $\Psi_{Z,\gamma}$, we obtain spatially diffeomorphism invariant states $\Psi^{\text{Diff}}_{Z,\gamma} := \eta_{\text{Diff}}(\Psi_{Z,\gamma})$ and physical states $\Psi^{\text{phys}}_{Z,\gamma} := \eta_{\text{phys}} \circ \eta_{\text{Diff}}(\Psi_{Z,\gamma})$ which can serve as *Ansätze* for semiclassical states in the corresponding Hilbert spaces. Whether they continue to have the desired semiclassical properties with respect to spatially diffeomorphism invariant or Dirac observables respectively is the subject of current research.

5.4 Quantum Black-hole Physics

The main achievement of LQG in this application is to provide a microscopic explanation of the Bekenstein–Hawking entropy, see [103] and references therein. The classical starting point is the theory of isolated and dynamical horizons [104] which is somehow a local[42] definition of an event horizon and captures the intuitive idea of a black hole in equilibrium. The notion of an isolated horizon uses, among other things, the classical field equations and therefore is a classical concept which is imported into the quantum theory by hand. In other words, the presence of the black hole is put in classically leading to an inner boundary of spacetime. It would be more desirable to have entirely quantum criteria at one's disposal, see, e.g., [105] for first steps; however, the following partly semiclassical framework is completely consistent and satisfactory.

The presence of the inner boundary leads to boundary conditions which, intuitively speaking, reduce the gauge freedom at the boundary and thus give rise to boundary degrees of freedom. Remarkably, their dynamics is described by a $U(1)$ quantum Chern Simons theory. On the other hand, the bulk is described by LQG. In order to compute the entropy of the black hole one counts the number of eigenstates of the area operator of the S^2 cross sections S of the horizon[43] whose eigenvalues fit into the interval $[\text{Ar}_0 - \ell_P^2, \text{Ar}_0 + \ell_P^2]$, where Ar_0 is some macroscopic area.

This number would be infinite if S would be an arbitrary surface. Namely a bulk state is described, near the horizon, by the ordered sets of punctures of the bulk graph with the surface S and at each such puncture p by the total

[42] The usual definition of a black-hole region as the complement of the past of future null infinity obviously requires the knowledge of the entire spacetime and is inappropriate to do local physics.

[43] Due to the boundary conditions this turns out to be a Dirac observable. In particular, different cross sections have the same area.

spin j_p to which the edges running into p couple. The area eigenvalue for such a configuration is given by [78, 98]

$$\lambda = \hbar\kappa\beta \sum_p \sqrt{j_p(j_p+1)} \tag{6}$$

For fixed j_p there are an infinite number of spin network states which couple to total j_p (for instance, let two edges run into p with spins $j_1,\ j_2 = j_1 + j_p$ where j_1 is arbitrary). Hence, if we would count bulk states, the entropy would diverge.

However, the physical reasoning is that what we must count are horizon states of the Chern Simons theory because the horizon degrees of freedom are the intrinsic description of the black hole and not the bulk degrees of freedom. Due to the quantum boundary conditions, the surface and bulk degrees of freedom are connected in the following way in the quantum theory: Around each puncture, the holonomy along the loop of the $U(1)$ Chern Simons connection must be, roughly speaking, equal to the signed area (flux) of the surface bounded by the loop. Hence what matters to the surface theory is the number j_p and not the detailed recoupling that created it. In other words, one ignores the multiplicities of the j_p.

With this in mind one can count now the number of eigenvalues. This would again be infinite if there would not be an area gap, that is, a smallest non-vanishing area eigenvalue which one can read off from (6). The result is [106–108]

$$\ln(N) = \frac{\beta}{\beta_0} \frac{\mathrm{Ar}_0}{4\ell_P^2} + O(\ln(\mathrm{Ar}_0/\ell_P^2)) \tag{7}$$

where β_0 is a numerical constant. This is the Bekenstein Hawking formula if we set $\beta = \beta_0$ which has been suggested to be one way to fix the Immirzi parameter. This would be inconsistent if β_0 would depend on the hair of the black hole. However, the constant β_0 is universal, all black holes of the Schwarzschild–Reissner–Nordstrom–Newman–Kerr family are allowed as well as Yang–Mills and dilatonic hair. Notice that these black holes are of astrophysical interest, they are non-supersymmetric and far from extremal, in contrast to the similar calculations in string theory which heavily depend on extremality.

In summary, there is an unexpected, consistent interplay between classical black-hole physics, quantum Chern Simons theory and LQG. Future improvements should include the development of quantum horizons and Hawking radiation.

5.5 Loop Quantum Cosmology (LQC)

Loop quantum cosmology (LQC) is not the cosmological sector of LQG.[44] Rather it is the usual homogeneous minisuperspace quantised by the methods

[44] So far there is no satisfactory derivation of LQC from LQG, LQC is constructed "by analogy".

of LQG. This has a kinematical and a dynamical side [109]. As far as the kinematics is concerned, although LQC has only a finite number of degrees of freedom one can circumvent the Stone–von Neumann uniqueness theorem for the representations of the canonical commutation relations by dropping the assumption of weak continuity of the Weyl operators. This is in complete analogy to LQG where holonomies but no connections are well defined as operators for precisely the same reason. The corresponding Hilbert space is then not of the Schrödinger type $L_2(\mathbb{R}, dx)$ but rather of the Bohr type $L_2(\overline{\mathbb{R}}, d\mu_0)$. Here $\overline{\mathbb{R}}$ is the Bohr compactification of the real line. It is the counterpart of $\overline{\mathcal{A}}$ and can be defined as the Gel'fand spectrum of the Abelean C* algebra generated by the functions $q \mapsto \exp(i\mu q)$. This algebra is called the algebra of quasiperiodic functions and is the counterpart of $\overline{\text{Cyl}}$. Finally μ_0 is the Haar measure on $\overline{\mathbb{R}}$ which is in complete analogy to the Ashtekar–Lewandowski measure of LQG.

On the dynamical side the situation in LQG is matched in the sense that in the Hamiltonian constraint one cannot work with connections but only with holonomies. Hence one has to modify the classical constraint by working with, say $\sin(\mu_0 q)/\mu_0$ rather than q where μ_0 is an arbitrarily small but finite constant.[45] Since also inverse powers of the momentum p conjugate to q appear in the Hamiltonian constraint one uses the same key kind of key identities as in LQG such as

$$i r \mu_0 \frac{\text{sgn}(p)}{|p|^{1-r}} = e^{-i\mu_0 q}\{|p|^r, e^{i\mu_0 q}\} \tag{8}$$

where $0 < r < 1$ is a rational number, in order to avoid negative powers of the p operator in the quantum theory.

The main advantage of this model is that one can carry out almost all steps of the quantisation programme and compare it with the conventional Schrödinger quantisation (Wheeler DeWitt theory). The predictions of the model are in fact quite intriguing: Avoidance of curvature singularities, deterministic quantum gauge flow through the would-be singularity, inflation from quantum geometry, avoidance of chaos in Bianchi IX cosmologies, recovery of conventional cosmology at large-scale factor etc. See [76] for a review. The most mathematically precise treatment can be found in [110].

Of course, one is never sure whether the simplifications that are made in toy models spoil its predictive power, that is, whether the results of the toy model continue to hold in the full theory. Partial confirmation of LQC singularity avoidance results within full LQG can be found in [112], although via a completly different mechanism. However, at least the model

[45] See [109] for arguments to fix the value of μ_0. In LQG the analogue of μ_0 would be the regulator ϵ in the loop attachment; however, in LQG all values of ϵ are equivalent due to spatial diffeomorphism invariance. This does not happen in LQC because in LQC the spatial diffeomorphism group is gauge fixed so that the appearance of μ_0 could be considered as an artefact of the simplicity of the model.

tests important aspects of the full theory, in particular the key identities of the type (27) without which these spectacular results of LQC would not have been possible.

Notice that LQG and in particular LQC can easily deal with de Sitter space kind of situations while this seems to be harder in superstring theory whose effective low energy limit should be supergravity on de Sitter space. However, the de Sitter algebra does not admit a positive Hamiltonian indicating that supergravity on de Sitter space is unstable. This is potentially alarming because recent observations indicate that the universe currently undergoes a de Sitter phase.

5.6 LQG Phenomenology

The field of LQG phenomenology has just started to develop, mostly because in the majority of cases there is no clear-cut derivation of the phenomenological assumptions made from full LQG. See [113, 114] for a review. One of the hottest topics in this fields are signatures of Lorentz invariance violation. A phenomenological description of this could be doubly special relativity (DSR) [115], a theory in which not only the speed of light but also the Planck energy is (inertial) frame independent. In 3D it turns out to be possible to connect DSR [116], non-commutative geometry [117] and LQG in the spin foam formulation but in 4D this is still elusive.

The intuitive idea behind Lorentz invariance violation in quantum gravity is the apparently Planck scale discreteness of LQG: If true, then quantum geometry looks more like a crystal than vacuum even if the gravitational vacuum state looks like Minkowski space on large scales. Hence there could be non-trivial dispersion relations for light propagation leading to energy-dependent time-of-arrival delays of photons of high energies that have travelled a long distance. One possible source of such signals are γ ray bursts at cosmological distances [118, 119] which would be detectable by the GLAST detector provided that the effect is linear in E/E_P where E is the photon energy and E_P the Planck energy. For first steps towards a systematic derivation from LQG see [120, 121]. Notice that for the five perturbative string theories on the Minkowski target space Lorentz invariance is built in axiomatically, hence Lorentz invariance violation could discriminate between LQG and string theory.

Another hot topic concerns cosmology. To this realm belong the prediction of deviations from the scale invariance of the power spectrum of the cosmic microwave background radiation (CMBR) [122, 123] using LQC (related) methods which might be detectable by the WMAP or PLANCK detectors.

Suffice it to say that this field is largely unexplored and that it needs more input both from experiments and theory.

6 Conclusions and Outlook

We hope to have given a brief but self-critical account of the status of LQG with special focus on the most important issue, the implementation of the quantum dynamics. In particular, we hope to have addressed most if not all issues that have been brought up in [12]. We presented them from an "inside" point of view and showed why the mostly technically correct description in [12], in our mind, is often unnecessarily sceptic, inconclusive or incomplete. Notice also that we only reported results well known in the LQG literature. We emphasise this because the unfamiliar reader may have the impression that only [12] unveiled the issues discussed there.

We have indicated why non-separable Hilbert spaces are no obstacle in LQG, they may even be welcome! There has been important progress recently on the frontiers of the semiclassical limit, the physical Hilbert space, physical (Dirac) observables, the physical Hamiltonian, the constraint algebra, the avoidance of anomalies and quantisation ambiguities, the covariant formulation [13] as well as physical applications which were insufficiently appreciated in [12]. The report given in [12] in many ways displays the field of LQG as it was a decade ago and thus ignores the progress made since then during which the field quadrupled in size. We hope to have clarified in this report that important developments were left out in [12] thus hopefully reducing the negative flavour conveyed there.

Let us discuss further issues touched upon in [12] which were not yet mentioned in this chapter:

1. A folklore statement that seems to have entered several physics blogs is that weakly discontinuous representations of the kind used in LQG do not work for the harmonic oscillator so why should they work for more complicated theories? This is the conclusion reached in [124]. As we will now show, while [124] is technically correct, its physical conclusion is *completely wrong*. In [124] one used a representation discussed first for QED [125] in order to avoid the negative norm states of the Gupta–Bleuler formulation. In this representation neither position q nor momentum p operators are well defined, only the Weyl operators $U(a) = \exp(iaq)$, $V(b) = \exp(ibp)$ exist. Hence the usual harmonic oscillator Hamiltonian $H = q^2 + p^2$ does not exist in this representation. Consider the substitute $H_\epsilon = [\sin^2(\epsilon q) + \sin^2(\epsilon p)]/\epsilon^2$. What is shown in [124] is that this operator is ill-defined as $\epsilon \to 0$. Is this a surprise? Of course not, we knew this without calculation because the representation is not weakly continuous. The divergence of H_ϵ as $\epsilon \to 0$ in discontinuous representations is therefore a trivial observation. However, what is physically much more interesting is the following. Fix an energy level E_0 above which the harmonic oscillator becomes relativistic and thus becomes inappropriate to model the correct physics. Let[46] $a_\epsilon^\dagger := [\sin(q\epsilon) + i\sin(p\epsilon)]/\epsilon$.

[46] Notice that classically $H_\epsilon = |a_\epsilon|^2$.

Consider the finite number of observables

$$b_{\epsilon,\mathrm{n}} := \frac{1}{n!}(a_\epsilon)^n\ (a_\epsilon^\dagger a_\epsilon)(a_\epsilon^\dagger)^n,\ n = 0,..,N = E_0/\hbar \tag{1}$$

Let Ω_0 be the Fock vacuum in the Schrödinger representation and ω the state underlying the discontinuous representation. Fix a finite measurement precision δ. Since the Fock representation is weakly continuous we find $\epsilon_0(N,\delta)$ such that $|<\Omega_0, b_{\epsilon,\mathrm{n}}\Omega_0> -n\hbar| < \delta/2$ for all $\epsilon \leq \epsilon_0$. On the other hand, by Fell's theorem[47] we find a trace class operator $\rho_{\mathrm{N},\delta}$ in the GNS representation determined by ω such that $|\mathrm{Tr}(\rho_{\mathrm{N},\delta} b_{\epsilon_0,\mathrm{n}}) - <\Omega_0, b_{\epsilon_0,\mathrm{n}}\Omega_0>| < \delta/2$ for all $n = 0,1,..,N$. It follows that with arbitrary, finite precision $\delta > 0$ we find states in the Fock and discontinuous representations respectively whose energy expectation values are given with precision δ by the usual value $n\hbar$. This implies that the two states cannot be physically distinguished.

In [127, 128] even more was shown:[48] There the spectrum of the operator H_ϵ was studied and the eigenvectors were determined explicitly. One could show that by tuning ϵ according to N, δ even the first N eigenvalues do not differ more than δ from $(n+1)\hbar$. Moreover, having fixed such an ϵ, the non-separable Hilbert space is a direct sum of separable H_ϵ invariant subspaces, and if we just consider the algebra generated by a_ϵ each of them is superselected. Hence we may restrict to any one of these irreducible subspaces and conclude that the physics of the discontinuous representation is indistinguishable from the physics of the Schrödinger representation within the error δ. This should be compared with the statement found in [124] that in discontinuous representations the physics of the harmonic oscillator is not correctly reproduced.

2. Actually the paper [124] was triggered by [129] where the following was shown:
   Using discontinuous representations one can quantise the closed bosonic string in any spacetime dimension without encountering anomalies, ghosts (negative norm states) or a tachyon state (instabilities). The representation-independent and purely algebraic no-go theorem of [130] that the Virasoro anomaly is unavoidable is circumvented by quantising the Witt group Diff(S^1)$\times$ Diff(S^1) rather than its algebra diff(S^1)$\oplus$dif(S^1). Since the representation of the Witt group is discontinuous, the infinitesimal generators do not exist and there is no Virasoro

47 The abstract statement is [6]: The folium of a faithful state on a C*−algebra is weakly dense in the set of all states. Here the folium of states are all trace class operators on the corresponding GNS Hilbert space. The theorem applies to the unique C* algebra [126] generated by the Weyl operators $U(a)$, $V(b)$ which we are considering here. The representations considered in [124, 125] are faithful.

48 In a representation which was continuous in one of p or q but discontinuous in the other. But similar results hold in this completely discontinuous representation considered here.

algebra in this discontinuous representation, exactly like in LQG. However, as in LQG, a unitary representation of the Witt group is sufficient in order to obtain the Hilbert space of physical states via group-averaging techniques and even a representation of the invariant charges [131] of the closed bosonic string.

Does this mean that the magical dimension $D = 26$ cannot be seen in this representation? Of course it can: One way to detect it in the usual Fock representation of the string is by considering the Poincaré algebra (in the lightcone gauge) and ask that it closes. For the LQG string [129] again the Poincaré group is represented unitarily but weakly discontinuously. However, we can approximate the generators as above in terms of the corresponding Weyl operators using some tiny but finite parameter ϵ. Since these are a finite number of operators in the corresponding C^* algebra, an appeal to Fell's theorem and using continuity of the Weyl operators in the Fock representation gurantees that we find a state in the folium of the LQG string with respect to which the expectation values of the approximate Poincaré generators coincides with their vacuum (or higher excited state) expectation values in the Fock representation to arbitrary precision δ.

Thus $D = 26$ is also hidden in this discontinuous representation, it is just that for no D there is a quantisation obstruction. Of course, much still has to be studied for the LQG string, e.g. a formulation of scattering theory; however, the purpose of [129] was not to propose a phenomenologically interesting model but rather to indicate that $D = 26$ is not necessarily sacred but rather a feature of the specific Fock quantisation used.

3. One sometimes reads the statement [6] that the instantaneous fields (smeared only in 3D rather than in 4D) are too singular in interacting quantum field theories. In fact, in Wightman field theories one can read Haag's theorem as saying that the representation of the interacting field algebra (which contains dynamical information) is never unitarily equivalent to the representation of the canonical commutation relations of the free field algebra (which lacks the information about the interaction). This seems to imply an obstruction to canonical quantisation where one precisely starts from a purely kinematical representation of the Poincaré algebra of the instantaneous fields. In LQG this no-go theorem is circumvented because the quantum field theory that one constructs is not a Wightman field theory: It is a QFT of a new kind, namely a background-independent QFT to which Haag's theorem as stated above does not apply because the Wightman axioms do not hold. In fact, in LQG the interaction is encoded in the self-adjoint master constraint which is well (densely) defined.
4. In [12] we find the question, where in LQG does one find the counter terms [7] of perturbative quantum gravity? More generally, how does one make contact with perturbative QFT and what role does the renormalisation group play in LQG, if any? These perturbative questions are hard to

answer in a theory which is formulated non-perturbatively; however, let us make a guess:[49]

Once a physical Hamiltonian such as the one of Sect. 4.4. has been successfully quantised one can in principle define scattering theory in the textbook way, that is, one would compute transition amplitudes between initial and final physical states. This may be hard to do technically but there is no obstruction in principle. In order to recover perturbation theory around Minkowski space one will consider a physical state (vacuum) which is a minimal energy state with respect to that Hamiltonian and peaked on Minkowski space. The physical excitations of that state can be considered as the analogues of the graviton excitations of the perturbative formulation. Now by construction the transition amplitudes (n point functions) are finite; however, there will be of course screening effects, that is, effectively a running of couplings where the energy scale at which one measures is fed in by the physical state by which one probes a given operator. This is the way we expect to recover renormalisation effects.

As far as counterterms are concerned, as we have frequently stated, there are correction terms in all semiclassical computations done so far which depend on the Planck mass, see e.g. [121] where a finite but large quantum gravity correction to the cosmological constant[50] is computed which results from photon field propagation on fluctuating spacetimes. Similar results are expected with respect to graviton propagation. These counterterm operators are formulated in terms of the canonical fields but using the field equations one can presumably recast their classical limits into covariant counterterm Lagrangeans.

Of course it is on the burden of LQG to show that this is really what happens, but it is not that there are no ideas for possible mechanisms.

We will now answer a number of frequently asked questions which one can find in [12]:

1. *Is there only mathematical progress in LQG?*
   A continuously updated and fairly complete list of all LQG publications to date can be found in [132]. A brief look at this list will show that there are papers of all levels of rigour and that mathematically more sophisticated papers were motivated and driven by less rigorous papers which started from a physical idea. It is true that in LQG one puts stress on mathematical rigour. The reason is that developing background-independent QFTs means walking on terra incognita. Hence, one does not have the luxury to be cavalier about mathematical details as in background-dependent QFTs

[49] Of course, one could ask whether the question is meaningful if quantum gravity, which is believed to be non-renormalisable, simply does not admit a perturbative formulation. However, it is believed that perturbative quantum gravity does make sense as an effective theory.

[50] The cosmological "constant" therefore becomes dynamical.

where well-established theorems ensure that there are rigorous versions of formal calculations.

Section 5 should have indicated that current research is focussed on hot research topics such as semiclassical quantum gravity (contact with QFT on curved spacetimes), quantum cosmology and quantum black-hole physics. These results together with the huge body of work on spin foam models were hardly mentioned in [12]; see, however, [13]. Ignoring this research performed over the past 10 years means giving an out-of-date presentation of LQG which would be similar to writing a review on string theory without mentioning D-branes, M-theory and the landscape.

Notice also that being a much smaller and younger field than string theory or high energy physics[51] which in addition cannot just use the techniques from ordinary particle physics but in fact must first develop its own mathematical framework from scratch, the amount of results obtained so far is naturally smaller due to lack of man power.

2. *Has there anything been gained as compared to the Wheeler–DeWitt framework?*
Any serious theoretical physicist will confirm that it is almost a miracle that one can tame mathematical monsters such as the area, volume, Hamiltonian constraint or master constraint operator *at all.* These operators are integrals over delicate, non-polynomial functions of operator-valued distributions evaluated at the same point which are hopelessely singular in usual background-dependent Fock representations. Moreover, not only can one give mathematcial sense to them, they are even free of UV divergences. This is the beauty of background independence and provides a precise implementation of the old physical idea that quantum gravity should provide the *natural regulator of ordinary QFT UV divergences.*

For the first time one can write down a concrete, mathematically well-defined proposal for the Hamiltonian or master constraint and study its physical properties. For the first time one can actually construct rigorous solutions thereof. For the first time one can precisely define a kinematical, spatially diffeomorphism invariant or physical Hilbert space. For the first time one could show that the semiclassical limit of at least the graph non-changing master constraint is the correct one with respect to rigorously defined, kinematical coherent states.[52]

It is true that not all questions have been answered in connection with the quantum dynamics and research on it will continue to occupy many researchers during many years to come. However, what is asked for in [12] is too much: Nobody expects that one can completely solve the theory. We cannot even solve classical general relativity completely and we will probably never be able to. General relativity and even more so LQG are not

[51] LQG is the focal point of only an order of 10^2 reserachers worldwide.

[52] Notice that it is meaningless the semiclassical limit of a constraint operator with physical coherent states which by definition are in its (generalised) kernel.

integrable systems such as string theory on Minkowski background target space which is mathematically relatively trivial as a field theory. Even today people working in classical general relativity struggle to get analytical results for the gravitational waves radiated by, say, a black-hole merger. The problem was posed almost half a century ago but recent progress is mostly due to increasing computing power. Gravitational waves is just a tiny sector of classical general relativity and in LQG we also must hope that we can at least analyse the theory in sufficient detail in those sectors. In [12] the authors ask for (approximately) physical semiclassical states that enable one to investigate the QFT on curved spacetimes limit of the theory. We claim that these exist: Kinematical coherent states have been introduced in [86]. We can choose to have them peaked on the constraint surface and then those states solve the master constraint approximately [77], that is, $||\widehat{M}\psi|| \approx 0$. These states will enable us to perform semiclassical perturbation theory as described in [77] and the non-diagonisability of the volume operator poses absolutely no problem here.

Finally, again a glance at [132] reveals that in LQG there is linear progress on the quantisation programme outlined in Sect. 3 over the past 20 years. One never changed the rules of that programme which means that the velocity of progress is naturally decreasing as one faces the ever tougher steps of that programme. We just mention that because from [12] one could sometimes get the impression that what is criticised is that researchers in LQG did not identify the open problems mentioned in [12]. They did as one can see from the publications, but some of the problems simply have not yet been solved and are topics of current research. That does not mean that they cannot be solved at all and so far every hurdle in LQG was taken.

3. *Is general covariance broken in LQG?*
When reading [12] one may get the impression that spacetime diffeomorphism invariance is broken right from the beginning just because one performs a 3+1 split of the action. This impression is wrong. As we have tried to explain in Sect. 2 the constraints require that physical observables are idependent of the foliation that one introduces in the canonical formulation. This is the same in string theory where the Witt constraints require worldsheet diffeomorphism invariance of physical observables. These constraints are implemented in the LQG Hilbert space and their kernel defines spacetime diffeomorphism invariant states. The question of off-shell versus on-shell closure is still open for the Hamiltonian constraint. This is why the master constraint programme was introduced as a possible alternative which seems to work in the sense that for the master constraint one could show that these constraints and their algebra have the correct classical limit [77]. They are maybe implemented anomalously in the master constraint but by subtracting from the master constraint the minimum of the spectrum the anomaly can be cancelled which corresponds to some kind of

normal ordering. Notice that this is an *off-shell* closure of the constraints as asked for in [12].

In [12] the authors illustrate the importance of on-shell versus off-shell closure by multiplying a given set of non-anomalous quantum constraints with structure constants with a central operator. These modified constraints still close on-shell while the off-shell algebra would now close only with structure functions. However, there are now possibly extra solutions, namely those in the kernel of the central operator, hence the physical Hilbert space would suffer from an infinite number of ambiguities. We find this example inconclusive for the following reason: In LQG one did not randomly multiply the Hamiltonian constraint by something else but rather just used the classical expression and directly quantised it by reasonable regularisation techniques. Furthermore, when modifying the classical constraints by multiplying it with a Dirac observable, the modified constraints define the same constraint surface as the original ones only where the Dirac observable does not vanish. Hence the extra solutions in the kernel of the central operator are physically not allowed and therefore, in this example at least, the modified quantum constraints in fact define the same physical Hilbert space as the original constraints.

4. *Does non-separability of the Hilbert space prevent the emergence of the continuum in the semiclassical limit?*
   In [12] the authors point out the non-separability of the kinematical Hilbert space which originates from the weak discontinuity of the holonomy operators. They call this the *pulverisation of the continuum* in the sense that all, even infinitesimally different, edges lead to orthogonal spin network states. The only topology on the set graphs with respect to which the scalar product is continuous is the *discrete topology* (every subset is open). They then ask whether the continuum can be recovered in the semiclassical limit. The answer is in the affirmative: The approximately physical states [86] (kinematical coherent states which are peaked on the constraint surface of the phase space) are labelled by *smooth* classical fields and the corresponding expectation values of physical operators such as the master constraint depend even smoothly on those fields, not only continuously, see, e.g., [77].
5. *Is the ambiguity in the Hamiltonian constraint comparable to non-renormlisability of perturbative quantum gravity?*
   Definitely not:
   First of all there is a crucial qualitative difference: Perturbative quantum gravity cannot make sense as a fundamental theory because the perturbation series divereges for all possible choices of the renormalisation constants. LQG is a finite theory for any choice of the ambiguity parameters. Next, the countably infinite number of renormalisation constants in perturbative quantum gravity take continuous values so that the number of ambiguities is uncountably infinite while the physical Hilbert space of LQG depends only on a discrete number of ambiguities. Finally, in

perturbative quantum gravity all values of the infinite number of renormalisation constants are, a priori, all equally natural while in LQG all of the discrete choices are pathological except for a finite number. As we have explained, without some notion of naturality, even ordinary QFT suffers from an infinite number of ambiguities (such as all possible discretisations of Yang–Mills theories on all possible lattices). Hence applying naturalness, the amount of ambiguity in LQG reduces to a finite number of ambiguities which is comparable to the degree of ambiguity of a renormalisable ordinary quantum field theory.

In the appendix the interested reader can find an example where the necessity of mathematical machinery is illustrated in a concrete physical question, namely whether the so-called "Kodama state" is a physical state of LQG. This would not be possible without it and therefore exemplifies "what has been gained".

Acknowledgements

We thank the members of the Algebraic quantum field theory community, especially Dorothea Bahns, Detlev Buchholz, Klaus Fredenhagen, Robert Schrader and Rainer Verch for their encouragement to write this chapter. Also the pressure on the author from the Editorial Board of *Classical and Quantum Gravity* to write an answer to [12] is gratefully acknowledged. The author has benefitted from fruitful discussions with Abhay Ashtekar, Carlo Rovelli and Lee Smolin and is grateful for many comments by Herman Nicolai, Kasper Peeters and Marija Zamaklar. Furthermore we want to thank Hans Kastrup very much for frequently pointing out the necessity to respond to [124]. Finally, we are grateful to Ion Olimpiu Stamatescu who invited and urged the author to write this chapter for the forthcoming book *Approaches to Fundamental Physics. An Assessment of Current Theoretical Ideas* (Springer Berlin Heidelberg 2007).

The part of the research performed at the Perimeter Institute for Theoretical Physics was supported in part by funds from the Government of Canada through NSERC and from the Province of Ontario through MEDT.

Appendix

The Kodama State

We end this chapter by displaying a concrete example which illustrates the necessity of all the mathematical machinery in order to reach a conclusive answer for precise questions. The example is the so-called *Kodama state* [133] which is frequently claimed to be an exact solution to all constraints of LQG [134]. In fact the Kodama state has attracted much attention in the early days of LQG (see [135] and references therein) because of its formal connection with the Jones polynomial [136], which would therefore seem to be an exact solution (in the loop representation) of all the constraints of LQG.

We will now show that this does not hold for various reasons. To see what is going on, consider pure gravity with a cosmological constant. After multiplying the Hamiltonian constraint by the factor $\sqrt{\det(q)}$ it is given by

$$\tilde{H} = \epsilon^{jkl} \epsilon_{\mathrm{abc}} E^a_j E^b_k [B^{\mathbb{C}c}_l + \Lambda E^c_l] \tag{1}$$

where Λ is the cosmological constant and $B^{\mathbb{C}c}_j = \epsilon^{abc} F^{\mathbb{C}j}_{\mathrm{ab}}/2$ the magnetic field of the complex connection $A^{\mathbb{C}}$ which is the pull-back to σ of the self-dual part of the spin connection (annihilating the tetrad).

The idea underlying the Kodama state is that the $SL(2,\mathbb{C})$ Chern–Simons action

$$S_{\mathrm{CS}}[A^{\mathbb{C}}] := \int_\sigma \mathrm{Tr}(F^{\mathbb{C}} \wedge A^{\mathbb{C}} - \frac{1}{3} A^{\mathbb{C}} \wedge A^{\mathbb{C}} \wedge A^{\mathbb{C}}) \tag{2}$$

is the generating functional of the magnetic field, that is, $\delta S_{\mathrm{CS}}/\delta A^{\mathbb{C}j}_a(x) = B^{\mathbb{C}a}_j(x)$ where the functional derivative is in the sense of the space of smooth $SL(2,\mathbb{C})$ connections $\mathcal{A}_{\mathbb{C}}$. Now the canonical brackets $\{E^a_j(x), A^{\mathbb{C}k}_b(y)\} = i\kappa\delta(x,y)$ suggest to formally define a Hilbert space $\mathcal{H}_{\mathbb{C}} = L_2(\mathcal{A}_{\mathbb{C}}, [dA_{\mathbb{C}}\, d\overline{A^{\mathbb{C}}}])$ of square integrable, holomorphic functions on $\mathcal{A}_{\mathbb{C}}$ with respect to formal Lebesgue measure and to represent $A^{\mathbb{C}j}_a(x)$ as a multiplication operator and $E^a_j(x)$ as $-\ell_P^2 \delta/\delta A^{\mathbb{C}j}_a(x)$. This formally satisfies the canonical commutation relations.

As is well known, the Chern–Simons action is invariant under infinitesimal gauge transformations, and as an integral of a three form over all of σ it is also spatially diffeomorphsim invariant. Moreover, in the ordering (1) the Kodama state

$$\Psi_{\mathrm{Kodama}} = e^{\frac{1}{\Lambda \ell_P^2} S_{\mathrm{CS}}} \tag{3}$$

is annihilated by the Hamiltonian constraint. This is exciting because the nine conditions $B = -\Lambda E$ satisfied by this state is easily seen to correspond to de Sitter space for the appropriate sign of Λ [134]. Hence, the Kodama state could be argued to correspond to the de Sitter vacuum of LQG.

There are several flaws with this formal calculation:

A. *Adjointness relations*
The formal representation of the canonical commutation relations just outlined is not a representation of the *algebra generated by E, $A^{\mathbb{C}}$, that is, the adjointness relations are not satisfied. These demand that E is self-adjoint and that $A^{\mathbb{C}} + \overline{A^{\mathbb{C}}} = 2\Gamma(E)$ where Γ is the spin connection of E. It is clear that the "measure" $[dA^{\mathbb{C}} d\overline{A^{\mathbb{C}}}]$ cannot implement these adjointness

relations, hence we have to incorporate a formal kernel $K(A^{\mathbb{C}}, \overline{A^{\mathbb{C}}})$. A formal Fourier transform calculation [2] reveals that

$$K(A^{\mathbb{C}}, \overline{A^{\mathbb{C}}}) = \int [dE] \exp(i \int_\sigma d^3x \, [\frac{A_a^{\mathbb{C}j} + \overline{A_a^{\mathbb{C}j}}}{2} - \Gamma_a^j] E_j^a \tag{4}$$

Even without specifying the details of this functional integral, with this kernel the inner product is no longer positive (semi) definite.

B. *Euclidean gravity*
Suppose we replace $A^{\mathbb{C}}$ by a real connection. This formally corresponds to Euclidean gravity and now the formal Hilbert space would be $\mathcal{H} = L_2(\mathcal{A}, [dA])$ which does give rise to a formal representation of the algebra underlying A, E if $E_j^a(x) = i\ell_P^2 \delta / \delta A_a^j(x)$. Now the Kodama state becomes

$$\Psi_{\text{Kodama}} = e^{\frac{i}{\Lambda \ell_P^2} S_{\text{CS}}} \tag{5}$$

Being a pure phase, it is not normalisable in that formal inner product.

C. *Measurability*
Now consider instead the rigorous Ashtekar–Isham–Lewandowski representation $L_2(\overline{\mathcal{A}}, d\mu_0)$. Certainly the operator corresponding to (1) does not exist in that representation but let us forget about the origin of Ψ_{Kodama} and just ask whether it defines an element of that Hilbert space. Being a pure phase it is formally normalisable because the measure μ_0 is normalised. However, the question is whether the Kodama state is a measurable function[53] in order that we can compute inner products between the Kodama state and, say, spin network functions. It is easy to see that this is not the case. For instance this follows from the fact that if we would triangulate the integral over the the Chern–Simons action in order to replace the integral by a Riemann sum over certain holonomies (these are measurable functions) and consider the infinite refinement limit, then in this limit the Kodama state has non-vanishing inner product with an uncountably infinite number of spin network functions. Thus, it is not normalisable when viewed as a proper L_2 function. This can also be interpreted differently: Recall that L_2 functions are only defined up to sets of measure zero. The Chern–Simons action is a priori defined only on the measure zero subset of smooth connections. The extension to $\overline{\mathcal{A}}$ that we just tried by representing it as a linear combination of spin network functions is no longer a phase.

D. *Distributional solution*
One could interpret the last item as saying that the Chern–Simons state defines a rigorous element of $\mathcal{D}^*$ and now the question is whether it is

53 A function is said to be measurable if the preimages of open subsets of $\mathbb{C}$ are measurable subsets. In our case, the measurable states are generated by the Borel sets of $\overline{\mathcal{A}}$.

annihilated by the rigorously defined dual of the Hamiltonian constraint–constructed in Sect. 4. It is easy to see that this is not the case because the Hamiltonian constraint with a cosmological constant term, although its dual formally acquires the ordering as in (1), is not proportional to $B+\Lambda E$ because the volume operator that enters the cosmological constant term is not quantised in the form $E^a_j e^j_a$ but rather as $\sqrt{\det(|E|)}$. One could of course write the smeared constraint in the form

$$H(N) = \int_\sigma N \,\mathrm{Tr}([F + \Lambda * E] \wedge \{A, V\}) \quad (6)$$

where V is the volume functional and proceed as in Sect. 4 although it would be somewhat awkward to define the volume operator in this way. However, even if this would work, this would still only define a solution to the Euclidean constraint.

This discussion hopefully illustrates the physical importance of the mathematical notions introduced and shows that LQG has been brought to a level of mathematcal rigour that allows to actually answer physical questions. Without it we could not have decided if and in which sense the Kodama state is a physical state.

References

1. C. Rovelli. *Quantum Gravity*, (Cambridge University Press, Cambridge, 2004).
2. T. Thiemann. *Modern Canonical Quantum General Relativity*, (Cambridge University Press, Cambridge, 2006) (at press). [gr-qc/0110034]
3. C. Rovelli. Loop quantum gravity. *Living Rev. Rel.* **1** (1998), 1. [gr-qc/9710008]
   C. Rovelli. Strings, loops and others: a critical survey of the present approaches to quantum gravity. plenary lecture given at 15th Intl. Conf. on Gen. Rel. and Gravitation (GR15), Pune, India, Dec 16–21, 1997. [gr-qc/9803024]
   M. Gaul and C. Rovelli, Loop quantum gravity and the meaning of diffeomorphism invariance. *Lect. Notes Phys.* **541** (2000), 277–324. [gr-qc/9910079]
   T. Thiemann. Lectures on loop quantum gravity. *Lect. Notes Phys.* **631** (2003), 41–135. [gr-qc/0210094]
   A. Ashtekar and J. Lewandowski. Background-independent quantum gravity: a status report. *Class. Quant. Grav.* **21** (2004), R53. [gr-qc/0404018]
   L. Smolin. An invitation to loop quantum gravity. [hep-th/0408048]
4. R. M. Wald. *Quantum field theory in curved space-time and black hole thermodynamics*, (Chicago University Press, Chicago, 1995).
5. R. Brunetti, K. Fredenhagen and R. Verch. The generally covariant locality principle: a new paradigm for local quantum field theory. *Commun. Math. Phys.* **237** (2003), 31–68. [math-ph/0112041]
6. R. Haag. *Local Quantum Physics*, 2nd ed., (Springer Verlag, Berlin, 1996).

7. N. Marcus and A. Sagnotti. The ultraviolet behavior of N=4 Yang-Mills nnd the power counting of extended superspace. *Nucl. Phys.* **B256** (1985), 77.
M. H. Goroff and A. Sagnotti. The ultraviolet behavior of Einstein gravity. *Nucl. Phys.* **B266** (1986), 709.
Non – renormalizability of (last hope) $D = 11$ supergravity with a terse survey of divergences in quantum gravities. [hep-th/9905017]
8. M. H. Goroff and A. Sagnotti. Quantum gravity at two loops. *Phys. Lett.* **B160** (1985), 81.
9. J. Polchinski. *String Theory*, Vol. 1: An introduction to the bosonic string, Vol. 2: Superstring theory and beyond, (Cambridge University Press, Cambridge, 1998).
10. E. D'Hoker and D.H. Phong. Lectures on Two Loop Superstrings. [hep-th/0211111]
11. G. Scharf. *Finite Quantum Electrodynamics: The Causal Approach*, (Springer Verlag, Berlin, 1995).
12. H. Nicolai, K. Peeters and M. Zamaklar. Loop quantum gravity: an outside view. *Class. Quant. Grav.* **22** (2005), R193. [hep-th/0501114]
13. H. Nicolai and K. Peeters. Loop and spin foam quantum gravity: a brief guide for beginners. [gr-qc/0601129]
14. F. Denef and M. Douglas. Distributions of flux vacua. *JHEP* **0405** (2004), 072. [hep-th/0404116]
J. Shelton, W. Taylor, B. Wecht. Generalized flux vacua. [hep-th/0607015]
15. L. Susskind. The anthropic landscape of string theory. [hep-th/0302219]
16. L. Smolin. Scientific alternatives to the anthropic principle. [hep-th/0407213]
17. L. Smolin. The case for background independence. [hep-th/0507235]
18. J. Maldacena. The large N limit of superconformal field theories and supergravity. *Adv. Theor. Math. Phys.* **2** (1998), 231–252. [hep-th/9711200]
19. T. Thiemann. The Phoenix project: master constraint programme for loop quantum gravity. *Class. Quant. Grav.* **23** (2006), 2211–2248. [gr-qc/0305080]
20. T. Thiemann. Quantum Spin Dynamics (QSD). *Class. Quantum Grav.* **15** (1998), 839–873. [gr-qc/9606089]
21. D.M. Gitman and I. V. Tyutin. *Quantization of Fields with Constraints*, (Springer-Verlag, Berlin, 1990).
22. R. P. Woodard. Avoiding dark energy with 1/r modifications of gravity. [astro-ph/0601672]
23. R. Geroch. The domain of dependence. *Journ. Math. Phys.*, **11** (1970), 437–509.
24. R. Beig and O. Murchadha. The Poincaré group as the symmetry group of canonical general relativity. *Ann. Phys.* **174** (1987), 463.
25. P. A. M. Dirac. *Lectures on Quantum Mechanics*, (Belfer Graduate School of Science, Yeshiva University Press, New York, 1964).
26. M. Henneaux and C. Teitelboim. *Quantization of Gauge Systems*, (Princeton University Press, Princeton, 1992).
27. S. A. Hojman, K. Kuchar and C. Teitelboim. Geometrodynamics regained. *Annals Phys.* **96** (1976), 88–135.
28. T. Thiemann. The LQG string: loop quantum gravity quantization of string theory I: Flat target space. *Class. Quant. Grav.* **23** (2006), 1923–1970. [hep-th/0401172]

29. N. M. J. Woodhouse. *Geometric Quantization*, 2nd. ed., (Clarendon Press, Oxford, 1991).
30. C. Rovelli. What is observable in classical and quantum gravity? *Class. Quantum Grav.* **8** (1991), 297–316.
    C. Rovelli. Quantum reference systems. *Class. Quantum Grav.* **8** (1991), 317–332.
    C. Rovelli. Time in quantum gravity: physics beyond the Schrodinger regime. *Phys. Rev.* **D43** (1991), 442–456.
    C. Rovelli. Quantum mechanics without time: a model. *Phys. Rev.* **D42** (1990), 2638–2646.
31. B. Dittrich. Partial and complete observables for Hamiltonian constrained systems. [gr-qc/0411013]
    B. Dittrich. Partial and complete observables for canonical general relativity. [gr-qc/0507106]
32. T. Thiemann. Reduced phase space quantization and Dirac observables. *Class. Quant. Grav.* **23** (2006), 1163–1180. [gr-qc/0411031]
33. T. Thiemann. Solving the problem of time in general relativity and cosmology with phantoms and k-essence. [astro-ph/0607380]
34. B. Dittrich and T. Thiemann. Testing the master constraint programme for loop quantum gravity: I. General framework. *Class. Quant. Grav.* **23** (2006), 1025–1066. [gr-qc/0411138]
35. B. Dittrich and T. Thiemann. Testing the master constraint programme for loop quantum gravity: II. Finite – dimensional systems. *Class. Quant. Grav.* **23** (2006), 1067–1088. [gr-qc/0411139]
    B. Dittrich and T. Thiemann. Testing the master constraint programme for loop quantum gravity: III. SL(2R) models. *Class. Quant. Grav.* **23** (2006), 1089–1120. [gr-qc/0411140]
    B. Dittrich and T. Thiemann. Testing the master constraint programme for loop quantum gravity: IV. Free field theories. *Class. Quant. Grav.* **23** (2006), 1121–1142. [gr-qc/0411141]
    B. Dittrich and T. Thiemann. Testing the master constraint programme for loop quantum gravity: V. Interacting field theories. *Class. Quant. Grav.* **23** (2006), 1143–1162. [gr-qc/0411142]
36. J. Klauder. Universal procedure for enforcing quantum constraints. *Nucl. Phys.* **B547** (1999), 397–412. [hep-th/9901010]
    A. Kempf and J. R. Klauder, On the implementation of constraints through projection operators, *J. Phys.* **A34** (2001), 1019–1036. [quant-ph/0009072]
37. D. Giulini and D. Marolf. On the generality of refined algebraic quantization. *Class. Quant. Grav.* **16** (1999), 2479–2488. [gr-qc/9812024]
38. T. Thiemann. Quantum Spin Dynamics (QSD): II. The kernel of the Wheeler-DeWitt constraint operator. *Class. Quantum Grav.* **15** (1998), 875–905. [gr-qc/9606090]
    T. Thiemann. Quantum Spin Dynamics (QSD): III. Quantum constraint algebra and physical scalar product in quantum general relativity. *Class. Quantum Grav.* **15** (1998), 1207–1247. [gr-qc/9705017]
    T. Thiemann. Quantum Spin Dynamics (QSD): IV. 2+1 Euclidean quantum gravity as a model to test 3+1 Lorentzian quantum gravity. *Class. Quantum Grav.* **15** (1998), 1249–1280. [gr-qc/9705018]

T. Thiemann. Quantum Spin Dynamics (QSD): V. Quantum gravity as the natural regulator of the Hamiltonian constraint of matter quantum field theories. *Class. Quantum Grav.* **15** (1998), 1281–1314. [gr-qc/9705019]
T. Thiemann. Quantum Spin Dynamics (QSD): VI. Quantum Poincaré algebra and a quantum positivity of energy theorem for canonical quantum gravity. *Class. Quantum Grav.* **15** (1998), 1463–1485. [gr-qc/9705020]
39. B. S. DeWitt. Quantum theory of gravity. I. The canonical theory. *Phys. Rev.* **160** (1967), 1113–1148.
B. S. DeWitt. Quantum theory of gravity. II. The manifestly covariant theory. *Phys. Rev.* **162** (1967), 1195–1238.
B. S. DeWitt. Quantum theory of gravity. III. Applications of the covariant theory. *Phys. Rev.* **162** (1967), 1239–1256.
40. A. Ashtekar. New variables for classical and quantum gravity. *Phys. Rev. Lett.* **57** (1986), 2244–2247.
A. Ashtekar. New Hamiltonian formulation of general relativity. *Phys. Rev.* **D36** (1987), 1587–1602.
41. A. Ashtekar and C.J. Isham. Representations of the holonomy algebras of gravity and non-Abelean gauge theories. *Class. Quantum Grav.* **9** (1992), 1433. [hep-th/9202053]
42. A. Ashtekar and J. Lewandowski. Representation theory of analytic holonomy $C^\star$ algebras. In *Knots and Quantum Gravity*, J. Baez (ed.), (Oxford University Press, Oxford 1994). [gr-qc/9311010]
43. A. Ashtekar, J. Lewandowski, D. Marolf, J. Mourão and T. Thiemann. Quantization of diffeomorphism invariant theories of connections with local degrees of freedom. *Journ. Math. Phys.* **36** (1995), 6456–6493. [gr-qc/9504018]
44. J.M. Mourão, T. Thiemann and J.M. Velhinho. Physical properties of quantum field theory measures. *J. Math. Phys.* **40** (1999), 2337–2353. [hep-th/9711139]
45. F. Barbero. Real Ashtekar variables for Lorentzian signature space times. *Phys. Rev.* **D51** (1995), 5507–5510.
F. Barbero. Reality conditions and Ashtekar variables: a different perspective. *Phys. Rev.* **D51** (1995), 5498–5506.
46. G. Immirzi. Quantum gravity and Regge calculus. *Nucl. Phys. Proc. Suppl.* **57** (1997), 65. [gr-qc/9701052]
C. Rovelli and T. Thiemann. The Immirzi parameter in quantum general relativity. *Phys. Rev.* **D57** (1998), 1009–1014. [gr-qc/9705059]
47. B. Brügmann and J. Pullin. Intersecting N loop solutions of the Hamiltonian constraint of quantum gravity. *Nucl. Phys.* **B363** (1991), 221–246.
B. Brügmann, J. Pullin and R. Gambini. Knot invariants as nondegenerate quantum geometries. *Phys. Rev. Lett.* **68** (1992), 431–434. B. Brügmann, J. Pullin and R. Gambini. Jones polynomials for intersecting knots as physical states of quantum gravity. *Nucl. Phys.* **B385** (1992), 587–603.
48. T. Thiemann. Anomaly-free formulation of non-perturbative, four-dimensional Lorentzian quantum gravity. *Physics Letters* **B380** (1996), 257–264. [gr-qc/9606088]
49. J. Samuel. Canonical gravity, diffeomorphisms and objective histories. *Class. Quant. Grav.* **17** (2000), 4645–4654. [gr-qc/0005094]

J. Samuel. Is Barbero's Hamiltonian formulation a gauge theory of Lorentzian gravity? *Class. Quantum Grav.* **17** (2000), L141. [gr-qc/00050095]
50. S. Alexandrov. SO(4,C) covariant Ashtekar-Barbero gravity and the Immirzi parameter. *Class. Quant. Grav.* **17** (2000), 4255–4268. [gr-qc/0005085]
51. R. Gambini and A. Trias. Second quantization of the free electromagnetic field as quantum mechanics in the loop space. *Phys. Rev.* **D22** (1980), 1380.
C. Di Bartolo, F. Nori, R. Gambini and A. Trias. Loop space quantum formulation of free electromagnetism. *Lett. Nuov. Cim.* **38** (1983), 497.
R. Gambini and A. Trias. Gauge dynamics in the C representation. *Nucl. Phys.* **B278** (1986), 436.
52. R. Giles. The reconstruction of gauge potentials from Wilson loops. *Phys. Rev.* **D8** (1981), 2160.
53. T. Jacobson and L. Smolin. Nonperturbative quantum geometries. *Nucl. Phys.* **B299** (1988), 295.
54. C. Rovelli and L. Smolin. Loop space representation of quantum general relativity. *Nucl. Phys.* **B331** (1990), 80.
55. A. Ashtekar, A. Corichi and J.A. Zapata. Quantum theory of geometry III: Non-commutativity of Riemannian structures. *Class. Quant. Grav.* **15** (1998), 2955–2972 [gr-qc/9806041]
56. H. Araki. Hamiltonian formalism and the canonical commutation relations in quantum field theory. *J. Math. Phys.* **1** (1960), 492.
57. J. Lewandowski, A. Okolow, H. Sahlmann and T. Thiemann. Uniqueness of diffeomorphism invariant states on holonomy – flux algebras. *Comm. Math. Phys.* **267** (2006), 703–733. [gr-qc/0504147]
58. C. Fleischhack. *Representations of the Weyl algebra in quantum geometry.* [math-ph/0407006]
59. O. Bratteli and D. W. Robinson. *Operator algebras and quantum statistical mechanics*, Vol. 1,2, (Springer Verlag, Berlin, 1997).
60. W. Rudin. *Real and complex analysis*, (McGraw-Hill, New York, 1987).
61. J. Velhinho. A groupoid approach to spaces of generalized connections. *J. Geom. Phys.* **41** (2002) 166–180. [hep-th/0011200]
62. A. Ashtekar and J. Lewandowski. Projective techniques and functional integration for gauge theories. *J. Math. Phys.* **36** (1995), 2170–2191. [gr-qc/9411046]
A. Ashtekar and J. Lewandowski. Differential geometry on the space of connections via graphs and projective limits. *Journ. Geo. Physics* **17** (1995), 191–230. [hep-th/9412073]
63. J. R. Munkres. *Toplogy: A First Course*, (Prentice Hall Inc., Englewood Cliffs (NJ), 1980).
64. Y. Yamasaki. *Measures on Infinite Dimensional Spaces*, (World Scientific, Singapore, 1985).
65. H. Sahlmann and T. Thiemann. Irreducibility of the Ashtekar – Isham – Lewandowski representation. *Class. Quant. Grav.* **23** (2006), 4453–4472. [gr-qc/0303074]
66. spin network basis C. Rovelli and L. Smolin. Spin networks and quantum gravity. *Phys. Rev.* **D53** (1995), 5743–5759. [gr-qc/9505006]

J. Baez. Spin networks in non-perturbative quantum gravity. In *The Interface of Knots and Physics*, L. Kauffman (ed.), (American Mathematical Society, Providence, Rhode Island, 1996). [gr-qc/9504036]

67. M. Reed, B. Simon. *Methods of Modern Mathematical Physics*, Vols. 1–4, (Academic Press, Boston, 1980).
68. N. Grot and C. Rovelli. Moduli space structure of knots with intersections. *J. Math. Phys.* **37** (1996), 3014–3021. [gr-qc/9604010]
69. J.-A. Zapata. A combinatorial approach to diffeomorphism invariant quantum gauge theories. *Journ. Math. Phys.* **38** (1997), 5663–5681. [gr-qc/9703037]
    J.-A. Zapata. A combinatorial space from loop quantum gravity. *Gen. Rel. Grav.* **30** (1998), 1229–1245. [gr-qc/9703038]
70. W. Fairbairn and C. Rovelli. Separable Hilbert space in loop quantum gravity. *J. Math. Phys.* **45** (2004), 2802–2814. [gr-qc/0403047]
71. J. Velhinho. Comments on the kinematical structure of loop quantum cosmology. *Class. Quant. Grav.* **21** (2004), L109. [gr-qc/0406008]
72. D. Marolf and J. Lewandowski. Loop constraints: A habitat and their algebra. *Int. J. Mod. Phys.* **D7** (1998), 299–330. [gr-qc/9710016]
73. R. Gambini, J. Lewandowski, D. Marolf and J. Pullin. On the consistency of the constraint algebra in spin network gravity. *Int. J. Mod. Phys.* **D7** (1998), 97–109. [gr-qc/9710018]
74. T. Thiemann. Quantum spin dynamics (QSD): VIII. The master constraint. *Class. Quant. Grav.* **23** (2006), 2249–2266. [gr-qc/0510011]
    M. Han and Y. Ma. Master constraint operator in loop quantum gravity. *Phys. Lett.* **B635** (2006), 225–231. [gr-qc/0510014]
75. T. Thiemann. Kinematical Hilbert spaces for fermionic and Higgs quantum field theories. *Class. Quantum Grav.* **15** (1998), 1487–1512. [gr-qc/9705021]
76. M. Bojowald and H. A. Morales-Tecotl. Cosmological applications of loop quantum gravity. *Lect. Notes Phys.* **646** (2004), 421–462. [gr-qc/0306008]
77. K. Giesel and T. Thiemann. Algebraic quantum gravity (AQG) I. Conceptual setup. [gr-qc/0607099]
    K. Giesel and T. Thiemann. Algebraic quantum gravity (AQG) II. Semiclassical analysis. [gr-qc/0607100]
    K. Giesel and T. Thiemann. Algebraic quantum gravity (AQG) III. Semiclassical perturbation theory. [gr-qc/0607101]
78. Rovelli and L. Smolin. Discreteness of volume and area in quantum gravity. *Nucl. Phys.* **B442** (1995), 593–622. Erratum: *Nucl. Phys.* **B456** (1995), 753. [gr-qc/9411005]
79. A. Ashtekar and J. Lewandowski. Quantum theory of geometry II: volume operators. *Adv. Theo. Math. Phys.* **1** (1997), 388–429. [gr-qc/9711031]
80. K. Giesel and T. Thiemann. Consistency check on volume and triad operator quantisation in loop quantum gravity. I. *Class. Quant. Grav.* **23** (2006), 5667–5691. [gr-qc/0507036]
    K. Giesel and T. Thiemann. Consistency check on volume and triad operator quantisation in loop quantum gravity. II. *Class. Quant. Grav.* **23**, (2006) 5693–5771. [gr-qc/0507037]
81. P. Hajíček and K. Kuchař. Constraint quantization of parametrized relativistic gauge systems in curved space-times. *Phys. Rev.* **D41** (1990), 1091–1104.

P. Hajíček and K. Kuchař. Transversal affine connection and quantization of constrained systems. *Journ. Math. Phys.* **31** (1990), 1723–1732.

82. L. Smolin. The classical limit and the form of the Hamiltonian constraint in non-perturbative quantum general relativity. [gr-qc/9609034]
83. E. Witten. (2+1)-dimensional gravity as an exactly solvable system. *Nucl. Phys.* **B311** (1988), 46.
84. M. Gaul and C. Rovelli. A generalized Hamiltonian contraint operator in loop quantum gravity and its simplest Euclidean matrix elements. *Class. Quant. Grav.* **18** (2001) 1593–1624. [gr-qc/0011106]
85. A. Perez. On the regularization ambiguities in loop quantum gravity. *Phys. Rev.* **D73** (2006), 044007. [gr-qc/0509118]
86. T. Thiemann. Quantum Spin Dynamics (QSD): VII. Symplectic structures and continuum lattice formulations of gauge field theories. *Class. Quant. Grav.* **18** (2001) 3293–3338. [hep-th/0005232]
   T. Thiemann. Complexifier coherent states for canonical quantum general relativity. *Class. Quant. Grav.* **23** (2006), 2063–2118. [gr-qc/0206037]
   T. Thiemann. Gauge Field Theory Coherent States (GCS): I. General properties. *Class. Quant. Grav.* **18** (2001), 2025–2064. [hep-th/0005233]
   T. Thiemann and O. Winkler. Gauge Field Theory Coherent States (GCS): II. Peakedness properties. *Class. Quant. Grav.* **18** (2001) 2561–2636. [hep-th/0005237]
   T. Thiemann and O. Winkler. Gauge Field Theory Coherent States (GCS): III. Ehrenfest theorems. *Class. Quantum Grav.* **18** (2001), 4629–4681. [hep-th/0005234]
   T. Thiemann and O. Winkler. Gauge field theory coherent states (GCS): IV. Infinite tensor product and thermodynamic limit. *Class. Quantum Grav.* **18** (2001), 4997–5033. [hep-th/0005235]
   H. Sahlmann, T. Thiemann and O. Winkler. Coherent states for canonical quantum general relativity and the infinite tensor product extension. *Nucl. Phys.* **B606** (2001) 401–440. [gr-qc/0102038]
87. P. Hasenfratz. The Theoretical Background and Properties of Perfect Actions. [hep-lat/9803027]
   S. Hauswith. Perfect Discretizations of Differential Operators. [hep-lat/0003007]; The Perfect Laplace Operator for Non-Trivial Boundaries. [hep-lat/0010033]
88. J. Brunnemann and T. Thiemann. Simplification of the spectral analysis of the volume operator in loop quantum gravity. *Class. Quant. Grav.* **23** (2006), 1289–1346. [gr-qc/0405060]
89. T. Thiemann. Closed formula for the matrix elements of the volume operator in canonical quantum gravity. *Journ. Math. Phys.* **39** (1998), 3347–3371. [gr-qc/9606091]
90. D. Stauffer and A. Aharony. *Introduction to Percolation Theory*, 2nd ed., (Taylor and Francis, London, 1994).
   D. M. Cvetovic, M. Doob and H. Sachs. *Spectra of Graphs*, (Academic Press, New York, 1979).
91. A. Perez. Spin foam models for quantum gravity. *Class. Quant. Grav.* **20** (2003), R43. [gr-qc/0301113]
92. M. Reisenberger and C. Rovelli. Sum over surfaces form of loop quantum gravity. *Phys. Rev.* **D56** (1997), 3490–3508. [gr-qc/9612035]

93. E. Buffenoir, M. Henneaux, K. Noui and Ph. Roche. Hamiltonian analysis of Plebanski theory. *Class. Quant. Grav.* **21** (2004), 5203–5220. [gr-qc/0404041]
94. J. W. Barrett and L. Crane. Relativistic spin networks and quantum gravity. *J. Math. Phys.* **39** (1998), 3296–3302. [gr-qc/9709028]
J. W. Barrett and L. Crane. A Lorentzian signature model for quantum general relativity. *Class. Quant. Grav.* **17** (2000) 3101–3118. [gr-qc/9904025]
95. J. C. Baez, J. D. Christensen, T. R. Halford and D. C. Tsang. Spin foam models of Riemannian quantum gravity. *Class. Quant. Grav.* **19** (2002), 4627–4648. [gr-qc/0202017]
J. C. Baez and J. D. Christensen. Positivity of spin foam amplitudes. *Class. Quant. Grav.* **19** (2002), 2291–2306. [gr-qc/0110044]
96. L. Freidel. Group field theory: an overview. *Int. J. Theor. Phys.* **44** (2005), 1769–1783. [hep-th/0505016]
97. J. Ambjorn, M. Carfora and A. Marzuoli. *The geometry of dynamical triangulations*, (Springer-Verlag, Berlin, 1998).
98. A. Ashtekar and J. Lewandowski. Quantum theory of geometry I: Area Operators. *Class. Quantum Grav.* **14** (1997), A55–A82. [gr-qc/9602046]
99. T. Thiemann. A length operator for canonical quantum gravity. *Journ. Math. Phys.* **39** (1998), 3372–3392. [gr-qc/9606092]
100. B. Dittrich and T. Thiemann. Facts and fiction about Dirac observables. (to appear)
101. M. Varadarajan. Fock representations from U(1) holonomy algebras. *Phys. Rev.* **D61** (2000), 104001. [gr-qc/0001050]
M. Varadarajan. Photons from quantized electric flux representations. *Phys. Rev.* **D64** (2001), 104003. [gr-qc/0104051]
M. Varadarajan. Gravitons from a loop representation of linearized gravity. *Phys. Rev.* **D66** (2002), 024017. [gr-qc/0204067]
M. Varadarajan. The Graviton vacuum as a distributional state in kinematic loop quantum gravity. *Class. Quant. Grav.* **22** (2005), 1207–1238. [gr-qc/0410120]
102. A. Ashtekar and J. Lewandowski. Relation between polymer and Fock excitations. *Class. Quant. Grav.* **18** (2001), L117–L128. [gr-qc/0107043]
103. A. Ashtekar. Classical and quantum physics of isolated horizons: a brief overview. *Lect. Notes Phys.* **541** (2000) 50–70.
104. S. Hayward. Marginal surfaces and apparent horizons. [gr-qc/9303006]
S. Hayward. On the definition of averagely trapped surfaces. *Class. Quant. Grav.* **10** (1993), L137–L140. [gr-qc/9304042]
S. Hayward. General laws of black hole dynamics. *Phys. Rev.* **D49** (1994), 6467–6474.
S. Hayward, S. Mukohyama and M.C. Ashworth. Dynamic black hole entropy. *Phys. Lett.* **A256** (1999), 347–350. [gr-qc/9810006]
A. Ashtekar and B. Krishnan. Dynamical horizons and their properties. *Phys. Rev.* **D68** (2003), 104030. [gr-qc/0308033]
105. V. Husain and O. Winkler. Quantum black holes. *Class. Quant. Grav.* **22** (2005), L135–L142. [gr-qc/0412039]

106. A. Ashtekar, J. C. Baez and K. Krasnov. Quantum geometry of isolated horizons and black hole entropy. *Adv. Theor. Math. Phys.* **4** (2001), 1–94. [gr-qc/0005126]
107. M. Domagala and J. Lewandowski. Black hole entropy from quantum geometry. *Class. Quant. Grav.* **21** (2004), 5233–5244. [gr-qc/0407051]
108. K. Meissner. Black hole entropy in loop quantum gravity. *Class. Quant. Grav.* **21** (2004), 5245–5252. [gr-qc/0407052]
109. A. Ashtekar, M. Bojowald and J. Lewandowski. Mathematical structure of loop quantum cosmology. *Adv. Theor. Math. Phys.* **7** (2003), 233. [gr-qc/0304074]
110. A. Ashtekar, T. Pawlowski and P. Singh. Quantum nature of the big bang. *Phys. Rev. Lett.* **96** (2006), 141301. [gr-qc/0602086]
111. A. Ashtekar, T. Pawlowski and P. Singh. Quantum nature of the big bang. *Phys. Rev. Lett.* **96** (2006), 141301. [gr-qc/0602086]
A. Ashtekar, T. Pawlowski and P. Singh. Quantum nature of the big bang: an analytical and numerical investigation. I. *Phys. Rev.* **D73** (2006), 124038. [gr-qc/0604013]
A. Ashtekar, T. Pawlowski and P. Singh. Quantum nature of the big bang: improved dynamics. [gr-qc/0607039]
112. J. Brunnemann and T. Thiemann. On (cosmological) singularity avoidance in loop quantum gravity. *Class. Quant. Grav.* **23** (2006), 1395–1428. [gr-qc/0505032]
J. Brunnemann and T. Thiemann. Unboundedness of triad – like operators in loop quantum gravity. *Class. Quant. Grav.* **23** (2006), 1429–1484. [gr-qc/0505033]
113. T. Jacobson, S. Liberati and D. Mattingly. Lorentz violation at high energy: concepts, phenomena and astrophysical constraints. *Annals Phys.* **321** (2006), 150–196. [astro-ph/0505267]
114. S. Hossenfelder. Interpretation of quantum field theories with a minimal length scale. *Phys. Rev.* **D73** (2006), 105013. [hep-th/0603032]
115. J. Kowalski-Glikman. Introduction to doubly special relativity. *Lect. Notes Phys.* **669** (2005), 131–159. [hep-th/0405273]
116. L. Freidel, J. Kowalski-Glikman and L. Smolin. 2+1 gravity and doubly special relativity. *Phys. Rev.* **D69** (2004), 044001. [hep-th/0307085]
117. L. Freidel and S. Majid. Noncommutative harmonic analysis, sampling theory and the Duflo map in 2+1 quantum gravity. [hep-th/0601004]
118. G. Amelino-Camelia, John R. Ellis, N.E. Mavromatos, D.V. Nanopoulos and Subir Sarkar. Potential sensitivity of gamma ray burster observations to wave dispersion in vacuo. *Nature.* **393** (1998) 763–765. [astro-ph/9712103]
119. S. D. Biller et al. Limits to quantum gravity effects from observations of TeV flares in active galaxies. *Phys. Rev. Lett.* **83** (1999), 2108–2111. [gr-qc/9810044]
120. R. Gambini and J. Pullin, Nonstandard optics from quantum spacetime. *Phys. Rev.* **D59** (1999), 124021. [gr-qc/9809038]
121. H. Sahlmann and T. Thiemann. Towards the QFT on curved spacetime limit of QGR. 1. A general scheme. *Class. Quant. Grav.* **23** (2006), 867–908. [gr-qc/0207030]
H. Sahlmann and T. Thiemann. Towards the QFT on curved spacetime limit of QGR. 2. A concrete implementation. *Class. Quant. Grav.* **23** (2006), 909–954. [gr-qc/0207031]

122. S. Hofmann and O. Winkler. The spectrum of fluctuations in inflationary cosmology. [astro-ph/0411124]
123. S. Tsujikawa, P. Singh and R. Maartens. Loop quantum gravity effects on inflation and the CMB. *Class. Quant. Grav.* **21** (2004), 5767–5775. [astro-ph/0311015]
124. Robert C. Helling, G. Policastro. String quantization: Fock vs. LQG representations. [hep-th/0409182]
125. H. Narnhofer and W. Thirring. Covariant QED without indefinite metric. *Rev. Math. Phys.* **SI1** (1992), 197–211.
126. J. Slawny. On factor representations and the C^*-algebra of canonical commutation relations. *Comm. Math. Phys.* **24** (1972), 151–170.
127. A. Ashtekar, S. Fairhurst and J. L. Willis. Quantum gravity, shadow states and quantum mechanics. *Class. Quant. Grav.* **20** (2003), 1031. [gr-qc/0207106]
128. K. Fredenhagen, F. Reszewski. Polymer state approximations of Schrodinger wave functions. [gr-qc/0606090]
129. T. Thiemann. The LQG string: loop quantum gravity quantization of string theory I: Flat target space. *Class. Quant. Grav.* **23** (2006), 1923–1970. [hep-th/0401172]
130. G. Mack, "Introduction to Conformal Invariant Quantum Field Theory in two and more Dimensions", in: Cargese 1987, "Nonperturbative Quantum Field Theory", 1987; Preprint DESY 88–120
131. K. Pohlmeyer. A group theoretical approach to the quantization of the free relativistic closed string. *Phys. Lett.* **B119** (1982), 100.
D. Bahns. The invariant charges of the Nambu – Goto string and canonical quantisation. *J. Math. Phys.* **45** (2004), 4640–4660. [hep-th/0403108]
132. A. Hauser and A. Corichi. Bibliography of publications related to classical self-dual variables and loop quantum gravity, [gr-qc/0509039]
133. H. Kodama. Holomorphic wave function of the universe. *Phys. Rev.* **D42** (1990), 2548–2565.
134. L. Freidel and L. Smolin. Linearization of the Kodama state. *Class. Quant. Grav.* **21** (2004), 3831–3844. [hep-th/0310224]
135. R. Gambini and J. Pullin. *Loops, Knots, Gauge Theories and Quantum Gravity,* (Cambridge University Press, Cambridge, 1996).
136. E. Witten. Quantum field theory and the Jones polynomial. *Comm. Math. Phys.* **121** (1989), 351–399.

Quantum Einstein Gravity: Towards an Asymptotically Safe Field Theory of Gravity

O. Lauscher[1] and M. Reuter[2]

[1] Institute of Theoretical Physics, University of Leipzig, Augustusplatz 10-11, 04109 Leipzig, Germany
lauscher@itp.uni-leipzig.de

[2] Institute of Physics, University of Mainz, Staudingerweg 7, 55099 Mainz, Germany
reuter@thep.physik.uni-mainz.de

1 Introduction

Quantized general relativity, based upon the Einstein–Hilbert action

$$S_{\text{EH}} = \frac{1}{16\pi G} \int d^4x \sqrt{-g} \, \{-R + 2\Lambda\} \ , \tag{1.1}$$

is well known to be perturbatively nonrenormalizable. This has led to the widespread belief that a straightforward quantization of the metric degrees of freedom cannot lead to a mathematically consistent and predictive *fundamental* theory valid down to arbitrarily small spacetime distances. Einstein gravity was rather considered merely an *effective* theory whose range of applicability is limited to a phenomenological description of gravitational effects at distances much larger than the Planck length.

In particle physics one usually considers a theory fundamental if it is perturbatively renormalizable. The virtue of such models is that one can "hide" their infinities in only finitely many basic parameters (masses, gauge couplings, etc.) which are intrinsically undetermined within the theory and whose value must be taken from the experiment. All other couplings are then well-defined computable functions of those few parameters. In nonrenormalizable effective theories, on the other hand, the divergence structure is such that increasing orders of the loop expansion require an increasing number of new counter terms and, as a consequence, of undetermined free parameters. Typically, at high energies, all these unknown parameters enter on an equal footing which threatens the predictivity of the theory.

However, there are examples of field theories which do "exist" as fundamental theories despite their perturbative nonrenormalizability [1, 2]. These models are "nonperturbatively renormalizable" along the lines of Wilson's modern formulation of renormalization theory [1]. They are constructed by

O. Lauscher and M. Reuter: *Quantum Einstein Gravity: Towards an Asymptotically Safe Field Theory of Gravity*, Lect. Notes Phys. **721**, 265–285 (2007)
DOI 10.1007/978-3-540-71117-9_11

performing the limit of infinite ultraviolet (UV) cutoff ("continuum limit") at a non-Gaussian renormalization group fixed point g_{*i} in the space $\{\mathrm{g}_i\}$ of all (dimensionless, essential) couplings g_i which parametrize a general action functional. This construction has to be contrasted with the standard perturbative renormalization which, at least implicitly, is based upon the Gaussian fixed point at which all couplings vanish, $\mathrm{g}_{*i} = 0$ [3, 4].

2 Asymptotic Safety

In his "asymptotic safety" scenario Weinberg [5] has put forward the idea that, perhaps, a quantum field theory of gravity can be constructed nonperturbatively by invoking a non-Gaussian UV fixed point ($\mathrm{g}_{*i} \neq 0$). The resulting theory would be "asymptotically safe" in the sense that at high energies unphysical singularities are likely to be absent.

The arena in which the idea is formulated is the so-called "theory space". By definition, it is the space of all action functionals $A[\,\cdot\,]$ which depend on a given set of fields and are invariant under certain symmetries. Hence the theory space $\{A[\,\cdot\,]\}$ is fixed once the field contents and the symmetries are fixed. The infinitely many generalized couplings g_i needed to parametrize a general action functional are local coordinates on theory space. In gravity one deals with functionals $A[g_{\mu\nu}, \cdots]$ which are required to depend on the metric in a diffeomorphism invariant way. (The dots represent matter fields and possibly background fields introduced for technical convenience.) Theory space carries a crucial geometric structure, namely a vector field which encodes the effect of a Kadanoff–Wilson-type block spin or "coarse graining" procedure, suitably reformulated in the continuum. The components $\boldsymbol{\beta}_i$ of this vector field are the beta-functions of the couplings g_i. They describe the dependence of $\mathrm{g}_i \equiv \mathrm{g}_i(k)$ on the coarse graining scale k:

$$k\,\partial_k \mathrm{g}_i = \boldsymbol{\beta}_i(\mathrm{g}_1, \mathrm{g}_2, \cdots) \tag{2.1}$$

By definition, k is taken to be a mass scale. Roughly speaking the running couplings $\mathrm{g}_i(k)$ describe the dynamics of field averages, the averaging volume having a linear extension of the order $1/k$. The $\mathrm{g}_i(k)$'s should be thought of as parametrizing a running action functional $\Gamma_k[g_{\mu\nu}, \cdots]$. By definition, the renormalization group (RG) trajectories, i.e. the solutions to the "exact renormalization group equation" (2.1) are the integral curves of the vector field $\boldsymbol{\beta} \equiv (\boldsymbol{\beta}_i)$ defining the "RG flow".

The asymptotic safety scenario assumes that $\boldsymbol{\beta}$ has a zero at a point with coordinates g_{*i} not all of which are zero. Given such a non-Gaussian fixed point (NGFP) of the RG flow one defines its UV critical surface, or unstable manifold $\mathcal{S}_{\mathrm{UV}}$ to consist of all points of theory space which are attracted into it in the limit $k \to \infty$. (Note that increasing k amounts to going in the direction *opposite* to the natural coarse graining flow.) The dimensionality

$\dim(\mathcal{S}_{\rm UV}) \equiv \Delta_{\rm UV}$ is given by the number of attractive (for *in*creasing cutoff k) directions in the space of couplings. The linearized flow near the fixed point is governed by the Jacobi matrix $\mathbf{B} = (B_{ij})$, $B_{ij} \equiv \partial_j \boldsymbol{\beta}_i(\mathrm{g}_*)$:

$$k\,\partial_k\,\mathrm{g}_i(k) = \sum_j B_{ij}\,\left(\mathrm{g}_j(k) - \mathrm{g}_{*j}\right)\ . \tag{2.2}$$

The general solution to this equation reads

$$\mathrm{g}_i(k) = \mathrm{g}_{*i} + \sum_I C_I\, V_i^I \left(\frac{k_0}{k}\right)^{\theta_I} \tag{2.3}$$

where the V^I's are the right-eigenvectors of $\mathbf{B}$ with (complex) eigenvalues $-\theta_I$. Furthermore, k_0 is a fixed reference scale, and the C_I's are constants of integration. If $\mathrm{g}_i(k)$ is to approach g_{*i} in the infinite cutoff limit $k \to \infty$ we must set $C_I = 0$ for all I with $\mathrm{Re}\,\theta_I < 0$. Hence the dimensionality $\Delta_{\rm UV}$ equals the number of $\mathbf{B}$-eigenvalues with a negative real part, i.e. the number of θ_I's with a positive real part.

A specific quantum field theory is defined by an RG trajectory which exists globally, i.e. is well behaved all the way down from "$k = \infty$" in the UV to $k = 0$ in the IR. The key idea of asymptotic safety is to base the theory upon one of the trajectories running inside the hypersurface $\mathcal{S}_{\rm UV}$ since these trajectories are manifestly well behaved and free from fatal singularities (blowing up couplings, etc.) in the large-k limit. Moreover, a theory based upon a trajectory inside a finite dimensional $\mathcal{S}_{\rm UV}$ has predictive power. The problem of an increasing number of counter terms and undetermined parameters which plagues effective theories does not arise.

In fact, in order to select a specific quantum theory we have to fix $\Delta_{\rm UV}$ free parameters which are not predicted by the theory and must be taken from experiment. When we *lower* the cutoff, only $\Delta_{\rm UV}$ parameters in the initial action are "relevant", and fixing these parameters amounts to picking a specific trajectory on $\mathcal{S}_{\rm UV}$; the remaining "irrelevant" parameters are all attracted towards $\mathcal{S}_{\rm UV}$ automatically. Therefore the theory has the more predictive power, the smaller the dimensionality of $\mathcal{S}_{\rm UV}$, i.e. the fewer UV attractive eigendirections the non-Gaussian fixed point has. If $\Delta_{\rm UV} < \infty$, the quantum field theory thus constructed is as predictive as a perturbatively renormalizable model with $\Delta_{\rm UV}$ "renormalizable couplings", i.e. couplings relevant at the Gaussian fixed point.

It is plausible that $\mathcal{S}_{\rm UV}$ is indeed finite dimensional. If the dimensionless g_i's arise as $\mathrm{g}_i(k) = k^{-d_i}\bar{\mathrm{g}}_i(k)$ by rescaling (with the cutoff k) the original couplings $\bar{\mathrm{g}}_i$ with mass dimensions d_i, then $\boldsymbol{\beta}_i = -d_i \mathrm{g}_i + \cdots$ and $B_{ij} = -d_i\delta_{ij} + \cdots$ where the dots stand for the quantum corrections. Ignoring them, $\theta_i = d_i + \cdots$, and $\Delta_{\rm UV}$ equals the number of positive d_i's. Since adding derivatives or powers of fields to a monomial in the action always lowers d_i, there can be at most a finite number of positive d_i's and, therefore, of negative eigenvalues of **B**. Thus, barring the presumably rather exotic possibility that the quantum corrections change the signs of infinitely many elements in **B**, the dimensionality of $\mathcal{S}_{\text{UV}}$ is finite [5]. Since asymptotic safety is necessarily linked to large anomalous dimensions this argument has to be taken with a grain of salt, of course. Nevertheless, the available calculations seem to support this picture.

We emphasize that, in general, the UV fixed point on which the above construction is based, if it exists, has no reason to be of the simple Einstein–Hilbert form (1.1). The initial point of the RG trajectory $\Gamma_{k\to\infty}$ is expected to contain many more invariants, both local (curvature polynomials) and nonlocal ones. For this reason the asymptotic safety scenario is *not* a quantization of general relativity, and it cannot be compared in this respect to the loop quantum gravity approach, for instance. In a conventional field theory setting the functional $\Gamma_{k\to\infty}$ corresponds to the bare (or "classical") action S which usually can be chosen (almost) freely. It is one of the many attractive features of the asymptotic safety scenario that the bare action is fixed by the theory itself and actually can be *computed*, namely by searching for zeros of $\boldsymbol{\beta}$. In this respect it has, almost by construction, a degree of predictivity which cannot be reached by any scheme trying to quantize a given classical action.

3 RG Flow of the Effective Average Action

During the past few years, the asymptotic safety scenario in Quantum Einstein Gravity (QEG) has been mostly investigated in the framework of the effective average action [6]-[21], [4], a specific formulation of the Wilsonian RG which originally was developed for theories in flat space [22–24] and has been first applied to gravity in [6].

Quite generally, the effective average action Γ_k is a coarse-grained free energy functional that describes the behavior of the theory at the mass scale k. It contains the quantum effects of all fluctuations of the dynamical variables with momenta larger than k, but not of those with momenta smaller than k. As k is decreased, an increasing number of degrees of freedom is integrated out. The method thus complies, at an intuitive level, with the coarse-graining picture of the previous section. The successive averaging of the fluctuation variable is achieved by a k-dependent IR cutoff term $\Delta_k S$ which is added to the classical action in the standard Euclidean functional integral. This term gives a momentum-dependent mass square $\mathcal{R}_k(p^2)$ to the field modes with momentum p. It is designed to vanish if $p^2 \gg k^2$, but suppresses the contributions of the modes with $p^2 < k^2$ to the path integral. When regarded as a function of k, Γ_k describes a curve in theory space that interpolates between the classical action $S = \Gamma_{k\to\infty}$ and the conventional effective action $\Gamma = \Gamma_{k=0}$. The change of Γ_k induced by an infinitesimal change of k is described by a functional differential equation, the exact RG equation. In a symbolic notation it reads

$$k\,\partial_k \Gamma_k = \frac{1}{2}\,\mathrm{STr}\left[\left(\Gamma_k^{(2)} + \mathcal{R}_k\right)^{-1} k\,\partial_k \mathcal{R}_k\right]\,. \tag{3.1}$$

For a detailed discussion of this equation we must refer to the literature [6]. Suffice it to say that, expanding $\Gamma_k[g_{\mu\nu},\cdots]$ in terms of diffeomorphism invariant field monomials $I_i[g_{\mu\nu},\cdots]$ with coefficients $\mathrm{g}_i(k)$, (3.1) assumes the component form (2.1).

In general it is impossible to find exact solutions to (3.1) and we are forced to rely upon approximations. A powerful nonperturbative approximation scheme is the truncation of theory space where the RG flow is projected onto a finite-dimensional subspace. In practice one makes an ansatz for Γ_k that comprises only a few couplings and inserts it into the RG equation. This leads to a, now finite, set of coupled differential equations of the form (2.1).

The simplest approximation one might try is the "Einstein–Hilbert truncation" [6, 8] defined by the ansatz

$$\Gamma_k[g_{\mu\nu}] = (16\pi G_k)^{-1}\int d^dx\,\sqrt{g}\,\{-R(g) + 2\bar{\lambda}_k\} \tag{3.2}$$

It applies to a d-dimensional Euclidean spacetime and involves only the cosmological constant $\bar{\lambda}_k$ and the Newton constant G_k as running parameters. Inserting (3.2) into the RG equation (3.1) one obtains a set of two $\boldsymbol{\beta}$-functions $(\boldsymbol{\beta}_\lambda, \boldsymbol{\beta}_g)$ for the dimensionless cosmological constant $\lambda_k \equiv k^{-2}\bar{\lambda}_k$ and the dimensionless Newton constant $g_k \equiv k^{d-2}G_k$, respectively. They describe a two-dimensional RG flow on the plane with coordinates $\mathrm{g}_1 \equiv \lambda$ and $\mathrm{g}_2 \equiv g$. At a fixed point (λ_*, g_*), both $\boldsymbol{\beta}$-functions vanish simultaneously. In the Einstein–Hilbert truncation there exists both a trivial Gaussian fixed point (GFP) at $\lambda_* = g_* = 0$ and, quite remarkably, also a UV attractive NGFP at $(\lambda_*, g_*) \neq (0,0)$. In Fig. 1 we show part of the g-λ theory space and the corresponding RG flow for $d = 4$. The trajectories are obtained by numerically integrating the differential equations $k\,\partial_k\lambda = \boldsymbol{\beta}_\lambda(\lambda, g)$ and $k\,\partial_k g = \boldsymbol{\beta}_g(\lambda, g)$. The arrows point in the direction of increasing coarse graining, i.e. from the UV towards the IR. We observe that three types of trajectories emanate from the NGFP: those of Type Ia (Type IIIa) run towards negative (positive) cosmological constants, while the "separatrix", the unique trajectory (of Type IIa) crossing over from the NGFP to the GFP, has a vanishing cosmological constant in the IR. The flow is defined on the half-plane $\lambda < 1/2$ only; it cannot be continued beyond $\lambda = 1/2$ as the $\boldsymbol{\beta}$-functions become singular there. In fact, the Type IIIa-trajectories cannot be integrated down to $k = 0$ within the Einstein–Hilbert approximation. They terminate at a nonzero k_{term} where they run into the $\lambda = 1/2$–singularity. Near k_{term} a more general truncation is needed in order to continue the flow.

In Weinberg's original paper [5] the asymptotic safety idea was tested in $d = 2 + \epsilon$ dimensions where $0 < \epsilon \ll 1$ was chosen so that the $\boldsymbol{\beta}$-functions

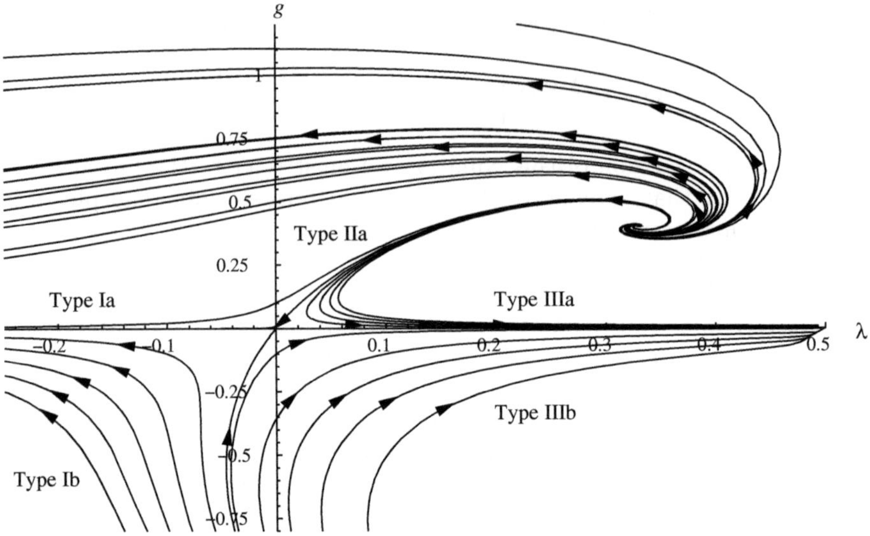

Fig. 1. Part of theory space of the Einstein–Hilbert truncation with its RG flow. The arrows point in the direction of decreasing values of k. The flow is dominated by an NGFP in the first quadrant and a trivial one at the origin. (From [9].)

(actually $\boldsymbol{\beta}_g$ only) could be found by an ϵ-expansion. Before the advent of the exact RG equations no practical tool was known which would have allowed a nonperturbative calculation of the $\boldsymbol{\beta}$-functions in the physically interesting case of $d = 4$ spacetime dimensions. However, as we saw above, the effective average action in the Einstein–Hilbert approximation does indeed predict the existence of an NGFP in a nonperturbative setting. It was first analyzed in [13, 8, 9], and also first investigations of its possible role in black-hole physics [25] and cosmology [26, 27] were performed already.

The detailed analyses of [8, 9] demonstrated that the NGFP found has all the properties necessary for asymptotic safety. In particular one has a pair of complex conjugate critical exponents $\theta' \pm \mathrm{i}\,\theta''$ with $\theta' > 0$, implying that the NGFP, for $k \to \infty$, attracts all trajectories in the half-plane $g > 0$. (The lower half-plane $g < 0$ is unphysical probably since it corresponds to a negative Newton constant.) Because of the nonvanishing imaginary part $\theta'' \neq 0$, all trajectories spiral around the NGFP before hitting it.

The question of crucial importance is whether the fixed point predicted by the Einstein–Hilbert truncation actually approximates a fixed point in the exact theory, or whether it is an artifact of the truncation. In [8–10] evidence was found, which, in our opinion, strongly supports the hypothesis that there does indeed exist an NGFP in the exact four-dimensional theory, with exactly the properties required for the asymptotic safety scenario. In these investigations the reliability of the Einstein–Hilbert truncation was tested both by analyzing the cutoff scheme-dependence within this truncation [8, 9] and by generalizing the truncation ansatz itself [10]. The idea behind the first method is as follows.

The cutoff operator $\mathcal{R}_k(p^2)$ is specified by a matrix in field space and a "shape function" $R^{(0)}(p^2/k^2)$ which describes the details of how the modes get suppressed in the IR when p^2 drops below k^2. We checked the cutoff scheme dependence of the various quantities of interest both by looking at their dependence on the function $R^{(0)}$ and comparing two different matrix structures. Universal quantities are particularly important in this respect because, by definition, they are strictly cutoff scheme-independent in the exact theory. Any truncation leads to a residual scheme dependence of these quantities, however. Its magnitude is a natural indicator for the quality of the truncation [28]. Typical examples of universal quantities are the critical exponents θ_I. The existence or nonexistence of a fixed point is also a universal, scheme-independent feature, but its precise location in parameter space is scheme dependent. Nevertheless it can be shown that, in $d = 4$, the product $g_*\lambda_*$ must be universal [8] while g_* and λ_* separately are not.

The detailed numerical analysis of the Einstein–Hilbert RG flow near the NGFP [8, 9] shows that the universal quantities, in particular the product $g_*\,\lambda_*$, are indeed scheme independent at a quite impressive level of accuracy. As the many numerical "miracles" which lead to the almost perfect cancellation of the $R^{(0)}$-dependence would have no reason to occur if there was not a fixed point in the exact theory as an organizing principle, the results of this analysis can be considered strong evidence in favor of a fixed point in the exact, untruncated theory.

The ultimate justification of any truncation is that when one adds further terms to it its physical predictions do not change significantly any more. As a first step towards testing the stability of the Einstein–Hilbert truncation against the inclusion of other invariants [10] we took a (curvature)^2-term into account:

$$\Gamma_k[g_{\mu\nu}] = \int d^dx\,\sqrt{g}\,\Big\{(16\pi G_k)^{-1}\left[-R(g)+2\bar{\lambda}_k\right]+\bar{\beta}_k\,R^2(g)\Big\} \tag{3.3}$$

Inserting (3.3) into the functional RG equation yields a set of $\boldsymbol{\beta}$-functions $(\boldsymbol{\beta}_\lambda, \boldsymbol{\beta}_g, \boldsymbol{\beta}_\beta)$ for the dimensionless couplings λ_k, g_k and $\beta_k \equiv k^{4-d}\bar{\beta}_k$. They describe the RG flow on the three-dimensional λ-g-β-space. Despite the extreme algebraic complexity of the three $\boldsymbol{\beta}$-functions it was possible to show [10–12] that they, too, admit an NGFP $(\lambda_*, g_*, \beta_*)$ with exactly the properties needed for asymptotic safety. In particular it turned out to be UV attractive in all three directions. The value of β_* is extremely tiny, and close to the NGFP the projection of the three-dimensional flow onto the λ-g-subspace is very well described by the Einstein–Hilbert truncation which ignores the third direction from the outset. The λ_*- and g_*-values and the critical exponents related to the flow in the λ-g-subspace, as predicted by the three-dimensional truncation, agree almost perfectly with those from the Einstein–Hilbert approximation. Analyzing the scheme dependence of the universal quantities one finds again a highly remarkable $R^{(0)}$-independence – which is truly amazing if one visualizes the huge amount of nontrivial numerical compensations and cancellations among several dozens of $R^{(0)}$-dependent terms which is necessary to make $g_* \lambda_*$, say, approximately independent of the shape function $R^{(0)}$.

On the basis of these results we believe that the NGFP occuring in the Einstein–Hilbert truncation is very unlikely to be an artifact of this truncation but rather may be considered the projection of an NGFP in the exact theory. The fixed point and all its qualitative properties are stable against variations of the cutoff and the inclusion of a further invariant in the truncation. It is particularly remarkable that within the scheme dependence the additional R^2-term has essentially no impact on the fixed point. These are certainly very nontrivial indications supporting the conjecture that four-dimensional QEG indeed possesses an RG fixed point with the properties needed for its nonperturbative renormalizability.

This view is further supported by two conceptually independent investigations. In [19] a proper time renormalization group equation rather than the flow equation of the average action has been used, and again a suitable NGFP was found. This framework is conceptually somewhat simpler than that of the effective average action; it amounts to an RG-improved 1-loop calculation with an IR cutoff. Furthermore, in [29] the functional integral over the subsector of metrics admitting two Killing vectors has been performed *exactly*, and again an NGFP was found, this time in a setting and an approximation which is *very* different from that of the truncated Γ_k-flows. As for the inclusion of matter fields, both in the average action [14–16, 20] and the symmetry reduction approach [29], a suitable NGFP has been established for a broad class of matter systems. For a more detailed review of the asymptotic safety scenario the reader is referred to [30].

4 Scale-Dependent Metrics and the Resolution Function $\ell(k)$

In the following we take the existence of a suitable NGFP on the full theory space for granted and explore some of the properties of asymptotic safety, in particular we try to gain some understanding of what a "quantum spacetime" is like. Unless stated otherwise we consider pure Euclidean gravity in $d = 4$.

The running effective average action $\Gamma_k[g_{\mu\nu}]$ defines an infinite set of effective field theories, valid near the scale k which we may vary between $k = 0$ and $k = \infty$. Intuitively speaking, the solution $\langle g_{\mu\nu}\rangle_k$ of the scale-dependent field equation

$$\frac{\delta \Gamma_k}{\delta g_{\mu\nu}(x)}[\langle g\rangle_k] = 0 \tag{4.1}$$

can be interpreted as the metric averaged over (Euclidean) spacetime volumes of a linear extension ℓ which typically is of the order of $1/k$. Knowing the scale dependence of Γ_k, i.e. the renormalization group trajectory $k \mapsto \Gamma_k$, we can derive the running effective Einstein equations (4.1) for any k and, after fixing appropriate boundary conditions and symmetry requirements, follow their solution $\langle g_{\mu\nu} \rangle_k$ from $k = \infty$ to $k = 0$.

The infinitely many equations of (4.1), one for each scale k, are valid *simultaneously.* They all refer *to the same* physical system, the "quantum spacetime". They describe its effective metric structure on different length scales. An observer using a "microscope" with a resolution $\approx k^{-1}$ will perceive the universe to be a Riemannian manifold with metric $\langle g_{\mu\nu} \rangle_k$. At every fixed k, $\langle g_{\mu\nu} \rangle_k$ is a smooth classical metric. But since the quantum spacetime is characterized by the infinity of metrics $\{\langle g_{\mu\nu} \rangle_k | k = 0, \cdots, \infty\}$ it can acquire very nonclassical and in particular fractal features. In fact, every proper distance calculated from $\langle g_{\mu\nu} \rangle_k$ is unavoidably scale dependent. This phenomenon is familiar from fractal geometry, a famous example being the coast line of England whose length depends on the size of the yardstick used to measure it [31].

Let us describe more precisely what it means to "average" over Euclidean spacetime volumes. The quantity we can freely tune is the IR cutoff scale k, and the "resolving power" of the microscope, henceforth denoted ℓ, is an a priori unknown function of k. (In flat space, $\ell \approx \pi/k$.) In order to understand the relationship between ℓ and k we must recall some more steps from the construction of $\Gamma_k[g_{\mu\nu}]$ in [6].

The effective average action is obtained by introducing an IR cutoff into the path-integral over all metrics, gauge fixed by means of a background gauge fixing condition. Even without a cutoff the resulting effective action $\Gamma[g_{\mu\nu}; \bar{g}_{\mu\nu}]$ depends on two metrics, the expectation value of the quantum field, $g_{\mu\nu}$, and the background field $\bar{g}_{\mu\nu}$. This is a standard technique, and it is well known [32] that the functional $\Gamma[g_{\mu\nu}] \equiv \Gamma[g_{\mu\nu}; \bar{g}_{\mu\nu} = g_{\mu\nu}]$ obtained by equating the two metrics can be used to generate the 1PI Green's functions of the theory.

(We emphasize, however, that the average action method is manifestly *background independent* despite the temporary use of $\bar{g}_{\mu\nu}$ at an intermediate level. At no stage in the derivation of the $\boldsymbol{\beta}$-functions it is necessary to assign a concrete metric to $\bar{g}_{\mu\nu}$, such as $\bar{g}_{\mu\nu} = \eta_{\mu\nu}$ in standard perturbation theory, say. The RG flow, i.e. the vector field $\boldsymbol{\beta}$, on the theory space of diffeomorphism invariant action functionals depending on $g_{\mu\nu}$ and $\bar{g}_{\mu\nu}$ is a highly universal object: it neither depends on any specific metric, nor on any specific action.)

The IR cutoff of the average action is implemented by first expressing the functional integral over all metric fluctuations in terms of eigenmodes of $\bar{D}^2$, the covariant Laplacian formed with the aid of the background metric $\bar{g}_{\mu\nu}$. Then the suppression term $\Delta_k S$ is introduced which damps the contribution of all $-\bar{D}^2$-modes with eigenvalues smaller than k^2. Coupling the dynamical fields to sources and Legendre-transforming leads to the scale-dependent functional $\Gamma_k[g_{\mu\nu}; \bar{g}_{\mu\nu}]$, and the action with one argument again obtains by equating the two metrics:

$$\Gamma_k[g_{\mu\nu}] \equiv \Gamma_k[g_{\mu\nu}; \bar{g}_{\mu\nu} = g_{\mu\nu}] \,. \tag{4.2}$$

It is this action which appears in the effective field equations (4.1).

A solution to those effective field equations represents a scale-dependent "one-point function" $\langle g_{\mu\nu}(x)\rangle_k$. Even though the metric operator is a highly singular object, as is any quantum field operator localized at a point, $\langle g_{\mu\nu}(x)\rangle_k$ is a smooth function of x in general, at most up to isolated singularities. In fact, within the Einstein–Hilbert truncation, (4.1) has the same form as the classical Einstein equation, and all its well-known solutions, with G and $\bar{\lambda}$ replaced by the k-dependent G_k and $\bar{\lambda}_k$, respectively, provide examples of such one-point functions. The smoothness of $\langle g_{\mu\nu}(x)\rangle_k$, at every *fixed* value of $k \in (0, \infty)$, is due to the averaging over infinitely many configurations of the microscopic (i.e., quantum) metric. This average is performed with the path integral containing the cutoff term $\Delta_k S$. The occurrence of a smooth one-point function is familiar from standard field theory. A well-known text book example from quantum electrodynamics is the Uehling potential, the radiatively corrected field of an electric point charge. In quantum gravity, in a formalism without gauge fixing, one might encounter additional problems due to the fact that it is impossible to specify any particular point "x" in the quantum ensemble so that observables would always have to contain an integration over x. In the average action approach this problem does not arise since the renormalization group equation pertains to an explicitly gauge fixed path integral. As a result, for every given k, the labels "x" are in a one-to-one correspondence with the points of spacetime. It is a nontrivial issue, however, to make sure that when one compares solutions $\langle g_{\mu\nu}(x)\rangle_k$ for different values of k the coordinates x refer to the same point always. This can be done, for instance, by deriving a flow equation directly for the solution: $k\,\partial_k \langle g_{\mu\nu}(x)\rangle_k = \cdots$ [33]. A simple example of an equation of this kind (or rather its solution) is the relation (5.4) below. Once we have found a family of metrics $\langle g_{\mu\nu}(x)\rangle_k$ where "x" refers to the same point for any value of k we may perform only k-independent diffeomorphisms on this family if we want to maintain this property. A priori we could have changed the coordinates at each level k separately, but clearly this would destroy the scale-independent one-to-one correspondence between points and coordinates.

In the spirit of effective field theory, and by the very construction of the effective average action [24], $\langle g_{\mu\nu}\rangle_k$ should be thought of as the metric relevant in any single-scale physical process involving momenta of the order k, in the sense that fluctuations about the average are smallest if $\langle g_{\mu\nu}\rangle_k$ is used for this particular, physically determined value of k. The concrete identification of k depends on the physical situation or process under consideration. A typical example of a quantity which potentially can act as an IR cutoff, well known from deep inelastic scattering, for instance, is the (4-momentum)^2, or virtuality, of a particle.

It is natural to ask how much of the spacetime structure is revealed by an experiment ("microscope") with a given characteristic scale k. Because of the identification of the two metrics in (4.2) we see that, in a sense, it is the eigenmodes of $\bar{D}^2 = D^2$, constructed from the argument of $\Gamma_k[g]$, which are cut off at k^2. This last observation is essential for the following algorithm

[23, 34] for the reconstruction of the averaging scale ℓ from the cutoff k. The input data is the set of metrics characterizing a quantum manifold, $\{\langle g_{\mu\nu}\rangle_k\}$. The idea is to deduce the relation $\ell = \ell(k)$ from the spectral properties of the scale-dependent Laplacian $\Delta(k) \equiv D^2(\langle g_{\mu\nu}\rangle_k)$ built with the solution of the effective field equation. More precisely, for every fixed value of k, one solves the eigenvalue problem of $-\Delta(k)$ and studies the properties of the special eigenfunctions whose eigenvalue is k^2, or nearest to k^2 in the case of a discrete spectrum. We shall refer to an eigenmode of $-\Delta(k)$ whose eigenvalue is (approximately) the square of the cutoff k as a "cutoff mode" (COM) and denote the set of all COMs by $\mathsf{COM}(k)$.

If we ignore the k-dependence of $\Delta(k)$ for a moment (as it would be appropriate for matter theories in flat space) the COMs are, for a sharp cutoff, precisely the last modes integrated out when lowering the cutoff, since the suppression term in the path integral cuts out all modes of the metric fluctuation with eigenvalue smaller than k^2.

For a non-gauge theory in flat space the coarse graining or averaging of fields is a well-defined procedure, based upon ordinary Fourier analysis, and one finds that in this case the length ℓ is essentially the wavelength of the last modes integrated out, the COMs.

This observation motivates the following definition of ℓ in quantum gravity. We determine the COMs of $-\Delta(k)$, analyze how fast these eigenfunctions vary on spacetime, and read off a typical coordinate distance Δx^μ characterizing the scale on which they vary. For an oscillatory COM, for example, Δx^μ would correspond to an oscillation period. (In general there is a certain freedom in the precise identification of the Δx^μ belonging to a specific cutoff mode. This ambiguity can be resolved by refining the definition of Δx^μ on a case-by-case basis only.) Finally we use the metric $\langle g_{\mu\nu}\rangle_k$ itself in order to convert Δx^μ to a proper length. This proper length, by definition, is ℓ. Repeating the above steps for all values of k, we end up with a function $\ell = \ell(k)$. In general one will find that ℓ depends on the position on the manifold as well as on the direction of Δx^μ.

Applying the above algorithm on a nondynamical flat spacetime one recovers the expected result $\ell(k) = \pi/k$. In [34] a specific example of a QEG spacetime has been constructed, the quantum S^4, which is an ordinary 4-sphere at every fixed scale, with a k-dependent radius, though. In this case, too, the resolution function was found to be $\ell(k) = \pi/k$.

Thus the construction and interpretation of a QEG spacetime proceeds, in a nutshell, as follows. We start from a fixed RG trajectory $k \mapsto \Gamma_k$, derive its effective field equations at each k, and solve them. The resulting quantum mechanical counterpart of a classical spacetime is equipped with the infinity of Riemannian metrics $\{\langle g_{\mu\nu}\rangle_k | k = 0, \cdots, \infty\}$ where the parameter k is only a book-keeping device a priori. It can be given a physical interpretation by relating it to the COM length scale characterizing the averaging procedure: One constructs the Laplacian $-D^2(\langle g_{\mu\nu}\rangle_k)$, diagonalizes it, looks how rapidly its k^2-eigenfunction varies, and "measures" the length of typical variations with the metric $\langle g_{\mu\nu} \rangle_k$ itself. In the ideal case one can solve the resulting $\ell = \ell(k)$ for $k = k(\ell)$ and reinterpret the metric $\langle g_{\mu\nu} \rangle_k$ as referring to a microscope with a known position and direction-dependent resolving power ℓ. The price we have to pay for the background independence is that we cannot freely choose ℓ directly but rather k only.

5 Microscopic Structure of the QEG Spacetimes

One of the intriguing conclusions we reached in [8, 10] was that the QEG spacetimes are fractals and that their effective dimensionality is scale dependent. It equals 4 at macroscopic distances ($\ell \gg \ell_{\rm Pl}$) but, near $\ell \approx \ell_{\rm Pl}$, it gets dynamically reduced to the value 2. For $\ell \ll \ell_{\rm Pl}$ spacetime is, in a precise sense [8], a 2-dimensional fractal.

In [26] the specific form of the graviton propagator on this fractal was applied in a cosmological context. It was argued that it gives rise to a Harrison–Zeldovich spectrum of primordial geometry fluctuations, perhaps responsible for the CMBR spectrum observed today. (In [25–27], [35]-[40] various types of "RG improvements" were used to explore possible physical manifestations of the scale dependence of the gravitational parameters.)

A priori there exist several plausible definitions of a fractal dimensionality of spacetime. In our original argument [8] we used the one based upon the anomalous dimension η_N at the NGFP. We shall review this argument in the rest of this section. Then, in Sect. 6, we evaluate the spectral dimension for the QEG spacetimes [41] and demonstrate that it displays the same dimensional reduction $4 \to 2$ as the one based upon η_N. The spectral dimension has also been determined in Monte Carlo simulations of causal (i.e. Lorentzian) dynamical triangulations [42]-[45] and it will be interesting to compare the results.

For simplicity we use the Einstein–Hilbert truncation to start with, and we consider spacetimes with classical dimensionality $d = 4$. The corresponding RG trajectories are shown in Fig. 1. For $k \to \infty$, all of them approach the NGFP (λ_*, g_*) so that the dimensionful quantities run according to

$$G_k \approx g_*/k^2 \,, \qquad \bar{\lambda}_k \approx \lambda_* \, k^2 \tag{5.1}$$

The behavior (5.1) is realized in the asymptotic scaling regime $k \gg m_{\rm Pl}$. Near $k = m_{\rm Pl}$ the trajectories cross over towards the GFP. Since we are interested only in the limiting cases of very small and very large distances the following caricature of an RG trajectory will be sufficient. We assume that G_k and $\bar{\lambda}_k$ behave as in (5.1) for $k \gg m_{\rm Pl}$, and that they assume constant values for $k \ll m_{\rm Pl}$. The precise interpolation between the two regimes could be obtained numerically [9] but will not be needed here.

The argument of [10] concerning the fractal nature of the QEG spacetimes is as follows. Within the Einstein–Hilbert truncation of theory space, the effective field equations (4.1) happen to coincide with the ordinary Einstein equation, but with G_k and $\bar{\lambda}_k$ replacing the classical constants. Without matter,

$$R_{\mu\nu}(\langle g\rangle_k) = \bar{\lambda}_k \langle g_{\mu\nu}\rangle_k \tag{5.2}$$

Since in the absence of dimensionful constants of integration $\bar{\lambda}_k$ is the only quantity in this equation which sets a scale, every solution to (5.2) has a typical radius of curvature $r_c(k) \propto 1/\sqrt{\bar{\lambda}_k}$. (For instance, the S^4-solution has the radius $r_c = \sqrt{3/\bar{\lambda}_k}$.) If we want to explore the spacetime structure at a fixed length scale ℓ we should use the action $\Gamma_k[g_{\mu\nu}]$ at $k \approx \pi/\ell$ because with this functional a tree-level analysis is sufficient to describe the essential physics at this scale, including the relevant quantum effects. Hence, when we observe the spacetime with a microscope of resolution ℓ, we will see an average radius of curvature given by $r_c(\ell) \equiv r_c(k = \pi/\ell)$. Once ℓ is smaller than the Planck length $\ell_{\rm Pl} \equiv m_{\rm Pl}^{-1}$ we are in the fixed point regime where $\bar{\lambda}_k \propto k^2$ so that $r_c(k) \propto 1/k$, or

$$r_c(\ell) \propto \ell \tag{5.3}$$

Thus, when we look at the structure of spacetime with a microscope of resolution $\ell \ll \ell_{\rm Pl}$, the average radius of curvature which we measure is proportional to the resolution itself. If we want to probe finer details and decrease ℓ we automatically decrease r_c and hence *in*crease the average curvature. Spacetime seems to be more strongly curved at small distances than at larger ones. The scale-free relation (5.3) suggests that at distances below the Planck length the QEG spacetime is a special kind of fractal with a self-similar structure. It has no intrinsic scale because in the fractal regime, i.e. when the RG trajectory is still close to the NGFP, the parameters which usually set the scales of the gravitational interaction, G and $\bar{\lambda}$, are not yet "frozen out". This happens only later on, somewhere halfway between the NGFP and the GFP, at a scale of the order of $m_{\rm Pl}$. Below this scale, G_k and $\bar{\lambda}_k$ stop running and, as a result, $r_c(k)$ becomes independent of k so that $r_c(\ell) = \text{const}$ for $\ell \gg \ell_{\rm Pl}$. In this regime $\langle g_{\mu\nu}\rangle_k$ is k-independent, indicating that the macroscopic spacetime is describable by a single smooth Riemannian manifold.

The above argument made essential use of the proportionality $\ell \propto 1/k$. In the fixed point regime it follows trivially from the fact that there exist no other relevant dimensionful parameters so that $1/k$ is the only length scale one can form. The algorithm for the determination of $\ell(k)$ described above yields the same answer.

It is easy to make the k-dependence of $\langle g_{\mu\nu}\rangle_k$ explicit. Picking an arbitrary reference scale k_0 we rewrite (5.2) as $[\bar{\lambda}_{k_0}/\bar{\lambda}_k]\, R^\mu{}_\nu(\langle g\rangle_k) = \bar{\lambda}_{k_0}\, \delta^\mu_\nu$. Since $R^\mu{}_\nu(c\, g) = c^{-1}\, R^\mu{}_\nu(g)$ for any constant $c > 0$, the average metric and its inverse scale as

$$\langle g_{\mu\nu}(x)\rangle_k = [\bar{\lambda}_{k_0}/\bar{\lambda}_k]\, \langle g_{\mu\nu}(x)\rangle_{k_0} \tag{5.4}$$

$$\langle g^{\mu\nu}(x)\rangle_k = [\bar{\lambda}_k/\bar{\lambda}_{k_0}]\, \langle g^{\mu\nu}(x)\rangle_{k_0} \tag{5.5}$$

These relations are valid provided the family of solutions considered exists for all scales between k_0 and k, and $\bar{\lambda}_k$ has the same sign always.

As we discussed in [8] the QEG spacetime has an effective dimensionality which is k-dependent and hence noninteger in general. The discussion was based upon the anomalous dimension η_N of the operator $\int \sqrt{g}\, R$. It is defined as $\eta_N \equiv -k\,\partial_k \ln Z_{Nk}$, where $Z_{Nk} \propto 1/G_k$ is the wavefunction renormalization of the metric [6]. In a sense which we shall make more precise in a moment, the effective dimensionality of spacetime equals $4 + \eta_N$. The RG trajectories of the Einstein–Hilbert truncation (within its domain of validity) have $\eta_N \approx 0$ for $k \to 0$[1] and $\eta_N \approx -2$ for $k \to \infty$, the smooth change by two units occuring near $k \approx m_{\rm Pl}$. As a consequence, the effective dimensionality is 4 for $\ell \gg \ell_{\rm Pl}$ and 2 for $\ell \ll \ell_{\rm Pl}$.

In the exact theory, and in any truncation, the UV fixed point has an anomalous dimension $\eta \equiv \eta_N(\lambda_*, g_*) = -2$ [8, 10]. We can use this information in order to determine the momentum dependence of the dressed graviton propagator for momenta $p^2 \gg m_{\rm Pl}^2$. Expanding the Γ_k of (3.2) about flat space and omitting the standard tensor structures we find the inverse propagator $\widetilde{\mathcal{G}}_k(p)^{-1} \propto Z_N(k)\, p^2$. The conventional dressed propagator $\widetilde{\mathcal{G}}(p)$, the one contained in $\Gamma \equiv \Gamma_{k=0}$, obtains from the exact $\widetilde{\mathcal{G}}_k$ by taking the limit $k \to 0$. For $p^2 > k^2 \gg m_{\rm Pl}^2$ the actual cutoff scale is the physical momentum p^2 itself[2] so that the k-evolution of $\widetilde{\mathcal{G}}_k(p)$ stops at the threshold $k = \sqrt{p^2}$. Therefore

$$\widetilde{\mathcal{G}}(p)^{-1} \propto Z_N\left(k = \sqrt{p^2}\right) p^2 \propto (p^2)^{1-\frac{\eta}{2}} \tag{5.6}$$

because $Z_N(k) \propto k^{-\eta}$ when $\eta \equiv -\partial_t \ln Z_N$ is (approximately) constant. In d dimensions, and for $\eta \neq 2 - d$, the Fourier transform of $\widetilde{\mathcal{G}}(p) \propto 1/(p^2)^{1-\eta/2}$ yields the following propagator in position space:

$$\mathcal{G}(x; y) \propto \frac{1}{|x - y|^{d-2+\eta}} \,. \tag{5.7}$$

This form of the propagator is well known from the theory of critical phenomena, for instance. (In the latter case it applies to large distances.) Equation (5.7) is not valid directly at the NGFP. For $d = 4$ and $\eta = -2$ the dressed propagator is $\widetilde{\mathcal{G}}(p) = 1/p^4$ which has the following representation in position space:

$$\mathcal{G}(x; y) = -\frac{1}{8\pi^2} \ln\left(\mu\,|x - y|\right) \,. \tag{5.8}$$

Here μ is an arbitrary constant with the dimension of a mass. Obviously (5.8) has the same form as a $1/p^2$-propagator in 2 dimensions.

[1] In the case of type IIIa trajectories [9, 39] the macroscopic k-value is still far above $k_{\rm term}$, i.e. in the "GR regime" described in [39].

[2] See Sect. 1 of [37] for a detailed discussion of "decoupling" phenomena of this kind.

Slightly away from the NGFP, before other physical scales intervene, the propagator is of the familiar type (5.7) which shows that the quantity η_N has the standard interpretation of an anomalous dimension in the sense that fluctuation effects modify the decay properties of $\mathcal{G}$ so as to correspond to a spacetime of effective dimensionality $4 + \eta_N$.

Thus the properties of the RG trajectories imply the following "dimensional reduction": Spacetime, probed by a "graviton" with $p^2 \ll m_{\mathrm{Pl}}^2$ is four-dimensional, but it appears to be two-dimensional for a graviton with $p^2 \gg m_{\mathrm{Pl}}^2$ [8].

It is interesting to note that in d classical dimensions, where the macroscopic spacetime is d-dimensional, the anomalous dimension at the fixed point is $\eta = 2 - d$. Therefore, for any d, the dimensionality of the fractal as implied by η_N is $d + \eta = 2$ [8, 10].

6 The Spectral Dimension

Next we turn to the spectral dimension $\mathcal{D}_s$ of the QEG spacetimes. This particular definition of a fractal dimension is borrowed from the theory of diffusion processes on fractals [46] and is easily adapted to the quantum gravity context [47, 45]. In particular it has been used in the Monte Carlo studies mentioned above.

Let us study the diffusion of a scalar test particle on a d-dimensional classical Euclidean manifold with a fixed smooth metric $g_{\mu\nu}(x)$. The corresponding heat-kernel $K_g(x, x'; T)$ giving the probability for the particle to diffuse from x' to x during the fictitious diffusion time T satisfies the heat equation $\partial_T K_g(x, x'; T) = \Delta_g K_g(x, x'; T)$ where $\Delta_g \equiv D^2$ denotes the scalar Laplacian: $\Delta_g \phi \equiv g^{-1/2}\, \partial_\mu (g^{1/2}\, g^{\mu\nu}\, \partial_\nu \phi)$. The heat-kernel is a matrix element of the operator $\exp(T\,\Delta_g)$. In the random walk picture its trace per unit volume, $P_g(T) \equiv V^{-1} \int d^d x\, \sqrt{g(x)}\, K_g(x, x; T) \equiv V^{-1}\, \mathrm{Tr}\, \exp(T\,\Delta_g)$, has the interpretation of an average return probability. (Here $V \equiv \int d^d x\, \sqrt{g}$ denotes the total volume.) It is well known that P_g possesses an asymptotic expansion (for $T \to 0$) of the form $P_g(T) = (4\pi T)^{-d/2} \sum_{n=0}^{\infty} A_n\, T^n$. For an infinite flat space, for instance, it reads $P_g(T) = (4\pi T)^{-d/2}$ for all T. Thus, knowing the function P_g, one can recover the dimensionality of the target manifold as the T-independent logarithmic derivative $d = -2\, d \ln P_g(T)/d \ln T$. This formula can also be used for curved spaces and spaces with finite volume V provided T is not taken too large [45].

In QEG where we functionally integrate over all metrics it is natural to replace $P_g(T)$ by its expectation value. Symbolically, $P(T) \equiv \langle P_\gamma(T) \rangle$ where $\gamma_{\mu\nu}$ denotes the microscopic metric (integration variable) and the expectation value is with respect to the ordinary path integral (without IR cutoff) containing the fixed point action.Given $P(T)$, we define the spectral dimension of the quantum spacetime in analogy with the classical formula:

$$\mathcal{D}_{\mathrm{s}} = -2\frac{d\ln P(T)}{d\ln T} \tag{6.1}$$

Let us now evaluate (6.1) using the average action method. The fictitious diffusion process takes place on a "manifold" which, at every fixed scale, is described by a smooth Riemannian metric $\langle g_{\mu\nu}\rangle_k$. While the situation appears to be classical at fixed k, nonclassical features emerge in the regime with nontrivial RG running since there the metric depends on the scale at which the spacetime structure is probed.

The nonclassical features are encoded in the properties of the diffusion operator. Denoting the covariant Laplacians corresponding to the metrics $\langle g_{\mu\nu}\rangle_k$ and $\langle g_{\mu\nu}\rangle_{k_0}$ by $\Delta(k)$ and $\Delta(k_0)$, respectively, (5.4) and (5.5) imply that they are related by

$$\Delta(k) = [\bar{\lambda}_k/\bar{\lambda}_{k_0}]\,\Delta(k_0) \tag{6.2}$$

When $k, k_0 \gg m_{\mathrm{Pl}}$ we have, for example,

$$\Delta(k) = (k/k_0)^2\,\Delta(k_0) \tag{6.3}$$

Recalling that the average action Γ_k defines an effective field theory at the scale k we have that $\langle\mathcal{O}(\gamma_{\mu\nu})\rangle \approx \mathcal{O}(\langle g_{\mu\nu}\rangle_k)$ if the operator $\mathcal{O}$ involves typical covariant momenta of the order k and $\langle g_{\mu\nu}\rangle_k$ solves (4.1). In the following we exploit this relationship for the RHS of the diffusion equation, $\mathcal{O} = \Delta_\gamma\,K_\gamma(x,x';T)$. It is crucial here to correctly identify the relevant scale k.

If the diffusion process involves only a small interval of scales near k over which $\bar{\lambda}_k$ does not change much the corresponding heat equation must contain the $\Delta(k)$ for this specific, fixed value of k:

$$\partial_T K(x,x';T) = \Delta(k)K(x,x';T) \tag{6.4}$$

Denoting the eigenvalues of $-\Delta(k_0)$ by $\mathcal{E}_n$ and the corresponding eigenfunctions by ϕ_n, this equation is solved by

$$K(x,x';T) = \sum_n \phi_n(x)\,\phi_n(x')\,\exp\Big(-F(k^2)\,\mathcal{E}_n\,T\Big) \tag{6.5}$$

Here we introduced the convenient notation $F(k^2) \equiv \bar{\lambda}_k/\bar{\lambda}_{k_0}$. Knowing this propagation kernel we can time-evolve any initial probability distribution $p(x;0)$ according to $p(x;T) = \int d^4x'\,\sqrt{g_0(x')}\,K(x,x';T)\,p(x';0)$ with g_0 the determinant of $\langle g_{\mu\nu}\rangle_{k_0}$. If the initial distribution has an eigenfunction expansion of the form $p(x;0) = \sum_n C_n\,\phi_n(x)$ we obtain

$$p(x;T) = \sum_n C_n\,\phi_n(x)\,\exp\Big(-F(k^2)\,\mathcal{E}_n\,T\Big) \tag{6.6}$$

If the C_n's are significantly different from zero only for a single eigenvalue $\mathcal{E}_N$, we are dealing with a single-scale problem. In the usual spirit of effective field theories we would then identify $k^2 = \mathcal{E}_N$ as the relevant scale at which the running couplings are to be evaluated. However, in general the C_n's are different from zero over a wide range of eigenvalues. In this case we face a multiscale problem where different modes ϕ_n probe the spacetime on different length scales.

If $\Delta(k_0)$ corresponds to flat space, say, the eigenfunctions $\phi_n \equiv \phi_p$ are plane waves with momentum p^μ, and they resolve structures on a length scale ℓ of order $\pi/|p|$. Hence, in terms of the eigenvalue $\mathcal{E}_n \equiv \mathcal{E}_p = p^2$ the resolution is $\ell \approx \pi/\sqrt{\mathcal{E}_n}$. This suggests that when the manifold is probed by a mode with eigenvalue $\mathcal{E}_n$ it "sees" the metric $\langle g_{\mu\nu}\rangle_k$ for the scale $k = \sqrt{\mathcal{E}_n}$. Actually the identification $k = \sqrt{\mathcal{E}_n}$ is correct also for curved space since, in the construction of Γ_k, the parameter k is introduced precisely as a cutoff in the spectrum of the covariant Laplacian.

Therefore we conclude that under the spectral sum of (6.6) we must use the scale $k^2 = \mathcal{E}_n$ which depends explicitly on the resolving power of the corresponding mode. Likewise, in (6.5), $F(k^2)$ is to be interpreted as $F(\mathcal{E}_n)$. Thus we obtain the traced propagation kernel

$$\begin{aligned} P(T) &= V^{-1} \sum_n \exp\Big[- F(\mathcal{E}_n)\, \mathcal{E}_n\, T\Big] \\ &= V^{-1}\, \mathrm{Tr}\, \exp\Big[F\Big(- \Delta(k_0)\Big)\, \Delta(k_0)\, T\Big] \end{aligned} \tag{6.7}$$

It is convenient to choose k_0 as a macroscopic scale in a regime where there are no strong RG effects any more.

Furthermore, let us assume for a moment that at k_0 the cosmological constant is tiny, $\bar{\lambda}_{k_0} \approx 0$, so that $\langle g_{\mu\nu}\rangle_{k_0}$ is an approximately flat metric. In this case the trace in (6.7) is easily evaluated in a plane wave basis:

$$P(T) = \int \frac{d^4p}{(2\pi)^4}\, \exp\left[-p^2\, F(p^2)\, T\right] \tag{6.8}$$

The T-dependence of (6.8) determines the fractal dimensionality of spacetime via (6.1). In the limits $T \to \infty$ and $T \to 0$ where the random walks probe very large and small distances, respectively, we obtain the dimensionalities corresponding to the largest and smallest length scales possible. The limits $T \to \infty$ and $T \to 0$ of $P(T)$ are determined by the behavior of $F(p^2) \equiv \bar{\lambda}(k = \sqrt{p^2})/\bar{\lambda}_{k_0}$ for $p^2 \to 0$ and $p^2 \to \infty$, respectively.

For an RG trajectory where the renormalization effects stop below some threshold we have $F(p^2 \to 0) = 1$. In this case (6.8) yields $P(T) \propto 1/T^2$, and we conclude that the macroscopic spectral dimension is $\mathcal{D}_\mathrm{s} = 4$.

In the fixed point regime we have $\bar{\lambda}_k \propto k^2$, and therefore $F(p^2) \propto p^2$. As a result, the exponent in (6.8) is proportional to p^4 now. This implies the $T \to 0$–behavior $P(T) \propto 1/T$. It corresponds to the spectral dimension $\mathcal{D}_\mathrm{s} = 2$.

This result holds for all RG trajectories since only the fixed-point properties were used. In particular it is independent of $\bar{\lambda}_{k_0}$ on macroscopic scales. Indeed, the above assumption that $\langle g_{\mu\nu}\rangle_{k_0}$ is flat was not necessary for obtaining $\mathcal{D}_s = 2$. This follows from the fact that even for a curved metric the spectral sum (6.7) can be represented by an Euler–MacLaurin series which always implies (6.8) as the leading term for $T \to 0$.

Thus we may conclude that on very large and very small length scales the spectral dimensions of the QEG spacetimes are

$$\begin{aligned} \mathcal{D}_s(T \to \infty) &= 4 \\ \mathcal{D}_s(T \to 0) &= 2 \end{aligned} \tag{6.9}$$

The dimensionality of the fractal at sub-Planckian distances is found to be 2 again, as in the first argument based upon η_N. Remarkably, the equality of $4 + \eta$ and $\mathcal{D}_s$ is a special feature of 4 classical dimensions. Generalizing for d classical dimensions, the fixed point running of Newton's constant becomes $G_k \propto k^{2-d}$ with a dimension-dependent exponent, while $\bar{\lambda}_k \propto k^2$ continues to have a quadratic k-dependence. As a result, the $\underset{\sim}{\mathcal{G}}(k)$ of (5.6) is proportional to $1/p^d$ in general so that, for any d, the two-dimensional looking graviton propagator (5.8) is obtained. (This is equivalent to saying that $\eta = 2 - d$, or $d + \eta = 2$, for arbitrary d.)

On the other hand, the impact of the RG effects on the diffusion process is to replace the operator Δ by Δ^2, for any d, since the cosmological constant always runs quadratically. Hence, in the fixed point regime, (6.8) becomes $P(T) \propto \int d^d p \, \exp\left[-p^4 T\right] \propto T^{-d/4}$. This T-dependence implies the spectral dimension

$$\mathcal{D}_s(d) = d/2 \tag{6.10}$$

This value coincides with $d + \eta$ if, and only if, $d = 4$. It is an intriguing speculation that this could have something to do with the observed macroscopic dimensionality of spacetime.

For the sake of clarity and to be as explicit as possible we described the computation of $\mathcal{D}_s$ within the Einstein–Hilbert truncation. However, it is easy to see [41] that the only nontrivial ingredient of this computation, the scaling behavior $\Delta(k) \propto k^2$, is in fact an exact consequence of asymptotic safety. If the fixed point exists, simple dimensional analysis implies $\Delta(k) \propto k^2$ at the untruncated level, and this in turn gives rise to (6.10). If QEG is asymptotically safe, $\mathcal{D}_s = 2$ at sub-Planckian distances is an *exact* nonperturbative result for all of its spacetimes. To be as explicit as possible, we described the arguments leading to $\mathcal{D}_s = 2$ in the context of the average action. They are, however, to a large extent independent of the concrete formalism used; see [30] for further details.

It is interesting to compare the result (6.9) to the spectral dimensions which were recently obtained by Monte Carlo simulations of the causal dynamical triangulation model of quantum gravity [45]:

$$\begin{aligned} \mathcal{D}_s(T \to \infty) &= 4.02 \pm 0.1 \\ \mathcal{D}_s(T \to 0) &= 1.80 \pm 0.25 \end{aligned} \qquad (6.11)$$

These figures, too, suggest that the long-distance and short-distance spectral dimension should be 4 and 2, respectively. The dimensional reduction from 4 to 2 dimensions is a highly nontrivial dynamical phenomenon which seems to occur in both QEG and the discrete triangulation model. We find it quite remarkable that the discrete and the continuum approach lead to essentially identical conclusions in this respect. This could be a first hint indicating that the discrete model and QEG in the average action formulation describe the same physics.

7 Concluding Remarks

In the light of the RG properties of the effective average action it is indeed rather likely that four-dimensional Quantum Einstein Gravity can be defined ("renormalized") nonperturbatively along the lines of asymptotic safety. It seems quite possible to construct a quantum field theory of the spacetime metric which is not only an effective one, but rather a fundamental one and which is mathematically consistent and predictive on the smallest possible length scales even. If so, it is not necessary to leave the realm of quantum field theory in order to construct a satisfactory quantum gravity. This is at variance with the basic credo of string theory, for instance, which is also claimed to provide a consistent gravity theory. Here a very high price has to be paid for curing the problems of perturbative gravity, however: one has to live with infinitely many (unobserved) matter fields.

The average action approach has led to specific predictions for the spacetime structure in nonperturbative, asymptotically safe gravity. The general picture of the QEG spacetimes which emerged is as follows. At sub-Planckian distances spacetime is a fractal of dimensionality $\mathcal{D}_s = 4 + \eta = 2$. It can be thought of as a self-similar hierarchy of superimposed Riemannian manifolds of any curvature. As one considers larger length scales where the RG running of the gravitational parameters comes to a halt, the "ripples" in the spacetime gradually disappear and the structure of a classical four-dimensional manifold is recovered.

References

1. K.G. Wilson, J. Kogut: Phys. Rept. **12**, 75 (1974);
   K.G. Wilson: Rev. Mod. Phys. **47**, 773 (1975).
2. G. Parisi: Nucl. Phys. B **100**, 368 (1975), Nucl. Phys. B **254**, 58 (1985);
   K. Gawedzki, A. Kupiainen: Nucl. Phys. B **262**, 33 (1985), Phys. Rev. Lett.

54, 2191 (1985), Phys. Rev. Lett. **55**, 363 (1985);
B. Rosenstein, B.J. Warr, S.H. Park: Phys. Rept. **205**, 59 (1991);
C. de Calan, P.A. Faria da Veiga, J. Magnen, R. Sénéor: Phys. Rev. Lett. **66**, 3233 (1991).

3. J. Polchinski: Nucl. Phys. B **231**, 269 (1984).
4. For a review see: C. Bagnuls and C. Bervillier: Phys. Rept. **348**, 91 (2001);
T.R. Morris: Prog. Theor. Phys. Suppl. **131**, 395 (1998);
J. Polonyi: Central Eur. J. Phys. **1**, 1 (2004).
5. S. Weinberg: Ultraviolet divergences in quantum theories of gravitation. In: *General Relativity, an Einstein Centenary Survey*, ed by S.W. Hawking and W. Israel (Cambridge University Press 1979) pp. 790–831;
S. Weinberg: preprint hep-th/9702027.
6. M. Reuter: Phys. Rev. D **57**, 971 (1998) [hep-th/9605030].
7. D. Dou and R. Percacci: Class. Quant. Grav. **15**, 3449 (1998).
8. O. Lauscher and M. Reuter: Phys. Rev. D **65**, 025013 (2002) [hep-th /0108040].
9. M. Reuter and F. Saueressig: Phys. Rev. D **65**, 065016 (2002) [hep-th/0110054].
10. O. Lauscher and M. Reuter: Phys. Rev. D **66**, 025026 (2002) [hep-th/ 0205062].
11. O. Lauscher and M. Reuter: Class. Quant. Grav. **19**, 483 (2002) [hep-th/0110021].
12. O. Lauscher and M. Reuter: Int. J. Mod. Phys. A **17**, 993 (2002) [hep-th/0112089].
13. W. Souma: Prog. Theor. Phys. **102**, 181 (1999).
14. R. Percacci and D. Perini: Phys. Rev. D **67**, 081503 (2003).
15. R. Percacci and D. Perini: Phys. Rev. D **68**, 044018 (2003).
16. D. Perini: Nucl. Phys. Proc. Suppl. **127** C, 185 (2004).
17. M. Reuter and F. Saueressig: Phys. Rev. D **66**, 125001 (2002) [hep-th/0206145];
Fortschr. Phys. **52**, 650 (2004) [hep-th/0311056].
18. D. Litim: Phys. Rev. Lett. **92**, 201301 (2004).
19. A. Bonanno, M. Reuter: JHEP **02**, 035 (2005) [hep-th/0410191].
20. R. Percacci and D. Perini: Class. Quant. Grav. **21**, 5035 (2004).
21. R. Percacci: preprint hep-th/0409199.
22. C. Wetterich: Phys. Lett. B **301**, 90 (1993).
23. M. Reuter and C. Wetterich: Nucl. Phys. B **417**, 181 (1994), Nucl. Phys. B **427**, 291 (1994), Nucl. Phys. B **391**, 147 (1993), Nucl. Phys. B **408**, 91 (1993);
M. Reuter: Phys. Rev. D **53**, 4430 (1996), Mod. Phys. Lett. A **12**, 2777 (1997).
24. For a review see: J. Berges, N. Tetradis and C. Wetterich: Phys. Rept. **363**, 223 (2002);
C. Wetterich: Int. J. Mod. Phys. A **16**, 1951 (2001).
25. A. Bonanno and M. Reuter: Phys. Rev. D **62**, 043008 (2000) [hep-th/0002196];
Phys. Rev. D **60**, 084011 (1999) [gr-qc/9811026].
26. A. Bonanno and M. Reuter: Phys. Rev. D **65**, 043508 (2002) [hep-th/0106133];
M. Reuter and F. Saueressig: JCAP **09**, 012 (2005) [hep-th/0507167].
27. A. Bonanno and M. Reuter: Phys. Lett. B **527**, 9 (2002) [astro-ph/0106468];
Int. J. Mod. Phys. D **13**, 107 (2004) [astro-ph/0210472].
28. J.-I. Sumi, W. Souma, K.-I. Aoki, H. Terao, K. Morikawa: preprint hep-th/0002231.
29. P. Forgács and M. Niedermaier: preprint hep-th/0207028;
M. Niedermaier: JHEP **12**, 066 (2002); Nucl. Phys. B **673**, 131 (2003); preprint gr-qc/0610018.

30. M. Niedermaier and M. Reuter: Living Rev. Relativity **9**, 5 (2006).
31. B. Mandelbrot: *The Fractal Geometry of Nature* (Freeman, New York 1977).
32. L.F. Abbott: Nucl. Phys. B **185**, 189 (1981);
B.S. DeWitt: Phys. Rev. **162**, 1195 (1967);
M.T. Grisaru, P. van Nieuwenhuizen and C.C. Wu: Phys. Rev. D **12**, 3203 (1975);
D.M. Capper, J.J. Dulwich and M. Ramon Medrano: Nucl. Phys. B **254**, 737 (1985);
S.L. Adler: Rev. Mod. Phys. **54**, 729 (1982).
33. M. Reuter and J.-M. Schwindt: JHEP 07**01**, 049 (2007) [hep-th/0611294].
34. M. Reuter and J.-M. Schwindt: JHEP **01**, 070 (2006) [hep-th/0511021].
35. E. Bentivegna, A. Bonanno and M. Reuter: JCAP **01**, 001 (2004) [astro-ph/0303150].
36. A. Bonanno, G. Esposito and C. Rubano: Gen. Rel. Grav. **35**, 1899 (2003); Class. Quant. Grav. **21**, 5005 (2004).
37. M. Reuter and H. Weyer: Phys. Rev. D **69**, 104022 (2004) [hep-th/0311196].
38. M. Reuter and H. Weyer: Phys. Rev. D **70**, 124028 (2004) [hep-th/0410117].
39. M. Reuter and H. Weyer: JCAP **12**, 001 (2004) [hep-th/0410119].
40. J. Moffat: JCAP **05**, 003 (2005) [astro-ph/0412195];
J.R. Brownstein and J. Moffat: Astrophys. J. **636**, 721 (2006); Mon. Not. Roy. Astron. Soc. **367**, 527 (2006).
41. O. Lauscher and M. Reuter: JHEP **10**, 050 (2005) [hep-th/0508202].
42. A. Dasgupta and R. Loll: Nucl. Phys. B **606**, 357 (2001);
J. Ambjørn, J. Jurkiewicz and R. Loll: Nucl. Phys. B **610**, 347 (2001), Phys. Rev. Lett. **85**, 924 (2000);
R. Loll. Nucl. Phys. Proc. Suppl. **94**, 96 (2001);
J. Ambjørn: preprint gr-qc/0201028.
43. J. Ambjørn, J. Jurkiewicz and R. Loll: Phys. Rev. Lett. **93**, 131301 (2004).
44. J. Ambjørn, J. Jurkiewicz and R. Loll: Phys. Lett. B **607**, 205 (2005).
45. J. Ambjørn, J. Jurkiewicz and R. Loll: Phys. Rev. Lett. **95**, 171301 (2005); Phys. Rev. D **72**, 064014 (2005).
46. D. ben-Avraham and S. Havlin: *Diffusion and reactions in fractals and disordered systems* (Cambridge University Press, Cambridge 2004).
47. H. Kawai, M. Ninomiya: Nucl. Phys. B **336**, 115 (1990);
R. Floreanini and R. Percacci: Nucl. Phys. B **436**, 141 (1995);
I. Antoniadis, P.O. Mazur and E. Mottola: Phys. Lett. B **444**, 284 (1998).

Part VI

String Theory

String Theory: An Overview

J. Louis[1,2], T. Mohaupt[3], and S. Theisen[4]

[1] II. Institut für Theoretische Physik, Universität Hamburg, 22761 Hamburg, Germany
[2] Zentrum für Mathematische Physik, Universität Hamburg, 20146 Hamburg, Germany
jan.louis@desy.de
[3] Theoretical Physics Division, Department of Mathematical Sciences, University of Liverpool, Peach Street, Liverpool L69 7ZL, U.K.
Thomas.Mohaupt@liverpool.ac.uk
[4] Max-Planck-Institut für Gravitationsphysik, Albert-Einstein-Institut, Am Mühlenberg 1, 14476 Golm, Germany
theisen@aei.mpg.de

1 Introduction

String theory is not, in contrast to general relativity and quantum field theory, a theory in the strict sense. There is, e.g., no axiomatic formulation and there is no set of defining equations of motion. Instead there is a set of rules which have been developed over the years. They have led to rather spectacular results and have passed all conceivable consistency checks. As has become clear, string theory is more than a theory of strings. This is most apparent through the rôle played by D-branes. They are additional extended dynamical objects whose existence within the theory can be inferred from a variety of arguments. D-branes and other types of p-branes (p-dimensional membranes) are essential for the web of non-perturbative dualities between the known perturbative string theories. Since it is not clear whether strings will remain the fundamental degrees of freedom in the final form of the theory, the term M-theory is frequently used instead of non-perturbative string theory.

However, both notions are programmatic, as the underlying dynamical principle and, closely related to this, the symmetries of string theory have not yet been found. It would thus be more appropriate to speak about a 'theory under construction'; nevertheless, following common usage, we will always speak of string theory or M-theory, the later being understood as the working title for the non-perturbative completion of string theory. At the moment it is not possible to present a well-rounded-off view of string theory. All this non-technical overview is able to accomplish is to recall some of the successes of the theory, mention some of the current activities and some of the open challenges.

J. Louis et al.: *String Theory: An Overview*, Lect. Notes Phys. **721**, 289–323 (2007)
DOI 10.1007/978-3-540-71117-9_12

2 Beyond the Standard Model

String theory is a proposal for a unifying framework of high energy physics. At currently achievable accelerator energies of $\mathcal{O}(1\text{TeV})$ or, equivalently, at distance scales $> 10^{-19}$ m, the standard model (SM) of Particle Physics (amended with appropriate neutrino masses) provides a successful and predictive theoretical description. It is based on the mathematical framework of a local quantum gauge field theory with gauge group $SU(3) \times SU(2) \times U(1)$. Nevertheless it is believed that the SM is merely a low-energy effective description of a more fundamental theory. There are several reasons for this belief.

(1) The standard model has many free parameters (coupling constants, mixing angles, etc.) which have to be fixed by experiments. They could, a priori, take any value (within a given range such that the effective description is still valid and that perturbative calculations, on which the comparisons with experiment are based, are justified). In addition there is no explanation of the particle spectrum and its symmetries; they are also experimental input, the only theoretical restriction being the requirement of cancellation of gauge anomalies.
(2) The standard model does not include the gravitational interactions. While this can be safely ignored at laboratory energies, there is nevertheless an energy range where gravity competes with the gauge interactions. This is a consequence of the fact that the gravitational coupling constant G_N has dimension of (length)^2. One should therefore consider the dimensionless quantity $G_N E^2$, where E is the energy scale of the experiment. With $G_N \sim M_{\text{Pl}}^{-2}$ this means that the gravitational interaction becomes large at energies comparable with the Planck scale. It is in this regime that a quantum theory of gravity is needed, provided that we assume that gravity is a fundamental interaction and not an effective one (which would not be quantized).
(3) In the standard model space-time is non-dynamical and smooth. The number of dimensions and the geometry (four-dimensional Minkowski space-time) are fixed and the back-reaction on the geometry is neglected. However, at very high energies this is no longer appropriate; for instance, if a mass m is squeezed into a volume of a size smaller than its Schwarzschild radius $r_s \sim l_p^2\,\text{m}$, then we expect that it will collapse into a black hole.

These points make it highly desirable to have a unified quantum theory of all interactions. Here 'unification' can be understood in two ways. The broader, conceptual meaning of unification is to have a consistent framework which includes both quantum theory and gravity. The predominant belief among particle physicists is that this mainly requires to 'quantize gravity', i.e., to reformulate Einstein's theory of gravity as a quantum theory. However, one should bear in mind that 'quantization' is only an, albeit successful, formal device for formulating quantum theories. It is quite plausible that the introduction of a dynamical space-time requires significant modifications of quantum theory as well. We will briefly comment on the stringy perspective on this below.

The 'conceptual' type of unification is not only aesthetically appealing, but also mandatory if we want to address the physics of the early universe and of black holes in a consistent way. The second, more narrow, meaning of unification is that, qualitatively speaking, all forces in nature are manifestations of one single force (as, e.g., in the Kaluza–Klein scenario). The popular though somewhat over-ambitious terms 'theory of everything' or 'Weltformel' have been coined to illustrate this idea. More concretely, one can put forward the working hypothesis that there is a symmetry principle, such that (i) the coupling constants of all interactions can be expressed in terms of one fundamental coupling constant, and such that (ii) all particles organize into irreducible representations. While this idea is aesthetically pleasing, it is by no means a necessary requirement for having a consistent fundamental theory of nature. However, the idea of unification (in the narrow sense) has great heuristic value, as it has stimulated the formulation of interesting theories.

While the unification of the non-gravitational interactions within the framework of quantum gauge theories does not meet fundamental problems, this changes once one attempts to include gravity. One way to see this is due to the perturbative non-renormalizability of the gravitational interaction.[1]

In this context, the main differences between the gravitational and the renormalizable Yang–Mills gauge interactions are (1) the graviton has spin two while Yang–Mills gauge bosons have spin one; (2) the gravitational coupling constant has negative mass-dimension while the gauge coupling constants are dimensionless. Difference (2) renders the theory of gravity, based on the Einstein-Hilbert action, perturbatively non-renormalizable: the UV infinities in Feynman diagrams cannot be absorbed by a finite number of local counter-terms. The cure for the Fermi theory of weak interactions, that is pulling the four-fermion interaction apart by inserting a propagator line of (massive) gauge bosons, does not work for the theory of gravity with its infinity of interaction vertices. However, if one expands the individual lines of a Feynman diagram into tubes or strips, thus replacing the world-lines by world-sheets of closed or open strings, one solves, in one go, the problem with UV infinities and replaces the infinitely many interaction vertices by a finite number of basic three-point interactions (cf. Fig. 1).

In addition, all elementary particles, gauge fields and matter fields, correspond to vibration modes of the string, which is the only fundamental object (in contrast to a quantum field for each particle species as, e.g., in grand unification models). Since every consistent quantum theory of strings necessarily contains a massless spin two particle (which has the kinematical and dynamical properties of a graviton), it automatically includes gravity. Therefore string theory is a unified theory in both meanings of the word.

[1] We will not discuss conceptual problems but refer instead to the contributions on quantum gravity in this book.

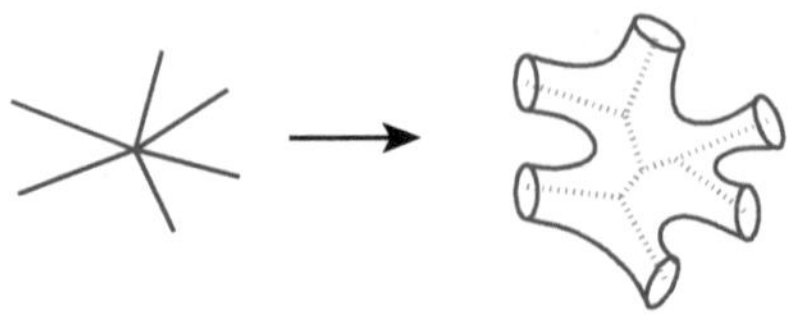

Fig. 1. 'Thickening' a field theory interaction vertex

While unification in the narrow sense is manifest in the formalism, the conceptual unification of quantum theory and gravity has not yet been achieved in a satisfactory way. The reason for this is that the formalism of 'perturbative string theory', to be reviewed in Sects. 2 and 3, only gives a set of rules for computing on-shell scattering amplitudes in an on-shell background. While the UV finiteness of string amplitudes, which has been made highly plausible (though not been proved rigorously), is clearly relevant for the conceptual unification, it mainly takes care of a technical aspect. Given that the amplitudes are indeed finite, string theory provides only a perturbative theory of quantum gravity. Note, however, that this allows to do more than just to compute graviton scattering in a fixed background. It also allows to compute an effective action, which encodes quantum corrections to the Einstein–Hilbert term. This in turn has implications for black hole physics, which will be the subject of Sect. 8. At this point we just emphasize that conceptual issues of quantum gravity, such as black hole entropy, have been addressed successfully in string theory. This said, it must also be stressed that the range of conceptual points which can be addressed today is quite limited. The main reason is that as a starting point one always has to specify a reference background space-time. We will come back to this when discussing open questions in Sect. 10.

Next, let us briefly come back to the question whether the conceptual unification of quantum theory and gravity mainly requires to 'quantize gravity', or whether both quantum theory and gravity need to be modified in a more drastic way. In perturbative string theory quantization is applied in the same pragmatic spirit as in theoretical particle physics. Actually, the approach is at first glance a bit more naive, as one quantizes the relativistic string and thus does quantum mechanics rather than quantum field theory. The fact that this procedure results in a consistent perturbative theory of quantum gravity is a surprising discovery, and the deeper reason behind this remains to be understood. Heuristically, the improved UV behaviour can be understood in terms of the 'thickening' of propagators and vertices, which we mentioned above. As a consequence, classical physics is modified in two ways, not only by quantum corrections, but also by stringy corrections related to the finite size of strings. As we will see later, the string length replaces the Planck length as the fundamental scale (at least in perturbative string theory), while there are also transformations ('dualities') in the theory, which mutually exchange quantum corrections and stringy corrections. While the deeper implications of these observations remain to be explored, it indicates that the full theory does more than just 'quantize gravity'. The relation between string theory and quantum field theory is more complicated than suggested by the naive picture of 'thickening Feynman graphs'. While programmatically string theory intends to supersede quantum field theory, in its current state it is deeply entangled with it. The first point which makes this obvious is the rôle of two-dimensional conformal field theories, which we will elaborate on in Sects. 3 and 4. While string scattering amplitudes are finite, the two-dimensional field theories used in the Polyakov approach are just renormalizable. Therefore the concept of renormalization still plays a rôle. The second point, to be discussed in Sect. 7, is the AdS/CFT correspondence, which claims that string theory in certain backgrounds is equivalent to specific quantum field theories. Here, and in related proposals such as the so-called 'M(atrix) theory', one even contemplates to define string theory in terms of quantum field theory. Let us further note a trend shared by string theory and quantum field theory, namely the importance of effective field theories. Here again renormalization (understood in a Wilsonian spirit) plays an important rôle. Of course, the concept of string effective field theories is by itself consistent with the idea that string theory supersedes quantum field theory. However, the two previous examples show that in its present state string theory has a more complicated relationship with quantum field theory. The only systematic approach to go beyond local quantum field theory is string field theory, which aims to be a full-fledged quantum field theory of extended objects. Unfortunately, string field theory has proved to be complicated that progress was very slow. Moreover, it is not clear how the non-perturbative dualities, to be discussed in Sect. 6, which nowadays hold a central position in our understanding of string theory, fit together with string field theory.

The optimists hope, of course, that all the exciting observations made during the last years will ultimately condense into a new principle, which supersedes and conceptually unifies quantum field theory and gravity. But so far only some clues have been found, while the 'big picture' is still far from clear.

3 The Free String

The dynamics of the bosonic string in d-dimensional Minkowski space-time is governed by the Nambu–Goto action

$$S_{\text{NG}} = \frac{1}{2\pi\alpha'}\int_{\Sigma} d\sigma d\tau \sqrt{|\det G(X)|}.$$

The world-sheet Σ which is swept out by the string is parametrized by $\sigma^{\alpha} = (\sigma, \tau)$. The integral is the area of Σ measured with the induced metric $G_{\alpha\beta} = \partial_{\alpha}X^{\mu}\partial_{\beta}X^{\nu}\eta_{\mu\nu}$. $X^{\mu}(\sigma,\tau) : \Sigma \hookrightarrow M$ is the embedding of Σ into the d-dimensional space-time with Minkowski-metric $\eta_{\mu\nu}$, $\mu,\nu = 0,\ldots,d-1$. $T = \frac{1}{2\pi\alpha'}$ is the string tension. $l_{\text{s}} = \sqrt{\alpha'}$ is the string scale, a length scale characteristic for string theory. It replaces the Planck length l_{P} as the fundamental length scale.

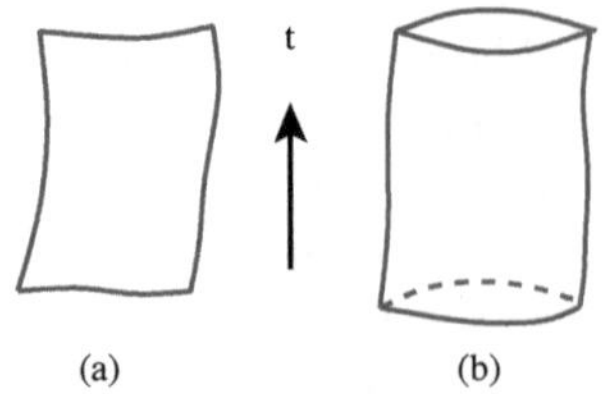

Fig. 2. World-sheet of a free (**a**) open and (**b**) closed string

$S_{\rm NG}$ is the direct generalization of the action of a point-like (i.e. zero-dimensional) relativistic particle to one-dimensional strings. The world-sheet of a freely propagating open and closed string has the topology of a strip and cylinder, respectively (Fig. 2). For the latter, $X^\mu(\sigma,\tau)$ is periodic (in σ) on the cylinder.

For the open string one can impose either Dirichlet (D) or Neumann (N) boundary conditions for each of the d fields X^μ at each of the two ends of the string. The physical meaning of Neumann boundary conditions is that space-time momentum does not flow off the ends of the string. With Dirichlet boundary conditions the position of the end of the string is fixed while space-time momentum flows off. d-dimensional Poincaré invariance demands that the total space-time momentum is conserved. This means that momentum must be absorbed by other dynamical objects. These objects, on which open strings end, are called Dirichlet branes, or D-branes, for short (cf. Fig. 3). If p of the spatial components of X^μ at one end of the string have Neumann boundary conditions and the remaining $d-p-1$ components have Dirichlet boundary conditions, this string ends on a Dp-brane. A D0-brane is also called D-particle and a D1-brane is called D-string. Fundamental strings (F-strings) and D-string are quite different objects. One difference is that an open F-string must end on a D-brane (but not vice versa). Other differences will be discussed below.

Propagation of the particles which correspond to the excitations of the open string is restricted to the world-volume of the D-brane while excitations of the closed string propagate in the full d-dimensional space-time.

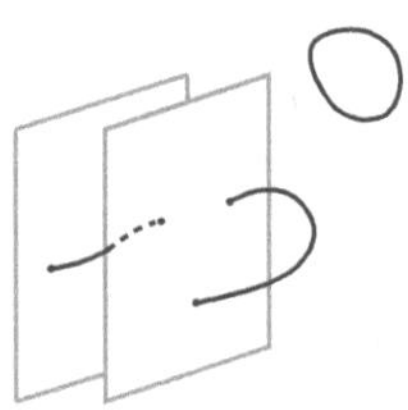

Fig. 3. Open strings end on D-branes; closed strings explore the d-dimensional space-time

Quantization of string theory simplifies if one uses the Polyakov action $S_{\rm P}$ which is classically equivalent to the Nambu–Goto action:

$$S_{\rm P} = -\frac{1}{4\pi\alpha'}\int_\Sigma d\sigma d\tau\sqrt{|\det h|}h^{\alpha\beta}\partial_\alpha X^\mu\partial_\beta X^\nu\eta_{\mu\nu}.$$

$S_{\rm P}$ is the action of d scalar fields $X^\mu(\sigma,\tau)$ which are coupled to two-dimensional gravity with metric $h_{\alpha\beta}(\sigma,\tau)$. $S_{\rm P}$ is invariant under (i) global space-time Poincaré transformations $X^\mu \to a^\mu{}_\nu X^\nu + b^\mu$, $a^\mu{}_\rho a^\nu{}_\sigma\eta_{\mu\nu} = \eta_{\rho\sigma}$; (ii) local reparametrizations of the world-sheet $\sigma^\alpha \to \tilde{\sigma}^\alpha(\sigma,\tau)$; and (iii) under local Weyl-rescalings of the metric $h_{\alpha\beta} \to \Omega^2(\sigma,\tau)h_{\alpha\beta}$. Local Weyl invariance implies tracelessness of the energy–momentum tensor $T_{\alpha\beta} = -\frac{4\pi}{\sqrt{|\det h|}}\frac{\delta}{\delta h^{\alpha\beta}}S_p$ of the world-sheet field theory, i.e. $h^{\alpha\beta}T_{\alpha\beta} = 0$. Reparametrization invariance can be used to go to conformal gauge[2] where $h_{\alpha\beta} = e^{2\varphi(\sigma,\tau)}\eta_{\alpha\beta}$. In the classical theory the Weyl degree of freedom φ decouples. Violation of the local Weyl invariance in the quantized theory is signalled by a conformal anomaly. It is measured by the central charge of the Virasoro algebra, the algebra of constraints ($T_{\alpha\beta} = 0$) in the quantized theory.

In the (1,1) supersymmetric version of the Polyakov action, every bosonic field X^μ and its two Majorana–Weyl superpartners $\psi^\mu_\pm$ of positive and negative chirality are coupled to two-dimensional world-sheet supergravity $(h_{\alpha\beta}, \chi^\pm_\alpha)$. In the classical theory, in addition to the metric degrees of freedom also those of the two world-sheet gravitini $\chi^\pm_\alpha$ are unphysical. This is a consequence of two-dimensional local world-sheet supersymmetry.

The fermions $\psi^\mu_\pm$ on the world-sheet of the closed string can be periodic (Ramond) or anti-periodic (Neveu–Schwarz), where the periodicity condition can be chosen independently for each chirality. This leads to four different sectors of the closed string theory.[3] Excitations in the (NS,NS) and the (R,R) sectors are space-time bosons while excitations in the two mixed sectors, (R,NS) and (NS,R), are space-time fermions. For the open string the boundary conditions couple the two chiralities to each other. This leads to two sectors: the NS sector with space-time bosons and the R sector with space-time fermions.

Quantization of string theory in Minkowski space-time is only possible in the critical dimension $d_{\rm crit}$, unless one is willing to accept that the quantum theory of strings contains an additional degree of freedom, the Liouville mode.

---

[2] Going to conformal gauge does not fix the reparametrization invariance completely. The remaining transformations are (in Euclidean signature on Σ) conformal transformations and the two-dimensional field theory on Σ in conformal gauge is a so-called 'conformal field theory'.

[3] Classically, one could define different theories by taking any subset of the possible boundary definitions. However, the quantum theory includes multiply connected world-sheets, and the theory must be invariant under the so-called 'modular transformations', to be discussed below. This in turn implies that all combinations of boundary conditions have to be included, and thus each type of boundary condition defines a sector of the quantum theory.

The status of Liouville string theory, also known as non-critical string theory, is not completely understood. Throughout this chapter we will fix the number of space-time dimensions to be d_{crit}. The critical dimension is $d_{\text{crit}} = 26$ for the bosonic string and $d_{\text{crit}} = 10$ for the fermionic string.[4] One obtains a positive-definite Hilbert space (no-ghost-theorem) and space-time Poincaré invariance as well as (super)Weyl invariance on the world-sheet are anomaly free. In the covariant BRST quantization the gauge fixing of the local symmetries leads to ghost fields, the reparametrization ghosts (b, c), and their superpartners (β, γ). In the critical dimension their contribution to the conformal anomaly is compensated by X^μ and ψ^μ.

The resulting spectrum of the theory contains a finite number of massless and infinitely many massive excitations with mass$^2 = \frac{n}{2\alpha'}$ with $n \in \mathbb{N}$. Among the states there are also tachyons with negative mass^2. They imply an instability of the vacuum. This is unavoidable in the bosonic string. However, the spectrum of the fermionic string must be truncated by an additional projection, the Gliozzi–Scherk–Olive (GSO) projection. This projection can be chosen such that the tachyon is projected out and the remaining spectrum is space-time supersymmetric. The GSO projection is necessary and can be understood as a consistency condition (modular invariance, locality of the CFT operator products) which must be imposed on the quantum mechanical scattering amplitudes, to be discussed in the next section. In a theory with only closed strings the spectrum has $\mathcal{N} = 2$ space-time supersymmetry. Two possible, inequivalent GSO projections lead to the non-chiral type IIA and to the chiral type IIB theory. Their massless spectra are those of ten-dimensional type IIA and type IIB supergravity, respectively.

The spectrum of type I theory with both open and closed strings is $\mathcal{N} = 1$ supersymmetric. Its massless spectrum is that of supersymmetric Yang–Mills theory, coupled to supergravity. The degrees of freedom of the Yang–Mills theory are excitations of the open string. The two ends of the open string carry charges in the fundamental representation of the gauge group (Chan–Paton factors) such that the open string has the quantum numbers of a gauge boson. The supergravity degrees of freedom are, as in the type II theories, the massless excitations of the closed string. Consistency requires the gauge group to be $SO(32)$. Only in this case gauge and gravitational anomalies vanish.

Type I and type II theories are also called superstring theories. In addition to the type I theory, there are two further string theories with $\mathcal{N} = 1$ space-time supersymmetry. These are the heterotic $E_8 \times E_8$ and $SO(32)$ theories. These theories have, like the type II theories, only closed strings. In contrast to the local (1,1) world-sheet supersymmetry of the superstring,

[4] Strictly speaking, one does not need to fix the number of space-time dimensions, but the number of degrees of freedom, as measured by the central charge of the world-sheet conformal field theory, which must be $c = 26$ and $c = 10$ in order to cancel the contribution from the reparametrization ghosts. The surplus degrees of freedom need not have the interpretation of string coordinates along extra dimensions.

heterotic theories have local (1,0) supersymmetry. The superpartner of X^μ is a single Majorana–Weyl fermion ψ^μ_+. Absence of gravitational anomalies on the world-sheet requires, in the fermionic formulation of the theory, 32 additional Majorana–Weyl fermions λ^a_-, $a = 1, \ldots, 32$. In the bosonic formulation these 32 fermions are replaced by 16 periodic chiral scalars $\Phi^I(\tau + \sigma)$, $I = 1, \ldots, 16$ which are the coordinates of a 16-dimensional torus. Modular invariance restricts the allowed tori to those which are generated by a 16-dimensional self-dual even lattice Λ via $T^{16} = \mathbb{R}^{16}/\Lambda$. There are precisely two such lattices which lead to the two allowed gauge groups $E_8 \times E_8$ and $SO(32)$. The massless spectra of the two heterotic theories are again those of supersymmetric Yang–Mills theory, coupled to supergravity, now with gauge group $E_8 \times E_8$ or $SO(32)$.

4 The Interacting String

The discussion in Sect. 3 was based on the free string theory. Interactions are introduced through the inclusion of topologically non trivial world-sheets. Figure 4 shows the decay of a closed string into two closed strings, while Fig. 5 shows the joining of two open strings into a closed string.

The strength of the interaction is controlled by the value of the dimensionless string coupling constant g, which is dynamically determined through the background value (vacuum expectation value) Φ_0 of the dilaton Φ, $g = e^{\Phi_0}$. The dilaton, as the graviton, is part of the massless spectrum of every string theory. Different values of g do not correspond to different theories but they parametrize ground states[5] of a given theory. The coupling constants of the different string theories are, however, a priori independent.

In string theory, the quantum field theoretical computation of scattering amplitudes by summation over Feynman diagrams is replaced by the summation over world-sheets of different topologies (cf. Fig. 6). Which topologies are

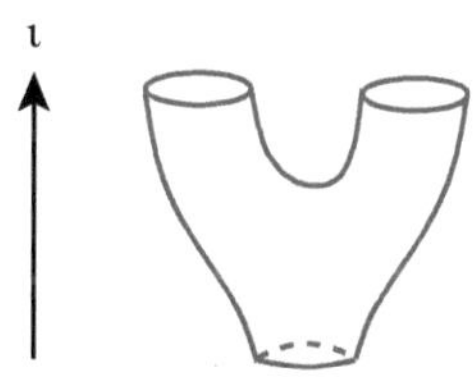

Fig. 4. Decay of a closed string into two closed strings

[5] By a (perturbative) string ground state ('string vacuum') we mean a conformal field theory with the correct properties, i.e. the correct central charge, modular invariant partition function, etc. A geometric realization can be provided by specific background configurations of the massless fields. In this section we choose $G_{\mu\nu}(X) = \eta_{\mu\nu}$, $\Phi = \Phi_0$ with all others set to zero. More general backgrounds will be mentioned in later sections.

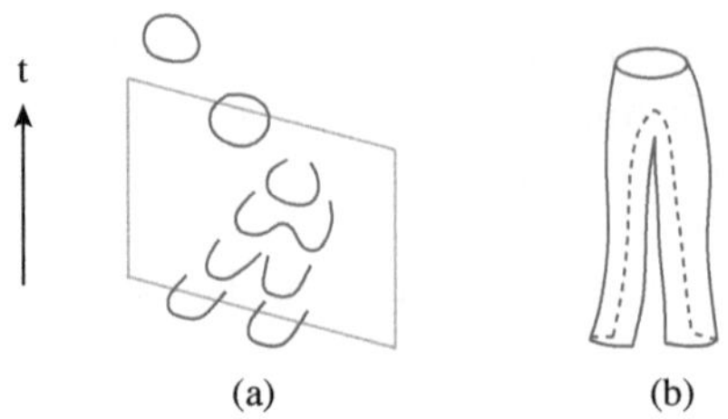

Fig. 5. Two open strings join to a closed string (**a**) in a space-time diagram and (**b**) in a world-sheet diagram

allowed depends on the string theory. In particular, in type I both orientable and non-orientable world-sheets must be summed over while the world-sheets of the four other theories must be orientable.

Scattering amplitudes $\mathcal{A}$ can be computed in perturbation theory. $\mathcal{A}$ is expanded in a power series in the string coupling g

$$\mathcal{A} = \sum_n g^n \mathcal{A}^{(n)}$$

where each term $\mathcal{A}^{(n)}$ is computed separately. The validity of perturbation theory requires $g \ll 1$. The power of g with which a given world-sheet contributes is the negative of its Euler number, i.e. it is determined by its topology. The scattering amplitudes $\mathcal{A}^{(n)}$ of physical states are correlation functions of BRST-invariant vertex operators of the quantum field theory on Σ as specified by the Polyakov action. In the path-integral formulation one has to sum over all metrics $h_{\alpha\beta}$ on Σ and over all embeddings X^μ of Σ in space-time M. The computation of scattering amplitudes is most easily done with the methods of conformal field theory. Using the local symmetries on Σ one goes to (super) conformal gauge and the infinite dimensional integration over $h_{\alpha\beta}$ ($\chi_\alpha^\pm$) is reduced to the finite dimensional integration over the (super) moduli of Σ and the integration over the Faddeev–Popov ghosts. For closed strings, requiring invariance of the amplitudes under those reparametrizations which are not continuously connected to the identity transformation (modular invariance) restricts the range of integration of the modular parameters to a fundamental region. Requiring modular invariance for the one-loop amplitudes guarantees

Fig. 6. One-loop quantum correction to the propagation of a closed string

the anomaly freedom of the space-time spectrum. It also is equivalent to imposing the GSO projection on the spectrum. For open strings one has to impose, in addition, a tadpole cancellation condition for the (R,R)-fields.

For external momenta which are small compared to the characteristic scale $l_s^{-1} = 1/\sqrt{\alpha'}$, the scattering amplitudes coincide with those of an effective field theory. Expanding the amplitudes in powers of α' corresponds to an expansion in powers of derivatives in the effective field theory. In leading order the low-energy effective action of each of the five string theories coincides with the action of the appropriate classical supergravity theory. The relations between the coupling constants of supergravity (the ten-dimensional Newton's constant $G_{\mathrm{N}}^{(10)}$ and the ten-dimensional Yang–Mills coupling g_{YM}) and g and l_s are $G_{\mathrm{N}}^{(10)} \sim g^2 l_s^8$, $g_{\mathrm{YM}}^2 \sim g^2 l_s^6$ (heterotic), and $g_{\mathrm{YM}}^2 \sim g l_s^6$ (type I). (Computable) corrections in α' result, e.g., in a modification of the Einstein–Hilbert action by terms which contain higher powers of the Riemann tensor.

The finite extent of the string becomes relevant for external momenta $\mathcal{O}(1/\sqrt{\alpha'})$ where deviations from quantum field theory become noticeable. For instance, in contrast to quantum field theory, string scattering amplitudes are UV finite. Heuristically this can be understood as follows: the point-like interaction vertices of QFT are now smeared. Perturbative finiteness of string theory[6] is a consequence of modular invariance which restricts the analogue of the Schwinger proper time-integral to a so-called 'fundamental region'. (Modular invariance is a property of string theory and requires, as a necessary condition, the excistence of an infinite number of excitations.) Within the framework of string theory one can compute perturbative corrections to the gravitational interaction. It is in this sense that string theory (in its supersymmetric version) is an UV finite and unitary perturbative quantum theory of gravity.

5 Compactification

So far the discussion was restricted to string theories in a given d_{crit}-dimensional Minkowski space-time with metric $\eta_{\mu\nu}$. However, it is possible to formulate string theory in topologically and metrically non-trivial space-times where, e.g., only d dimensions are infinitely extended and the remaining $d_{\mathrm{int}} = d_{\mathrm{crit}} - d$ are curled up and compact. One possible realization of such a compactification starts with a direct product Ansatz $M^d \times K^{\mathrm{int}}$ for the ten-dimensional space-time. Here M^d is d-dimensional Minkowski space and K^{int}

[6] A complete all-order proof of perturbative finiteness has not been worked out yet. While there are technical difficulties related to gauge fixing of the supermoduli at higher loops, there is no apparent obstruction to extending the existing finiteness results to all loops. (These difficulties seem to be absent in Berkovits' covariant and manifestly space-time supersymmetric quantization of the superstring.) This is different from the situation in maximally extended supergravity, where counterterms are possible, and, hence, to be expected, at higher loop level.

a d_{int}-dimensional compact manifold. In the Polyakov action this corresponds to replacing $\eta_{\mu\nu}$ by the metric $g_{\mu\nu}(X)$ of the product space. This leads to a two-dimensional non-linear σ-model with target space $M^d \times K^{\text{int}}$. Consistency of the compactification requires a conformally invariant σ-model and thus implies strong restrictions on K^{int}. The resulting d-dimensional theory depends on the geometry and topology of the compact manifold. For instance d-dimensional supersymmetric theories require that the manifold K^{10-d} admits Killing spinors.

The simplest consistent compactification of a closed string is on a circle S^1 with radius R. One requires $X(\sigma + 2\pi, \tau) = X(\sigma, \tau) + 2\pi w R$, $w \in \mathbb{Z}$, for one of the coordinates. This leads to additional massless and massive states: On the one hand, the Kaluza–Klein excitations with mass^2 $= (n/R)^2$, $n \in \mathbb{Z}$ which decouple for $R \to 0$ and, on the other hand, winding states with mass^2 $= (wR/l_s^2)^2$, which become massless for $R \to 0$. These winding states are characteristic for string theory and are not present in compactified field theories. They lead to a symmetry of the spectrum and the scattering amplitudes of the bosonic string under the T-duality transformation $R \to l_s^2/R$, $g \to g l_s/R$ under which Kaluza–Klein and winding states are exchanged. In other words geometrically different compactifications correspond to physically identically ground states of string theory. This symmetry implies to regard l_s as minimal length: Compactifications on a large circle is indistinguishable form compactification on a small circle. In both cases the limits $R \to \infty$ and $R \to 0$, respectively, lead to a continuum of massless states which is interpreted as the decompactification of an additional dimension. The compactification on S^1 leads to an additional free parameter, the radius R of the circle. Similar to the string coupling g it can be interpreted as the vacuum expectation value of a massless scalar field (modulus); e.g. $G_{25,25} = R^2$ for the bosonic string compactified on a circle in the X^{25} direction.

The ground states of the compactified theory are restricted via T-duality to either one of the two fundamental regions $R \in [l_s, \infty)$ or $R \in (0, l_s,]$. While T-duality is a symmetry of the bosonic string this is not the case for type II strings: T-duality transforms type IIA theory on S^1_R to type IIB theory on $S^1_{l_s^2/R}$.

A simple generalization of a compactification on a circle is the compactification on a d_{int}-dimensional torus $T^{d_{\text{int}}}$. Here T-duality is a non-Abelian discrete symmetry on the parameter space (moduli space) of the compactification whose local coordinates are, among others, the components of the metric on $T^{d_{\text{int}}}$.

Of physical interest is the case $d = 4$. For type II theories compactification on T^6 leads, at the level of the low-energy effective action, to $\mathcal{N} = 8$ supergravity and for the type I and the two heterotic theories to $\mathcal{N} = 4$ Super–Yang–Mills (SYM) theory coupled to $\mathcal{N} = 4$ supergravity. $\mathcal{N} = 1(2)$ supergravity is obtained by compactification of the heterotic (type II) string on six-dimensional Calabi–Yau manifolds.[7] The large number of topologically different Calabi–Yau manifolds leads to many different four-dimensional theories which differ in their spectra of string excitations. This, in turn, leads to different low-energy effective actions which differ from each other in gauge group, spectrum of massless particles, and interactions. Size and shape of the Calabi–Yau manifold are parametrized by the (perturbatively) undetermined vacuum expectation values of neutral (under the gauge group) scalar fields, the moduli fields.

Discrete symmetries which act on the moduli space of a given compactification and which are exact in every order of string perturbation theory are called T-duality. Mirror symmetry of Calabi–Yau compactifications is a non-trivial example of a T-duality. It states that compactifications on a pair of topologically different Calabi–Yau manifolds, a so-called 'mirror pair', are completely equivalent and undistinguishable. This, as already the simple example of the compactification on a circle, demonstrates that strings probe the geometry of a manifolds quite differently than point particle probes. One therefore speaks of 'string geometry'.

In the language of conformal field theory, compactification of a superstring theory means that one replaces $(10 - d)$ of the free superfields (X^i, ψ^i) by a superconformal non-linear sigma-model with target space K^{int} and the same central charge $c^{\text{int}} = 3d_{\text{int}}/2$. More generally one can take an 'internal' superconformal field theory of the same central charge as long as it satisfies consistency conditions such as modular invariance. Such a theory has, in general, no formulation as a sigma-model and does thus not admit a geometric interpretation. An analogous discussion also holds for heterotic theories where the contributions of the additional fields λ^a or Φ^I have to be taken into account.

More general compactifications than the one discussed so far are not only specified by the metric on K^{int} but by additional non-trivial background values of other massless bosonic fields. For example, a consistent compactification of type IIB on $AdS_5 \times S^5$ needs a non-trivial background value for the self-dual five-form field strength F_5 which provides the necessary vacuum energy density to balance the curvature of each factor. As for Calabi–Yau manifolds this compactification is an exact conformal field theory and it plays a prominent rôle in the AdS/CFT correspondence which we discuss in Sect. 7.

For more general compactifications with background fields, their backreaction on the geometry demands that one gives up the (geometric) direct product structure of the Ansatz and replaces it by a warped product where the metric of the infinitely extended space-time depends on an overall scale factor – the warp factor – which can be a non-trivial function of the coordinates of the compact space. Examples are compactifications where the Calabi–Yau manifold is replaced by a manifold with $SU(3)$-structure (rather than $SU(3)$ holonomy). Such generalized compactifications arise when localized sources

[7] Calabi–Yau manifolds are compact Kähler manifolds with $SU(3)$ holonomy.

for the background fields (D-branes, orientifold planes) and/or background fluxes are present, i.e. non-vanishing VEVs for the (R,R) and (NS,NS) anti-symmetric tensor fields (see also Sect. 6).

6 Duality and M-Theory

So far we have only discussed the perturbative quantization of strings which propagate through a fixed classical background space-time. A complete theory of quantum gravity should, however, dynamically generate the background space-time. At this time, string theory has not yet achieved this, but there has been recent progress within the AdS/CFT correspondence.

The main problem in taking space-time to be dynamical ab initio is that a non-perturbative formulation of the theory does not yet exists. This situation is quite unusual. One often encounters that a theory is, at least in principle, known but in order to compute quantities of interest one must develop perturbative methods which allow approximate computations. In string theory the situation is quite different: only the perturbation series is known while the fundamental formulation from which it can be derived is still lacking.

One possible way to access the non-perturbative regime is via the duality between weakly and strongly coupled theories, a concept which is well known for supersymmetric quantum field theories. It provides control over the strongly coupled regime of a given theory via perturbative methods applied to the dual theory. The two theories which comprise a duality pair are often very different perturbatively; they might differ, e.g., in their degrees of freedom and their symmetries. The perturbative degrees of freedom of one theory might be solitons, i.e. localized solutions of the classical equations of motion of the weakly coupled dual theory. These solitons are not part of the perturbative spectrum since their masses diverge as the coupling constant g approaches zero. If the solitons become very light and weakly coupled as $g \to \infty$, they might play the rôle of the elementary degrees of freedom of the dual theory. A duality between a weakly and a strongly coupled theory is called S-duality.

S-duality in string theory is non-perturbative in the power series expansion in the coupling constant g, but it is perturbative in the expansion in l_s. For T-duality the situation is reversed. The non-perturbative nature in the expansion in l_s manifests itself, e.g. in mirror symmetry, through the contribution of world-sheet instantons $\sim e^{-R^2/l_s^2}$, where R is the overall size of the Calabi–Yau manifold. A discrete symmetry which is neither perturbative in g nor in l_s is called U-duality.

To prove S-duality (or U-duality) is difficult, since it presupposes a non-pertubative formulation of the theory. However, one can check the duality hypothesis on those solitonic states whose quantum corrections are controllable and whose masses, as functions of the coupling constants, can be exactly determined at weak coupling. For these states an extrapolation to strong coupling is allowed and the comparison with perturbative states of the dual theory is then possible. Such Bogomolny–Prasad–Sommerfield (BPS) states are present in field and string theories with extended supersymmetry. They have the distinctive property that they preserve some of the supersymmetries, i.e. they are annihilated by some of the supercharges, the generators of the supersymmetry algebra.

The BPS-spectrum of string theory contains, in particular, the D-branes. In analogy to the coupling of an electrically charged particle to the Maxwell potential $A^{(1)}$, an 'electric' p-dimensional Dp-brane couples to a $(p+1)$-form potential $C^{(p+1)}$. In addition to the electrically charged branes there are also 'magnetically charged' branes. They are characterized through the field strength $(*H)^{(8-p)}$ which is dual to $H^{(p+2)} = dC^{(p+1)}$. This means that the object which is dual (in the sense of Hodge duality) to an electrically charged Dp-brane is a magnetically charged $(6-p)$-dimensional D$(6-p)$-brane. The potentials C to which branes couple are the massless fields in the (R,R) sectors of superstring theories.

The (NS,NS) sector of the type II and the heterotic string theories also contains an anti-symmetric tensor field $B_{\mu\nu}$ to which their fundamental string (F1) couples. The dual magnetic object is the five-dimensional NS5-brane. The massless bosonic fields and the D-brane spectra of the different string theories are summarized in Tables 1 and 2. The massless fermionic fields are determined by space-time supersymmetry.

Branes were first discovered as classical solutions of the effective supergravity theories. The supergravity solutions describe extended objects and contain, in addition to a non-trivial space-time metric and the dilaton, a non-vanishing $(p+2)$-form field strength $H^{(p+2)}$. Subsequently the solutions which couple to (R,R) fields got their string theoretic interpretation as D-branes, namely the dynamical objects on which open strings end and

Table 1. Bosonic massless fields in type II theories, the closed string sector of type I and in the heterotic theories. $G_{\mu\nu}$ is the space-time metric, $B_{\mu\nu}$ an anti-symmetric tensor field (Kalb-Ramond field), and Φ the dilaton. A_μ is the vector potential of the gauge groups $E_8 \times E_8$ and $SO(32)$, respectively. $C^{(p)}$ is a p-form field with field strength $H^{(p+1)} = dC^{(p)}$. The field strength of $C^{(4)+}$ is self-dual, $H^{(5)} = *H^{(5)}$, and $H^{(0)}$ is a non-propagating 0-form field strength. (The type I string also has SO(32) gauge bosons from the open string sector)

Sektor	(NS,NS)	(R,R)
Type IIA	$G_{\mu\nu}, B_{\mu\nu}, \Phi$	$H^{(0)}, C^{(1)}, C^{(3)}$
Type IIB	$G_{\mu\nu}, B_{\mu\nu}, \Phi$	$C^{(0)}, C^{(2)}, C^{(4)+}$
Type I	$G_{\mu\nu}, \Phi$	$C^{(2)}$
Heterotic	$G_{\mu\nu}, B_{\mu\nu}, \Phi, A_\mu$	

Table 2. D-brane spectra of superstring theories. The D(−1) brane of type IIB is a D-instanton. The D9 brane in type I is degenerate. It implies that open strings can move freely in the ten-dimensional space-time. All remaining D-branes are in one-to-one correspondence to 'electric' (R,R)-potentials and their 'magnetic' duals. D-branes couple to these potentials as sources. T-duality changes the boundary conditions of open strings, N↔D. This means that T-duality maps Dp-branes to D($p \pm 1$)-branes, depending on whether the T-duality direction is along (−) or perpendicular (+) to the world-volume of the brane. Type II and heterotic theories also have a NS5 brane

Dp-branes	p
type IIA	0,2,4,6,8
type IIB	-1,1 3 5 7
type I	1 5 9

to which they transfer space-time momentum. Those which couple to $B_{\mu\nu}$ or its dual are identified with the fundamental string and the NS5 brane, respectively. If one computes the tension (energy density) of the F1 solution (which carries 'electric' B-charge) one finds that it is independent of the string coupling constant. The tension for the NS5-brane (which carries 'magnetic' B-charge) however, behaves as $\tau_{\text{NS5}} \sim 1/g^2$ while that of D-branes depends on the string coupling as $\tau_{\text{D}} \sim 1/g$. This means that the NS5- and the D-branes are heavy and decouple in the weak coupling limit $g \to 0$. They are part of the non-perturbative sector of the respective pertubatively defined string theory. At strong coupling, $g \gg 1$, the BPS-p-branes become light. In some cases they can be viewed as the fundamental objects of a dual theory which possesses a perturbative expansion in powers of $g_{\text{dual}} = 1/g$.

An example of this in $d = 10$ is the S-duality between the heterotic SO(32)-string and the type I string. The coupling constants of these two theories are inverse of each other, and the D-string of type I is mapped, in the limit of strong coupling, to the fundamental heterotic string.

The type IIB theory in $d = 10$ possesses both an F-string and a D-string. The relation between their tensions is $\tau_{\text{F1}}/\tau_{\text{D1}} = g$, i.e. at strong coupling the D-string is much lighter than the F-string. The type IIB theory is self-dual under S-duality, i.e. it is invariant under $g \to 1/g$ and simultaneous exchange of D- and F-strings and their dual magnetic objects, the D5 and NS5 branes. T-duality relates the type IIB theory with the type IIA theory. T-duality also relates the two heterotic theories with each other.

The type IIA theory has BPS bound states of n D0-branes with mass $m \sim n/(gl_s)$. These states can be interpreted as Kaluza–Klein excitations of an 11-dimensional theory which has been compactified on an S^1 with radius $R_{11} = gl_s$ (cf. the discussion of S^1 compactification in Sect. 5). In the strong coupling limit $g \to \infty$, the type IIA theory possesses 11-dimensional Poincaré invariance. At low energies the massless excitations and their interactions are described by the unique 11-dimensional supergravity theory. Using $G_N^{(11)} \sim R_{11} G_N^{(10)} \sim M_{11}^{-9}$, the characteristic mass scale of the 11-dimensional theory is $M_{11} = g^{-1/3} l_s^{-1}$. At energies $\mathcal{O}(M_{11})$, neither string theory nor supergravity are adequate descriptions. Both have to be superseded by an as yet unknown theory which has been given the name M-theory. The strongly coupled $E_8 \times E_8$ heterotic string can also be interpreted as a compactification of M-theory, namely on an interval. The gauge degrees of freedom of one E_8-factor are located on each of the two ten-dimensional boundaries at the end of the interval.

The duality relations imply that the five string theories are merely different perturbative approximations of one and the same fundamental theory. The fact that 11-dimensional supergravity also appears indicates that the five string theories cannot provide a complete description in the strong coupling regime. The hypothetical theory, from which the five string theories and 11-dimensional supergravity can be derived in different approximations, is called M-theory. The elementary excitations of this theory depend on the approximation. As 11-dimensional theory it possesses membranes, i.e. M2-branes, and their dual objects, M5-branes. The fundamental string of type IIA arises upon compactification on a circle of radius R_{11} where the M2 branes is wrapped around the circle.

In addition to the duality relations which was we have discussed here, there are other connections between various string theories, in the critical dimension as well as in the compactified theory. In all non-perturbative dualities branes play an essential rôle.

In the presence of D-branes one has, besides the excitation modes of the closed string, also those of the open string whose endpoints move along the world-volume of the branes. For instance, at low energies ($l_s \to 0$), the dynamics of the massless modes of N coincident D3-branes is described by a four-dimensional $\mathcal{N} = 4$ SYM-theory with gauge group $U(N)$. This gauge theory is localized on the world-volume of the D3-branes. Its gauge coupling constant is $g_{\rm YM}^2 = g$. In the limit $l_s \to 0$ the modes of the open string and gravity decouple. Many different theories can be constructed by an appropriate choice of D-brane configurations and e.g. many features of the vacuum structure of the supersymmetric extension of QCD (SQSD) can be 'understood' in the brane picture.

7 AdS/CFT

String theory dates back to the pre-QCD era, as an attempt to understand the scattering data of hadrons. Veneziano 'guessed' a formula (known as the Veneziano formula) which correctly incorporates the empirically motivated duality hypothesis, which states that the complete four-point amplitude can be written either as a sum over only s-channel poles or as a sum over only t-channel poles. It was soon realized that the Veneziano amplitude can be derived from a theory of (bosonic) strings. Serious problems related to the high-energy behaviour of the Veneziano amplitude and, in particular, the discovery of QCD as a renormalizable QFT made string theory as a theory of the strong interaction obsolete. Furthermore, the discovery of a critical dimension and the presence of a massless spin-two particle were considered as indications that string theory might be the correct framework for a theory of quantum gravity. This has become the prevailing point of view.

However, more recently, based on the AdS/CFT conjecture of Maldacena, string theory has become a powerful analytic tool for studying strongly coupled gauge field theories. The most interesting such theory is QCD at low energies. While no gravity dual has yet been found, many (supersymmetric) generalizations have been studied using the so-called 'gauge theory - gravity duality'.

In its simplest version, the AdS/CFT correspondence arises from analysing a system of N coincident D3 branes. For small $gN = g_{\rm YM}^2 N$, i.e. for small 't Hooft coupling, the world-volume theory on the branes is the conformally invariant $U(N)$ $\mathcal{N} = 4$ supersymmetric gauge theory. Its degrees of freedom arise from the massless excitations of the open strings ending on the branes. This theory is coupled to supergravity in the ten-dimensional space-time. The supergravity fields arise from massless excitations on the closed strings. As long as gN is small, one can neglect the backreaction of the branes on the geometry and the assumption of the D3 branes embedded in ten-dimensional Minkowski space-time is appropriate. In the limit $l_s \to 0$ the gauge theory on the brane decouples from the gravity theory in the bulk. If gN becomes large, the backreaction can no longer be neglected and the system is better described by the geometry of the brane solutions of type IIB supergravity. The above decoupling limit now leads to a decoupling of the region close to the branes, the so-called 'near-horizon region' which has $AdS_5 \times S^5$ geometry, from the asymptotic region, where one obtains a theory of free gravitons in ten-dimensional Minkowski space-time. Comparison then suggest a correspondence between four-dimensional $\mathcal{N} = 4$ SYM theory and type IIB string theory compactified on $AdS_5 \times S^5$. It also implies the relations $(R/l_s)^4 = 4\pi g N = 4\pi g_{\rm YM}^2 N$ between the string scale l_s, the curvature radius R of the background geometry, the string coupling constant g, the rank of the gauge group N, and the gauge coupling constant $g_{\rm YM}^2$. One can think of the gauge theory degrees of freedom to be located at the conformal boundary of AdS_5 which is four-dimensional Minkowski space-time (up to global issues). In this sense, the AdS/CFT correpondence is a very concrete realization of the holographic principle (see also Sect. 8). One can further interpret the radial coordinate as the energy scale in the field theory.

As long as the radius of curvature is large and the string coupling constant is small, one can approximate the type IIB string theory by IIB supergravity on this background. One then obtains a duality between a *quantum* field theory – $\mathcal{N} = 4$ SYM in the large-N limit – and a *classical* gravity theory. Evidence for this duality is provided by a matching of the symmetries: the isometry group of the space-time coincides with the global symmetries of the gauge theory (conformal invariance and $\mathcal{R}$-symmetry) and this extends to the supergroups. In particular the *AdS*-factor of the geometry indicates that the dual field theory is conformally invariant (which $\mathcal{N} = 4$ SYM is).[8] More detailed checks, which do not rely entirely on the symmetries, have been performed. For instance, the conformal anomaly of $\mathcal{N} = 4$ SYM, which is clearly a quantum effect of a four-dimensional field theory, can be computed via a classical gravity calculation. A precise matching between Kalazu–Klein states of the supergravity theory and gauge invariant operators is possible and many of their dynamical properties can be computed on both sides of the correspondence. Needless to say that they match precisely.

One obstacle to go beyond the supergravity approximation on the string theory side is that this requires the quantization of string theory compactified on $AdS_5 \times S^5$, which consists, in addition to a background metric, of a non-vanishing value of the self-dual (R,R) five-form field strength. At present, quantization in (R,R) backgrounds (as opposed to (NS,NS) backgrounds) is still an unsolved problem, at least in the so-called 'RNS' (Ramond–Neveu–Schwarz) formalism on which most of the string literature is based. But it has been shown that $AdS_5 \times S^5$ is a consistent background for string compactification to all orders in string pertubation theory.

Considerable progress has been made in the so-called 'BMN' (Berenstein–Nastase–Maldacena) limit where relevant configurations on the string side are classical solutions of the string sigma-model which correspond to macroscopically large strings rotating in the background geometry. On the gauge theory side the dual operators are those with large conformal dimension and R-charge (which is dual to the $SO(6)$ isometry of S^5).

Many generalizations of the correspondence have been constructed. For instance, in order to reduce the amount of supersymmetry one replaces S^5 by a five-dimensional compact manifold X^5 which can serve as the base of a six-dimensional Ricci-flat Kähler cone, i.e. X^5 must be a Einstein–Sasaki manifold (e.g. for $X^5 = S^5$ the Kähler cone is simply $\mathbb{R}^6$). Generalizations to non-conformal theories are neccessary if one wants a dual description of confining gauge theories (such as QCD). In fact, one can give a rather general criterium which the background geometry has to satisfy in order that the dual gauge theory is confining. This relies on the picture of the QCD string as a fundamental string which connects two quarks which are located on the 'boundary' of the dual geometry, but which plunges into the bulk (cf. Fig. 7) as this is the geodesic which connects its two endpoints. The expectation value of the Wilson loop $\langle W[\mathcal{C}]\rangle \sim e^{-TE(L)}$, where $E(L)$ is the potential energy, is the exponentiated area of the world-sheet of the open string which boundary $\mathcal{C}$, computed with the Nambu–Goto action in the given background geometry. If one does this in the $AdS_5 \times S^5$ geometry one finds $E(L) \sim T/L$, i.e. the Coulomb law. In a confining dual geometry one finds instead $E(L) = \sigma T L$,

[8] More generally, asymptotic *AdS* geometries are dual to field theories which are conformally invariant (fixed point of the beta-function) in the UV.

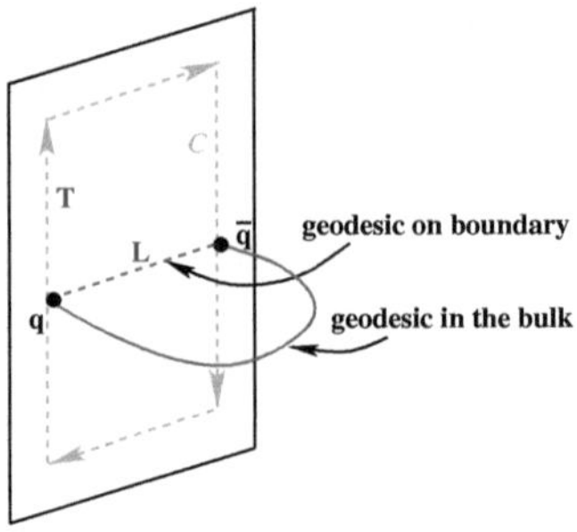

Fig. 7. 'QCD' string

where σ is the QCD string tension. Such a confining geometry must contain an additional scale which breaks conformal invariance, in terms of which σ can be expressed.

However, the regions in parameter space where, on the one hand, the gauge theory is weakly coupled and perturbation theory appropriate and, on the other hand, where the string coupling is weak and the space-time curvature small, i.e. where the supergravity approximation of string theory is good, do not overlap. Therefore, a direct comparison is only possible for protected operators which exists in field theories with extended supersymmetry (BPS states). But one can use the conjectured correpondence to arrive at 'predictions' about strongly coupled gauge theories, e.g. about their phase structure, the spectrum of mesons, chiral symmetry breaking, etc. in regimes where other analytical methods are not available. All these concepts have a geometric analogue within the dual gravity description. Also, the interpretation of the radial coordinate as the energy scale has been made precise, e.g., in extracting the gauge theory beta-function from geometrical data.

Other generalizations of the Maldacena conjecture lead to holographic descriptions of theories in other than four dimensions. For instance, compactification of 11-dimensional supergravity on $AdS_7 \times S^4$ leads to a six-dimensional theory of interacting tensor multiplets on the world-volume of coincident M5-branes. Again, the number of branes N is related to the radius of curvature of the geometry ($R \sim l_p N^{1/3}$). Nothing is known about this theory from the field theory side but definite predictions, e.g. about the number of degrees of freedom, i.e. that it grows as N^3, can be obtained from its dual gravity description.

Perhaps the most important lesson from these developments is the duality between quantum field theories (without gravity) and higher-dimensional gravitational theories (such as supergravity or string theory). A dual description of real QCD (four-dimensional, non-supersymmetric, $SU(3)$ gauge group, etc.) has not yet been found. But it has been demonstrated that the high energy behaviour of the Veneziano amplitude, when interpreted within AdS/CFT context (generalized to non-conformal backgrounds) where the radial coordinate serves as the energy scale, changes and that it is no longer in disagreement with experiments.

8 Black-Hole Entropy

Black holes are a major testing ground for ideas about quantum gravity. They are subject to the laws of black-hole mechanics which formally have the same structure as the laws of thermodynamics. Combining this with the Hawking effect, which allows to assign a temperature to a black hole, this leads to the identifications

$$T = \frac{\kappa_S}{2\pi}\left(\frac{\hbar}{c}\right) , \quad \mathcal{S}_{\text{thermo}} = \frac{A}{4}\left(\frac{c^3}{G_N \hbar}\right) . \tag{8.1}$$

Here T is the Hawking temperature, $\mathcal{S}_{\text{thermo}}$ is the Bekenstein–Hawking entropy, κ_S the surface gravity, and A the area of the event horizon.[9] The occurrence of $\hbar$ and G_N clearly shows that black-hole entropy can only be described within the framework of a theory of quantum gravity. Below we will often use Planckian units and set $c = \hbar = G_N = 1$. In analogy to the relation between thermodynamics and statistical mechanics, this suggests that while Einstein gravity describes black holes at the macroscopic level, a theory of quantum gravity should provide the microscopic description. In particular, it should be possible to relate the thermodynamical entropy to a statistical entropy, which measures the degeneracy of microscopic states for a given macroscopic state,

$$\mathcal{S}_{\text{stat}} = \log N(M, Q, J) . \tag{8.2}$$

Here the macroscopic state of a black hole is characterized by its mass M, charge Q, and angular momentum J, and $N(M, Q, J)$ denotes the number of microscopic black-hole states with given values for these quantities. It is a benchmark for any candidate theory of quantum gravity whether such microscopic states can be identified and counted, and whether the thermodynamical and statistical entropies agree.

The simple fact that the entropy is proportional to an area (the area of the black-hole's event horizon) rather than a volume, leads to the concept of holography. The information contained inside the region enclosed by the horizon is represented as a hologram on the horizon: all information about the inside is stored on the holographic screen.[10] This is in sharp contrast with

---

[9] For a Schwarzschild black-hole $\kappa_S = c^4/(4G_N M)$ and $A = 4\pi\left(\frac{2MG_N}{c^2}\right)^2$.

[10] More generally, the holographic principle asserts that the information contained in some region of space can be represented as a 'hologram' – a theory which 'lives' on the boundary of that region. It furthermore asserts that the theory on the boundary of the region of space in questions should contain at most one bit of information per Planck area l^2_{Planck}.

what we expect from statistical mechanics and local quantum field theory where the entropy is an extensive quantity and should thus be proportional to the volume of the system. The lesson we learn from this is that the nature of the degrees of freedom of quantum gravity is quite different from that of a local quantum field theory.

In this section we discuss some of the research done on black-hole entropy in string theory. Except for the initial discussion of the heuristic string–black hole correspondence, we restrict ourselves to BPS black holes, i.e. black holes which are invariant under a subset of the supersymmetry transformations of the underlying string theory. For BPS black-holes, string theory provides a quantitative explanation of black-hole entropy. The agreement between thermodynamical and statistical entropy extends beyond the leading term in the semiclassical limit. Moreover, the calculations show that at subleading level the entropy of stringy black holes follows Wald's generalized formula for black-hole entropy, which deviates from the simple area law once quantum corrections (higher derivative corrections) to the Einstein–Hilbert action are taken into account.

We first discuss the heuristic string–black hole correspondence which, while qualitative, has the virtue to apply to Schwarzschild-type black holes. The basic idea is that 'heavy strings states are black holes'. Let us start with string perturbation theory in flat space. Free strings have an infinite tower of states of ever increasing mass, $m^2 \sim \frac{n}{\alpha'}$, $n \in \mathbb{N}$. If we take the string coupling g to be finite, but small, the feedback of a sufficiently light string state on its ambient space-time is negligible. A rough way of estimating this feedback is to compare the characteristic length scale of string theory, $l_S = \sqrt{\alpha'}$, to the gravitational scale of a string state of mass m, i.e. its Schwarzschild radius $r_S \sim G_N m \sim \sqrt{n\alpha'}\, g^2$. Here we used the relation $G_N \sim g^2\alpha'$ between the four-dimensional Newton constant G_N, the string scale $\sqrt{\alpha'}$, and the string coupling constant, together with the mass formula.[11] The string length $\sqrt{\alpha'}$ is the smallest length scale which can be resolved by scattering string states. Since the feedback of a string state on the space-time geometry is estimated by r_S, it is negligible if $\sqrt{\alpha'} \gg r_S$. For given coupling this requires that the mass of the state is sufficiently small, while for given mass the coupling must be sufficiently small. In this regime the number of string states of given mass can be counted, since we know the spectrum of free strings in flat space-time. The asymptotic number of states is governed by the Hardy–Ramanujan formula and grows like $e^{\sqrt{n}}$. In other words the statistical entropy of string states grows like

$$\mathcal{S}_{\text{stat}} \sim \sqrt{n} \tag{8.3}$$

for large n.

Let us now either increase the mass, at fixed coupling, or increase the coupling, at fixed mass. Then r_S will grow relative to $\sqrt{\alpha'}$. While we do not

[11] We assume that the additional dimensions required for consistency have been compactified on a manifold of size $(\alpha')^3$.

know what happens in detail, we know qualitatively what happens in the opposite extreme regime $r_S \gg \sqrt{\alpha'}$, where the gravitational scale is much larger than the string scale. Here we can use that the long-wavelength approximation of string theory is provided by an effective field theory, which can be constructed using string perturbation theory. The effective action has an expansion in string loops, controlled by g, so that we need to keep the string coupling small enough. Moreover, it has an expansion in the string length $\sqrt{\alpha'}$. The long-wavelength or low energy expansion of the action is an expansion in derivatives. σ-model loop corrections (i.e. higher orders in α') which appear in each order in the genus expansion (higher orders in g) give rise to higher derivative terms.

In this regime gravity is described by the Einstein–Hilbert action plus an infinite series of higher curvature terms. For the time being, we only take into account the leading Einstein–Hilbert term. Then we are in the realm of general relativity and expect that an object which sits within its Schwarzschild radius forms a black hole. Therefore our original string state should correspond to a black-hole solution of the effective field theory. The associated thermodynamical entropy is the Bekenstein–Hawking entropy. For a Schwarzschild-type black hole carrying the mass of the string state we obtain

$$\mathcal{S}_{\text{thermo}} = \frac{A}{4G_N} \simeq G_N m^2 \simeq g^2 n \ .$$

Comparing this to the statistical entropy of string states (8.3), we see that both entropies are different, in general. However, they agree, up to a numerical constant of order unity, when the Schwarzschild radius is of the order of the string scale, $r_S \simeq \sqrt{\alpha'}$, or, equivalently, for a string coupling of order $g^2\sqrt{n} \simeq 1$. The observation that the entropies of strings and black holes match here support the idea of a phase transition (or maybe a smooth crossover) from a perturbative string regime to a black-hole regime. In particular, this provides a scenario for the endpoint of the decay of black holes through Hawking radiation: once the black hole has shrunk to a size of order $\sqrt{\alpha'}$, it converts into a highly excited string state, which then decays according to the rules of string perturbation theory. It is encouraging that a string has the right number of states to account for the states of a black hole with equal mass. This scenario is compatible with unitarity, and elaborates on the old idea of a correspondence between black holes and elementary particles. The idea of a phase transition is further supported by the observation that the Hawking temperature of a black hole of size $\sqrt{\alpha'}$ equals the Hagedorn temperature, which is interpreted as the limiting temperature for a grand canonical ensemble of strings.

While this scenario is broad and appealing, it is very qualitative. In particular, the string and black-hole entropy only match up to a multiplicative factor of order unity, and the interpolation between the perturbative string regime and the black-hole regime is bold, because one has no control over the intermediate regime. There is no a priori argument which connects the number of states in the two extreme regimes, which in principle could change drastically. And indeed, we saw that the two entropies are different in general,[12] indicating that the number of states changes when going from one extreme regime to the other.

Therefore we will now focus on a subset of states which are under much better control. Here, the entropies of strings and black holes will not just match for some particular value of the coupling, but they will be equal. Here 'equal' means equality up to additive terms which are subleading in the semiclassical limit, corresponding to large mass. In particular, there is no undetermined or mismatching factor between the leading terms of the two entropies. The relevant subclass of states are the BPS states, which sit in special, so-called 'short' or BPS representations of the supersymmetry algebra. BPS states carry central charges under the supersymmetry algebra, and have the minimal mass compatible with their central charges. In supergravity the central charges are determined by the electric and magnetic charges under gauge interactions which are mediated by the gauge bosons in the supergravity multiplet (graviphotons).

In string perturbation theory, BPS states appear as a special subset of the string states. In the corresponding effective supergravity theory, BPS states are realized as supersymmetric solitons, more specifically as extremal black-hole solutions with Killing spinors. The comparison of black hole and string entropy proceeds by constructing BPS black-hole solutions with given charges and by comparing the resulting entropy to the number of string BPS states with the same mass and the same charges. In various examples where both entropies have been computed in their respective regimes, it has been found that they agree, even when including subleading corrections.

Let us discuss an explicit example for the quantitative version of the string–black hole correspondence. We consider four-dimensional string compactifications with $\mathcal{N}=4$ supersymmetry. For concreteness, we employ the realization through the heterotic string, compactified on a six-torus. For generic moduli the gauge group of this compactification is $U(1)^{28}$, and the electric charges carried by elementary string states can be combined into a vector $\mathbf{q}$ which takes values in a 28-dimensional lattice $\Gamma_{22,6}$, which comes equipped with an indefinite bilinear form of signature $(22,6)$. Incidentally, the problem of counting BPS states of charge $\mathbf{q}$ is equivalent to counting the number of states for the open bosonic string in 26 dimensions. Hence the result follows from the Hardy–Ramanujan formula. The corresponding entropy is

$$\mathcal{S}_{\text{stat}} = 4\pi\sqrt{\frac{|\mathbf{q}^2|}{2}} + \cdots \,. \tag{8.4}$$

Here $\mathbf{q}^2$ is the (indefinite) scalar product of the charge vector $\mathbf{q}\in\Gamma_{22,6}$ with itself. We have displayed the leading contribution in the limit of large charges

[12] Note that the black-hole entropy is bigger than the string entropy if we are in the black-hole regime, and vice versa.

$|\mathbf{q}^2| \gg 1$ (which, through the BPS condition, implies large mass). There are corrections, starting with a term proportional to $\log |\mathbf{q}|^2$, followed by an infinite series of terms which involve negative powers of $|\mathbf{q}^2|$, plus further corrections which are exponentially suppressed for large charges.

The corresponding effective field theory is, to leading order in derivatives, an $\mathcal{N} = 4$ supergravity theory coupled to 22 vector multiplets. It turns out that BPS solutions with charges $\mathbf{q}$ always have a null singularity, i.e. the event horizon coincides with the singularity and has vanishing area. As a consequence, the Bekenstein–Hawking entropy is zero

$$\mathcal{S}_{\text{thermo}} = \frac{A}{4} = 0 \,,$$

and disagrees with the statistical entropy of string states. This is, however, not the end of the story. Since space-time curvature becomes large close to the horizon, one cannot trust the two-derivative effective action. Once the leading curvature-squared terms are taken into account, the null singularity is replaced by a smooth horizon of area $A = 8\pi\sqrt{\frac{1}{2}|\mathbf{q}^2|}$. The corresponding Bekenstein–Hawking entropy $\mathcal{S}_{\text{Bekenstein-Hawking}} = \frac{A}{4} = 2\pi\sqrt{\frac{1}{2}|\mathbf{q}^2|}$ is finite, but disagrees with the statistical entropy by a factor 2. However, as pointed out some time ago by R. Wald, the area law has to be replaced by a more refined formula, once the gravitational action contains higher derivative terms. In contrast to the naive area law, Wald's modified law assures that the first law of black hole mechanics remains valid. For the case at hand, Wald's modified formula amounts to an additive correction term $\frac{A}{4}$, which leads to precise agreement between the leading term of the thermodynamical entropy

$$\mathcal{S}_{\text{thermo}} = 4\pi\sqrt{\frac{|\mathbf{q}^2|}{2}} + \cdots \tag{8.5}$$

and the statistical entropy. Like the statistical entropy, the thermodynamical entropy is further modified if subleading corrections are taken into account. For the thermodynamical entropy, the corrections come from further subleading terms in the effective action. As for the statistical entropy, these corrections are logarithms, inverse powers, and exponentials in $|\mathbf{q}^2|$.

The next step is therefore to compare the subleading contributions to both entropies. In the above example, no full agreement between statistical and thermodynamical entropy has been achieved to date. The problem seems to be related to the fact that for BPS black holes in $\mathcal{N} = 4$ compactifications, which carry only electric charge, the scalar fields take values in a particular subspace of the moduli space, which is singular unless instanton corrections are taken into account. This reflects itself in the fact that black-hole solution has a vanishing horizon area at leading order. While further work is needed to better understand this class of BPS black holes, the situtation is much better for generic BPS black holes, which carry both electric and magnetic charges.

The most general BPS black-hole solution of an $\mathcal{N} = 4$ compactification carries 28 electric charges $\mathbf{q}$, but also 28 magnetic charges $\mathbf{p}$, which lie on a lattice of the form $\Gamma_{22,6}$.[13] When using the two-derivative effective action, the entropy of such dyonic BPS black holes is

$$\mathcal{S}_{\text{thermo}} = \pi\sqrt{\mathbf{p}^2\mathbf{q}^2 - (\mathbf{p}\cdot\mathbf{q})^2}\,. \tag{8.6}$$

Observe that for purely electric charge the entropy vanishes. This is the subcase we discussed above. What are the corresponding string theory microstates? Fundamental strings do not carry magnetic charges with respect to the gauge group $U(1)^{28}$. However, magnetic charges are carried by heterotic five-branes, which are solitonic objects occurring in the heterotic string theory. Dyonic BPS states with arbitrary electric and magnetic charge correspond to bound states of fundamental heterotic strings and heterotic five-branes. The number of BPS states with given charges is known in terms of an integral representation. When evaluating this integral at its leading saddle point, one recovers (8.6). But as in the case (8.4) there are subleading corrections to both the statistical and thermodynamical entropy. This time the corrections agree even when including contributions which are exponentially suppressed for large charges. At the level of the effective action, this corresponds to including the contribution of an infinite series of instanton corrections to the higher-derivative terms. The agreement crucially depends on using Wald's modified formula instead of the naive area law.

There are several other types of brane configurations where a quantitative agreement between statistical and thermodynamical entropy, including subleading corrections, has been found. In particular, the first examples of such a matching involved D-branes, rather than fundamental strings and solitonic five-branes. With this amount of evidence, it is fair to say that string theory can account quantitatively for the entropy of BPS black holes. String theory is unrivaled in that the matching of statistical and thermodynamical entropy does not involve the tuning of free parameters, and that the matching extends to subleading corrections and is sensitive to the distinction between Wald's law and the area law. This success also illustrates that a consistent perturbative theory of quantum gravity accounts for much more than 'graviton scattering in a fixed background'. In particular, string perturbation theory can be used to derive higher curvature corrections to the Einstein–Hilbert action. These in turn modify black-hole solutions, smooth singularities, and give contributions to the entropy. These are genuine quantum gravity effects, as the higher-derivative terms are generated by quantum corrections.[14] While the

[13] By Dirac quantization, electric and magnetic charges lie on dual lattices. However, the charge lattice turns out to be self-dual, so that one has two copies of the same lattice.

[14] To be precise, the terms relevant for the $\mathcal{N} = 4$ compactifications discussed above are 'tree-level plus instantons' (in the string coupling g) for the heterotic string and 'one-loop' for the dual description by the type-II string.

agreement of statistical and thermodynamic entropy strongly suggests that string theory has the right number of degrees of freedom to account for the microstates of BPS black holes, a more direct understanding of these states as states of black holes is certainly needed. Recently, an intriguing proposal has been put forward by H. Ooguri, A. Strominger, and C. Vafa, which defines a 'black hole partition function' and relates it to the partition function of the topological string. This could be a major step forward in this direction.

A clear limitation of the approach described here is that it relies on supersymmetry, or, to be precise, that it applies to supersymmetric states only. However, there are other approaches to black holes within string theory, which we are not able to discuss here for lack of space. But let us mention that recently there has been considerable interest in studying non-supersymmetric extremal black holes. It turns out that many features of supersymmetric black holes carry over, and in particular that higher derivative corrections can be taken into account. Moreover, black-hole entropy has been studied extensively in the context of the AdS/CFT correspondence, which can be viewed as a concrete realization of the 'holographic principle'.[15] Finally, a new line of thought is the 'fuzzball proposal', which views BPS black holes as superpositions of smooth geometries, one for each black-hole microstate. This approach might be a first step towards a detailed understanding of the interpolation between the string perturbative regime and the black-hole regime.

So far, string theory does not yet provide a complete account of black-hole physics. Nevertheless, black-holes are clearly the most successful application of string theory in the gravitational realm. They will continue to be a major subject of interest in the string community, and, maybe, the results will even reshape our understanding of what string theory is.

9 Approaches to Phenomenology

As we discussed in Sect. 5 the spectrum of excitations of a string compactified, e.g., on a Calabi–Yau manifold, contains a finite number of massless excitations L and an infinite number of massive modes H. Their mass is of the order of the characteristic scale of the string M_s. Among the massless modes one finds generically a spin–2 degree of freedom which is identified with Einstein's graviton. In addition massless spin–1 gauge bosons of some gauge group G, families of massless chiral fermions in fundamental and anomaly free representations of G and elementary spin–0 bosons which can serve as candidates for Higgs-like fields can appear among the massless modes. Such string backgrounds are not only a candidate for a consistent quantum gravity but also a candidate for a unified theory of all known particles and their interactions.

In order to check this proposal, it is necessary to identify the standard model (SM) as the low energy limit. This amounts to the identification of the

[15] The duality with a unitary quantum field theory strengthens the claim that there is no information loss during black-hole evaporation via Hawking radiation.

particle spectrum of the standard model (or some generalization thereof) as well as their couplings in a low-energy effective Lagrangian L_{eff}. The effective Lagrangian can be computed systematically in perturbation theory by studying string scattering processes at energy scales E far below the characteristic scale M_s. Demanding that the S-matrix of the effective field theory coincides with the string S-matrix for energy scales, $E \ll M_s$ determines the effective Lagrangian.

However, the programme just outlined has a number of serious drawbacks. First of all, the S-matrix elements in string theory can currently only be reliably computed as a perturbative expansion in the string coupling g. Second of all, a large class of consistent S-matrices, each corresponding to a two-dimensional conformal field theory, do exist. This in turn leads to a large number of different effective theories with different L_{eff}. Every set of S-matrix elements (or equivalently every consistent CFT) can be viewed as a different vacuum of the same string theory. Each string vacuum is as good as any other or, in other words, the vacuum is degenerate and there is presently no understanding what selects one vacuum over another and lifts the vacuum degeneracy. Finally, string theory only contains one scale M_s and one dimensionless coupling g and hence all light modes L are exactly massless. This is reminiscent of a standard model without the Higgs mechanism where all fermions and gauge bosons are also exactly massless. Thus, one has to understand what mechanism generates the weak scale M_Z (and why it is so small).

Given this state of affairs there are a number of possible strategies to make further progress. One approach – commonly called 'string phenomenology' – does not attempt to explain the mechanism which lifts the vacuum degeneracy and chooses the true vacuum. Rather it surveys the whole space of string ground states and looks for particularly 'promising' candidate vacua. The criteria of what is a 'promising' string vacuum is of course ambiguous and different aspects have dominated this field over the years. After the discovery of the heterotic string and its Calabi–Yau compactification in 1984/85 all of string phenomenology focused on vacua of the $E_8 \times E_8$ heterotic string with four flat space-time dimensions with Minkowskian signature, a gauge group $G \subset E_8 \times E_8$ which is big enough to contain the $SU(3) \times SU(2) \times U(1)$ of the SM and at least three light chiral generations. In addition, $\mathcal{N} = 1$ local space-time supersymmetry was imposed at M_s since it seems very difficult to understand how the hierarchy M_Z/M_s can be generated and kept stable without supersymmetry.[16] Furthermore, most of the known consistent string vacua with space-time fermions are already supersymmetric and within our current understanding supersymmetry appears to be a plausible (if not necessary) symmetry of string theory.

[16] $\mathcal{N} = 1$ is chosen since such supersymmetric theories can easily have chiral fermions. This is not possible for $\mathcal{N} > 1$.

Almost all string vacua contain gauge neutral scalar fields M^i ('moduli of the compactification') which are flat directions of the perturbative effective potential. Thus their vacuum expectation values (VEVs) are undetermined in perturbation theory and therefore they are additional free parameters of a given string vacuum. They set the (inverse) gauge couplings $g_{\rm YM}^{-2}$ and the Yukawa couplings Y of the theory. Thus, as in any QFT, both the gauge couplings and the Yukawa couplings are free parameters of the effective low-energy string theory. However, the situation here is slightly better than in a QFT. First of all, the fact that the couplings depend on scalar field VEVs opens up the possibility of a dynamical determination of the couplings. If we understood what mechanism lifts the flat directions and induces a $\langle M^i\rangle$ we would have a dynamical way of understanding the values of the dimensionless couplings $g_{\rm YM}$ and Y. Furthermore, it is quite possible that for a given vacuum the ratio of Yukawa couplings displays some special properties which can be tested experimentally.

The perturbative heterotic string has the additional feature that the gauge coupling is universal at the tree level. The generic gauge group is a product of simple factors $G = \prod_a G_a$ with gauge couplings g_a for each factor G_a which are identical even without the existence of a covering GUT group

$$g_a^{-2} = \mathrm{Re}\langle \Phi\rangle \qquad \text{for all } a\,, \tag{9.1}$$

where Φ is the dilaton field. (Strictly speaking there is an integer normalization factor k_a in (9.1) which we have omitted here for simplicity.) Thus the perturbative heterotic string very generically predicts a universal gauge coupling. It also predicts the scale at which the coupling constants unify to be $M_s \approx 5\cdot 10^{17}$ GeV. Current electro-weak precision data seem to favour a unification of the gauge couplings at approximately $3\cdot 10^{16}$ GeV which is indeed remarkably close to the string value. However, given the present precision the mismatch of a factor of 20 cannot be simply ignored.

Despite some of the successes of the heterotic string vacua there are a number of questions left unanswered in the perturbative approach outlined above. We still have to understand how the light modes get their masses, how M_Z and the hierarchy is generated, what lifts the vacuum degeneracy and induces VEVs for the M^i, and finally how supersymmetry is broken at low energies. The belief (and hope) is that all of these problems are just an artefact of string perturbation theory and that once we understand the non-perturbative phase of string theory these problems will have a (hopefully realistic) solution.

Since we lack a fully developed non-perturbative formulation of string theory there are various ways to argue the structure of possible non-perturbative corrections. First, one might assume that the dominant non-perturbative effects arise at energy scales well below M_s and therefore can be described by field-theoretic means. Clearly these non-perturbative effects are part of string theory and the real assumption is that they dominate over the 'stringy' effects. This assumption is partly motivated by the fact that in order to generate a hierarchy these non-perturbative effects have to occur at an energy scale well below M_s. An example of such a non-perturbative effect is gaugino condensation in a hidden sector which can be analysed already in supergravity. The hidden sector which has no renormalizable interactions with the observable sector is taken to be an asymptotically free non-abelian gauge theory which is weakly coupled at M_s but becomes strongly coupled at

$$\Lambda_c = M_{\rm Pl}\, e^{-\frac{8\pi^2}{b g_{\rm YM}^2}} \ll M_{\rm Pl}\,, \tag{9.2}$$

where b is the coefficient of the one-loop β-function. Such hidden sectors do indeed exist in string theory, the matterless E_8 of Calabi–Yau compactification of the heterotic string is only one example. As we already discussed, the gauge couplings are field dependent in string theory and thus a non-trivial potential for M^i is generated

$$g_{\rm YM}^{-2}(M^i) \to \Lambda_c(M^i) \to V_{\rm np}(M^i)\,. \tag{9.3}$$

At the minimum of V supersymmetry can be spontaneously broken and non-trivial VEVs for M^i can be generated. However, generically a large cosmological constant arises in almost all of the models considered so far and no realistic scenario satisfying all phenomenological constraints has been constructed. As a consequence a more detailed low-energy phenomenology of such models has not been developed.

In recent years a slight variation on this setup has been studied which goes under the name of 'Brane World Scenarios'. Here the standard model or its generalization lives on a stack of space-time filling D-branes in a type II bulk. Supersymmetry is spontaneously broken by additionally turning on background fluxes in the bulk already at the tree level. The fluxes generate a potential which fixes some of the moduli but in general additional non-perturbative effects have to be employed in order to fix all of them and to obtain a (meta-stable) ground state. This aspect is particularly important if one attempts to construct de Sitter vacua with a small cosmological constant. A detailed analysis of these 'Brane World Scenarios' is currently under way.

If the gauge degrees of freedom of the standard model arise as excitations of a D-brane they can be viewed as localized on a three-dimensional plane within a higher-dimensional space. This implies that the 'extra' dimensions can only be probed by the gravitational interaction. Currently Newtons $\frac{1}{r}$ law is experimentally established down to the sub-millimeter range while the Coulomb $\frac{1}{r}$ law has been established in Bhabha-scattering at LEP down to 10^{-18} m. This opens up the theoretical possibility of 'large extra dimension' which are only transparent for gravity. If they are large enough they can be seen as deviations from Newtons $\frac{1}{r}$ law in gravitational torsion experiments or at LHC by producing appropriate Kaluza–Klein excitations. The phenomenological signatures of such scenarios have been studied in detail.

10 Open Questions

The most obvious shortcoming of 'perturbative string theory' is that it is limited to a set of rules for computing on-shell scattering amplitudes in an on-shell background. If we want to address conceptual issues of quantum gravity this is a double handicap: quantities can only be computed as formal power series in the string coupling, and one has to fix an on-shell background in advance. Perturbative and non-perturbative dualities have certainly enhanced the range of quantities which can be computed, but without changing these points fundamentally. Direct non-perturbative methods, such as an instanton calculus, are in a very early state of development. One might hope that string field theory supersedes the (conceptually) cumbersome 'first quantized' formalism which is still mostly used. However, string field theory is very complicated to work with. With the notable exception of tachyon condensation, string field theory has mainly been used to reproduce results obtained before in the 'first quantized' approach.

Manifest background independence is certainly a desirable feature of any theory of gravity. String theory is background independent, in the sense that different on-shell backgrounds are different solutions of one underlying theory. Formally, this is clear from the fact that deformations of on-shell backgrounds correspond to marginal deformations of the world-sheet action, which in turn are equivalent to inserting the vertex operator for a coherent string state into correlators. However, background independence is not manifest, as one needs to fix a reference background, or equivalently a world-sheet conformal theory before being able to deform it. Therefore there is always an, albeit conventional, cut between the space-time geometry (plus other background fields) and the dynamics in the background. Compared to approaches to quantum gravity which focus on quantizing four-dimensional Einstein gravity, string theory faces additional challenges. The various perturbative and non-perturbative dualities clearly indicate that there is a huge redundancy between the consistent string backgrounds. In particular, since dualities mix the gravitational with other degrees of freedom, one gets identifications between space-time geometries with different topologies. The most prominent example of this is mirror symmetry, which relates Calabi–Yau threefolds with opposite Euler numbers (and reflected Hodge diamonds), and 'second quantized mirror symmetry', which relates type-II string theory on certain Calabi–Yau threefolds to the heterotic string on $K3 \times T^2$ (together with a certain choice of gauge fields inside the $K3$-surface). While this appears to be a deep observation, what is lacking so far is a sufficiently abstract and general concept of 'state', which allows one to understand why these apparently different space-times (amended with other background fields) represent the same state. One closely related question is, what is the geometry underlying string theory?

Both quantum corrections, controlled by g and stringy corrections, controlled by α', have consequences for space-time geometry. Since dualities can exchange quantum effects and stringy effects, both kinds of modifications are related, and the distinction between them depends on the 'duality frame' one is using. So far, Calabi–Yau compactifications, in particular in the setting of the topological string, have been the major playground for exploring 'string geometry'. More work needs to be done in this framework before addressing these questions within the full theory. Note that topological string theory is rich enough to address issues such as background independence, quantum space-time structure ('space-time foam'), and the quest for a non-perturbative formulation.

The problem of 'string geometry' can also be rephrased from another perspective, by highlighting the distinction between 'geometrical backgrounds' and 'observed geometry'. Geometrical backgrounds are classical data, which are used to define the world-sheet conformal field theory, while observed geometry is the geometry one infers by probing space with string or brane states. This is illustrated by the example of strings in Minkowski space-time. While Minkowski space-time does not have a minimal length scale, the shortest length resolved by scattering string states is the string length $\sqrt{\alpha'}$. Thus there is a qualitative difference between the classical background used to define the world-sheet theory, and the geometry 'seen' by strings. Things become more complicated if we probe the same geometry using different objects. In particular, D-particles (D0-branes) resolve a different minimal length scale, the 11-dimensional Planck scale, which is related to the string length scale through the vacuum expectation value of the dilaton. This again illustrates the high redundancy in the description of observable quantities. What is needed here is a disentanglement between observables and gauge symmetries. While all this is 'well known' within the string community, and has been discussed in several publications, a more focused effort might be needed to make progress in these important conceptual questions.

The simplest consistent string backgrounds are ten-dimensional Minkowski space, populated by either of the five supersymmetric perturbative string theories, and 11-dimensional Minkowski space, for which only the massless sector and the BPS states are known. This clearly presses the question why we live in a four-dimensional universe. Moreover, even when taking the attitude to impose that the additional space dimensions are unobservable at the presently realizable energy scales, one still meets the problem that there is a huge number of ways to compactify the theory to four dimensions. This is, first of all, a serious obstacle for testing the theory empirically based on its predictions. And, second, it leaves us with the question whether the particular solution which describes our universe (assuming that such a solution really exists) has been chosen by a historical accident, or whether there is a dynamical explanation. Since currently no convincing dynamical explanation is at hand, an eloquent group within the string community advocates the use of the anthropic principle within the context of eternal inflation. Not surprisingly, this move has provoked harsh criticism, which in its most pointed form discards anthropic reasoning as being unscientific. Before commenting on the anthropic principle, let us point out that it is not clear a priori which properties of our universe can be explained by recourse to laws and which properties are just historical facts. This is, of course, closely related to the distinction between 'equations' and 'initial conditions' which is the key point of a famous essay by E. Wigner. While for most branches of physics it is not controversial that science explains 'regularities among events', while initial conditions are contingent (true but not necessarily true), we nowadays tend to expect more than this of quantum cosmology. However, the idea that a theory could dispense itself from initial conditions, or could, in some sense, explain them, might just be wrong. This said, we need to stress that there are alternatives to anthropic considerations, which are worth exploring. As a matter of fact, the present state of string theory does not allow us to study generic time-dependent space-times in the full theory. Therefore the main problem with anthropic reasoning is that it could prevent us from further developing the theory. A moderate goal, which has been subject of some recent activity, is to use effective fields theories, which incorporate some relevant stringy features, to show that string vacua with rich spectra of light (compared to string or Planck scale), stable, charged particles are preferred dynamically. Another, more demanding question is why four large space-time dimensions should be preferred. Ultimately, one needs to develop the formalism of string theory beyond the framework of fixed background on-shell amplitudes before these questions can be addressed properly.

11 Some Concluding Remarks

String theory has to a large extent been developed by exploring internal requirements of consistency, often in a formal rather than mathematically rigorous way. The underlying mathematical structure is very rich, and has lead to very non-trivial predictions, insights, and developments, which in turn have stimulated work by pure mathematicians and a vivid exchange of ideas between physicists and mathematicians. Some observations made in string theory, such as mirror symmetry, have already been put on firm ground. Topological string theory, which is not only a toy version of string theory but also a tool which allows to compute various quantities relevant for particle phenomenology, is well understood perturbatively, while a non-perturbative formulation is currently in the center of interest and might be within reach. This supports the expectation that a mathematically satisfactory formulation of the full string theory will be found eventually, although it is hard to estimate how long this will take. While there are good indications that the theory is consistent, its relevance for physics is less clear. Certainly, many ideas which have grown out of string theory, notably the AdS/CFT correspondence and the idea of extra dimensions, have had considerable influence on quantum field theory, particle physics, and gravitational physics. While some of these ideas are purely technical, like methods for the computation of amplitudes which are now commonly used in QCD (helicity amplitudes), other ideas are conceptual.

In particular, string theory makes it very natural to express physical phenomena, including non-perturbative quantum phenomena in terms of geometry. Two prominent examples are the geometrical realization of strong–weak coupling dualities, as in the Seiberg–Witten solution of $\mathcal{N} = 2$ gauge theories and its lifts to type-II string theory and 11-dimensional M-theory, and the holographic renormalization group, where the energy scale of a four-dimensional quantum field theory is literally treated as an extra dimension. But without direct empirical evidence, string theory might just be a technical tool, or a catalyser for ideas which one could also have developed independently. While this would not necessarily be bad in the sense of invalidating the work done in this field, most people working on string theory do certainly hope that it captures fundamental features of space, time, and matter. Then, finding ways of testing the theory through experiment or observation is indispensable.

From the perspective of the standard model one is eagerly waiting for experimental signatures which lead us beyond its domain of validity and, hopefully, indicate a particular type of extension. Whether signals of new physics will give us clues about the relevance of string theory will strongly depend on what kind of new physics will be found. Low-energy supersymmetry, with a rich spectrum of supersymmetric particles, would certainly be very attractive for particle phenomenology. It would also fit with the idea that physics at higher-energy scales is organized by higher symmetries, and would thus indirectly support string theory as the ultimate form of unified theory. However, it would also indicate that string effects only become relevant at the (four-dimensional) Planck scale, and then it will be very difficult to distill direct evidence for string theory out of the data.

The situation would be much better in the alternative scenarios with 'large' (TeV-scale rather than Planck scale) extra dimensions. This would involve gravity and it would also take us beyond the realm of renormalizable QFTs. In this case new concepts, such as those offered by string theory, become relevant already at the TeV scale.

Considerations of string theory have led to the discovery of very non-trivial mathematical structures, which might hold key for formulating a unified quantum theory of all interactions. This also gives confidence that there is a mathematically consistent and physically relevant theory underlying all the facets of what we today mean by string theory, whose complete fundamental structure and symmetries are still to be uncovered.

Selected References

Almost all papers on string theory since 1991 are available on http://de.arxiv.org/archive/hep-th. Below we will only give a few references which will guide the reader to more specific publications.

Popular Accounts
B. Greene, "The Elegant Universe: Superstring, Hidden Dimensions, and the Quest for the Ultimate Theory," W. W. Norton & Company, 1999.
L. Randall, "Warped Passages: Unraveling the Mysteries of the Universe's Hidden Dimensions," HarperCollins Publishers, 2005.
L. Susskind, "The Cosmic Landscape: String Theory and the Illusion of Intelligent Design," Little Brown and Company, 2005.

Textbooks
M. B. Green, J. H. Schwarz and E. Witten, "Superstring Theory. Vol. 1: Introduction, Vol. 2: Loop Amplitudes, Anomalies And Phenomenology," Cambridge University Press 1987.
D. Lüst and S. Theisen "Lectures on String Theory", Lect. Notes Phys. **346** (1989).
J. Polchinski, "String theory. Vol. 1: An introduction to the bosonic string, Vol. 2: Superstring theory and beyond," Cambridge University Press 1998.
B. Zwiebach, "A first course in string theory," Cambridge University Press 2004.
L. Susskind and J. Lindesay, "An Introduction to Black Holes, Information and the String Theory Revolution: The Holographic Universe," World Scientific Publishing 2004.

Review Articles
T. Mohaupt, "Black hole entropy, special geometry and strings," Fortsch. Phys. **49** (2001) 3, hep-th/0007195.
B. Pioline, "Lectures on on Black Holes, Topological Strings and Quantum Attractors,"
hep-th/0607227.
O. Aharony, S. S. Gubser, J. M. Maldacena, H. Ooguri and Y. Oz, "Large N field theories, string theory and gravity," Phys. Rept. **323**, 183 (2000), hep-th/9905111.
A. Sen, "An introduction to non-perturbative string theory," hep-th/9802051.
M. Grana, "Flux compactifications in string theory: A comprehensive review," Phys. Rept. **423**, 91 (2006), hep-th/0509003 .
R. Bousso, "The holographic principle," Rev. Mod. Phys. **74**, 825 (2002), hep-th/0203101
C. P. Bachas, "Lectures on D-branes," hep-th/9806199.
J. M. Maldacena, "Black Holes in String Theory," hep-th/9607235.
G. Horowitz, "Spacetime in string theory," New J. Phys. **7** (2005) 198, gr-qc/0410049.

Part VII

Cosmology

Dark Energy

N. Straumann

Institute for Theoretical Physics, University of Zurich, Winterthurerstrasse 180, 8057 Zurich, Switzerland
norbert.straumann@freesurf.ch

1 Introduction

Cosmology is going through a fruitful and exciting period. Some of the developments are definitely also of interest to physicists outside the fields of astrophysics and cosmology.

This chapter covers some particularly fascinating and topical subjects. A central theme will be the current evidence that the recent ($z < 1$) Universe is dominated by an exotic nearly homogeneous dark energy density with *negative* pressure. The simplest candidate for this unknown so-called *dark energy* is a cosmological term in Einstein's field equations, a possibility that has been considered during all the history of relativistic cosmology. Independently of what this exotic energy density is, one thing is certain since a long time: The energy density belonging to the cosmological constant is not larger than the cosmological critical density, and thus *incredibly small by particle physics standards.* This is a profound mystery, since we expect that all sorts of *vacuum energies* contribute to the effective cosmological constant.

Since this is such an important issue it should be of interest to indicate how convincing the evidence for this finding really is, or whether one should remain skeptical. Much of this is based on the observed temperature fluctuations of the cosmic microwave background radiation (CMB), and large-scale structure formation. The first evidence for an accelerating expansion of the Universe, and still the only direct one, came from the Hubble diagram for Type Ia supernovae. When combined with other measurements a cosmological world model of the Friedmann–Lemaître variety has emerged that is spatially almost flat, with about 70% of its energy contained in the form dark energy. A detailed analysis of the existing data requires a considerable amount of theoretical machinery that is beyond the scope of this contribution. For interested readers we shall refer to some books, reviews, and articles that may be most convenient to penetrate deeper into various topics.

N. Straumann: *Dark Energy*, Lect. Notes Phys. **721**, 327–397 (2007)
DOI 10.1007/978-3-540-71117-9_13

Since this book addresses mostly readers whose main interests are outside astrophysics and cosmology, I do not presuppose a serious training in cosmology. However, I do assume some working knowledge of general relativity (GR). As a source, and for references, I usually quote my recent textbook [1]. The essentials of the Friedmann–Lemaître models will be summarized in Appendices A and B. Appendix C provides a brief introduction to *inflation*, a key idea of modern cosmology.

2 Einstein's Original Motivation of the Λ-Term

One of the contributions in the famous book *Albert Einstein: Philosopher–Scientist* [2] is a chapter by George E. Lemaître entitled "The Cosmological Constant". In the introduction he says: *"The history of science provides many instances of discoveries which have been made for reasons which are no longer considered satisfactory. It may be that the discovery of the cosmological constant is such a case."* When the book appeared in 1949 – at the occasion of Einstein's seventieth birthday – Lemaître could not be fully aware of how right he was, how profound the cosmological constant problem really is, especially since he was not a quantum physicist.

We begin this contribution in reviewing the main aspects of the history of the Λ-term, from its introduction in 1917 up to the point when it became widely clear that we are facing a deep mystery. (See also [3] and [4].) I describe first the *classical* aspect of the historical development.

Einstein introduced the cosmological term when he applied GR the first time to cosmology [5]. Presumably the main reason why Einstein turned so soon after the completion of GR to cosmology had much to do with Machian ideas on the origin of inertia, which played in those years an important role in Einstein's thinking. His intention was to eliminate all vestiges of absolute space. He was, in particular, convinced that isolated masses cannot impose a structure on space at infinity. Einstein was actually thinking about the problem regarding the choice of boundary conditions at infinity already in spring 1916. In a letter to Michele Besso on 14 May 1916 he also mentions the possibility of the world being finite. A few months later he expanded on this in letters to Willem de Sitter. It is along these lines that he postulated a Universe that is spatially finite and closed, a Universe in which no boundary conditions are needed. He then believed that this was the only way to satisfy what he later [7] named *Mach's principle*, in the sense that the metric field should be determined uniquely by the energy-momentum tensor.

In addition, Einstein assumed that the Universe was *static*. This was not unreasonable at the time, because the relative velocities of the stars as observed were small. (Recall that astronomers only learned later that spiral nebulae are independent star systems outside the Milky Way. This was definitely established when in 1924 Hubble found that there were Cepheid variables in Andromeda and also in other galaxies.)

These two assumptions were, however, not compatible with Einstein's original field equations. For this reason, Einstein added the famous Λ-term, which is compatible with the principles of GR, in particular with the energy–momentum law $\nabla_\nu T^{\mu\nu} = 0$ for matter. The modified field equations in standard notation and signature $(- + ++)$ are

$$G_{\mu\nu} = 8\pi G T_{\mu\nu} - \Lambda g_{\mu\nu} \ . \tag{1}$$

The cosmological term is, in four dimensions, the only possible complication of the field equations if no higher than second-order derivatives of the metric are allowed (*Lovelock theorem*). This remarkable uniqueness is one of the most attractive features of GR. (In higher dimensions additional terms satisfying this requirement are allowed.)

For the static Einstein universe the field equations (1) imply the two relations

$$4\pi G\rho = \frac{1}{a^2} = \Lambda \ , \tag{2}$$

where ρ is the mass density of the dust-filled universe (zero pressure) and a is the radius of curvature. (We remark, in passing, that the Einstein universe is the only static dust solution; one does not have to assume isotropy or homogeneity. Its instability was demonstrated by Lemaître in 1927.) Einstein was very pleased by this direct connection between the mass density and geometry, because he thought that this was in accord with Mach's philosophy.

Einstein concludes with the following sentences:

> *In order to arrive at this consistent view, we admittedly had to introduce an extension of the field equations of gravitation which is not justified by our actual knowledge of gravitation. It has to be emphasized, however, that a positive curvature of space is given by our results, even if the supplementary term is not introduced. That term is necessary only for the purpose of making possible a quasi-static distribution of matter, as required by the fact of the small velocities of the stars.*

To de Sitter, Einstein emphasized in a letter on 12 March 1917 that his cosmological model was intended primarily to settle the question "whether the basic idea of relativity can be followed through its completion, or whether it leads to contradictions". And he adds whether the model corresponds to reality was another matter.

Only later Einstein came to realize that Mach's philosophy is predicated on an antiquated ontology that seeks to reduce the metric field to an epiphenomenon of matter. It became increasingly clear to him that the metric field has an independent existence, and his enthusiasm for what he called Mach's principle later decreased. In a letter to F. Pirani he wrote in 1954: *"As a matter of fact, one should no longer speak of Mach's principle at all"* [8]. GR still preserves some remnant of Newton's absolute space and time.

3 From Static to Expanding World Models

Surprisingly to Einstein, de Sitter discovered in the same year, 1917, a completely different static cosmological model which also incorporated the cosmological constant, but was *anti-Machian*, because it contained no matter [9]. For this reason, Einstein tried to discard it on various grounds (more on this below). The original form of the metric was

$$g = -\Big[1 - (\frac{r}{R})^2\Big] dt^2 + \frac{dr^2}{1 - (\frac{r}{R})^2} + r^2 (d\vartheta^2 + \sin^2 \vartheta d\varphi^2) \, .$$

Here, the spatial part is the standard metric of a three-sphere of radius R, with $R = (3/\Lambda)^{1/2}$. The model had one very interesting property: For light sources moving along static worldlines there is a gravitational redshift, which became known as the *de Sitter effect.* This was thought to have some bearing on the redshift results obtained by Slipher. Because the fundamental (static) worldlines in this model are not geodesic, a freely falling object released by any static observer will be seen by him to accelerate away, generating also local velocity (Doppler) redshifts corresponding to *peculiar velocities.* In the second edition of his book [10], published in 1924, Eddington writes about this

> *de Sitter's theory gives a double explanation for this motion of recession; first there is a general tendency to scatter (...); second there is a general displacement of spectral lines to the red in distant objects owing to the slowing down of atomic vibrations (...), which would erroneously be interpreted as a motion of recession.*

I do not want to enter into all the confusion over the de Sitter universe. One source of this was the apparent singularity at $r = R = (3/\Lambda)^{1/2}$. This was at first thoroughly misunderstood even by Einstein and Weyl. ('The Einstein–de Sitter–Weyl–Klein Debate' is now published in Vol. 8 of the *Collected Papers* [6].) At the end, Einstein had to acknowledge that de Sitter's solution is fully regular and matter-free and thus indeed a counter example to Mach's principle. But he still discarded the solution as physically irrelevant because it is not globally static. This is clearly expressed in a letter from Weyl to Klein, after he had discussed the issue during a visit of Einstein in Zurich [11]. An important discussion of the redshift of galaxies in de Sitter's model by H. Weyl in 1923 should be mentioned. Weyl introduced an expanding version[1] of the de Sitter model [12]. For *small* distances his result reduced to what later became known as the Hubble law. Independently of Weyl, Cornelius Lanczos introduced in 1922 also a non-stationary interpretation of de Sitter's solution in the form of a Friedmann spacetime with a positive spatial curvature

[1] I recall that the de Sitter model has many different interpretations, depending on the class of fundamental observers that is singled out.

[13]. In a second paper he also derived the redshift for the non-stationary interpretation [14].

Until about 1930 almost everybody believed that the Universe was static, in spite of the two fundamental papers by Friedmann [15] in 1922 and 1924 and Lemaître's independent work [16] in 1927. These path-breaking papers were in fact largely ignored. The history of this early period has – as is often the case – been distorted by some widely read documents. Einstein too accepted the idea of an expanding Universe only much later. After the first paper of Friedmann, he published a brief note claiming an error in Friedmann's work; when it was pointed out to him that it was his error, Einstein published a retraction of his comment, with a sentence that luckily was deleted before publication: *"[Friedmann's paper] while mathematically correct is of no physical significance"*. In comments to Lemaître during the Solvay meeting in 1927, Einstein again rejected the expanding universe solutions as physically unacceptable. According to Lemaître, Einstein was telling him, *"Vos calculs sont corrects, mais votre physique est abominable."* It appears astonishing that Einstein – after having studied carefully Friedmann's papers – did not realize that his static model is unstable, and hence that the Universe has to be expanding or contracting. On the other hand, I found in the archive of the ETH many years ago a postcard of Einstein to Weyl from 1923, related to Weyl's reinterpretation of de Sitter's solution, with the following interesting sentence: *"If there is no quasi-static world, then away with the cosmological term."*

It also is not well known that Hubble interpreted his famous results on the redshift of the radiation emitted by distant "nebulae" in the framework of the de Sitter model, as was suggested by Eddington.

The general attitude is well illustrated by the following remark of Eddington at a Royal Astronomical Society meeting in January 1930: *"One puzzling question is why there should be only two solutions. I suppose the trouble is that people look for static solutions."*

Lemaître, who had been for a short time a post-doctoral student of Eddington, read this remark in a report to the meeting published in *Observatory*, and wrote to Eddington pointing out his 1927 paper. Eddington had seen that paper, but had completely forgotten about it. But now he was greatly impressed and recommended Lemaître's work in a letter to *Nature*. He also arranged for a translation which appeared in MNRAS [17]. Eddington also "pointed out that it was immediately deducible from his [Lemaître's] formulae that Einstein's world is unstable, so that an expanding or a contracting universe is an inevitable result of Einstein's law of gravitation."

Lemaître's successful explanation of Hubble's discovery finally changed the viewpoint of the majority of workers in the field. At this point, Einstein *rejected the cosmological term as superfluous and no longer justified* [18]. At the end of the paper, in which he published his new view, Einstein adds some remarks about the age problem which was quite severe without the Λ-term, since Hubble's value of the Hubble parameter was then about seven times too large. Einstein is, however, not very worried and suggests two ways out. First he says that the matter distribution is in reality inhomogeneous and that the approximate treatment may be illusionary. Then he adds that in astronomy one should be cautious with large extrapolations in time.

Einstein repeated his new standpoint also much later [19], and this was adopted by many other influential workers, e.g. by Pauli [20]. Whether Einstein really considered the introduction of the Λ-term as "the biggest blunder of his life" appears doubtful to me. In his published work and letters I never found such a strong statement. Einstein discarded the cosmological term just for simplicity reasons. For a minority of cosmologists (O. Heckmann, for example [21]), this was not sufficient reason. Paraphrasing Rabi, one might ask, "who ordered it away"?

Einstein published his new view in the *Sitzungsberichte der Preussischen Akademie der Wissenschaften.* The correct citation is,

Einstein. A. (1931). Sitzungsber. Preuss. Akad. Wiss. 235–37.

Many authors have quoted this paper but never read it. As a result, the quotations gradually changed in an interesting, quite systematic fashion. Some steps are shown in the following sequence:

- A. Einstein. 1931. Sitzsber. Preuss. Akad. Wiss. ...
- A. Einstein. Sitzber. Preuss. Akad. Wiss. ... (1931)
- A. Einstein (1931). Sber. preuss. Akad. Wiss. ...
- Einstein. A .. 1931. Sb. Preuss. Akad. Wiss. ...
- A. Einstein. S.-B. Preuss. Akad. Wis. ...1931
- A. Einstein. S.B. Preuss. Akad. Wiss. (1931) ...
- Einstein, A., and Preuss, S.B. (1931). Akad. Wiss. **235**

Presumably, one day some historian of science will try to find out what happened with the young physicist S.B. Preuss, who apparently wrote just one important paper and then disappeared from the scene.

After the Λ-force was rejected by its inventor, other cosmologists, like Eddington, retained it. One major reason was that it solved the problem of the age of the Universe when the Hubble time scale was thought to be only 2 billion years (corresponding to the value $H_0 \sim 500$ km $s^{-1}Mpc^{-1}$ of the Hubble constant). This was even shorter than the age of the Earth. In addition, Eddington and others overestimated the age of stars and stellar systems.

For this reason, the Λ-term was employed again and a model was revived which Lemaître had singled out from the many solutions of the Friedmann–Lemaître equations.[2] This so-called "Lemaître hesitation universe" is closed and has a repulsive Λ-force ($\Lambda > 0$), which is slightly greater than the value

[2] I recall that Friedmann included the Λ-term in his basic equations. I find it remarkable that for the negatively curved solutions he pointed out that these may be open or compact (but not simply connected).

chosen by Einstein. It begins with a big bang and has the following two stages of expansion. In the first the Λ-force is not important, the expansion is decelerated due to gravity and slowly approaches the radius of the Einstein universe. At about the same time, the repulsion becomes stronger than gravity and a second stage of expansion begins which eventually inflates. In this way a positive Λ was employed to reconcile the expansion of the Universe with the age of stars.

Repulsive Effect of a Positive Cosmological Constant

The *repulsive* effect of a positive cosmological constant can be seen from the following consequence of Einstein's field equations for the time-dependent scale factor $a(t)$ (see Appendix A):

$$\ddot{a} = -\frac{4\pi G}{3}(\rho + 3p)a + \frac{\Lambda}{3}a \,, \tag{3}$$

where p is the pressure of all forms of matter.

Historically, the Newtonian analog of the cosmological term was regarded by Einstein, Weyl, Pauli, and others as a *Yukawa term*. This is not correct, as I now show.

For a better understanding of the action of the Λ-term it may be helpful to consider a general static spacetime with the metric (in adapted coordinates)

$$ds^2 = -\varphi^2 dt^2 + g_{ik} dx^i dx^k \,, \tag{4}$$

where φ and g_{ik} depend only on the spatial coordinates x^i. The component R_{00} of the Ricci tensor is given by $R_{00} = \bar{\Delta}\varphi/\varphi$, where $\bar{\Delta}$ is the three-dimensional Laplace operator for the spatial metric g_{ik} in (4) (see, e.g., [1]). Let us write (1) in the form

$$G_{\mu\nu} = \kappa(T_{\mu\nu} + T^{\Lambda}_{\mu\nu}) \qquad (\kappa = 8\pi G) \,, \tag{5}$$

with

$$T^{\Lambda}_{\mu\nu} = -\frac{\Lambda}{8\pi G} g_{\mu\nu} \,. \tag{6}$$

This has the form of the energy–momentum tensor of an ideal fluid, with energy density $\rho_\Lambda = \Lambda/8\pi G$ and pressure $p_\Lambda = -\rho_\Lambda$.[3] For an ideal fluid at rest Einstein's field equation implies

$$\frac{1}{\varphi}\bar{\Delta}\varphi = 4\pi G\Big[(\rho + 3p) + \underbrace{(\rho_\Lambda + 3p_\Lambda)}_{-2\rho_\Lambda}\Big] \,. \tag{7}$$

Since the energy density and the pressure appear in the combination $\rho + 3p$, we understand that a positive ρ_Λ leads to a repulsion (as in (3)). In the

[3] This way of looking at the cosmological term was soon (in 1918) emphasized by Schrödinger and also by F. Klein.

Newtonian limit we have $\varphi \simeq 1 + \phi$ (ϕ : Newtonian potential) and $p \ll \rho$, hence we obtain the modified Poisson equation

$$\Delta\phi = 4\pi G(\rho - 2\rho_\Lambda) \ . \tag{8}$$

This is the correct Newtonian limit.

As a result of revised values of the Hubble parameter and the development of the modern theory of stellar evolution in the 1950s, the controversy over ages was resolved and the Λ-term became again unnecessary. (Some tension remained for values of the Hubble parameter at the higher end of published values.)

However, in 1967 it was revived again in order to explain why quasars appeared to have redshifts that concentrated near the value $z = 2$. The idea was that quasars were born in the hesitation era [22]. Then quasars at greatly different distances can have almost the same redshift, because the universe was almost static during that period. Other arguments in favor of this interpretation were based on the following peculiarity. When the redshifts of emission lines in quasar spectra exceed 1.95, then redshifts of absorption lines in the same spectra were, as a rule, equal to 1.95. This was then quite understandable, because quasar light would most likely have crossed intervening galaxies during the epoch of suspended expansion, which would result in almost identical redshifts of the absorption lines. However, with more observational data evidence for the Λ-term dispersed for the third time.

4 The Mystery of the Λ-Problem

At this point I want to leave the classical discussion of the Λ-term, and turn to the quantum aspect of the Λ-problem, where it really becomes very serious.

4.1 Historical Remarks

Since quantum physicists had so many other problems, it is not astonishing that in the early years they did not worry about this subject. An exception was Pauli, who wondered in the early 1920s whether the zero-point energy of the radiation field could be gravitationally effective.

As background I recall that Planck had introduced the zero-point energy with somewhat strange arguments in 1911. The physical role of the zero-point energy was much discussed in the early years of quantum theory. There was, for instance, a paper by Einstein and Stern in 1913 [Collected Papers, Vol. 4, Doc. 11; see also the Editorial Note, p. 270] that aroused widespread interest. In this, two arguments in favor of the zero-point energy were given. The first had to do with the specific heat of rotating (diatomic) molecules. The authors developed an approximate theory of the energy of rotating molecules and came to the conclusion that the resulting specific heat agreed much better with recent experimental results by Arnold Eucken, if they included the zero-point energy. The second argument was based on a new derivation of Planck's radiation formula. In both the arguments, Einstein and Stern made a number of problematic assumptions, and in fall 1913, Einstein retracted their results. At the second Solvay Congress in late October 1913, Einstein said that he no longer believed in the zero-point energy, and in a letter to Ehrenfest [Vol. 5, Doc. 481] he wrote that the zero-point energy was "dead as a doornail".

From Charly Enz and Armin Thellung – Pauli's last two assistants – I have learned that Pauli had discussed this issue extensively with O. Stern in Hamburg. Stern had calculated, but never published, the vapor pressure difference between the isotopes 20 and 22 of Neon (using Debye theory). He came to the conclusion that without zero-point energy this difference would be large enough for easy separation of the isotopes, which is not the case in reality. These considerations penetrated into Pauli's lectures on statistical mechanics [23] (which I attended). The theme was taken up in an article by Enz and Thellung [24]. This was originally written as a birthday gift for Pauli, but because of Pauli's early death this appeared in a memorial volume of Helv.Phys.Acta.

From Pauli's discussions with Enz and Thellung we know that Pauli estimated the influence of the zero-point energy of the radiation field – cutoff at the classical electron radius – on the radius of the universe, and came to the conclusion that it *"could not even reach to the moon"*.

When, as a student, I heard about this, I checked Pauli's unpublished[4] remark by doing the following little calculation (which Pauli must have done):

In units with $\hbar = c = 1$ the vacuum energy density of the radiation field is

$$\langle \rho \rangle_{vac} = \frac{8\pi}{(2\pi)^3} \int_0^{\omega_{max}} \frac{\omega}{2} \omega^2 d\omega = \frac{1}{8\pi^2} \omega_{max}^4 ,$$

with

$$\omega_{max} = \frac{2\pi}{\lambda_{max}} = \frac{2\pi m_e}{\alpha} .$$

The corresponding radius of the Einstein universe in (2) would then be ($M_{pl} \equiv 1/\sqrt{G}$)

$$a = \frac{\alpha^2}{(2\pi)^{\frac{2}{3}}} \frac{M_{pl}}{m_e} \frac{1}{m_e} \sim 31 \text{ km} .$$

This is indeed less than the distance to the moon. (It would be more consistent to use the curvature radius of the static de Sitter solution; the result is the same, up to the factor $\sqrt{3/2}$.)

For decades nobody else seems to have worried about contributions of quantum fluctuations to the cosmological constant, although physicists

[4] A trace of this is in Pauli's Handbuch article [25] on wave mechanics in the section where he discusses the meaning of the zero-point energy of the quantized radiation field.

learned after Dirac's hole theory that the vacuum state in quantum field theory is not an empty medium, but has interesting physical properties. As an important example I mention the papers by Heisenberg and Euler [26] in which they calculated the modifications of Maxwell's equations due to the polarization of the vacuum. Shortly afterward, Weisskopf [27] not only simplified their calculations but also gave a thorough discussion of the physics involved in charge renormalization. Weisskopf related the modification of Maxwell's Lagrangian to the change of the energy of the Dirac sea as a function of slowly varying external electromagnetic fields. (Avoiding the old-fashioned Dirac sea, this effective Lagrangian is due to the interaction of a classical electromagnetic field with the vacuum fluctuations of the electron positron field.) After a charge renormalization this change is finite and gives rise to electric and magnetic polarization vectors of the vacuum. In particular, the refraction index for light propagating perpendicular to a static homogeneous magnetic field depends on the polarization direction. This is the vacuum analog of the well-known Cotton–Mouton effect in optics. As a result, an initially linearly polarized light beam becomes elliptic. (In spite of great efforts it has not yet been possible to observe this effect.)

Another beautiful example for the importance of vacuum energies as a function of varying external conditions is the *Casimir effect.* This is the most widely cited example of how vacuum fluctuations can have observable consequences.

The presence of conducting plates modifies the vacuum energy density in a manner which depends on the separation of the plates. This leads to an attractive force between the two plates.

Historically, this was a byproduct of some applied industrial research in the stability of colloidal suspensions used to deposit films in the manufacture of lamps and cathode tubes. This lead Casimir and Polder to reconsider the theory of van der Waals interaction with *retardation* included. They found that this causes the interaction to vary at large intermolecular separations as r^{-7}. Casimir mentioned his result to Niels Bohr during a walk, and told him that he was puzzled by the extreme simplicity of the result at large distance. According to Casimir, Bohr mumbled something about zero-point energy. That was all, but it put him on the right track.

Precision experiments have recently confirmed the theoretical prediction to about 1%. By now the literature related to the Casimir effect is enormous. For further information we refer to the recent book [28].

4.2 Has Dark Energy been Discovered in the Lab?

It has been suggested by Beck and Mackey [29] that part of the zero-point energy of the radiation field that is gravitationally active can be determined from noise measurements of Josephson junctions. This caused some widespread attention. In a reaction we [30] showed that there is no basis for this claim, by following the reasoning in [29] for a much simpler model, for which it is very obvious that the authors misinterpreted their formulae. Quite generally, the absolute value of the zero-point energy of a quantum mechanical system has no physical meaning when gravitational coupling is ignored. All that is measurable are *changes* of the zero-point energy under variations of system parameters or of external couplings, like an applied voltage. For further information on the controversy, see [31] and [32].

4.3 Vacuum Energy and Gravity

When we consider the coupling to gravity, the vacuum energy density acts like a cosmological constant. In order to see this, first consider the vacuum expectation value of the energy–momentum tensor in Minkowski spacetime. Since the vacuum state is Lorentz invariant, this expectation value is an invariant symmetric tensor, hence proportional to the metric tensor. For a curved metric this is still the case, up to higher curvature terms:

$$\langle T_{\mu\nu}\rangle_{vac} = -g_{\mu\nu}\rho_{vac} + \textit{higher curvature terms}\,. \tag{9}$$

The *effective* cosmological constant, which controls the large-scale behavior of the Universe, is given by

$$\Lambda = 8\pi G\rho_{vac} + \Lambda_0\,, \tag{10}$$

where Λ_0 is a bare cosmological constant in Einstein's field equations.

We know from astronomical observations that $\rho_\Lambda \equiv \Lambda/8\pi G$ cannot be larger than about the critical density:

$$\begin{aligned}\rho_{crit} &= \frac{3H_0^2}{8\pi G} \\ &= 1.88\times 10^{-29} h_0^2 g\mathrm{cm}^{-3} \\ &\simeq (3\times 10^{-3} eV)^4\,,\end{aligned} \tag{11}$$

where h_0 is the *reduced Hubble parameter*

$$h_0 = H_0/(100\mathrm{kms}^{-1}\mathrm{Mpc}^{-1}) \tag{12}$$

that is close to 0.7.

It is a complete mystery as to why the two terms in (10) should almost exactly cancel. This is – more precisely stated – the famous Λ-problem.

As far as I know, the first who came back to possible contributions of the vacuum energy density to the cosmological constant was Zel'dovich. He discussed this issue in two papers [33] during the third renaissance period of the Λ-term, but before the advent of spontaneously broken gauge theories. The following remark by him is particularly interesting. Even if one assumes completely ad hoc that the zero-point contributions to the vacuum energy density are exactly cancelled by a bare term, there still remain higher-order effects.

In particular, *gravitational* interactions between the particles in the vacuum fluctuations are expected on dimensional grounds to lead to a gravitational self-energy density of order $G\mu^6$, where μ is some cutoff scale. Even for μ as low as 1 GeV (for no good reason) this is about 9 orders of magnitude larger than the observational bound.

This illustrates that there is something profound that we do not understand at all, certainly not in quantum field theory (so far also not in string theory). We are unable to calculate the vacuum energy density in quantum field theories, like the standard model of particle physics. But we can attempt to make what appear to be reasonable order-of-magnitude estimates for the various contributions. All expectations are *in gigantic conflict with the facts* (see below). Trying to arrange the cosmological constant to be zero is unnatural in a technical sense. It is like enforcing a particle to be massless, by fine-tuning the parameters of the theory when there is no symmetry principle which implies a vanishing mass. The vacuum energy density is unprotected from large quantum corrections. This problem is particularly severe in field theories with spontaneous symmetry breaking. In such models there are usually several possible vacuum states with different energy densities. Furthermore, the energy density is determined by what is called the effective potential, and this is a *dynamical* object. Nobody can see any reason why the vacuum of the standard model we ended up as the Universe cooled has – for particle physics standards – an almost vanishing energy density. Most probably, we will only have a satisfactory answer once we shall have a theory which successfully combines the concepts and laws of GR about gravity and spacetime structure with those of quantum theory.

4.4 Simple Estimates of Vacuum Energy Contributions

If we take into account the contributions to the vacuum energy from vacuum fluctuations in the fields of the standard model up to the currently explored energy, i.e., about the electroweak scale $M_F = G_F^{-1/2} \approx 300 GeV$ (G_F : Fermi coupling constant), we cannot expect an almost complete cancellation, because there is *no symmetry principle* in this energy range that could require this. The only symmetry principle which would imply this is *supersymmetry*, but supersymmetry is broken (if it is realized in nature). Hence we can at best expect a very imperfect cancellation below the electroweak scale, leaving a contribution of the order of M_F^4. (The contributions at higher energies may largely cancel if supersymmetry holds in the real world.)

We would reasonably expect that the vacuum energy density is at least as large as the condensation energy density of the QCD phase transition to the broken phase of chiral symmetry. Already this is far too large: $\sim \Lambda_{QCD}^4/16\pi^2 \sim 10^{-4} GeV^4$; this is *more than 40 orders of magnitude larger* than ρ_{crit}. Beside the formation of quark condensates $< \bar{q}q >$ in the QCD vacuum which break chirality, one also expects a gluon condensate $< G_a^{\mu\nu} G_{a\mu\nu} > \sim \Lambda_{QCD}^4$. This produces a significant vacuum energy density as a result of a dilatation anomaly: If Θ^μ_μ denotes the "classical" trace of the energy–momentum tensor, we have [34]

$$T^\mu_\mu = \Theta^\mu_\mu - \frac{\beta(g_s)}{2g_s} G^{\mu\nu}_a G_{a\mu\nu} , \tag{13}$$

where the second term is the QCD piece of the trace anomaly. $\beta(g_s)$ is the β-function of QCD that determines the running of the strong coupling constant g_s (see the contribution of Dosch to this book). I recall that this anomaly arises because a scale transformation is no more a symmetry if quantum corrections are included. Taking the vacuum expectation value of (13), we would again naively expect that $< \Theta^\mu_\mu >$ is of the order M_F^4. Even if this should vanish for some unknown reason, the anomalous piece is cosmologically gigantic. The expectation value $< G^{\mu\nu}_a G_{a\mu\nu} >$ can be estimated with QCD sum rules [35], and gives

$$< T^\mu_\mu >^{anom} \sim -(350MeV)^4 , \tag{14}$$

about 45 orders of magnitude larger than ρ_{crit}. This reasoning should show convincingly that the cosmological constant problem is indeed a profound one. (Note that there is some analogy with the (much milder) strong CP problem of QCD. However, in contrast to the Λ-problem, Peccei and Quinn [36] have shown that in this case there is a way to resolve the conundrum.)

Let us also have a look at the Higgs condensate of the electroweak theory. Recall that in the standard model we have for the Higgs doublet Φ in the broken phase for $< \Phi^* \Phi > \equiv \frac{1}{2}\phi^2$ the potential

$$V(\phi) = -\frac{1}{2} m^2 \phi^2 + \frac{\lambda}{8} \phi^4 . \tag{15}$$

Setting as usual $\phi = v + H$, where v is the value of ϕ where V has its minimum,

$$v = \sqrt{\frac{2m^2}{\lambda}} = 2^{-1/4} G_F^{-1/2} \sim 246GeV , \tag{16}$$

we find that the Higgs mass is related to λ by $\lambda = M_H^2/v^2$. For $\phi = v$ we obtain the energy density of the Higgs condensate

$$V(\phi = v) = -\frac{m^4}{2\lambda} = -\frac{1}{8\sqrt{2}} M_F^2 M_H^2 = \mathcal{O}(M_F^4) . \tag{17}$$

We can, of course, add a constant V_0 to the potential (15) such that it cancels the Higgs vacuum energy in the broken phase – including higher-order corrections. This again requires an extreme fine tuning. A remainder of only $\mathcal{O}(m_e^4)$, say, would be catastrophic. This remark is also highly relevant for models of inflation and quintessence.

In attempts beyond the standard model the vacuum energy problem so far remains, and often becomes even worse. For instance, in supergravity theories with spontaneously broken supersymmetry there is the following simple relation between the gravitino mass m_g and the vacuum energy density

$$\rho_{vac} = \frac{3}{8\pi G} m_g^2 .$$

Comparing this with (11) we find

$$\frac{\rho_{vac}}{\rho_{crit}} \simeq 10^{122} \left(\frac{m_g}{m_{Pl}} \right)^2 .$$

Even for $m_g \sim 1\ eV$ this ratio becomes 10^{66}. (m_g is related to the parameter F characterizing the strength of the supersymmetry breaking by $m_g = (4\pi G/3)^{1/2} F$, so $m_g \sim 1\ eV$ corresponds to $F^{1/2} \sim 100\ TeV$.)

Also string theory has not yet offered convincing clues why the cosmological constant is so extremely small. The main reason is that a *low energy mechanism* is required, and since supersymmetry is broken, one again expects a magnitude of order M_F^4, which is *at least 50 orders of magnitude too large* (see also [37]). However, non-supersymmetric physics in string theory is at the very beginning and workers in the field hope that further progress might eventually lead to an understanding of the cosmological constant problem.

I hope I have convinced the reader that we are indeed facing a profound mystery. (For other recent reviews, see also [38–41]. These contain more extended lists of references.)

5 Luminosity–Redshift Relation for Type Ia Supernovae

A few years ago the Hubble diagram for Type Ia supernovae gave, as a big surprise, the first serious evidence for a currently accelerating Universe. Before presenting and discussing critically these exciting results, we develop on the basis of Appendix A some theoretical background.

5.1 Theoretical Redshift–Luminosity Relation

In cosmology several different distance measures are in use, which are all related by simple redshift factors (see Sect. A.4). The one which is relevant in this section is the *luminosity distance* D_L. We recall that this is defined by

$$D_L = (\mathcal{L}/4\pi\mathcal{F})^{1/2} , \tag{18}$$

where $\mathcal{L}$ is the intrinsic luminosity of the source and $\mathcal{F}$ the observed energy flux.

We want to express this in terms of the redshift z of the source and some of the cosmological parameters. If the comoving radial coordinate r is chosen such that the Friedmann–Lemaître metric takes the form

$$g = -dt^2 + a^2(t)\left[\frac{dr^2}{1-kr^2} + r^2 d\Omega^2\right], \quad k = 0, \pm 1 , \tag{19}$$

then we have

$$\mathcal{F} dt_0 = \mathcal{L} dt_e \cdot \frac{1}{1+z} \cdot \frac{1}{4\pi (r_e a(t_0))^2} .$$

The second factor on the right is due to the redshift of the photon energy; the indices $0, e$ refer to the present and emission times, respectively. Using also $1 + z = a(t_0)/a(t_e)$, we find in a first step:

$$D_L(z) = a_0(1+z) r(z) \quad (a_0 \equiv a(t_0)) . \tag{20}$$

We need the function $r(z)$. From

$$dz = -\frac{a_0}{a}\frac{\dot{a}}{a} dt , \qquad dt = -a(t)\frac{dr}{\sqrt{1-kr^2}}$$

for light rays, we see that

$$\frac{dr}{\sqrt{1-kr^2}} = \frac{1}{a_0}\frac{dz}{H(z)} \qquad (H(z) = \frac{\dot{a}}{a}) . \tag{21}$$

Now, we make use of the Friedmann equation

$$H^2 + \frac{k}{a^2} = \frac{8\pi G}{3}\rho . \tag{22}$$

Let us decompose the total energy–mass density ρ into non-relativistic (NR), relativistic (R), Λ, quintessence (Q), and possibly other contributions

$$\rho = \rho_{NR} + \rho_R + \rho_\Lambda + \rho_Q + \cdots . \tag{23}$$

For the relevant cosmic period we can assume that the "energy equation"

$$\frac{d}{da}(\rho a^3) = -3pa^2 \tag{24}$$

also holds for the individual components $X = NR, R, \Lambda, Q, \cdots$. If $w_X \equiv p_X/\rho_X$ is constant, this implies that

$$\rho_X a^{3(1+w_X)} = \text{const} . \tag{25}$$

Therefore,

$$\rho = \sum_X \left(\rho_X a^{3(1+w_X)}\right)_0 \frac{1}{a^{3(1+w_X)}} = \sum_X (\rho_X)_0 (1+z)^{3(1+w_X)} . \tag{26}$$

Hence the Friedmann equation (22) can be written as

$$\frac{H^2(z)}{H_0^2} + \frac{k}{H_0^2 a_0^2}(1+z)^2 = \sum_X \Omega_X (1+z)^{3(1+w_X)} , \tag{27}$$

where Ω_X is the dimensionless density parameter for the species X,

$$\Omega_X = \frac{(\rho_X)_0}{\rho_{crit}} , \tag{28}$$

where ρ_{crit} is the critical density:

$$\begin{aligned} \rho_{crit} &= \frac{3H_0^2}{8\pi G} \\ &= 1.88 \times 10^{-29} \; h_0^2 \; g \; cm^{-3} \\ &= 8 \times 10^{-47} h_0^2 \; GeV^{\,4} . \end{aligned} \tag{29}$$

Here h_0 denotes the *reduced Hubble parameter*

$$h_0 = H_0/(100 \; km \; s^{-1} \; Mpc^{-1}) \simeq 0.7 . \tag{30}$$

Using also the curvature parameter $\Omega_K \equiv -k/H_0^2 a_0^2$, we obtain the useful form

$$\boxed{H^2(z) = H_0^2 E^2(z; \Omega_K, \Omega_X) ,} \tag{31}$$

with

$$E^2(z; \Omega_K, \Omega_X) = \Omega_K (1+z)^2 + \sum_X \Omega_X (1+z)^{3(1+w_X)} . \tag{32}$$

Especially for $z = 0$ this gives

$$\Omega_K + \Omega_0 = 1, \quad \Omega_0 \equiv \sum_X \Omega_X . \tag{33}$$

If we use (31) in (21), we get

$$\int_0^{r(z)} \frac{dr}{\sqrt{1-kr^2}} = \frac{1}{H_0 a_0} \int_0^z \frac{dz'}{E(z')} \tag{34}$$

and thus

$$r(z) = \mathcal{S}(\chi(z)) , \tag{35}$$

where

$$\chi(z) = \frac{1}{H_0 a_0} \int_0^z \frac{dz'}{E(z')} \tag{36}$$

and

$$\mathcal{S}(\chi) = \begin{cases} \sin\chi & : \quad k = 1 \\ \chi & : \quad k = 0 \\ \sinh\chi & : \quad k = 1 . \end{cases} \tag{37}$$

Inserting this in (20) gives finally the relation we were looking for

$$D_L(z) = \frac{1}{H_0}\mathcal{D}_L(z;\Omega_K,\Omega_X) \; , \qquad (38)$$

with

$$\mathcal{D}_L(z;\Omega_K,\Omega_X) = (1+z)\frac{1}{|\Omega_K|^{1/2}}\mathcal{S}\left(|\Omega_K|^{1/2}\int_0^z \frac{dz'}{E(z')}\right) \qquad (39)$$

for $k = \pm 1$. For a flat universe, $\Omega_K = 0$ or equivalently $\Omega_0 = 1$, the "Hubble-constant-free" luminosity distance is

$$\mathcal{D}_L(z) = (1+z)\int_0^z \frac{dz'}{E(z')} \; . \qquad (40)$$

Astronomers use as logarithmic measures of $\mathcal{L}$ and $\mathcal{F}$ the *absolute and apparent magnitudes*,[5] denoted by M and m, respectively. The conventions are chosen such that the *distance modulus* $m - M$ is related to D_L as follows

$$m - M = 5\log\left(\frac{D_L}{1\ Mpc}\right) + 25 \; . \qquad (41)$$

Inserting the representation (38), we obtain the following relation between the apparent magnitude m and the redshift z:

$$m = \mathcal{M} + 5\log\mathcal{D}_L(z;\Omega_K,\Omega_X) \; , \qquad (42)$$

where, for our purpose, $\mathcal{M} = M - 5\log H_0 + 25$ is an uninteresting fit parameter. The comparison of this theoretical *magnitude redshift relation* with data will lead to interesting restrictions for the cosmological Ω-parameters. In practice often only Ω_M and Ω_Λ are kept as independent parameters, where from now on the subscript M denotes (as in most papers) non-relativistic matter.

The following remark about *degeneracy curves* in the Ω-plane is important in this context. For a fixed z in the presently explored interval, the contours defined by the equations $\mathcal{D}_L(z;\Omega_M,\Omega_\Lambda) = const$ have little curvature, and thus we can associate an approximate slope to them. For $z = 0.4$ the slope is about 1 and increases to 1.5-2 by $z = 0.8$ over the interesting range of Ω_M and Ω_Λ. Hence even quite accurate data can at best select a strip in the Ω-plane, with a slope in the range just discussed. This is the reason behind the shape of the likelihood regions shown later (Fig. 2).

In this context it is also interesting to determine the dependence of the *deceleration parameter*

$$q_0 = -\left(\frac{a\ddot{a}}{\dot{a}^2}\right)_0 \qquad (43)$$

[5] Beside the (bolometric) magnitudes m, M, astronomers also use magnitudes m_B, m_V, ... referring to certain wavelength bands B (blue), V (visual), and so on.

on Ω_M and Ω_Λ. At an any cosmic time we obtain from (107) and (26)

$$-\frac{\ddot{a}a}{\dot{a}^2} = \frac{1}{2}\frac{1}{E^2(z)}\sum_X \Omega_X(1+z)^{3(1+w_X)}(1+3w_X) . \tag{44}$$

For $z = 0$ this gives

$$q_0 = \frac{1}{2}\sum_X \Omega_X(1+3w_X) = \frac{1}{2}(\Omega_M - 2\Omega_\Lambda + \cdots) . \tag{45}$$

The line $q_0 = 0$ ($\Omega_\Lambda = \Omega_M/2$) separates decelerating from accelerating universes at the present time. For given values of Ω_M, Ω_Λ, etc., (44) vanishes for z determined by

$$\Omega_M(1+z)^3 - 2\Omega_\Lambda + \cdots = 0 . \tag{46}$$

This equation gives the redshift at which the deceleration period ends (coasting redshift).

Generalization for Dynamical Models of Dark Energy

If the vacuum energy constitutes the missing two-thirds of the average energy density of the *present* Universe, we would be confronted with the following *cosmic coincidence* problem: Since the vacuum energy density is constant in time – at least after the QCD phase transition – while the matter energy density decreases as the Universe expands, it would be more than surprising if the two are comparable just at about the present time, while their ratio was tiny in the early Universe and would become very large in the distant future. The goal of dynamical models of dark energy is to avoid such an extreme fine-tuning. The ratio p/ρ of this component then becomes a function of redshift, which we denote by $w_Q(z)$ (because the so-called "quintessence models" are particular examples). Then the function $E(z)$ in (32) gets modified.

To see how, we start from the energy equation (24) and write this as

$$\frac{d\ln(\rho_Q a^3)}{d\ln(1+z)} = 3w_Q .$$

This gives

$$\rho_Q(z) = \rho_{Q0}(1+z)^3 \exp\left(\int_0^{\ln(1+z)} 3w_Q(z')d\ln(1+z')\right)$$

or

$$\rho_Q(z) = \rho_{Q0} \exp\left(3\int_0^{\ln(1+z)} (1+w_Q(z'))d\ln(1+z')\right) . \tag{47}$$

Hence, we have to perform on the right of (32) the following substitution:

$$\Omega_Q(1+z)^{3(1+w_Q)} \to \Omega_Q \exp\left(3\int_0^{\ln(1+z)}(1+w_Q(z'))d\ln(1+z')\right) . \quad (48)$$

As indicated above, a much discussed class of dynamical models for dark energy are *quintessence models*. In many ways people thereby repeat what has been done in inflationary cosmology. The main motivation there was (see Appendix C) to avoid excessive fine tunings of standard big bang cosmology (horizon and flatness problems). In Appendix D we give a brief discussion of this class of models. It has to be emphasized, however, that quintessence models do *not* solve the vacuum energy problem, so far also not the coincidence puzzle.

5.2 Type Ia Supernovas as Standard Candles

It has long been recognized that supernovas of type Ia are excellent standard candles and are visible to cosmic distances [42] (the record is at present at a redshift of about 1.7). At relatively closed distances they can be used to measure the Hubble constant, by calibrating the absolute magnitude of nearby supernovas with various distance determinations (e.g., Cepheids). There is still some dispute over these calibration resulting in differences of about 10% for H_0. (For recent papers and references, see [43].)

In 1979, Tammann [44] and Colgate [45] independently suggested that at higher redshifts this subclass of supernovas can be used to determine also the deceleration parameter. In recent years this program became feasible, thanks to the development of new technologies which made it possible to obtain digital images of faint objects over sizable angular scales, and by making use of big telescopes such as Hubble and Keck.

There are two major teams investigating high-redshift SNe Ia, namely the "Supernova Cosmology Project" (SCP) and the "High-Z Supernova search Team" (HZT). Each team has found a large number of SNe, and both groups have published almost identical results. (For up-to-date information, see the home pages [46] and [47].)

Before discussing the most recent results, a few remarks about the nature and properties of type Ia SNe should be made. Observationally, they are characterized by the absence of hydrogen in their spectra, and the presence of some strong silicon lines near maximum. The immediate progenitors are most probably carbon–oxygen white dwarfs in close binary systems, but it must be said that these have not yet been clearly identified.[6]

In the standard scenario a white dwarf accretes matter from a non-degenerate companion until it approaches the critical Chandrasekhar mass

[6] This is perhaps not so astonishing, because the progenitors are presumably faint compact dwarf stars.

and ignites carbon burning deep in its interior of highly degenerate matter. This is followed by an outward-propagating nuclear flame leading to a total disruption of the white dwarf. Within a few seconds the star is converted largely into nickel and iron. The dispersed nickel radioactively decays to cobalt and then to iron in a few hundred days. A lot of effort has been invested to simulate these complicated processes. Clearly, the physics of thermonuclear runaway burning in degenerate matter is complex. In particular, since the thermonuclear combustion is highly turbulent, multidimensional simulations are required. This is an important subject of current research. (One gets a good impression of the present status from several articles in [48]. See also the review [49].) The theoretical uncertainties are such that, for instance, predictions for possible evolutionary changes are not reliable.

It is conceivable that in some cases a type Ia supernova is the result of a merging of two carbon–oxygen-rich white dwarfs with a combined mass surpassing the Chandrasekhar limit. Theoretical modeling indicates, however, that such a merging would lead to a collapse, rather than an SN Ia explosion. But this issue is still debated.

In view of the complex physics involved, it is not astonishing that type Ia supernovas are not perfect standard candles. Their peak absolute magnitudes have a dispersion of 0.3–0.5 mag, depending on the sample. Astronomers have, however, learned in recent years to reduce this dispersion by making use of empirical correlations between the absolute peak luminosity and light curve shapes. Examination of nearby SNe showed that the peak brightness is correlated with the time scale of their brightening and fading: slow decliners tend to be brighter than rapid ones. There are also some correlations with spectral properties. Using these correlations it became possible to reduce the remaining intrinsic dispersion, at least in the average, to $\simeq 0.15\ mag$. (For the various methods in use, and how they compare, see [50, 56], and references therein.) Other corrections, such as Galactic extinction, have been applied, resulting for each supernova in a corrected (rest-frame) magnitude. The redshift dependence of this quantity is compared with the theoretical expectation given by (41) and (39).

5.3 Results

After the classic papers [51–53] on the Hubble diagram for high-redshift type Ia supernovas, published by the SCP and HZT teams, significant progress has been made (for reviews, see [54] and [55]). I discuss first the main results presented in [56]. These are based on additional new data for $z > 1$, obtained in conjunction with the Great Observatories Origins Deep Survey (GOODS) Treasury program, conducted with the Advanced Camera for Surveys (ACS) aboard the Hubble Space Telescope (HST).

The quality of the data and some of the main results of the analysis are shown in Fig. 1. The data points in the top panel are the distance moduli relative to an empty uniformly expanding universe, $\Delta(m - M)$, and the redshifts

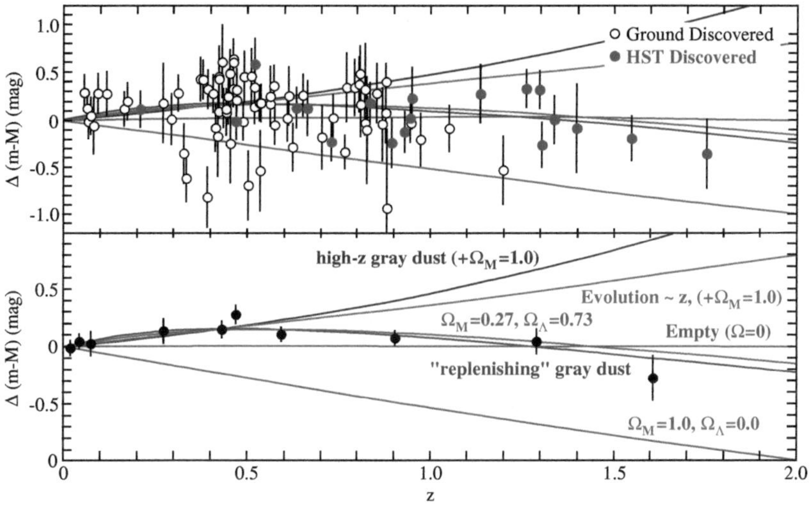

Fig. 1. Distance moduli relative to an empty uniformly expanding universe (residual Hubble diagram) for SNe Ia; see text for further explanations (Adapted from [56], Fig. 7.)

of a "gold" set of 157 SNe Ia. In this "reduced" Hubble diagram the filled symbols are the HST-discovered SNe Ia. The bottom panel shows weighted averages in fixed redshift bins.

These data are consistent with the "cosmic concordance" model ($\Omega_M = 0.3$, $\Omega_\Lambda = 0.7$), with $\chi^2_{dof} = 1.06$. For a flat universe with a cosmological constant, the fit gives $\Omega_M = 0.29\pm^{0.13}_{0.19}$ (equivalently, $\Omega_\Lambda = 0.71$). The other model curves will be discussed below. Likelihood regions in the $(\Omega_M, \Omega_\Lambda)$-plane, keeping only these parameters in (39) and averaging H_0, are shown in Fig. 2. To demonstrate the progress, old results from 1998 are also included. It will turn out that this information is largely complementary to the restrictions we shall obtain from the CMB anisotropies.

In the meantime new results have been published. Perhaps the best high-z SN Ia compilation to date are the results from the Supernova Legacy Survey (SNLS) of the first year [57]. The other main research group has also published new data at about the same time [58].

5.4 Systematic Uncertainties

Possible systematic uncertainties due to astrophysical effects have been discussed extensively in the literature. The most serious ones are (i) *dimming* by intergalactic dust, and (ii) *evolution* of SNe Ia over cosmic time, due to changes in progenitor mass, metallicity, and C/O ratio. I discuss these concerns only briefly (see also [54, 56]).

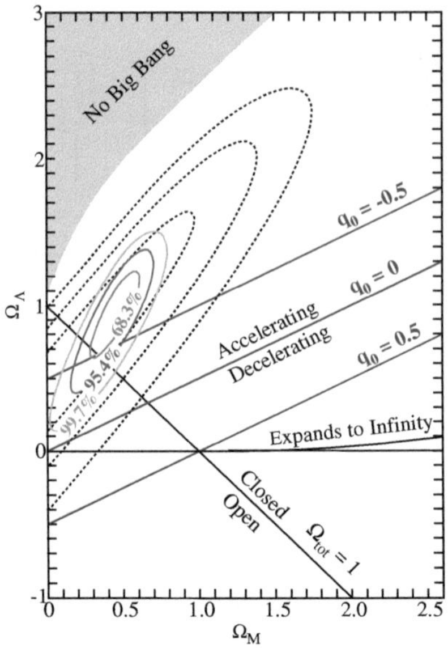

Fig. 2. Likelihood regions in the $(\Omega_M, \Omega_\Lambda)$-plane. The dotted contours are old results from 1998. (Adapted from [56], Fig. 8.)

Concerning extinction, detailed studies show that high-redshift SN Ia suffer little reddening; their B-V colors at maximum brightness are normal. However, it can a priori not be excluded that we see distant SNe through a grey dust with grain sizes large enough as to not imprint the reddening signature of typical interstellar extinction. One argument against this hypothesis is that this would also imply a larger dispersion than is observed. In Fig. 1 the expectation of a simple grey dust model is also shown. The new high-redshift data reject this monotonic model of astrophysical dimming. Equation (46) shows that at redshifts $z \geq (2\Omega_\Lambda/\Omega_M)^{1/3} - 1 \simeq 1.2$ the Universe is *decelerating*, and this provides an almost unambiguous signature for Λ, or some effective equivalent. There is now strong evidence for a transition from a deceleration to acceleration at a redshift $z = 0.46 \pm 0.13$.

The same data provide also some evidence against a simple luminosity evolution that could mimic an accelerating Universe. Other empirical constraints are obtained by comparing subsamples of low-redshift SN Ia believed to arise from old and young progenitors. It turns out that there is no difference within the measuring errors, *after* the correction based on the light-curve shape has been applied. Moreover, spectra of high-redshift SNe appear remarkably similar to those at low redshift. This is very reassuring. On the other hand, there seems to be a trend that more distant supernovas are bluer. It would, of course, be helpful if evolution could be predicted theoretically, but in view of what has been said earlier, this is not (yet) possible.

In conclusion, none of the investigated systematic errors appear to reconcile the data with $\Omega_\Lambda = 0$ and $q_0 \geq 0$. But further work is necessary before we can declare this as a really established fact.

To improve the observational situation a satellite mission called SNAP ("Supernovas Acceleration Probe") has been proposed [59]. According to the plans this satellite would observe about 2000 SNe within a year and much more detailed studies could then be performed. For the time being some scepticism with regard to the results that have been obtained is still not out of place, but the situation is steadily improving.

Finally, I mention a more theoretical complication. In the analysis of the data the luminosity distance for an ideal Friedmann universe was always used. But the data were taken in the real inhomogeneous Universe. This may perhaps not be good enough, especially for high-redshift standard candles. The simplest way to take this into account is to introduce a filling parameter which, roughly speaking, represents matter that exists in galaxies but not in the intergalactic medium. For a constant filling parameter one can determine the luminosity distance by solving the Dyer–Roeder equation. But now one has an additional parameter in fitting the data. For a flat universe this was investigated in [60]. We shall come back to this issue in Sect. 8.2.

6 Microwave Background Anisotropies

Investigations of the cosmic microwave background have presumably contributed most to the remarkable progress in cosmology during recent years (For a review, see [61]). Beside its spectrum, which is Planckian to an incredible degree, we also can study the temperature fluctuations over the "cosmic photosphere" at a redshift $z \approx 1100$. Through these we get access to crucial cosmological information (primordial density spectrum, cosmological parameters, etc.). A major reason for why this is possible relies on the fortunate circumstance that the fluctuations are tiny ($\sim 10^{-5}$) at the time of recombination. This allows us to treat the deviations from homogeneity and isotropy for an extended period of time perturbatively, i.e., by linearizing the Einstein and matter equations about solutions of the idealized Friedmann–Lemaître models. Since the physics is effectively *linear*, we can accurately work out the *evolution* of the perturbations during the early phases of the Universe, given a set of cosmological parameters. Confronting this with observations tells us a lot about the cosmological parameters as well as the initial conditions, and thus about the physics of the very early Universe. Through this window to the earliest phases of cosmic evolution we can, for instance, test general ideas and specific models of inflation.

6.1 Qualitative Remarks

Let me begin with some qualitative remarks, before I go into more technical details. Long before recombination (at temperatures $T > 6000$ K, say) photons, electrons, and baryons were so strongly coupled that these components may be treated together as a single fluid. In addition to this there is also a dark matter component. For all practical purposes the two interact only gravitationally. The investigation of such a two-component fluid for small deviations from an idealized Friedmann behavior is a well-studied application of cosmological perturbation theory (see, e.g., [63]).

At a later stage, when decoupling is approached, this approximate treatment breaks down because the mean free path of the photons becomes longer (and finally "infinite" after recombination). While the electrons and baryons can still be treated as a single fluid, the photons and their coupling to the electrons have to be described by the general relativistic Boltzmann equation. The latter is, of course, again linearized about the idealized Friedmann solution. Together with the linearized fluid equations (for baryons and cold dark matter, say) and the linearized Einstein equations one arrives at a complete system of equations for the various perturbation amplitudes of the metric and matter variables. There exist widely used codes, e.g. CMBFAST [62], that provide the CMB anisotropies – for given initial conditions – to a precision of about 1%. A lot of qualitative and semi-quantitative insight into the relevant physics can, however, be gained by looking at various approximations of the basic dynamical system.

Let us first discuss the temperature fluctuations. What is observed is the temperature autocorrelation:

$$C(\vartheta) := \left\langle \frac{\Delta T(\mathbf{n})}{T} \cdot \frac{\Delta T(\mathbf{n}')}{T} \right\rangle = \sum_{l=2}^{\infty} \frac{2l+1}{4\pi} C_l P_l(\cos\vartheta) \, , \tag{49}$$

where ϑ is the angle between the two directions of observation $\mathbf{n}, \mathbf{n}'$, and the average is taken ideally over all sky. The *angular power spectrum* is by definition $\frac{l(l+1)}{2\pi} C_l$ *versus* l $(\vartheta \simeq \pi/l)$.

A characteristic scale, which is reflected in the observed CMB anisotropies, is the sound horizon at last scattering, i.e., the distance over which a pressure wave can propagate until decoupling. This can be computed within the unperturbed model and subtends about half a degree on the sky for typical cosmological parameters. For scales larger than this sound horizon the fluctuations have been laid down in the very early Universe. These have been detected by the COBE satellite. The (gauge invariant brightness) temperature perturbation $\Theta = \Delta T/T$ is dominated by the combination of the intrinsic temperature fluctuations and gravitational redshift or blueshift effects. For example, photons that have to climb out of potential wells for high-density regions are redshifted. One can show that these effects combine for adiabatic initial conditions to $\frac{1}{3}\Psi$, where Ψ is one of the two gravitational Bardeen potentials. The latter, in turn, is directly related to the density perturbations. For scale-free initial perturbations and almost vanishing spatial curvature the corresponding angular power spectrum of the temperature fluctuations turns out to be nearly flat (Sachs–Wolfe plateau in Fig. 3).

On the other hand, inside the sound horizon before decoupling, acoustic, Doppler, gravitational redshift, and photon diffusion effects combine to the spectrum of small angle anisotropies shown in Fig. 3. These result from gravitationally driven synchronized acoustic oscillations of the photon–baryon fluid, which are damped by photon diffusion.

A particular realization of $\Theta(\mathbf{n})$, such as the one accessible to us (all sky map from our location), cannot be predicted. Theoretically, Θ is a random field $\Theta(\mathbf{x}, \eta, \mathbf{n})$, depending on the conformal time η, the spatial coordinates, and the observing direction $\mathbf{n}$. Its correlation functions should be rotationally invariant in $\mathbf{n}$, and respect the symmetries of the background time slices. If we expand Θ in terms of spherical harmonics,

$$\Theta(\mathbf{n}) = \sum_{lm} a_{lm} Y_{lm}(\mathbf{n}) \ , \tag{50}$$

the random variables a_{lm} have to satisfy

$$\langle a_{lm} \rangle = 0, \quad \langle a^\star_{lm} a_{l'm'} \rangle = \delta_{ll'} \delta_{mm'} C_l(\eta) \ , \tag{51}$$

where the $C_l(\eta)$ depend only on η. Hence the correlation function at the present time η_0 is given by (49), where $C_l = C_l(\eta_0)$, and the bracket now denotes the statistical average. Thus,

$$C_l = \frac{1}{2l+1} \left\langle \sum_{m=-l}^{l} a^\star_{lm} a_{lm} \right\rangle \ . \tag{52}$$

The standard deviations $\sigma(C_l)$ measure a fundamental uncertainty in the knowledge we can get about the C_l's. These are called *cosmic variances*, and are most pronounced for low l. In simple inflationary models the a_{lm} are Gaussian distributed, hence

$$\frac{\sigma(C_l)}{C_l} = \sqrt{\frac{2}{2l+1}} \ . \tag{53}$$

Therefore, the limitation imposed on us (only one sky in one universe) is small for large l.

6.2 Boltzmann Hierarchy

The brightness temperature fluctuation can be obtained from the perturbation of the photon distribution function by integrating over the magnitude of the photon momenta. The linearized Botzmann equation can then be translated into an equation for Θ, which we now regard as a function of η, x^i, and γ^j, where the γ^j are the directional cosines of the momentum vector relative to an orthonormal triad field of the unperturbed spatial metric with curvature

K. Next one performs a harmonic decomposition of Θ, which reads for the spatially flat case ($K = 0$)

$$\Theta(\eta, \boldsymbol{x}, \gamma) = (2\pi)^{-3/2} \int d^3k \sum_l \theta_l(\eta, k) G_l(\boldsymbol{x}, \gamma; \boldsymbol{k}) \;, \tag{54}$$

where

$$G_l(\boldsymbol{x}, \gamma; \boldsymbol{k}) = (-i)^l P_l(\hat{\boldsymbol{k}} \cdot \gamma) \exp(i\boldsymbol{k} \cdot \boldsymbol{x}) \;. \tag{55}$$

The dynamical variables $\theta_l(\eta)$ are the *brightness moments*, and should be regarded as random variables. Boltzmann's equation implies the following hierarchy of ordinary differential equations for the brightness moments[7] $\theta_l(\eta)$ (if polarization effects are neglected):

$$\theta_0' = -\frac{1}{3}k\theta_1 - \Phi' \;, \tag{56}$$

$$\theta_1' = k\Big(\theta_0 + \Psi - \frac{2}{5}\theta_2\Big) - \dot{\tau}(\theta_1 - V_b) \;, \tag{57}$$

$$\theta_2' = k\Big(\frac{2}{3}\theta_1 - \frac{3}{7}\theta_3\Big) - \dot{\tau}\frac{9}{10}\theta_2 \;, \tag{58}$$

$$\theta_l' = k\Big(\frac{l}{2l-1}\theta_{l-1} - \frac{l+1}{2l+3}\theta_{l+1}\Big) \;, \quad l > 2 \;. \tag{59}$$

Here, V_b is the gauge invariant scalar velocity perturbation of the baryons, $\dot{\tau} = x_e n_e \sigma_T a/a_0$, where a is the scale factor, $x_e n_e$ the unperturbed free electron density (x_e = ionization fraction), and σ_T the Thomson cross section. Moreover, Φ and Ψ denote the Bardeen potentials. (For further details, see, e.g., Sect. 6 of [3] or [63], where cosmological perturbation theory is developed in great detail.)

The C_l are determined by an integral over k, involving a primordial power spectrum (of curvature perturbations) and the $|\theta_l(\eta)|^2$, for the corresponding initial conditions (their transfer functions).

This system of equations is completed by the linearized fluid and Einstein equations. Various approximations for the Boltzmann hierarchy provide already a lot of insight. In particular, one can very nicely understand how damped acoustic oscillations are generated, and in which way they are influenced by the baryon fraction (again, see [3] or [63]). A typical theoretical CMB spectrum is shown in Fig. 3. (Beside the scalar contribution in the sense of cosmological perturbation theory, considered so far, the tensor contribution due to gravity waves is also shown there.)

6.3 Polarization

A polarization map of the CMB radiation provides important additional information to that obtainable from the temperature anisotropies. For example,

[7] In the literature the normalization of the θ_l is sometimes chosen differently: $\theta_l \to (2l+1)\theta_l$.

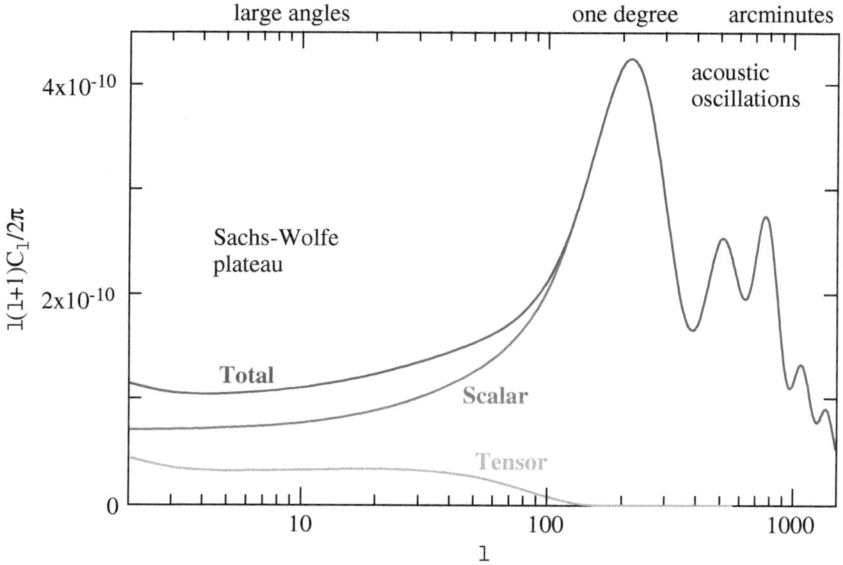

Fig. 3. Theoretical angular temperature–temperature (TT) power spectrum for adiabatic initial perturbations and typical cosmological parameters. The scalar and tensor contributions to the anisotropies are also shown

we can get constraints about the epoch of reionization. Most importantly, future polarization observations may reveal a stochastic background of gravity waves, generated in the very early Universe. In this section we give a brief introduction to the study of CMB polarization.

The mechanism which partially polarizes the CMB radiation is similar to that for the scattered light from the sky. Consider first scattering at a single electron of unpolarized radiation coming in from all directions. Due to the familiar polarization dependence of the differential Thomson cross section, the scattered radiation is, in general, polarized. It is easy to compute the corresponding Stokes parameters. Not surprisingly, they are not all equal to zero if and only if the intensity distribution of the incoming radiation has a non-vanishing quadrupole moment. The Stokes parameters Q and U are proportional to the overlap integral with the combinations $Y_{2,2} \pm Y_{2,-2}$ of the spherical harmonics, while V vanishes. This is basically the reason why a CMB polarization map traces (in the tight coupling limit) the quadrupole temperature distribution on the last scattering surface.

The polarization tensor of an all sky map of the CMB radiation can be parametrized in temperature fluctuation units, relative to the orthonormal basis $\{d\vartheta, \sin\vartheta\ d\varphi\}$ of the two sphere, in terms of the Pauli matrices as $\Theta \cdot 1 + Q\sigma_3 + U\sigma_1 + V\sigma_2$. The Stokes parameter V vanishes (no circular polarization). Therefore, the polarization properties can be described by the following symmetric trace-free tensor on S^2:

$$(\mathcal{P}_{ab}) = \begin{pmatrix} Q & U \\ U & -Q \end{pmatrix} . \tag{60}$$

As for gravity waves, the components Q and U transform under a rotation of the 2-bein by an angle α as

$$Q \pm iU \to e^{\pm 2i\alpha}(Q \pm iU) , \tag{61}$$

and are thus of spin-weight 2. $\mathcal{P}_{ab}$ can be decomposed uniquely into *electric* and *magnetic* parts:

$$\mathcal{P}_{ab} = E_{;ab} - \frac{1}{2} g_{ab} \Delta E + \frac{1}{2}(\varepsilon_a{}^c B_{;bc} + \varepsilon_b{}^c B_{;ac}) . \tag{62}$$

Expanding here the scalar functions E and B in terms of spherical harmonics, we obtain an expansion of the form

$$\mathcal{P}_{ab} = \sum_{l=2}^{\infty} \sum_m \left[a^E_{(lm)} Y^E_{(lm)ab} + a^B_{(lm)} Y^B_{(lm)ab} \right] \tag{63}$$

in terms of the tensor harmonics:

$$Y^E_{(lm)ab} := N_l (Y_{(lm);ab} - \frac{1}{2} g_{ab} Y_{(lm);c}{}^c), \quad Y^B_{(lm)ab} := \frac{1}{2} N_l (Y_{(lm);ac} \varepsilon^c{}_b + a \leftrightarrow b) , \tag{64}$$

where $l \geq 2$ and

$$N_l \equiv \left(\frac{2(l-2)!}{(l+2)!} \right)^{1/2} .$$

Equivalently, one can write this as

$$Q + iU = \sqrt{2} \sum_{l=2}^{\infty} \sum_m \left[a^E_{(lm)} + i a^B_{(lm)} \right] {}_2Y_l^m , \tag{65}$$

where ${}_sY_l^m$ are the spin-s harmonics.

As in (50) the multipole moments $a^E_{(lm)}$ and $a^B_{(lm)}$ are random variables, and we have equations analogous to (52):

$$C_l^{TE} = \frac{1}{2l+1} \sum_m \langle a^{\Theta\star}_{lm} a^E_{lm} \rangle, \text{ etc} . \tag{66}$$

(We have now put the superscript Θ on the a_{lm} of the temperature fluctuations.) The C_l's determine the various angular correlation functions. For example, one easily finds

$$\langle \Theta(\boldsymbol{n}) Q(\boldsymbol{n}') \rangle = \sum_l C_l^{TE} \frac{2l+1}{4\pi} N_l P_l^2(\cos\vartheta) . \tag{67}$$

For the spacetime-dependent Stokes parameters Q and U of the radiation field we can perform a normal mode decomposition analogous to (54). If, for simplicity, we again consider only scalar perturbations this reads

$$Q \pm iU = (2\pi)^{-3/2} \int d^3k \sum_l (E_l \pm iB_l)\, {}_{\pm 2}G_l^0 \,, \tag{68}$$

where

$${}_sG_l^m(\boldsymbol{x}, \gamma; \boldsymbol{k}) = (-i)^l \left(\frac{2l+1}{4\pi}\right)^{1/2} {}_sY_l^m(\gamma) \exp(i\boldsymbol{k} \cdot \boldsymbol{x}) \,, \tag{69}$$

if the mode vector $\mathbf{k}$ is chosen as the polar axis. (Note that G_l in (55) is equal to ${}_0G_l^0$.)

The Boltzmann equation implies a coupled hierarchy for the moments θ_l, E_l, and B_l [64, 65]. It turns out that the B_l vanish for scalar perturbations. Non-vanishing magnetic multipoles would be a unique signature for a spectrum of gravity waves. In a sudden decoupling approximation, the present electric multipole moments can be expressed in terms of the brightness quadrupole moment on the last scattering surface and spherical Bessel functions as

$$\frac{E_l(\eta_0, k)}{2l+1} \simeq \frac{3}{8}\theta_2(\eta_{dec}, k) \frac{l^2 j_l(k\eta_0)}{(k\eta_o)^2} \,. \tag{70}$$

Here one sees how the observable E_l's trace the quadrupole temperature anisotropy on the last scattering surface. In the tight coupling approximation the latter is proportional to the dipole moment θ_1.

7 Observational Results and Cosmological Parameters

In recent years several experiments gave clear evidence for multiple peaks in the angular temperature power spectrum at positions expected on the basis of the simplest inflationary models and big bang nucleosynthesis [66]. These results have been confirmed and substantially improved by the first-year WMAP data [67, 68, 72]. Fortunately, the improved data after three years of integration are now available [69]. Below we give a brief summary of some of the most important results.

Figure 4 shows the 3-year data of WMAP for the TT angular power spectrum, and the best fit (power law) ΛCDM model. The latter is a spatially flat model and involves the following six parameters: $\Omega_b h_0^2$, $\Omega_M h_0^2$, H_0, amplitude of fluctuations,[8] σ_8, optical depth τ, and the spectral index, n_s, of the primordial scalar power spectrum (see Appendix C.7). Figure 5 shows in addition the TE polarization data [70]. There are now also EE data that lead to a further reduction of the allowed parameter space. The first column in Table 1 shows the best fit values of the six parameters, using only the WMAP data.

[8] σ_8^2 is the variance of mass fluctuations in spheres of radius 8 h_0^{-1} Mpc. (For a precise definition, see, e.g., Appendix A of [63].)

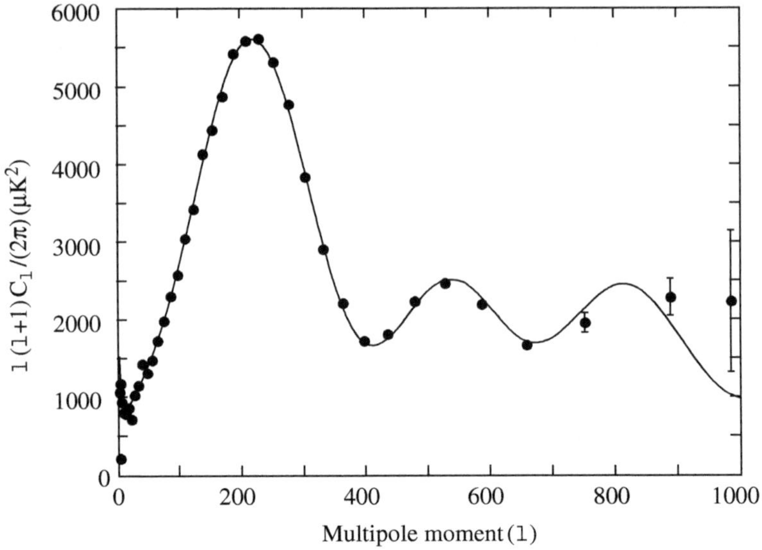

Fig. 4. Three-year WMAP data for the temperature–temperature (TT) power spectrum. The black line is the best fit ΛCDM model for the 3-year WMAP data. (Adapted from Fig. 2 of [69])

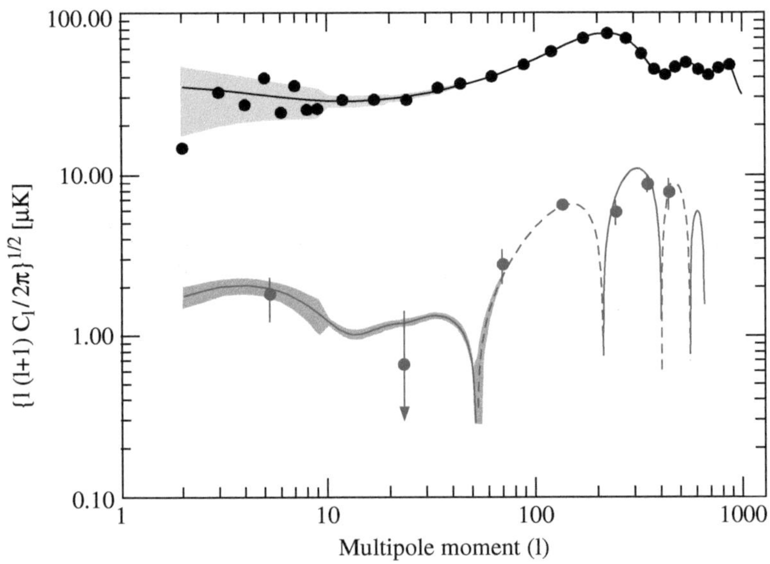

Fig. 5. WMAP data for the temperature-polarization TE power from Fig. 25 of [70])

Table 1.

Parameter	WMAP alone	WMAP + 2dFGRS
$100\Omega_b h_0^2$	$2.233^{+0.072}_{-0.0.091}$	$2.223^{+0.066}_{-0.083}$
$\Omega_M h_0^2$	$0.1268^{+0.0073}_{-0.0128}$	$0.1262^{+0.0050}_{-0.0103}$
h_0	$0.734^{+0.028}_{-0.038}$	$0.732^{+0.018}_{-0.025}$
Ω_M	$0.238^{+0.027}_{-0.045}$	$0.236^{+0.016}_{-0.029}$
σ_8	$0.744^{+0.050}_{-0.060}$	$0.737^{+0.033}_{-0.045}$
τ	$0.088^{+0.028}_{-0.034}$	$0.083^{+0.027}_{-0.031}$
n_s	$0.951^{+0.015}_{-0.019}$	$0.948^{+0.014}_{-0.018}$

Figure 6 shows the prediction of the model for the luminosity-redshift relation, together with the SLNS data [57] mentioned in Sect. 5.3. For other predictions and corresponding data sets, see [69].

Combining the WMAP results with other astronomical data reduces the uncertainties for some of the six parameters. This is illustrated in the second column which shows the 68% confidence ranges of a joint likelihood analysis when the power spectrum from the completed 2dFGRS [73] is added. In [69] other joint constraints are listed (see their Tables 5, 6). In Fig. 7 we reproduce one of many plots in [69] that shows the joint marginalized contours in the (Ω_M, h_0)-plane.

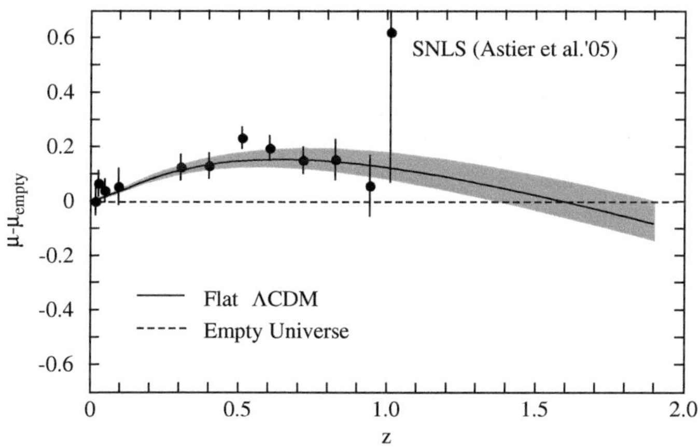

Fig. 6. Prediction for the luminosity-redshift relation from the ΛCDM model model fit to the WMAP data only. The ordinate is the deviation of the distance modulous from the empty universe model. The prediction is compared to the SNLS data [57]. (From Fig. 8 of [69])

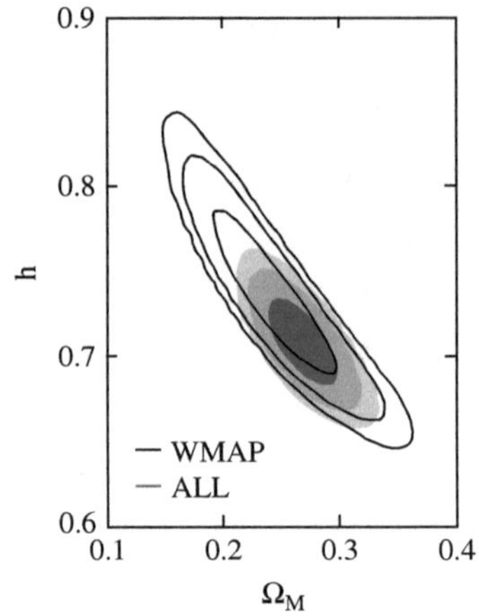

Fig. 7. Joint marginalized contours (68% and 95% confidence levels) in the (Ω_M, h_0)-plane for WMAP only (*solid lines*) and additional data (*filled red*) for the power-law ΛCDM model. (From Fig. 10 in [69])

The parameter space of the cosmological model can be extended in various ways. Because of intrinsic degeneracies, the CMB data alone are no more sufficient to determine unambiguously the cosmological model parameters. We illustrate this for non-flat models. For these the WMAP data (in particular, the position of the first acoustic peak) restricts the curvature parameter Ω_K to a narrow region around the degeneracy line $\Omega_K = -0.3040 + 0.4067$, $\Omega_\Lambda = 0.758^{+0.035}_{-0.058}$. This does not exclude models with $\Omega_\Lambda = 0$. However, when, for instance, the Hubble constant is restricted to an acceptable range, the universe must be nearly flat. For example, the restriction $h_0 = 0.72 \pm 0.08$ implies that $\Omega_K = -0.003^{+0.013}_{-0.017}$. Other strong limits are given in Table 11 of [69], assuming that $w = -1$. But even when this is relaxed, the combined data constrain Ω_K and w significantly (see Fig. 17 of [69]). The marginalized best fit values are $w = -1.062^{+0.128}_{-0.079}$, $\Omega_K = -0.024^{+0.016}_{-0.013}$ at the 68% confidence level.

The restrictions on w – assumed to have no z-dependence – for a flat model are illustrated in Fig. 8.

Another interesting result is that reionization of the Universe has set in at a redshift of $z_r = 10.9^{+2.7}_{-2.3}$. Later we shall add some remarks on what has been learnt about the primordial power spectrum.

It is most remarkable that a six parameter cosmological model is able to fit such a rich body of astronomical observations. There seems to be little room for significant modifications of the successful ΛCDM model. In spite of this we discuss in the next section some proposed attempts to explain the observations without dark energy.

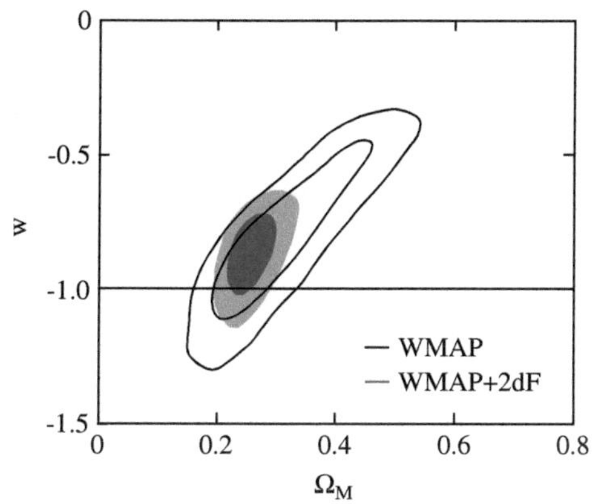

Fig. 8. Constraints on the equation of state parameter w in a flat universe model when WMAP data are combined with the 2dFGRS data. (From Fig. 15 in [69])

8 Alternatives to Dark Energy

In the previous two sections we have discussed some of the wide range of astronomical data that support the following 'concordance model': The Universe is spatially flat and dominated by a dark energy component and weakly interacting cold dark matter. Furthermore, the primordial fluctuations are adiabatic, nearly scale invariant and Gaussian, as predicted in simple inflationary models (see Sect. C.7). It is very likely that the present concordance model will survive phenomenologically.

A dominant dark energy component with density parameter $\simeq 0.7$ is so surprising that it should be examined whether this conclusion is really unavoidable. In what follows I shall briefly discuss some alternatives that have been proposed.

8.1 Changes in the Initial Conditions

Since we do not have a tested theory predicting the spectrum of primordial fluctuations, it appears reasonable to consider a wider range of possibilities than simple power laws. An instructive attempt in this direction was made some time ago [74], by constructing an Einstein–de Sitter model with $\Omega_\Lambda = 0$, fitting the CMB data as well as the power spectrum of 2dFGRS. In this the Hubble constant is, however, required to be rather low: $H_0 \simeq 46$ km/s/Mpc. The authors argued that this cannot definitely be excluded, because 'physical' methods lead mostly to relatively low values of H_0. In order to be consistent with matter fluctuations on cluster scales they added relic neutrinos with degenerate masses of order eV or a small contribution of quintessence with zero pressure ($w = 0$). In addition, they ignored the direct evidence for an accelerating Universe from the Hubble-diagram for distant Type Ia supernovae, on the basis of remaining systematic uncertainties. In the meantime, significant improvements in astronomical data sets have been made. In particular, the analysis of the 3-year WMAP data showed that there are no significant features in the primordial curvature fluctuation spectrum (see Sect. 5 of [69]). With the larger samples of high redshift supernovae and more precise information on large-scale galaxy clustering, such models with vanishing dark energy are no more possible [75].

8.2 Inhomogeneous Models

Backreaction

It has recently been suggested [76, 77] that perturbations on scales larger than the Hubble length, likely generated in the context of inflation, could mimic dark energy and cause acceleration. This suggestion caused a lot of discussion, and several papers addressed the question whether this is really possible. We repeat below a simple general argument given in [78] that the originally proposed mechanism cannot lead to acceleration, under the assumptions made in the cited papers. These include that the 4-velocity field u^μ of the CDM particles is geodesic and has zero vorticity $\omega_{\mu\nu}$. It is easy to see that these assumptions imply that the 1-form $\mathbf{u}$, belonging to the velocity field, has a vanishing exterior derivative. Hence we have locally $\mathbf{u} = dt$, thus u^μ is perpendicular to the slices $\{t = const\}$. Moreover the metric and the velocity have the form

$$g = -dt^2 + \bar{g}_t, \quad u = \partial_t \, , \tag{71}$$

where $\bar{g}_t$ is a t-dependent metric on slices of constant time t.

For such an inhomogeneous cosmological model one can introduce various definitions of the deceleration parameter which reduce to the familiar one for Friedmann models. We adopt here the one used in [77]. To motivate this, consider for some initial time t_{in} a spatial domain D and let this evolve according to the flow of u. If ω_t denotes the volume form belonging to $\bar{g}_t$, then we have for the volume $|D_t|$ and its time derivatives

$$|D_t| = \int \omega_t, \quad |\dot{D}_t| = \int \theta \omega_t, \quad |\ddot{D}_t| = \int (\dot{\theta} + \theta^2)\omega_t \, , \tag{72}$$

where $\theta = \nabla \cdot u$ denotes the expansion. If $l := |D_t|^{1/3}$, a natural definition of the deceleration parameter is $q = -(l\ddot{l})/\dot{l}^2$. This can be expressed as follows

$$\frac{1}{3}\frac{(|\dot{D}_t|)^2}{|D_t|^2} q = -\left(\frac{|\ddot{D}_t|}{|D_t|} - \frac{2}{3}\frac{(|\dot{D}_t|)^2}{|D_t|^2}\right) . \tag{73}$$

For an *infinitesimal* $|D_t|$ we obtain from the previous equations

$$\frac{1}{3}\theta^2 q = -(\dot{\theta} + \frac{1}{3}\theta^2) \, . \tag{74}$$

For the right-hand side we can now use the *Raychaudhuri equation*

$$\dot{\theta} + \frac{1}{3}\theta^2 = -\sigma_{\mu\nu}\sigma^{\mu\nu} + \omega_{\mu\nu}\omega^{\mu\nu} - R_{\mu\nu}u^\mu u^\nu , \tag{75}$$

where $\sigma_{\mu\nu}$ is the shear. For a vanishing vorticity, and imposing the strong energy condition (assumed in [77]), we see that $q \geq 0$. In this sense there is no acceleration.

A priori, a way out proposed in [76], is to argue that q as defined above is not what is measured in SN Ia observations. To analyze these one has to generalize the redshift–luminosity distance relation to inhomogeneous models. In doing this, two possible definitions for the deceleration parameter arise. One of them (q_4 in [78]) again has to be non-negative if the strong energy condition holds. The other (q_3 in [78]) may be negative, but in this case the supernova data would have to show acceleration in certain directions and deceleration in others. This is, however, not observed.

Kolb et al. have reacted to these considerations [79]. They admit that super-Hubble modes cannot lead to an acceleration, but they maintain that sub-Hubble modes may cause a large backreaction that may imply an effective acceleration. The authors stress that for investigating the effective dynamics averaging over a volume of size comparable with the present-day Hubble volume is essential. Let me add a few remarks on this. Adopting the notation

$$\langle\theta\rangle = \frac{\int \theta\omega_t}{\int \omega_t}, \quad \text{etc} , \tag{76}$$

and using the Raychaudhuri equation, we can write

$$\begin{aligned} \frac{1}{3}\frac{(|D_t|^{\cdot})^2}{|D_t|^2} q &= -\langle\dot{\theta} + \theta^2\rangle + \frac{2}{3}\langle\theta\rangle^2 \\ &= -\langle\dot{\theta} + \frac{1}{3}\theta^2\rangle - \frac{2}{3}(\langle\theta^2\rangle - \langle\theta\rangle^2) \\ &= \langle\sigma_{\mu\nu}\sigma^{\mu\nu} + R_{\mu\nu}u^\mu u^\nu\rangle - \frac{2}{3}(\langle\theta^2\rangle - \langle\theta\rangle^2) . \end{aligned} \tag{77}$$

The first term in the last equation, is non-negative if the strong energy condition holds, while the second term is non-positive.

The authors of [79] suggest that the second term may win and make q negative. To decide on the basis of detailed calculations whether this is indeed possible is a very difficult task. From what we know about the CMB radiation it appears, however, unlikely that there are such sizable perturbations out to very large scales. We shall say more about this in the next section.

The work by Kolb et al. triggered a lot of activity. (For a review, see [80].) We add some remarks about the ongoing discussion.

Power Spectrum of the Luminosity Distance

The deceleration parameter, defined in (73), has a simple geometrical meaning, but is not a directly measurable quantity. From an observational point of view, a more satisfactory approach is to generalize the magnitude–redshift relation, and study the fluctuations of the luminosity distance.

The magnitude–redshift relation in a perturbed Friedmann model has been derived in [81], and was later used to determine the angular power spectrum of the luminosity distance (the C_l's defined in analogy to (49)) [82]. One of the numerical results was that the uncertainties in determining cosmological parameters via the magnitude–redshift relation caused by fluctuations are small compared with the intrinsic dispersion in the absolute magnitude of Type Ia supernovae.

This subject was recently taken up in [83], as part of a program to develop the tools for extracting cosmological parameters, when much extended supernovae data become available.

Exact Inhomogeneous Model Studies

Effects of inhomogeneous matter distribution on light propagation were recently studied in the Lemaître–Tolman (LT) model, in order to see whether these can mimic an accelerated expansion.

The LT model is a family of spherically symmetric dust solutions of Einstein's equations, with a metric of the form

$$g = -dt^2 + \frac{R_{,r}^2(r,t)}{1+2E(r)}dr^2 + R^2(r,t)(d\vartheta^2 + \sin^2\vartheta d\varphi^2) . \tag{78}$$

The metric functions $E(r)$, $R(r,t)$, and a matter function $M(r)$ satisfy, as a consequence of Einstein's equations, the differential equations

$$M_{,r} = 4\pi\rho R^2 R_{,r}, \quad R_{,t}^2 = 2E + \frac{2GM}{R} + \frac{1}{3}\Lambda R^2 . \tag{79}$$

For these models the magnitude–redshift relation can be worked out exactly.

As an example we mention [84], where it was shown that for $\Lambda = 0$ the observed behavior of supernovae brightness cannot be fitted, unless our position in the model universe is very special. In that case one has to analyze also other data, in particular the CMB angular power spectrum. At the time of writing, this has not yet been done, but is certainly underway.

8.3 Modifications of Gravity

Since no satisfactory explanation of dark energy has emerged so far, possible modifications of GR, which would change the late expansion rate of the universe, have recently come into the focus of attention. The cosmic speed-up might, for instance, be explained by sub-dominant terms (like $1/R$) that become essential at small curvature. Modified gravity models have to be devised such that to pass the stringent Solar System tests, and are compatible with the observational data that support the concordance model.

Generalizations of the Einstein–Hilbert Action

The simplest generalization consists in replacing the Ricci scalar, R, in the Einstein–Hilbert action by a function $f(R)$. Note that this gives rise to fourth-order field equations.[9] Applying a suitable conformal transformation of the metric, the action becomes equivalent to a scalar-tensor theory. In detail, if we define a new metric $\tilde{g}_{\mu\nu} = \left[\exp(2\kappa/3)^{1/2\varphi}\right] g_{\mu\nu}$, then the action becomes

$$S = \int \left[\frac{1}{2\kappa}R[\tilde{g}] - \frac{1}{2}\tilde{g}^{\alpha\beta}\partial_\alpha\varphi\partial_\beta\varphi - V(\varphi) + L_{matter}\right]\sqrt{-\tilde{g}}d^4x \,, \tag{80}$$

where the potential V is determined by the function f. With this formulation one can, for instance, show that an arbitrary evolution of the scale factor $a(t)$ can be obtained with an appropriate choice of $f(R)$. It is also useful to check whether a particular model passes Solar System tests (acceptable Brans-Dicke parameter). One should, however, bear in mind that the two mathematically equivalent descriptions lead to physically different properties, for instance with regard to stability. These issues and the application for specific functions f to Friedmann spacetimes have recently been reviewed in [85].

A class of models that lead to cosmic acceleration is of the form $f(R) = R + \alpha/R^n$, $n > 0$. There has been a debate on whether such models (especially for $n = 1$) are consistent with Solar System tests. Some authors argued that this is the case, because they admit as a static spherically symmetric solution the Schwarzschild–de Sitter metric. This is, however, by no means sufficient. As already emphasized, this vacuum solution is far from unique. The correct one must match onto a physically acceptable solution for the interior of the star. In [86] it was shown for $n = 1$, i.e., for $f(R) = R - \mu^4/R$, that this requirement implies for the PPN parameter γ the value 1/2, in gross violation of the measured value $\gamma = 1 + (2.1 \pm 2.3) \times 10^{-5}$. This confirms an earlier claim by Chiba [87] that was based on the scalar-tensor reformulation (80).

Presumably, similar statements can be made for a large class of $f(R)$ models. Apart from their ad hoc nature, it has not yet been demonstrated that there are examples which satisfy all the constraints stressed above. The same can be said on generalizations [88], which include other curvature invariants, such as $R_{\mu\nu}R^{\mu\nu}$, $R_{\alpha\beta\gamma\delta}R^{\alpha\beta\gamma\delta}$. In addition, such models are in most cases *unstable*, like mechanical Lagrangian systems with higher derivatives [89].[10] An exception seem to be Lagrangians which are functions of

[9] Spherically symmetric vacuum solutions are, therefore, far from unique. Connected with this is that Birkhoff's theorem fails. So, on the basis of the vacuum equations the perihelion motion (for example) is no more predicted, but at best compatible with the theory. This is an enormous loss. (The reader may reflect about other drawbacks.)

[10] This paper contains a discussion of a generic instability of Lagrangian systems in mechanics with higher derivatives, which was discovered by M. Ostrogradski in 1850.

R and the Gauss–Bonnet invariant $G = R^2 - 4R_{\mu\nu}R^{\mu\nu} + R_{\alpha\beta\gamma\delta}R^{\alpha\beta\gamma\delta}$. By introducing two scalar fields such models can be written as an Einstein–Hilbert term plus a particular extra piece, containing a linear coupling to G. Because the Gauss–Bonnet invariant is a total divergence the corresponding field equations are of second order. This does, however, not guarantee that the theory is ghost-free. In [90] this question was studied for a class of models [88] for which there exist accelerating late-time power-law attractors and which satisfy the solar system constraints. It turned out that in a Friedmann background there are no ghosts, but there is instead *superluminal propagation* for a wide range of parameter space. This acausality is reminiscent of the Velo-Zwanziger phenomenon [92] for higher (> 1) spin fields coupled to external fields. It may very well be that it can only be avoided if very special conditions are satisfied. This issue deserves further investigations. See also [91].

First-Order Modifications of GR

The disadvantage of complicated fourth-order equations can be avoided by using the *Palatini variational principle*, in which the metric and the symmetric affine connection (the Christoffel symbols $\Gamma^\alpha{}_{\mu\nu}$) are considered to be independent fields.[11]

For GR the 'Palatini formulation' is equivalent to the Einstein–Hilbert variational principle, because the variational equation with respect to $\Gamma^\alpha{}_{\mu\nu}$ implies that the affine connection has to be the Levi–Civita connection. Things are no more that simple for $f(R)$ models:

$$S = \int \left[\frac{1}{2\kappa} f(R) + L_{matter}\right] \sqrt{-g}\, d^4x \ , \tag{81}$$

where $R[g,\Gamma] = g^{\alpha\beta}R_{\alpha\beta}[\Gamma]$, $R_{\alpha\beta}[\Gamma]$ being the Ricci tensor of the independent torsionless connection Γ. The equations of motion are in obvious notation

$$f'(R)R_{(\mu\nu)}[\Gamma] - \frac{1}{2} f(R) g_{\mu\nu} = \kappa T_{\mu\nu} \ , \tag{82}$$

$$\nabla^{\Gamma}_{\alpha}\left(\sqrt{-g} f'(R) g^{\mu\nu}\right) = 0 \ . \tag{83}$$

For the second of these equations one has to assume that L_{matter} is functionally independent of Γ. (It may, however, contain metric covariant derivatives.)

Equation (83) implies that

$$\nabla^{\Gamma}_{\alpha}\left[\sqrt{-\hat{g}}\hat{g}^{\mu\nu}\right] = 0 \tag{84}$$

for the conformally equivalent metric $\hat{g}_{\mu\nu} = f'(R) g_{\mu\nu}$. Hence, the $\Gamma^\alpha{}_{\mu\nu}$ are equal to the Christoffel symbols for the metric $\hat{g}_{\mu\nu}$.

---

[11] This approach was actually first introduced by Einstein (S.B. Preuss. Akad. Wiss., 414 (1925)). This is correctly stated in Pauli's classical text, p. 215.

The trace of (82) gives

$$Rf'(R) - 2f(R) = \kappa T \ .$$

Thanks to this algebraic equation we may regard R as a function of T. In the matter-free case it is identically satisfied if $f(R)$ is proportional to R^2. In all other cases R is equal to a constant c (which is in general not unique). If $f'(c) \neq 0$, (83) implies that Γ is the Levi–Civita connection of $g_{\mu\nu}$, and (82) reduces to Einstein's vacuum equation with a cosmological constant. In general, one can rewrite the field equations in the form of Einstein gravity with non-standard matter couplings.[12] Because of this it is, for instance, straightforward to develop cosmological perturbation theory [94].

Koivisto [95] has applied this to study the resulting matter power spectrum, and showed that the comparison with observations leads to strong constraints. The allowed parameter space for a model of the form $f(R) = R - \alpha R^{\beta}$ ($\alpha > 0$, $\beta < 1$) is reduced to a tiny region around the ΛCDM cosmology. For a related investigation, see [96].

The literature on this type of generalized gravity models is rapidly increasing.

Brane-World Models

Certain brane-world models[13] lead to modifications of Friedmann cosmology at very large scales. An interesting example has been proposed by Dvali, Gabadadze, and Porrati (DGP), for which the theory remains four-dimensional at 'short' distances, but crosses over to higher-dimensional behavior of gravity at some very large distance [97]. This model has the same number of parameters as the successful ΛCDM cosmology, but contains no dark energy. The resulting modified Friedmann equations can give rise to universes with accelerated expansion, due to an infrared modification of gravity.

In [100] the predictions of the model have been confronted with latest supernovae data [57], and the position of the acoustic peak in the Sloan digital sky survey (SDSS) correlation function for a luminous red galaxy sample [101]. The result is that a flat DGP brane model is ruled out at 3σ. A similar analysis was more recently performed in [99], including also the CMB shift parameter that effectively determines the first acoustic peak (see Sect. 8.1). The authors arrive at the conclusion that the flat DGP models are within the 1σ contours, but that the flat ΛCDM model provides a better fit to the data. They also point out some level of uncertainty in the use of the data, and conservatively conclude that the flat DGP models are within joint 2σ contours.

---

12 It is shown in [93] that if the matter action is independent of Γ, the theory is dynamically equivalent to a Brans-Dicke theory with Brans-Dicke parameter $-3/2$, plus a potentiel term.

13 For a review, see [98].

This nicely illustrates that observational data are restricting theoretical speculations more and more.

The DGP models have, however, serious defects on a fundamental level. A detailed analysis of the excitations about the self-accelerating solution showed that there is a *ghost mode* (negative kinetic energy) [102, 103]. Furthermore, it has very recently been pointed out [104] that due to superluminal fluctuations around non-trivial backgrounds, there is *no local causal evolution*. This infrared breakdown also happens for other apparently consistent low-energy effective theories.

* * *

The previous discussion should have made it clear that it is extremely difficult to construct consistent modifications of GR that lead to an accelerated universe at late times. The dark energy problems will presumably stay with us for a long time. Understanding the nature of DE is widely considered as one of the main goals of cosmological research for the next decade and beyond.

A Essentials of Friedmann–Lemaître Models

In this Appendix those parts of the standard model of cosmology that are needed throughout the text will be briefly introduced. More extensive treatments can be found at many places, for instance in the recent textbooks on cosmology [105], [106], [107], [108], [109].

A.1 Friedmann–Lemaître Spacetimes

There is now good evidence that the (recent as well as the early) Universe[14] is – on large scales – surprisingly homogeneous and isotropic. The most impressive support for this comes from extended redshift surveys of galaxies and from the truly remarkable isotropy of the CMB. In the two degree field (2dF) galaxy redshift survey,[15] completed in 2003, the redshifts of about 250,000 galaxies have been measured. The distribution of galaxies out to 4 billion light years shows that there are huge clusters, long filaments, and empty voids measuring over 100 million light years across. But the map also shows

[14] By *Universe* I always mean that part of the world around us which is in principle accessible to observations. In my opinion the 'Universe as a whole' is not a scientific concept. When talking about *model universes*, we develop on paper or with the help of computers, I tend to use lower case letters. In this domain we are, of course, free to make extrapolations and venture into speculations, but one should always be aware that there is the danger to be drifted into a kind of 'cosmo-mythology'.

[15] Consult the Home Page: http://www.mso.anu.edu. au/2dFGRS.

that there are *no larger structures*. The more extended SDSS has already produced very similar results, and will in the end have spectra of about a million galaxies.[16]

One arrives at the Friedmann(–Lemaître–Robertson–Walker) spacetimes by postulating that for each observer, moving along an integral curve of a distinguished four-velocity field u, the Universe looks spatially isotropic. Mathematically, this means the following: Let $Iso_x(M)$ be the group of local isometries of a Lorentz manifold (M, g), with fixed point $x \in M$, and let $SO_3(u_x)$ be the group of all linear transformations of the tangent space $T_x(M)$ which leave the four-velocity u_x invariant and induce special orthogonal transformations in the subspace orthogonal to u_x, then

$$\{T_x\phi : \ \phi \in Iso_x(M), \ \phi_\star u = u\} \supseteq SO_3(u_x)$$

($\phi_\star$ denotes the push-forward belonging to ϕ; see [1], p. 550). In [110] it is shown that this requirement implies that (M, g) is a Friedmann spacetime, whose structure we now recall. Note that (M, g) is then automatically homogeneous.

A *Friedmann spacetime* (M, g) is a warped product of the form $M = I \times \Sigma$, where I is an interval of $\mathbb{R}$, and the metric g is of the form

$$g = -dt^2 + a^2(t)\gamma \ , \tag{85}$$

such that (Σ, γ) is a Riemannian space of constant curvature $k = 0, \pm 1$. The distinguished time t is the *cosmic time*, and $a(t)$ is the *scale factor* (it plays the role of the warp factor (see Appendix B of [1])). Instead of t we often use the *conformal time* η, defined by $d\eta = dt/a(t)$. The velocity field is perpendicular to the slices of constant cosmic time, $u = \partial/\partial t$.

Spaces of Constant Curvature

For the space (Σ, γ) of constant curvature[17] the curvature is given by

$$R^{(3)}(X, Y)Z = k\left[\gamma(Z, Y)X - \gamma(Z, X)Y\right] \ ; \tag{86}$$

in components

$$R^{(3)}_{ijkl} = k(\gamma_{ik}\gamma_{jl} - \gamma_{il}\gamma_{jk}) \ . \tag{87}$$

Hence, the Ricci tensor and the scalar curvature are

$$R^{(3)}_{jl} = 2k\gamma_{jl} \quad , \quad R^{(3)} = 6k \ . \tag{88}$$

[16] For a description and pictures, see the Home Page: http://www.sdss.org/sdss.html.

[17] For a detailed discussion of these spaces I refer – for readers knowing German – to [111] or [112].

For the curvature two-forms we obtain from (87) relative to an orthonormal triad $\{\theta^i\}$

$$\Omega^{(3)}_{ij} = \frac{1}{2} R^{(3)}_{ijkl}\ \theta^k \wedge \theta^l = k\ \theta_i \wedge \theta_j \tag{89}$$

($\theta_i = \gamma_{ik}\theta^k$). The simply connected constant curvature spaces are in n dimensions the (n+1)-sphere S^{n+1} ($k = 1$), the Euclidean space ($k = 0$), and the pseudo-sphere ($k = -1$). Non-simply connected constant curvature spaces are obtained from these by forming quotients with respect to discrete isometry groups. (For detailed derivations, see [111].)

Curvature of Friedmann Spacetimes

Let $\{\bar{\theta}^i\}$ be any orthonormal triad on (Σ, γ). On this Riemannian space the first-structure equations read (we use the notation in [1]; quantities referring to this three-dimensional space are indicated by bars)

$$d\bar{\theta}^i + \bar{\omega}^i{}_j \wedge \bar{\theta}^j = 0\ . \tag{90}$$

On (M, g) we introduce the following orthonormal tetrad:

$$\theta^0 = dt, \quad \theta^i = a(t)\bar{\theta}^i\ . \tag{91}$$

From this and (90) we get

$$d\theta^0 = 0, \quad d\theta^i = \frac{\dot{a}}{a}\theta^0 \wedge \theta^i - a\ \bar{\omega}^i{}_j \wedge \bar{\theta}^j\ . \tag{92}$$

Comparing this with the first-structure equation for the Friedmann manifold implies

$$\omega^0{}_i \wedge \theta^i = 0, \quad \omega^i{}_0 \wedge \theta^0 + \omega^i{}_j \wedge \theta^j = \frac{\dot{a}}{a}\theta^i \wedge \theta^0 + a\ \bar{\omega}^i{}_j \wedge \bar{\theta}^j\ , \tag{93}$$

whence

$$\boxed{\omega^0{}_i = \frac{\dot{a}}{a}\ \theta^i, \quad \omega^i{}_j = \bar{\omega}^i{}_j\ .} \tag{94}$$

The worldlines of *comoving observers* are integral curves of the four-velocity field $u = \partial_t$. We claim that these are geodesics, i.e., that

$$\nabla_u u = 0\ . \tag{95}$$

To show this (and for other purposes) we introduce the basis $\{e_\mu\}$ of vector fields dual to (91). Since $u = e_0$ we have, using the connection forms (94),

$$\nabla_u u = \nabla_{e_0} e_0 = \omega^\lambda{}_0(e_0) e_\lambda = \omega^i{}_0(e_0) e_i = 0\ .$$

A.2 Einstein Equations for Friedmann Spacetimes

Inserting the connection forms (94) into the second-structure equations we readily find for the curvature 2-forms $\Omega^{\mu}{}_{\nu}$:

$$\Omega^0{}_i = \frac{\ddot{a}}{a}\theta^0 \wedge \theta^i, \quad \Omega^i{}_j = \frac{k + \dot{a}^2}{a^2}\theta^i \wedge \theta^j . \tag{96}$$

A routine calculation leads to the following components of the Einstein tensor relative to the basis (91)

$$G_{00} = 3\left(\frac{\dot{a}^2}{a^2} + \frac{k}{a^2}\right) , \tag{97}$$

$$G_{11} = G_{22} = G_{33} = -2\frac{\ddot{a}}{a} - \frac{\dot{a}^2}{a^2} - \frac{k}{a^2} , \tag{98}$$

$$G_{\mu\nu} = 0 \quad (\mu \neq \nu) . \tag{99}$$

In order to satisfy the field equations, the symmetries of $G_{\mu\nu}$ imply that the energy–momentum tensor *must* have the perfect fluid form (see [1], Sect. 1.4.2):

$$T^{\mu\nu} = (\rho + p)u^{\mu}u^{\nu} + pg^{\mu\nu} , \tag{100}$$

where u is the comoving velocity field introduced above.

Now, we can write down the field equations (including the cosmological term),

$$3\left(\frac{\dot{a}^2}{a^2} + \frac{k}{a^2}\right) = 8\pi G\rho + \Lambda , \tag{101}$$

$$-2\frac{\ddot{a}}{a} - \frac{\dot{a}^2}{a^2} - \frac{k}{a^2} = 8\pi Gp - \Lambda . \tag{102}$$

Although the 'energy–momentum conservation' does not provide an independent equation, it is useful to work this out. As expected, the momentum 'conservation' is automatically satisfied. For the 'energy conservation' we use the general form (see (1.37) in [1])

$$\nabla_u \rho = -(\rho + p)\nabla \cdot u . \tag{103}$$

In our case we have for the *expansion rate*

$$\nabla \cdot u = \omega^{\lambda}{}_0(e_{\lambda})u^0 = \omega^i{}_0(e_i) ,$$

thus with (94)

$$\nabla \cdot u = 3\frac{\dot{a}}{a} . \tag{104}$$

Therefore, (103) becomes

$$\dot{\rho} + 3\frac{\dot{a}}{a}(\rho + p) = 0 . \tag{105}$$

For a given equation of state, $p = p(\rho)$, we can use (105) in the form

$$\frac{d}{da}(\rho a^3) = -3pa^2 \tag{106}$$

to determine ρ as a function of the scale factor a. Examples: (1) For free massless particles (radiation) we have $p = \rho/3$, thus $\rho \propto a^{-4}$. (2) For dust ($p = 0$) we get $\rho \propto a^{-3}$.

With this knowledge the *Friedmann equation* (101) determines the time evolution of $a(t)$. It is easy to see that (102) follows from (101) and (105).

As an important consequence of (101) and (102) we obtain for the acceleration of the expansion

$$\ddot{a} = -\frac{4\pi G}{3}(\rho + 3p)a + \frac{1}{3}\Lambda a \ . \tag{107}$$

This shows that as long as $\rho + 3p$ is positive, the first term in (107) is decelerating, while a positive cosmological constant is repulsive. This becomes understandable if one writes the field equation as

$$G_{\mu\nu} = \kappa(T_{\mu\nu} + T^{\Lambda}_{\mu\nu}) \qquad (\kappa = 8\pi G) \ , \tag{108}$$

with

$$T^{\Lambda}_{\mu\nu} = -\frac{\Lambda}{8\pi G} g_{\mu\nu} \ . \tag{109}$$

This vacuum contribution has the form of the energy–momentum tensor of an ideal fluid, with energy density $\rho_\Lambda = \Lambda/8\pi G$ and pressure $p_\Lambda = -\rho_\Lambda$. Hence the combination $\rho_\Lambda + 3p_\Lambda$ is equal to $-2\rho_\Lambda$, and is thus negative. In what follows we shall often include in ρ and p the vacuum pieces.

A.3 Redshift

As a result of the expansion of the Universe the light of distant sources appears redshifted. The amount of redshift can be simply expressed in terms of the scale factor $a(t)$.

Consider two integral curves of the average velocity field u. We imagine that one describes the worldline of a distant comoving source and the other that of an observer at a telescope (see Fig. 9). Since light is propagating along null geodesics, we conclude from (85) that along the worldline of a light ray $dt = a(t)d\sigma$, where $d\sigma$ is the line element on the three-dimensional space (Σ, γ) of constant curvature $k = 0, \pm 1$. Hence the integral on the left of

$$\int_{t_e}^{t_o} \frac{dt}{a(t)} = \int_{source}^{obs.} d\sigma \ , \tag{110}$$

between the time of emission (t_e) and the arrival time at the observer (t_o) is independent of t_e and t_o. Therefore, if we consider a second light ray that is

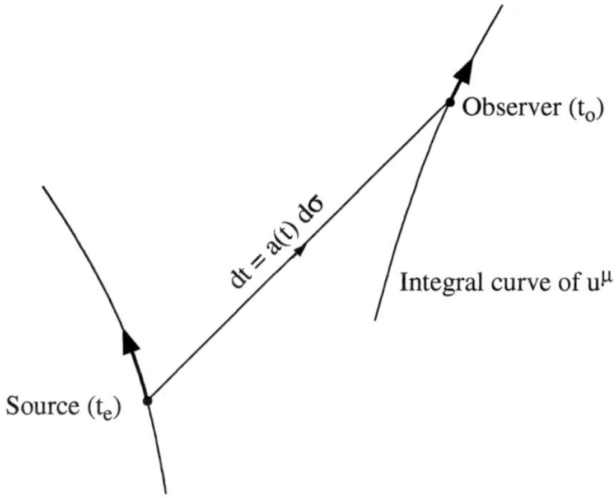

Fig. 9. Redshift for Friedmann models

emitted at the time $t_e + \Delta t_e$ and is received at the time $t_o + \Delta t_o$, we obtain from the last equation

$$\int_{t_e+\Delta t_e}^{t_o+\Delta t_o} \frac{dt}{a(t)} = \int_{t_e}^{t_o} \frac{dt}{a(t)} . \tag{111}$$

For a small Δt_e this gives

$$\frac{\Delta t_o}{a(t_o)} = \frac{\Delta t_e}{a(t_e)} .$$

The observed and the emitted frequences ν_o and ν_e, respectively, are thus related according to

$$\frac{\nu_o}{\nu_e} = \frac{\Delta t_e}{\Delta t_o} = \frac{a(t_e)}{a(t_o)} . \tag{112}$$

The redshift parameter z is defined by

$$z := \frac{\nu_e - \nu_o}{\nu_o} , \tag{113}$$

and is given by the key equation

$$\boxed{1 + z = \frac{a(t_o)}{a(t_e)}.} \tag{114}$$

One can also express this by the equation $\nu \cdot a = const$ along a null geodesic.

A.4 Cosmic Distance Measures

We now introduce a further important tool, namely operational definitions of three different distance measures, and show that they are related by simple redshift factors.

If D is the physical (proper) extension of a distant object and δ is its angle subtended, then the *angular diameter distance* D_A is defined by

$$D_A := D/\delta \,. \tag{115}$$

If the object is moving with the proper transversal velocity $V_\perp$ and with an apparent angular motion $d\delta/dt_0$, then the *proper-motion distance* is by definition

$$D_M := \frac{V_\perp}{d\delta/dt_0} \,. \tag{116}$$

Finally, if the object has the intrinsic luminosity $\mathcal{L}$ and $\mathcal{F}$ is the received energy flux then the *luminosity distance* is naturally defined as

$$D_L := (\mathcal{L}/4\pi\mathcal{F})^{1/2} \,. \tag{117}$$

Below we show that these three distances are related as follows

$$\boxed{D_L = (1+z)D_M = (1+z)^2 D_A.} \tag{118}$$

It will be useful to introduce on (Σ, γ) 'polar' coordinates (r, ϑ, φ), such that

$$\gamma = \frac{dr^2}{1-kr^2} + r^2 d\Omega^2, \quad d\Omega^2 = d\vartheta^2 + \sin^2\vartheta d\varphi^2 \,. \tag{119}$$

One easily verifies that the curvature forms of this metric satisfy (89). (This follows without doing any work by using in [1] the curvature forms (3.9) in the ansatz (3.3) for the Schwarzschild metric.)

To prove (118) we show that the three distances can be expressed as follows, if r_e denotes the comoving radial coordinate (in (119)) of the distant object and the observer is (without loss of generality) at $r = 0$.

$$D_A = r_e a(t_e), \quad D_M = r_e a(t_0), \quad D_L = r_e a(t_0)\frac{a(t_0)}{a(t_e)} \,. \tag{120}$$

Once this is established, (118) follows from (114).

From Fig. 10 and (119) we see that

$$D = a(t_e) r_e \delta \,, \tag{121}$$

hence the first equation in (120) holds.

To prove the second one we note that the source moves in a time dt_0 a proper transversal distance

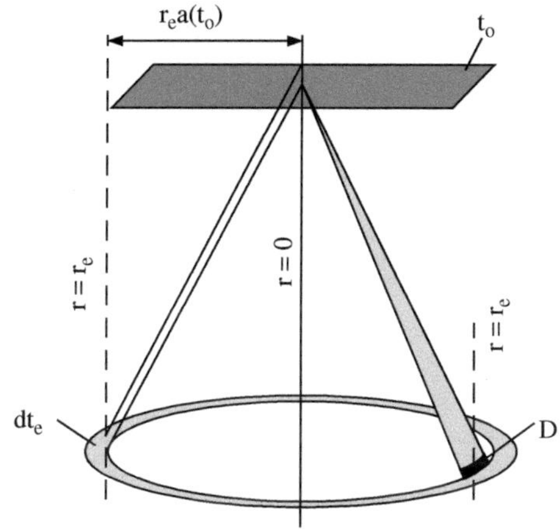

Fig. 10. Spacetime diagram for cosmic distance measures. The angular diameter distance $D_{ang} \equiv D_A$ and the luminosity distance $D_{lum} \equiv D_L$ have been introduced in this Appendix. The other two will be introduced in the Appendix C

$$dD = V_\perp dt_e = V_\perp dt_0 \frac{a(t_e)}{a(t_0)} .$$

Using again the metric (119) we see that the apparent angular motion is

$$d\delta = \frac{dD}{a(t_e) r_e} = \frac{V_\perp dt_0}{a(t_0) r_e} .$$

Inserting this into the definition (116) shows that the second equation in (120) holds. For the third equation we have to consider the observed energy flux. In a time dt_e the source emits an energy $\mathcal{L} dt_e$. This energy is redshifted to the present by a factor $a(t_e)/a(t_0)$, and is now distributed by (119) over a sphere with proper area $4\pi(r_e a(t_0))^2$ (see Fig. 10). Hence the received flux (*apparent luminosity*) is

$$\mathcal{F} = \mathcal{L} dt_e \frac{a(t_e)}{a(t_0)} \frac{1}{4\pi(r_e a(t_0))^2} \frac{1}{dt_0} ,$$

thus

$$\mathcal{F} = \frac{\mathcal{L} a^2(t_e)}{4\pi a^4(t_0) r_e^2} .$$

Inserting this into the definition (117) establishes the third equation in (120). For later applications we write the last equation in the more transparent form

$$\boxed{\mathcal{F} = \frac{\mathcal{L}}{4\pi(r_e a(t_0))^2} \frac{1}{(1+z)^2}.} \tag{122}$$

The last factor is due to redshift effects.

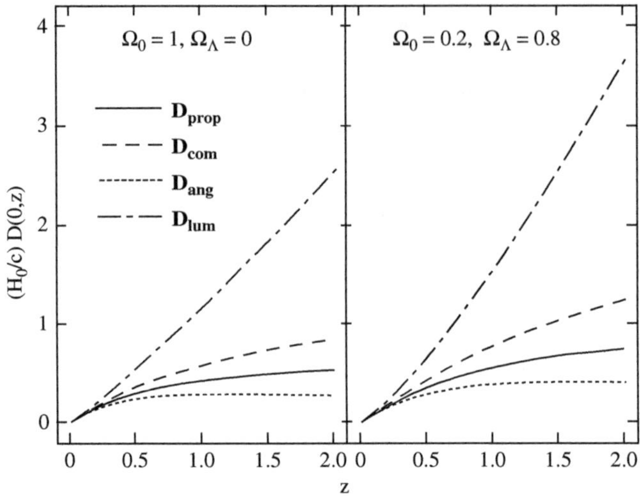

Fig. 11. Cosmological distance measures as a function of source redshift for two cosmological models

Two of the discussed distances as a function of z are shown in Fig. 11 for two Friedmann models with different cosmological parameters. The other two distance measures will be introduced in Appendix C.

B Thermal History below 100 MeV

B.1 Overview

Below the transition at about 200 MeV from a quark-gluon plasma to the confinement phase, the Universe was initially dominated by a complicated dense hadron soup. The abundance of pions, for example, was so high that they nearly overlapped. The pions, kaons, and other hadrons soon began to decay and most of the nucleons and antinucleons annihilated, leaving only a tiny baryon asymmetry. The energy density is then almost completely dominated by radiation and the stable leptons ($e^{\pm}$, the three neutrino flavors, and their antiparticles). For some time all these particles are in thermodynamic equilibrium. For this reason, only a few initial conditions have to be imposed. The Universe was never as simple as in this lepton era. (At this stage it is almost inconceivable that the complex world around us would eventually emerge.)

The first particles which freeze out of this equilibrium are the weakly interacting neutrinos. Let us estimate when this happened. The coupling of the neutrinos in the lepton era is dominated by the reactions:

$$e^- + e^+ \leftrightarrow \nu + \bar{\nu}\,, \quad e^{\pm} + \nu \to e^{\pm} + \nu\,, \quad e^{\pm} + \bar{\nu} \to e^{\pm} + \bar{\nu}\,.$$

For dimensional reasons, the cross sections are all of magnitude

$$\sigma \simeq G_F^2 T^2 \ , \tag{123}$$

where G_F is the Fermi coupling constant ($\hbar = c = k_B = 1$). Numerically, $G_F m_p^2 \simeq 10^{-5}$. On the other hand, the electron and neutrino densities n_e, n_ν are about T^3. For this reason, the reaction rates Γ for ν-scattering and ν-production per electron are of magnitude $c \cdot v \cdot n_e \simeq G_F^2 T^5$. This has to be compared with the expansion rate of the Universe

$$H = \frac{\dot{a}}{a} \simeq (G\rho)^{1/2} \ .$$

Since $\rho \simeq T^4$ we get

$$H \simeq G^{1/2} T^2 \ , \tag{124}$$

and thus

$$\frac{\Gamma}{H} \simeq G^{-1/2} G_F^2 T^3 \simeq (T/10^{10}\ K)^3 \ . \tag{125}$$

This ration is larger than 1 for $T > 10^{10}\ K \simeq 1$ MeV, and the neutrinos thus remain in thermodynamic equilibrium until the temperature has decreased to about 1 MeV. But even below this temperature the neutrinos remain Fermi distributed,

$$n_\nu(p)dp = \frac{1}{2\pi^2} \frac{1}{e^{p/T_\nu} + 1} p^2 dp \ , \tag{126}$$

as long as they can be treated as massless. The reason is that the number density decreases as a^{-3} and the momenta with a^{-1}. Because of this we also see that the neutrino temperature T_ν decreases after decoupling as a^{-1}. The same is, of course, true for photons. The reader will easily find out how the distribution evolves when neutrino masses are taken into account. (Since neutrino masses are so small this is only relevant at very late times.)

B.2 Chemical Potentials of the Leptons

The equilibrium reactions below 100 MeV, say, conserve several additive quantum numbers,[18] namely the electric charge Q, the baryon number B, and the three lepton numbers L_e, L_μ, L_τ. Correspondingly, there are five independent chemical potentials. Since particles and antiparticles can annihilate to photons, their chemical potentials are oppositely equal: $\mu_{e^-} = -\mu_{e^+}$, etc. From the following reactions

$$e^- + \mu^+ \rightarrow \nu_e + \bar{\nu}_\mu, \quad e^- + p \rightarrow \nu_e + n, \quad \mu^- + p \rightarrow \nu_\mu + n$$

[18] Even if B, L_e, L_μ, L_τ should not be strictly conserved, this is not relevant within a Hubble time H_0^{-1}.

we infer the equilibrium conditions

$$\mu_{e^-} - \mu_{\nu_e} = \mu_{\mu^-} - \mu_{\nu_\mu} = \mu_n - \mu_p \, . \tag{127}$$

As independent chemical potentials we can thus choose

$$\boxed{\mu_p, \, \mu_{e^-}, \, \mu_{\nu_e}, \, \mu_{\nu_\mu}, \, \mu_{\nu_\tau} .} \tag{128}$$

Because of local electric charge neutrality, the charge number density n_Q vanishes. From observations (see subsection E) we also know that the baryon number density n_b is much smaller than the photon number density ($\sim$ entropy density s_γ). The ratio n_B/s_γ remains constant for adiabatic expansion (both decrease with a^{-3}; see the next section). Moreover, the lepton number densities are

$$n_{L_e} = n_{e^-} + n_{\nu_e} - n_{e^+} - n_{\bar{\nu}_e}, \quad n_{L_\mu} = n_{\mu^-} + n_{\nu_\mu} - n_{\mu^+} - n_{\bar{\nu}_\mu}, \quad etc \, . \tag{129}$$

Since in the present Universe the number density of electrons is equal to that of the protons (bound or free), we know that after the disappearance of the muons $n_{e^-} \simeq n_{e^+}$ (recall $n_B \ll n_\gamma$), thus μ_{e^-} $(= -\mu_{e^+}) \simeq 0$. It is conceivable that the chemical potentials of the neutrinos and antineutrinos cannot be neglected, i.e., that n_{L_e} is not much smaller than the photon number density. In analogy to what we know about the baryon density we make the reasonable *asumption* that the lepton number densities are also much smaller than s_γ. Then we can take the chemical potentials of the neutrinos equal to zero ($|\mu_\nu|/kT \ll 1$). With what we said before, we can then put the five chemical potentials (128) equal to zero, because the charge number densities are all odd in them. Of course, n_B does not really vanish (otherwise we would not be here), but for the thermal history in the era we are considering they can be ignored.

B.3 Constancy of Entropy

Let ρ_{eq}, p_{eq} denote (in this subsection only) the total energy density and pressure of all particles in thermodynamic equilibrium. Since the chemical potentials of the leptons vanish, these quantities are only functions of the temperature T. According to the second law, the differential of the entropy $S(V,T)$ is given by

$$dS(V,T) = \frac{1}{T}[d(\rho_{eq}(T)V) + p_{eq}(T)dV] \, . \tag{130}$$

This implies

$$\begin{aligned} d(dS) = 0 &= d\left(\frac{1}{T}\right) \wedge d(\rho_{eq}(T)V) + d\left(\frac{p_{eq}(I)}{T}\right) \wedge dV \\ &= -\frac{\rho_{eq}}{T^2} dT \wedge dV + \frac{d}{dT}\left(\frac{p_{eq}(T)}{T}\right) dT \wedge dV \, , \end{aligned}$$

i.e., the Maxwell relation

$$\boxed{\frac{dp_{eq}(T)}{dT} = \frac{1}{T}[\rho_{eq}(T) + p_{eq}(T)] \ .} \tag{131}$$

If we use this in (130), we get

$$dS = d\left[\frac{V}{T}(\rho_{eq} + p_{eq})\right] ,$$

so the entropy density of the particles in equilibrium is

$$\boxed{s = \frac{1}{T}[\rho_{eq}(T) + p_{eq}(T)] \ .} \tag{132}$$

For an adiabatic expansion the entropy in a comoving volume remains constant:

$$S = a^3 s = \text{const} \ . \tag{133}$$

This constancy is equivalent to the energy equation (105) for the equilibrium part. Indeed, the latter can be written as

$$a^3 \frac{dp_{eq}}{dt} = \frac{d}{dt}[a^3(\rho_{eq} + p_{eq})] ,$$

and by (132) this is equivalent to $dS/dt = 0$.

In particular, we obtain for massless particles ($p = \rho/3$) from (131) again $\rho \propto T^4$ and from (132) that $S =$ constant implies $T \propto a^{-1}$.

Once the electrons and positrons have annihilated below $T \sim m_e$, the equilibrium components consist of photons, electrons, protons, and – after the big bang nucleosynthesis – of some light nuclei (mostly He^4). Since the charged particle number densities are much smaller than the photon number density, the photon temperature T_γ still decreases as a^{-1}. Let us show this formally. For this we consider beside the photons an ideal gas in thermodynamic equilibrium with the black body radiation. The total pressure and energy density are then (we use units with $\hbar = c = k_B = 1$; n is the number density of the non-relativistic gas particles with mass m):

$$p = nT + \frac{\pi^2}{45}T^4, \quad \rho = nm + \frac{nT}{\gamma - 1} + \frac{\pi^2}{15}T^4 \tag{134}$$

($\gamma = 5/3$ for a monoatomic gas). The conservation of the gas particles, $na^3 =$ const., together with the energy equation (106) implies, if $\sigma := s_\gamma/n$,

$$\frac{d\ln T}{d\ln a} = -\left[\frac{\sigma + 1}{\sigma + 1/3(\gamma - 1)}\right] .$$

For $\sigma \ll 1$ this gives the well-known relation $T \propto a^{3(\gamma-1)}$ for an adiabatic expansion of an ideal gas.

We are, however, dealing with the opposite situation $\sigma \gg 1$, and then we obtain, as expected, $a \cdot T = \text{const.}$

Let us look more closely at the famous ratio n_B/s_γ. We need

$$s_\gamma = \frac{4}{3T}\rho_\gamma = \frac{4\pi^2}{45}T^3 = 3.60 n_\gamma, \quad n_B = \rho_B/m_p = \Omega_B \rho_{crit}/m_p \,. \tag{135}$$

From the present value of $T_\gamma \simeq 2.7\ K$ and (30), $\rho_{crit} = 1.12 \times 10^{-5}\ h_0^2 (m_p/\text{cm}^3)$, we obtain as a measure for the baryon asymmetry of the Universe

$$\boxed{\frac{n_B}{s_\gamma} = 0.75 \times 10^{-8} (\Omega_B h_0^2)\,.} \tag{136}$$

It is one of the great challenges to explain this tiny number. So far, this has been achieved at best qualitatively in the framework of grand unified theories (GUTs).

B.4 Neutrino Temperature

During the electron–positron annihilation below $T = m_e$ the a-dependence is complicated, since the electrons can no more be treated as massless. We want to know at this point what the ratio T_γ/T_ν is after the annihilation. This can easily be obtained by using the constancy of comoving entropy for the photon–electron–positron system, which is sufficiently strongly coupled to maintain thermodynamic equilibrium.

We need the entropy for the electrons and positrons at $T \gg m_e$, long before annihilation begins. To compute this note the identity

$$\int_0^\infty \frac{x^n}{e^x - 1}dx - \int_0^\infty \frac{x^n}{e^x + 1}dx = 2\int_0^\infty \frac{x^n}{e^{2x} - 1}dx = \frac{1}{2^n}\int_0^\infty \frac{x^n}{e^x - 1}dx \,,$$

whence

$$\int_0^\infty \frac{x^n}{e^x + 1}dx = (1 - 2^{-n})\int_0^\infty \frac{x^n}{e^x - 1}dx \,. \tag{137}$$

In particular, we obtain for the entropies s_e, s_γ the following relation

$$s_e = \frac{7}{8}s_\gamma \quad (T \gg m_e)\,. \tag{138}$$

Equating the entropies for $T_\gamma \gg m_e$ and $T_\gamma \ll m_e$ gives

$$(T_\gamma a)^3\big|_{before}\left[1 + 2 \times \frac{7}{8}\right] = (T_\gamma a)^3\big|_{after} \times 1\,,$$

because the neutrino entropy is conserved. Therefore, we obtain

$$(aT_\gamma)\big|_{after} = \left(\frac{11}{4}\right)^{1/3}(aT_\gamma)\big|_{before}\,. \tag{139}$$

But $(aT_\nu)|_{after} = (aT_\nu)|_{before} = (aT_\gamma)|_{before}$, hence we obtain the important relation

$$\boxed{\left(\frac{T_\gamma}{T_\nu}\right)\bigg|_{after} = \left(\frac{11}{4}\right)^{1/3} = 1.401 \, .} \tag{140}$$

B.5 Epoch of Matter–Radiation Equality

In the main parts of these lectures the epoch when radiation (photons and neutrinos) have about the same energy density as non-relativistic matter (dark matter and baryons) plays a very important role. Let us determine the red-shift, z_{eq}, when there is equality.

For the three neutrino and antineutrino flavors the energy density is according to (137)

$$\rho_\nu = 3 \times \frac{7}{8} \times \left(\frac{4}{11}\right)^{4/3} \rho_\gamma \, . \tag{141}$$

Using

$$\frac{\rho_\gamma}{\rho_{crit}} = 2.47 \times 10^{-5} h_0^{-2} (1+z)^4 \, , \tag{142}$$

we obtain for the total radiation energy density, ρ_r,

$$\frac{\rho_r}{\rho_{crit}} = 4.15 \times 10^{-5} h_0^{-2} (1+z)^4 \, . \tag{143}$$

Equating this to

$$\frac{\rho_M}{\rho_{crit}} = \Omega_M (1+z)^3 \tag{144}$$

we obtain

$$\boxed{1 + z_{eq} = 2.4 \times 10^4 \Omega_M h_0^2 \, .} \tag{145}$$

Only a small fraction of Ω_M is baryonic. There are several methods to determine the fraction Ω_B in baryons. A traditional one comes from the abundances of the light elements. This is treated in most texts on cosmology. (German-speaking readers find a detailed discussion in my lecture notes [112], which are available in the Internet.) The comparison of the straightforward theory with observation gives a value in the range $\Omega_B h_0^2 = 0.021 \pm 0.002$. Other determinations are all compatible with this value. In Sect. 8 we shall obtain Ω_B from the CMB anisotropies. The striking agreement of different methods, sensitive to different physics, strongly supports our standard big bang picture of the Universe.

C Inflation and Primordial Power Spectra

C.1 Introduction

The horizon and flatness problems of standard big bang cosmology are so serious that the proposal of a very early accelerated expansion, preceding the hot era dominated by relativistic fluids, appears quite plausible. This general qualitative aspect of 'inflation' is now widely accepted. However, when it comes to concrete model building the situation is not satisfactory. Since we do not know the fundamental physics at superhigh energies not too far from the Planck scale, models of inflation are usually of a phenomenological nature. Most models consist of a number of scalar fields, including a suitable form for their potential. Usually there is no direct link to fundamental theories, like supergravity; however, there have been many attempts in this direction. For the time being, inflationary cosmology should be regarded as an attractive scenario, and not yet as a theory.

The most important aspect of inflationary cosmology is that *the generation of perturbations on large scales from initial quantum fluctuations is unavoidable and predictable.* For a given model these fluctuations can be calculated accurately, because they are tiny and cosmological perturbation theory can be applied. And, most importantly, these predictions can be *confronted with the cosmic microwave anisotropy measurements.* We are in the fortunate position to witness rapid progress in this field. The results from various experiments, most recently from WMAP, give already strong support of the basic predictions of inflation. Further experimental progress can be expected in the coming years.

C.2 The Horizon Problem and the General Idea of Inflation

I begin by describing the famous horizon puzzle, which is a very serious causality problem of standard big bang cosmology.

Past and Future Light Cone Distances

Consider our past light cone for a Friedmann spacetime model (Fig. 12). For a radial light ray the differential relation $dt = a(t)dr/(1-kr^2)^{1/2}$ holds for the coordinates (t, r) of the metric (19). The proper radius of the past light sphere at time t (cross section of the light cone with the hypersurface $\{t = const\}$) is

$$l_p(t) = a(t) \int_0^{r(t)} \frac{dr}{\sqrt{1-kr^2}} , \tag{146}$$

where the coordinate radius is determined by

$$\int_0^{r(t)} \frac{dr}{\sqrt{1-kr^2}} = \int_t^{t_0} \frac{dt'}{a(t')} . \tag{147}$$

Hence,

$$l_p(t) = a(t) \int_t^{t_0} \frac{dt'}{a(t')} . \tag{148}$$

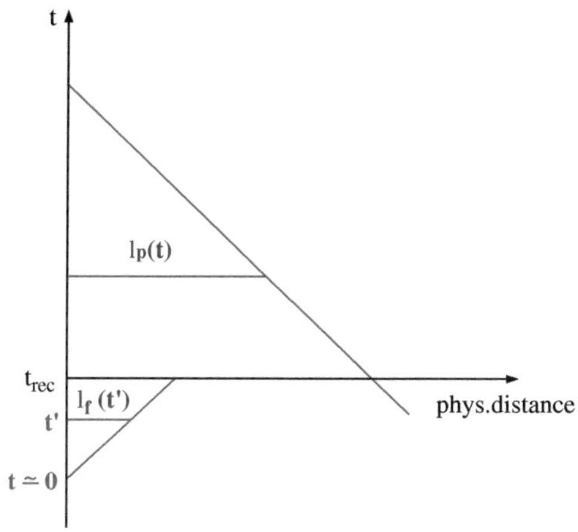

Fig. 12. Spacetime diagram illustrating the horizon problem

We rewrite this in terms of the redhift variable. From $1+z=a_0/a$ we get $dz=-(1+z)H\,dt$, so

$$\frac{dt}{dz}=-\frac{1}{H_0(1+z)E(z)}, \qquad H(z)=H_0E(z)\ .$$

Therefore,

$$l_p(z)=\frac{1}{H_0(1+z)}\int_0^z\frac{dz'}{E(z')}\ . \tag{149}$$

Similarly, the extension $l_f(t)$ of the forward light cone at time t of a very early event ($t\simeq 0,\ z\simeq\infty$) is

$$l_f(t)=a(t)\int_0^t\frac{dt'}{a(t')}=\frac{1}{H_0(1+z)}\int_z^\infty\frac{dz'}{E(z')}\ . \tag{150}$$

For the present Universe (t_0) this becomes what is called the *particle horizon distance*

$$D_{hor}=H_0^{-1}\int_0^\infty\frac{dz'}{E(z')}\ , \tag{151}$$

and gives the size of the *observable Universe.*

Analytical expressions for these distances are only available in special cases. For orientation we consider first the Einstein–de Sitter model ($K=0,\ \Omega_\Lambda=0,\ \Omega_M=1$), for which $a(t)=a_0(t/t_0)^{2/3}$ and thus

$$D_{hor}=3t_0=2H_0^{-1},\ \ l_f(t)=3t\ ,\ \ \frac{l_p}{l_f}=\left(\frac{t_0}{t}\right)^{1/3}-1=\sqrt{1+z}-1\ . \tag{152}$$

For a flat Universe a good fitting formula for cases of interest is (Hu and White)

$$D_{hor} \simeq 2H_0^{-1} \frac{1 + 0.084 \ln \Omega_M}{\sqrt{\Omega_M}} . \tag{153}$$

It is often convenient to work with 'comoving distances', by rescaling distances referring to time t (like $l_p(t), l_f(t)$) with the factor $a(t_0)/a(t) = 1 + z$ to the present. We indicate this by the superscript c. For instance,

$$l_p^c(z) = \frac{1}{H_0} \int_0^z \frac{dz'}{E(z')} . \tag{154}$$

This distance is plotted in Fig. 11 of Appendix A as $D_{com}(z)$. Note that for $a_0 = 1 : \ l_f^c(\eta) = \eta, \ l_p^c(\eta) = \eta_0 - \eta$. Hence (150) gives the following relation between η and z:

$$\eta = \frac{1}{H_0} \int_z^\infty \frac{dz'}{E(z')} .$$

The Number of Causality Distances on the Cosmic Photosphere

The number of causality distances at redshift z between two antipodal emission points is equal to $l_p(z)/l_f(z)$, and thus the ratio of the two integrals on the right of (149) and (150). We are particularly interested in this ratio at the time of last scattering with $z_{rec} \simeq 1100$. Then we can use for the numerator a flat Universe with non-relativistic matter, while for the denominator we can neglect in the standard hot big bang model Ω_K and Ω_Λ. A reasonable estimate is already obtained by using the simple expression in (152), i.e., $z_{rec}^{1/2} \approx 30$. A more accurate evaluation would increase this number to about 40. The length $l_f(z_{rec})$ subtends an angle of about 1 degree (exercise). How can it be that there is such a large number of causally disconnected regions we see on the microwave sky all having the same temperature? This is what is meant by the *horizon problem* and was a troublesome mystery before the invention of inflation.

Vacuum-Like Energy and Exponential Expansion

This causality problem is potentially avoided, if $l_f(t)$ would be increased in the very early Universe as a result of different physics. If a vacuum-like energy density would dominate, the Universe would undergo an *exponential expansion*. Indeed, in this case the Friedmann equation is

$$\left(\frac{\dot a}{a}\right)^2 + \frac{k}{a^2} = \frac{8\pi G}{3}\rho_{vac}, \quad \rho_{vac} \simeq \text{const} , \tag{155}$$

and has the solutions

$$a(t) \propto \begin{cases} \cosh H_{vac}t & : \quad k = 1 \\ e^{H_{vac}t} & : \quad k = 0 \\ \sinh H_{vac}t & : \quad k = 1 \,, \end{cases} \tag{156}$$

with

$$H_{vac} = \sqrt{\frac{8\pi G}{3}\rho_{vac}} \,. \tag{157}$$

Assume that such an exponential expansion starts for some reason at time t_i and ends at the *reheating time* t_e, after which standard expansion takes over. From

$$a(t) = a(t_i)e^{H_{vac}(t-t_i)} \quad (t_i < t < t_e) \,, \tag{158}$$

for $k = 0$ we get

$$l_f^c(t_e) \simeq a_0 \int_{t_i}^{t_e} \frac{dt}{a(t)} = \frac{a_0}{H_{vac}a(t_i)}\left(1 - e^{-H_{vac}\Delta t}\right) \simeq \frac{a_0}{H_{vac}a(t_i)} \,,$$

where $\Delta t := t_e - t_i$. We want to satisfy the condition $l_f^c(t_e) \gg l_p^c(t_e) \simeq H_0^{-1}$ (see (153)), i.e.,

$$a_i H_{vac} \ll a_0 H_0 \quad \Leftrightarrow \frac{a_i}{a_e} \ll \frac{a_0 H_0}{a_e H_{vac}} \tag{159}$$

or

$$e^{H_{vac}\Delta t} \gg \frac{a_e H_{vac}}{a_0 H_0} = \frac{H_{eq}a_{eq}}{H_0 a_0}\frac{H_{vac}a_e}{H_{eq}a_{eq}} \,.$$

Here, *eq* indicates the values at the time t_{eq} when the energy densities of non-relativistic and relativistic matter were equal. We now use the Friedmann equation for $k = 0$ and $w := p/\rho = \text{const}$. From (25) it follows that in this case

$$Ha \propto a^{-(1+3w)/2} \,,$$

and hence we arrive at

$$e^{H_{vac}\Delta t} \gg \left(\frac{a_0}{a_{eq}}\right)^{1/2}\left(\frac{a_{eq}}{a_e}\right) = (1+z_{eq})^{1/2}\left(\frac{T_e}{T_{eq}}\right) = (1+z_{eq})^{-1/2}\frac{T_{Pl}}{T_0}\frac{T_e}{T_{Pl}} \,, \tag{160}$$

where we used $aT = \text{const}$. So the number of e-folding periods during the inflationary period, $\mathcal{N} = H_{vac}\Delta t$, should satisfy

$$\mathcal{N} \gg \ln\left(\frac{T_{Pl}}{T_0}\right) - \frac{1}{2}\ln z_{eq} + \ln\left(\frac{T_e}{T_{Pl}}\right) \simeq 70 + \ln\left(\frac{T_e}{T_{Pl}}\right) \,. \tag{161}$$

For a typical GUT scale, $T_e \sim 10^{14}\ GeV$, we arrive at the condition $\mathcal{N} \gg 60$.

Such an exponential expansion would also solve the *flatness problem*, another worry of standard big bang cosmology. Let me recall how this problem arises.

The Friedmann equation (101) can be written as

$$(\Omega^{-1}-1)\rho a^2 = -\frac{3k}{8\pi G} = \text{const.}\,,$$

where

$$\Omega(t) := \frac{\rho(t)}{3H^2/8\pi G} \tag{162}$$

(ρ includes vacuum energy contributions). Thus

$$\Omega^{-1}-1 = (\Omega_0^{-1}-1)\frac{\rho_0 a_0^2}{\rho a^2}\,. \tag{163}$$

Without inflation we have

$$\rho = \rho_{eq}\left(\frac{a_{eq}}{a}\right)^4 \qquad (z > z_{eq})\,, \tag{164}$$

$$\rho = \rho_0\left(\frac{a_0}{a}\right)^3 \qquad (z < z_{eq})\,. \tag{165}$$

According to (26) z_{eq} is given by

$$1+z_{eq} = \frac{\Omega_M}{\Omega_R} \simeq 10^4\ \Omega_0 h_0^2\,. \tag{166}$$

For $z > z_{eq}$ we obtain from (163) and (164)

$$\Omega^{-1}-1 = (\Omega_0^{-1}-1)\frac{\rho_0 a_0^2}{\rho_{eq}a_{eq}^2}\frac{\rho_{eq}a_{eq}^2}{\rho a^2} = (\Omega_0^{-1}-1)(1+z_{eq})^{-1}\left(\frac{a}{a_{eq}}\right)^2 \tag{167}$$

or

$$\Omega^{-1}-1 = (\Omega_0^{-1}-1)(1+z_{eq})^{-1}\left(\frac{T_{eq}}{T}\right)^2 \simeq 10^{-60}(\Omega_0^{-1}-1)\left(\frac{T_{Pl}}{T}\right)^2\,. \tag{168}$$

Let us apply this equation for $T = 1\,\text{MeV}$, $\Omega_0 \simeq 0.2-0.3$. Then $|\,\Omega-1\,| \leq 10^{-15}$, thus the Universe was already incredibly flat at modest temperatures, not much higher than at the time of nucleosynthesis.

Such a fine tuning must have a physical reason. This is naturally provided by inflation, because our observable Universe could originate from a small patch at t_e. (A tiny part of the Earth surface is also practically flat.)

Beside the horizon scale $l_f(t)$, the *Hubble length* $H^{-1}(t) = a(t)/\dot{a}(t)$ plays also an important role. One might call this the "microphysics horizon", because this is the maximal distance microphysics can operate coherently in one expansion time. It is this length scale which enters in basic evolution equations, such as the equation of motion for a scalar field (see (175) below).

We sketch in Figs. 13–15 the various length scales in inflationary models, that is for models with a period of accelerated (e.g., exponential) expansion.

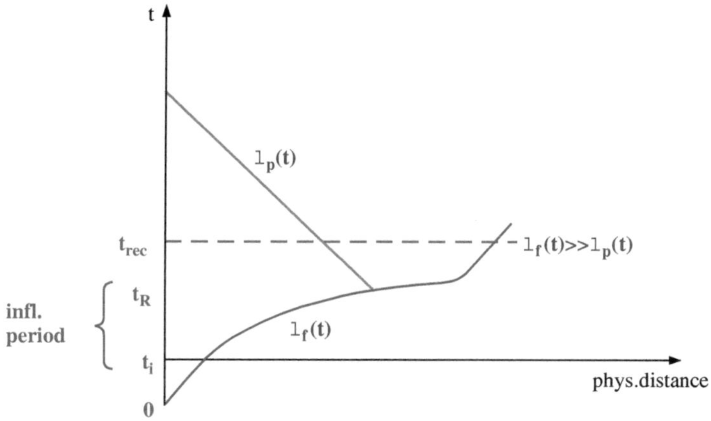

Fig. 13. Past and future light cones in models with an inflationary period

From these it is obvious that there can be – at least in principle – a *causal generation mechanism for perturbations*. This topic will be discussed in great detail in later parts of these lectures.

Exponential inflation is just an example. What we really need is an early phase during which the *comoving Hubble length decreases* (Fig. 15). This means that (for Friedmann spacetimes)

$$\boxed{\left(H^{-1}(t)/a\right)^{\cdot} < 0\,.} \tag{169}$$

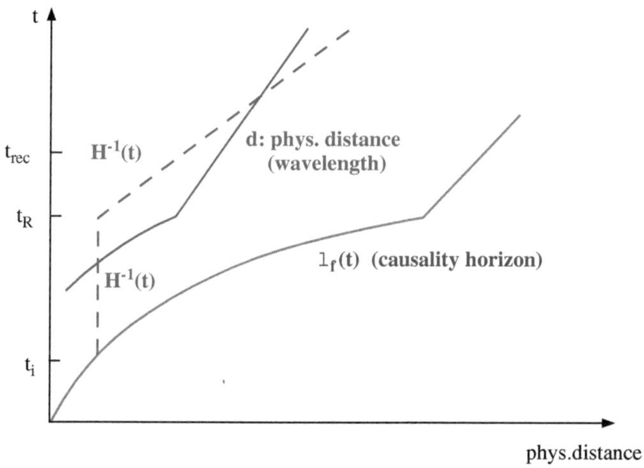

Fig. 14. Physical distance (e.g., between clusters of galaxies) and Hubble distance, and causality horizon in inflationary models

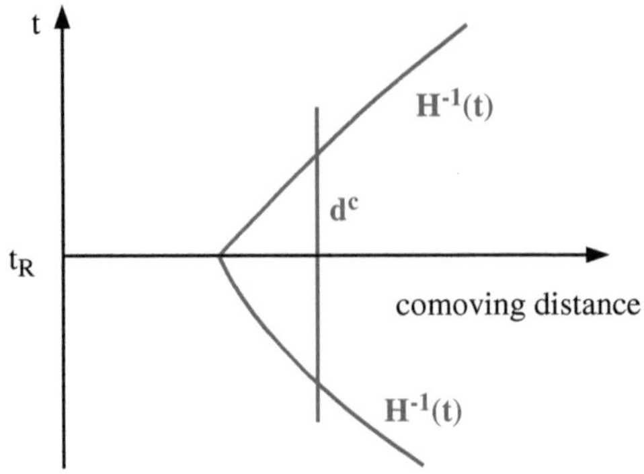

Fig. 15. Part of Fig. 14 expressed in terms of comoving distances

This is the *general definition of inflation*; equivalently, $\ddot{a} > 0$ (accelerated expansion). For a Friedmann model (107) tells us that

$$\ddot{a} > 0 \Leftrightarrow p < -\rho/3 \, . \tag{170}$$

This is, of course, not satisfied for 'ordinary' fluids.

Assume, as another example, *power-law inflation*: $a \propto t^p$. Then $\ddot{a} > 0 \Leftrightarrow p > 1$.

C.3 Scalar Field Models

Models with $p < -\rho/3$ are naturally obtained in scalar field theories. Most of the time we shall consider the simplest case of *one* neutral scalar field φ minimally coupled to gravity. Thus the Lagrangian density is assumed to be

$$\mathcal{L} = \frac{M_{pl}^2}{16\pi} R[g] - \frac{1}{2} \nabla_\mu \varphi \nabla^\mu \varphi - V(\varphi \, , \tag{171}$$

where $R[g]$ is the Ricci scalar for the metric g. The scalar field equation is

$$\Box \varphi = V_{,\varphi} \, , \tag{172}$$

and the energy–momentum tensor in the Einstein equation

$$G_{\mu\nu} = \frac{8\pi}{M_{Pl}^2} T_{\mu\nu} \tag{173}$$

is

$$T_{\mu\nu} = \nabla_\mu \varphi \nabla_\nu \varphi + g_{\mu\nu} \mathcal{L}_\varphi \tag{174}$$

($\mathcal{L}_\varphi$ is the scalar field part of (171)).

We consider first Friedmann spacetimes. Using previous notation, we obtain from (85)

$$\sqrt{-g} = a^3\sqrt{\gamma}, \quad \Box\varphi = \frac{1}{\sqrt{-g}}\partial_\mu(\sqrt{-g}g^{\mu\nu}\partial_\nu\varphi) = -\frac{1}{a^3}(a^3\dot{\varphi})^{\cdot} + \frac{1}{a^2}\triangle_\gamma\varphi\,.$$

The field equation (172) becomes

$$\boxed{\ddot{\varphi} + 3H\dot{\varphi} - \frac{1}{a^2}\triangle_\gamma\varphi = -V_{,\varphi}(\varphi)\,.} \tag{175}$$

Note that the expansion of the Universe induces a 'friction' term. In this basic equation one also sees the appearance of the Hubble length. From (174) we obtain for the energy density and the pressure of the scalar field

$$\rho_\varphi = T_{00} = \frac{1}{2}\dot{\varphi}^2 + V + \frac{1}{2a^2}(\nabla\varphi)^2\,, \tag{176}$$

$$p_\varphi = \frac{1}{3}T^i{}_i = \frac{1}{2}\dot{\varphi}^2 - V - \frac{1}{6a^2}(\nabla\varphi)^2\,. \tag{177}$$

(Here, $(\nabla\varphi)^2$ denotes the squared gradient on the 3-space (Σ,γ).)

Suppose the gradient terms can be neglected, and that φ is during a certain phase slowly varying in time, then we get

$$\rho_\varphi \approx V, \quad p_\varphi \approx -V\,. \tag{178}$$

Thus $p_\varphi \approx -\rho_\varphi$, as for a cosmological term.

Let us ignore for the time being the spatial inhomogeneities in the previous equations. Then these reduce to

$$\ddot{\varphi} + 3H\dot{\varphi} + V_{,\varphi}(\varphi) = 0\,; \tag{179}$$

$$\rho_\varphi = \frac{1}{2}\dot{\varphi}^2 + V, \quad p_\varphi = \frac{1}{2}\dot{\varphi}^2 - V\,. \tag{180}$$

Beside (179) the other dynamical equation is the Friedmann equation

$$\boxed{H^2 + \frac{K}{a^2} = \frac{8\pi}{3M_{Pl}^2}\left[\frac{1}{2}\dot{\varphi}^2 + V(\varphi)\right]\,.} \tag{181}$$

Equations (179) and (181) define a non-linear dynamical system for the dynamical variables $a(t), \varphi(t)$, which can be studied in detail (see, e.g., [113]).

Let us ignore the curvature term K/a^2 in (181). Differentiating this equation and using (179) shows that

$$\dot{H} = -\frac{4\pi}{M_{Pl}^2}\dot{\varphi}^2\,. \tag{182}$$

Regard H as a function of φ, then

$$\frac{dH}{d\varphi} = -\frac{4\pi}{M_{Pl}^2}\dot{\varphi}\,. \tag{183}$$

This allows us to write the Friedmann equation as

$$\left(\frac{dH}{d\varphi}\right)^2 - \frac{12\pi}{M_{Pl}^2}H^2(\varphi) = -\frac{32\pi^2}{M_{Pl}^4}V(\varphi) . \tag{184}$$

For a given potential $V(\varphi)$ this is a differential equation for $H(\varphi)$. Once this function is known, we obtain $\varphi(t)$ from (183) and $a(t)$ from (182).

C.4 Power-Law Inflation

We now proceed in the reverse order, assuming that $a(t)$ follows a power law

$$a(t) = \text{const. } t^p . \tag{185}$$

Then $H = p/t$, so by (182)

$$\dot{\varphi} = \sqrt{\frac{p}{4\pi}}M_{Pl}\frac{1}{t}, \quad \varphi(t) = \sqrt{\frac{p}{4\pi}}M_{Pl}\ln(t) + \text{const.} ,$$

hence

$$H \propto \exp\left(-\sqrt{\frac{4\pi}{p}}\frac{\varphi}{M_{Pl}}\right) . \tag{186}$$

Using this in (184) leads to an exponential potential

$$V(\varphi) = V_0 \exp\left(-4\sqrt{\frac{\pi}{p}}\frac{\varphi}{M_{Pl}}\right) . \tag{187}$$

C.5 Slow-Roll Approximation

An important class of solutions is obtained in the slow-roll approximation (SLA), in which the basic (179) and (181) can be replaced by

$$H^2 = \frac{8\pi}{3M_{Pl}^2}V(\varphi) , \tag{188}$$

$$3H\dot{\varphi} = -V_{,\varphi} . \tag{189}$$

A necessary condition for their validity is that the *slow-roll parameters*

$$\varepsilon_V(\varphi) := \frac{M_{Pl}^2}{16\pi}\left(\frac{V_{,\varphi}}{V}\right)^2 , \tag{190}$$

$$\eta_V(\varphi) := \frac{M_{Pl}^2}{8\pi}\frac{V_{,\varphi\varphi}}{V} \tag{191}$$

are small:

$$\varepsilon_V \ll 1, \quad |\, \eta_V \,| \ll 1 . \tag{192}$$

These conditions, which guarantee that the potential is flat, are, however, not sufficient.

The simplified system (188) and (189) implies

$$\boxed{\dot{\varphi}^2 = \frac{M_{Pl}^2}{24\pi}\frac{1}{V}\left(V_{,\varphi}\right)^2 .} \tag{193}$$

This is a differential equation for $\varphi(t)$.

Let us consider potentials of the form

$$V(\varphi) = \frac{\lambda}{n}\varphi^n . \tag{194}$$

Then (193) becomes

$$\boxed{\dot{\varphi}^2 = \frac{n^2 M_{Pl}^2}{24\pi}\frac{1}{\varphi^2}V .} \tag{195}$$

Hence, (188) implies

$$\frac{\dot{a}}{a} = -\frac{4\pi}{nM_{Pl}^2}(\varphi^2)^{\cdot} ,$$

and so

$$\boxed{a(t) = a_0 \exp\left[\frac{4\pi}{nM_{Pl}^2}(\varphi_0^2 - \varphi^2(t))\right] .} \tag{196}$$

We see from (195) that $\frac{1}{2}\dot{\varphi}^2 \ll V(\varphi)$ for

$$\varphi \gg \frac{n}{4\sqrt{3\pi}}M_{Pl} . \tag{197}$$

Consider first the example $n = 4$. Then (195) implies

$$\frac{\dot{\varphi}}{\varphi} = \sqrt{\frac{\lambda}{6\pi}}M_{Pl} \Rightarrow \varphi(t) = \varphi_0 \exp\left(-\sqrt{\frac{\lambda}{6\pi}}M_{Pl}\, t\right) . \tag{198}$$

For $n \neq 4$:

$$\varphi(t)^{2-n/2} = \varphi_0^{2-n/2} + t\left(2 - \frac{n}{2}\right)\sqrt{\frac{n\lambda}{24\pi}}M_{Pl}^{3-n/2} . \tag{199}$$

For the special case $n = 2$ this gives, using the notation $V = \frac{1}{2}m^2\varphi^2$, the simple result

$$\varphi(t) = \varphi_0 - \frac{mM_{Pl}}{2\sqrt{3\pi}}t . \tag{200}$$

Inserting this into (196) provides the time dependence of $a(t)$.

C.6 Why Did Inflation Start?

Attempts to answer this and related questions are *very speculative* indeed. A reasonable direction is to imagine random initial conditions and try to understand how inflation can emerge, perhaps generically, from such a state of matter. A. Linde first discussed a scenario along these lines which he called *chaotic inflation.* In the context of a single scalar field model he argued that typical initial conditions correspond to $\frac{1}{2}\dot{\varphi}^2 \sim \frac{1}{2}(\partial_i\varphi)^2 \sim V(\varphi) \sim 1$ (in Planckian units). The chance that the potential energy dominates in some domain of size $> \mathcal{O}(1)$ is presumably not very small. In this situation inflation could begin and $V(\varphi)$ would rapidly become even more dominant, which ensures continuation of inflation. Linde concluded from such considerations that chaotic inflation occurs under rather natural initial conditions. For this to happen, the form of the potential $V(\varphi)$ can even be a simple power law of the form (194). Many questions remain, however, open.

The chaotic inflationary Universe will look on very large scales – much larger than the present Hubble radius – extremely inhomogeneous. For a review of this scenario I refer to [114]. A much more extended discussion of inflationary models, including references, can be found in [107].

C.7 Inflation and Primordial Power Spectra

For a detailed derivation of the primordial power spectra that are generated as a result of quantum fluctuations during an inflationary period, I refer to my Combo-lectures [63].

The main steps are quite straightforward. First, one studies classical perturbations of the scalar field and the metric. For the scalar field one can reduce the problem to a Klein–Gordon equation with a time-dependent mass for a suitable gauge invariant perturbation amplitude. The quantization of this field follows standard rules. The quantization of the scalar part of the metric (Bardeen potentials) is then also fixed. Of particular interest is the power spectrum, $P_{\mathcal{R}}(k)$, of the so-called "curvature perturbation amplitude" $\mathcal{R}$. This is proportional to the Fourier transform of the two-point correlation function. More precisely, if

$$\mathcal{R}(\eta, \mathbf{x}) = (2\pi)^{-3/2} \int \mathcal{R}_{\mathbf{k}}(\eta) e^{i\mathbf{k}\cdot\mathbf{x}} d^3k \,,$$

then

$$\langle 0|\mathcal{R}_{\mathbf{k}}\mathcal{R}^{\dagger}_{\mathbf{k}'}|0\rangle =: \frac{2\pi^2}{k^3} P_{\mathcal{R}}(k)\delta^{(3)}(\mathbf{k}-\mathbf{k}') \,.$$

In the slow-roll approximation, this can be worked out explicitly, with the result

$$\mathcal{P}_{\mathcal{R}}(k) = \frac{4}{M_{Pl}^4} \frac{H^4}{(dH/d\varphi)^2}\Big|_{k=aH} \tag{201}$$

$$\simeq \frac{128\pi}{3} \frac{1}{M_{Pl}^6} \frac{U^3}{(U_{,\varphi})^2}\Big|_{k=aH} \,. \tag{202}$$

The expression on the right is evaluated at *horizon crossing* $k = aH$.

It is even simpler to determine the *power spectrum of gravitational waves* (tensor modes). In the same approximation one finds

$$\mathcal{P}_g(k) = \frac{16}{\pi} \frac{H^2}{M_{Pl}^2}\bigg|_{k=aH}, \quad H^2 \simeq \frac{8\pi}{3M_{Pl}^2} U . \tag{203}$$

For a given inflationary model, the power spectra are uniquely determined.

There is one delicate question, namely why we have chosen in the definition of the power spectrum the Fock state, relative to modes that at very short distances ($k/aH \to \infty$) approach the plane waves of the gravity free case with positive frequences. A priori, the initial state could contain all kinds of excitations. These would, however, be redshifted away by the enormous inflationary expansion, and the final power spectrum on interesting scales, much larger than the Hubble length, should be largely independent of possible initial excitations.

For a comparison with observations the power index, n_s, for scalar perturbations, defined by

$$n_s - 1 := \frac{d \ln P_{\mathcal{R}(k)}}{d \ln k} \tag{204}$$

is of particular interest. In terms of the slow-roll parameters (190) and (191) it is given by

$$n_s - 1 = -6\varepsilon_U + 2\eta_U , \tag{205}$$

whence the spectrum is nearly scale-free. For the ratio r of the amplitudes of P_g and $P_{\mathcal{R}}$ one finds $r = 16\varepsilon_U$. The WMAP data match the basic inflationary predictions, and are even well fit by the simplest model $U \propto \varphi^2$.

D Quintessence Models

In quintessence models the exotic missing energy with negative pressure is again described by a scalar field, whose potential is chosen such that the energy density of the homogeneous scalar field adjusts itself to be comparable to the matter density today for quite generic initial conditions, and is dominated by the potential energy. This ensures that the pressure becomes sufficiently negative. It is not simple to implement this general idea such that the model is phenomenologically viable. For instance, the success of BBN should not be spoiled. CMB and large-scale structure impose other constraints. One also would like to understand why cosmological acceleration started at about $z \sim 1$, and not much earlier or in the far future. There have been attempts to connect this with some characteristic events in the post-recombination Universe. On a fundamental level, the origin of a quintessence field that must be extremely weakly coupled to ordinary matter remains in the dark.

Let me briefly describe a simple model of this kind [115]. For the dynamics of the scalar field ϕ we adopt an exponential potential

$$V = V_0\, e^{-\lambda\phi/M_P} .$$

Such potentials often arise in Kaluza–Klein and string theories. Matter is described by a fluid with a baryotropic equation of state: $p_f = (\gamma - 1)\rho_f$.

For a Friedmann model with zero space-curvature, one can cast the basic equations into an autonomous two-dimensional dynamical system for the quantities

$$x(\tau) = \frac{\kappa\dot{\phi}}{\sqrt{6}H}, \quad y(\tau) = \frac{\kappa\sqrt{V}}{\sqrt{3}H} ,$$

where

$$H = \dot{a}/a, \quad \tau = \log a, \quad \kappa^2 = 8\pi G$$

($a(t)$ is the scalar factor). This system of autonomous differential equations has the form

$$\frac{dx}{d\tau} = f(x, y; \lambda, \gamma), \quad \frac{dy}{d\tau} = g(x, y; \lambda, \gamma) ,$$

where f and g are polynomials in x and y of third degree, which depend parametrically on λ and γ. The density parameters Ω_ϕ and Ω_f for the field ϕ and the fluid are given by

$$\Omega_\phi = x^2 + y^2, \quad \Omega_\phi + \Omega_f = 1 .$$

The interesting fact is that, for a large domain of the parameters λ, γ, the phase portrait has qualitatively the shape of Fig. 16. Therefore, under generic

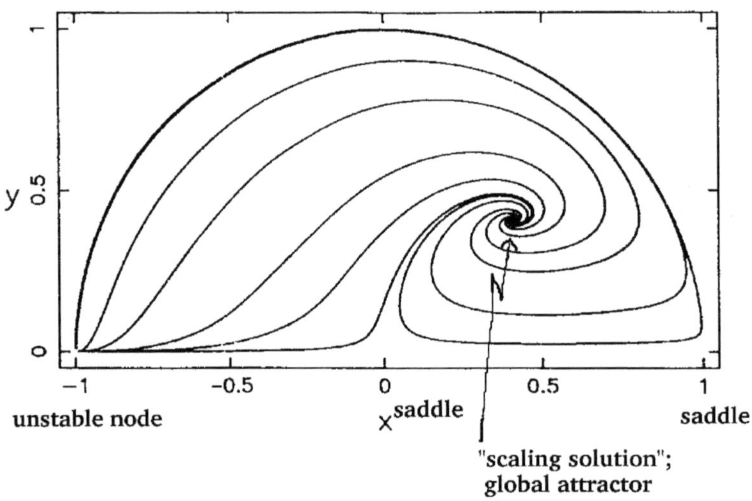

Fig. 16. Phase plane for $\gamma = 1$, $\lambda = 3$. The late-time attractor is the scaling solution with $x = y = 1/\sqrt{6}$ (from [115])

initial conditions, there is a global attractor (a node or a spiral) for which $\Omega_\phi = 3\gamma/\lambda^2$. For this "scaling solution" Ω_ϕ/Ω_f remains fixed, and for any other solution this ration is finally approached.

Unfortunately, if we set $p_\phi = (\gamma_\phi - 1)$ we find that $\gamma_\phi = 2x^2/(x^2+y^2)$, and this is equal to γ for the scaling solution. Thus this does *not* correspond to a quintessence solution. Moreover, the condition that ρ_ϕ should be subdominant during nucleosynthesis implies a small value for Ω_ϕ.

A more successful example of a so-called "tracker potential", with the property that the scalar field approaches a common evolutionary path from a wide range of initial conditions, has the form of an inverse power law, $V(\phi) = V_0/\phi^\alpha$ [117]. There is an extended literature on the subject. References [116]–[121] give a small selection of important early papers. For a recent review that describes also other scalar field models, see [122]. I emphasize once more that on the basis of the vacuum energy problem we would expect a huge additive constant for the quintessence potential that would destroy the whole picture. Thus, assuming for instance that the potential approaches zero as the scalar field goes to infinity has (so far) no basis. Apart of this and other fine tuning problems, I doubt that this kind of phenomenological models – with no natural field theoretical justification – will lead to an understanding of dark energy at a deeper level.

Fortunately, future more precise observations will allow us to decide whether the presently dominating exotic energy density satisfies $p/\rho = -1$ or whether this ratio is somewhere between -1 and $-1/3$. Recent studies (see [71, 72], and references therein), which make use of existing cosmological data, do not yet support quintessence. The restrictions for a possible redshift dependence are, so far, rather weak.

If convincing evidence for such a dependence should be established, we will not be able to predict the distant future of the Universe. Eventually, the dark energy density may perhaps become negative. This illustrates that we may *never be able to predict* the asymptotic behavior of the most grandiose of all dynamical systems. Other conclusions are left to the reader.

References

1. N. Straumann, *General Relativity, With Applications to Astrophysics*, Texts and Monographs in Physics, Springer Verlag, 2004.
2. G.E. Lemaître, in *Albert Einstein: Philosopher-Scientist*, P.A. Schilpp, ed., Illinois: The Library of Living Philosophers (1949).
3. N. Straumann, *On the Cosmological Constant Problems and the Astronomical Evidence for a Homogeneous Energy Density with Negative Pressure*, in *Poincarè Seminar 2002, Vacuum Energy – Renormalization*, B. Duplantier, and V. Rivasseau, eds.; Birkhäuser-Verlag 2003, pp. 7–51; astro-ph/0203330.
4. N. Straumann, *The History of the Cosmological Constant Problem*, in *On the Nature of Dark Energy*, IAP Astrophysics Colloquium 2002, Frontier Group, 2003, p. 17; gr-qc/0208027.

5. A. Einstein, *Sitzungsber. Preuss. Akad. Wiss.* phys.-math. Klasse VI, 142 (1917). See also: [6], Vol. 6, p. 540, Doc. 43.
6. A. Einstein, *The Collected Papers of Albert Einstein*, Vols. 1–9, Princeton University Press, 1987–. See also: [http:www. einstein.caltech.edu/].
7. A. Einstein, *On the Foundations of the General Theory of Relativity.* Reference [4], Vol. 7, Doc. 4.
8. A. Pais, *'Subtle is the Lord...': The Science and the Life of Albert Einstein.* Oxford University Press (1982). See especially Sect. 15e.
9. W. de Sitter, Proc. Acad. Sci., **19**, 1217 (1917); and **20**, 229 (1917).
10. A.S. Eddington, *The Mathematical Theory of Relativity.*Chelsea Publishing Company (1924). Third (unaltered) Edition (1975). See especially Sect. 70.
11. Letter from Hermann Weyl to Felix Klein, 7 February 1919; see also [5], Vol. 8, Part B, Doc. 567.
12. H. Weyl, Phys. Zeits. **24**, 230, (1923); Phil. Mag. **9**, 923 (1930).
13. C. Lanczos, Phys. Zeits. **23**, 539 (1922).
14. C. Lanczos, Zeits. f. Physik **17**, 168 (1923).
15. A. Friedmann, Z.Phys. **10**, 377 (1922); **21**, 326 (1924).
16. G.E. Lemaître, Ann. Soc. Sci. Brux. A **47**, 49 (1927).
17. G.E. Lemaître, Monthly Not. Roy. Astron. Soc. **91**, 483 (1931).
18. A. Einstein, S.B. Preuss. Akad. Wiss. (1931), 235.
19. A. Einstein, Appendix to the 2nd edn. of *The Meaning of Relativity*, (1945); reprinted in all later editions.
20. W. Pauli, *Theory of Relativity.* Pergamon Press (1958); Supplementary Note **19**.
21. O. Heckmann, *Theorien der Kosmologie*, berichtigter Nachdruck, Springer-Verlag (1968).
22. V. Petrosian, E.E. Salpeter, and P. Szekeres, Astrophys. J. **147**, 1222 (1967).
23. W. Pauli, *Pauli Lectures on Physics*; Ed. C.P. Enz. MIT Press (1973); Vol. 4, especially Sect. 20.
24. C.P. Enz, and A. Thellung, Helv. Phys. Acta **33**, 839 (1960).
25. W. Pauli, *Die allgemeinen Prinzipien der Wellenmechanik.* Handbuch der Physik, Vol. XXIV (1933). New edition by N. Straumann, Springer-Verlag (1990); see Appendix III, p. 202.
26. W. Heisenberg and H. Euler, Z. Phys. **38**, 714 (1936).
27. V.S. Weisskopf, Kongelige Danske Videnskabernes Selskab, Mathematisk-fysiske Meddelelser XIV, No.6 (1936).
28. M. Bordag, U. Mohideen, and V.M. Mostepanenko *New Developments in the Casimir Effect*, quant-ph/0106045.
29. C. Beck and M.C. Mackey, Phys. Lett. **B 605**, 295 (2005); astro-ph/0406504.
30. Ph. Jetzer and N. Straumann, Phys. Lett. **B 606**, 77 (2005); astro-ph/0411034.
31. C. Beck and M. C. Mackey, astro-ph/0603397.
32. Ph. Jetzer and N. Straumann, Phys. Lett. **B 639**, 57 (2006); [astro-ph/0604522].
33. Y.B. Zel'dovich, JETP letters **6**, 316 (1967); Soviet Physics Uspekhi **11**, 381 (1968).
34. C.G. Callan, S. Coleman, and R. Jackiw, Ann. Phys. **59**, 42 (1970).
35. T. Schäfer and E.V. Shuryak, Rev. Mod. Phys. **70**, 323 (1998).
36. R.D. Peccei and H. Quinn, Phys. Rev. Lett, **38**, 1440 (1977); Phys. Rev. **D16**, 1791 (1977).

37. E. Witten, hep-ph/0002297.
38. S.M. Caroll, *Living Reviews in Relativity*, astro-ph/0004075.
39. N. Straumann, in *Dark Matter in Astro- and Particle Physics*, Edited by H.V.Klapdor-Kleingrothaus, Springer (2001), p. 110.
40. S.E. Rugh and H. Zinkernagel, hep-th/0012253.
41. T. Padmanabhan, Phys. Repts. **380**, 235 (2003).
42. W. Baade, Astrophys. J. **88**, 285 (1938).
43. A. Sandage, G.A. Tammann, A. Saha, B. Reindl, F.D. Machetto and Panagia, astro-ph/0603647; A. Saha, F. Thim, G.A. Tammann, B. Reindl, A. Sandage, astro-ph/0602572.
44. G.A. Tammann. In *Astronomical Uses of the Space Telescope*, eds. F. Macchetto, F. Pacini and M. Tarenghi, p. 329. Garching: ESO.
45. S. Colgate, Astrophys. J. **232**,404 (1979).
46. SCP-Homepage: http://www-supernova.LBL.gov
47. HZT-Homepage: http://cfa-www.harvard.edu/cfa/oir/Research/supernova/HighZ.html
48. *Cosmic Explosions*, eds. S. Holt and W. Zhang, AIP Conference Proceedings 522, American Institute of Physics (New York) (2000).
49. W. Hillebrandt and J.C. Niemeyer, Ann. Rev. Astron. Astrophys. **38**, 191–230 (2000).
50. B. Leibundgut, Astron. Astrophys. **10**, 179 (2000).
51. S. Perlmutter, et al., Astrophys. J. **517**, 565 (1999).
52. B. Schmidt, et al., Astrophys. J. **507**, 46 (1998).
53. A.G. Riess, et al., Astron. J. **116**, 1009 (1998).
54. B. Leibundgut, Ann. Rev. Astron. Astrophys. **39**, 67 (2001).
55. A.V. Filippenko, *Measuring and Modeling the Universe*, ed. W.L. Freedman, Cambridge University Press, (2004); astro-ph/0307139.
56. A.G. Riess, et al., Astrophys. J. **607**, 665 (2004); astro-ph/0402512.
57. P. Astier, et al., Astron. Astrophys. **447**, 31 (2006) [astro-ph/0510447].
58. A. Clocchiatti, et al., Astron. Astrophys. **447**, 31 (2006); [astro-ph/0510155].
59. Snap-Homepage: http://snap.lbl.gov
60. R. Kantowski and R.C. Thomas, Astrophys. J. **561**, 491 (2001); astro-ph/0011176.
61. W. Hu and S. Dodelson, Annu. Rev. Astron. Astrophys. **40**, 171–216 (2002).
62. U. Seljak, and M. Zaldarriaga, Astrophys. J. **469**, 437 (1996). (See also *http://www.sns.ias.edu/matiasz/CMBFAST/cmbfast.html*)
63. N. Straumann, *From primordial quantum fluctuations to the anisotropies of the cosmic microwave background radiation*, Ann. Phys. (Leipzig) **15**, No. 10–11, 701–847 (2006); [hep-ph/0505249].
64. W. Hu and M. White, Phys. Rev. D **56**, 596(1997).
65. W. Hu, U. Seljak, M. White, and M. Zaldarriaga, Phys. Rev. D **57**, 3290 (1998).
66. G. Steigman, Int.J.Mod.Phys. **E15**, 1 (2006); astro-ph/0511534.
67. C.L. Bennett, et al., ApJS **148**, 1 (2003); ApJS **148**, 97 (2003).
68. D.N. Spergel, et al., ApJS **148** 175 (2003).
69. D.N. Spergel, et al., astro-ph/0603449.
70. L. Page et al., astro-ph/0603450.
71. M. Tegmark et al., Phys. Rev. **D69**, 103501 (2004); astro-ph/0310723.
72. U. Seljak, et al., Phys. Rev. **D71**, 103515 (2005); astro-ph/0407372; astro-ph/0604335.

73. S. Cole, et al., MNRAS, **362**, 505 (2005).
74. A. Blanchard, M. Douspis, M. Rowan-Robinson, and S. Sarkar, Astron.Astrophys. **412**, 35 (2003); astro-ph/0304237.
75. A. Blanchard, M. Douspis, M. Rowan-Robinson, and S. Sarkar, astro-ph/0512085.
76. E. Barausse, S. Matarrese, and A. Riotto, Phys. Rev. **D71**,0635537 (2005); astro-ph/0501152.
77. E. W. Kolb, S. Matarrese, A. Notari, and A. Riotto, hep-th/0503117.
78. Ch. M. Hirata and U. Seljak, Phys. Rev. **D72**, 083501 (2005); astro-ph/0503582.
79. E. W. Kolb, S. Matarrese, and A. Riotto, astro-ph/0506534.
80. S. Räsänen, astro-ph/0607626.
81. M. Sasaki, Mon. Not. R. Astron. Soc. **228**, 653 (1987).
82. N. Sigiura, N. Sugiyama, and M. Sasaki, Prog.Theo. Phys. **101**, 903 (1999).
83. C. Bonvin, R. Durrer and M.A. Gasparini, Phys. Rev. **D73**, 023523 (2006); astro-ph/0511183.
84. K. Bolejko, astro-ph/0512103.
85. S. Nojiri and S.D. Odintsov, hep-th/0601213.
86. A.L. Erickcek, T.L. Smith, and M. Kamiokowski, astro-ph/0610483.
87. T. Chiba, Phys. Lett. **B575**, 1 (2003); astro-ph/0307338.
88. S. Caroll, A. De Felice, V. Duvvuri, D. Easson, M. Trodden and M. Turner, Phys. Rev. **D70**, 063513 (2005).
89. R.P. Woodard, astro-ph/0601672.
90. A. De Felice, M. Hindmarsh, and M. Trodden, astro-ph/0604154.
91. G. Calcagni, B. de Carlos and A. De Felice, hep-th/0604201.
92. G. Velo and D. Zwanziger, Phys. Rev. **186**, 1337–41 (1969) ; Phys. Rev. **188**, 2218–22 (1969).
93. T.P. Sotiriou, gr-qc/0604028.
94. T. Koivisto and H. Kurki-Suonio, astro-ph/0509422.
95. T. Koivisto, astro-ph/0602031.
96. M. Amarzguioui, O. Elgaroy, D.F. Mota and T. Multmaki, astro-ph/0510519.
97. G.R. Dvali, G. Gabadadze and M. Porrati, Phys. Lett. B **485**, 208 (2000); hep-th/0005016.
98. R. Maartens, *Living Reviews*, gr-qc/0312059.
99. R. Maartens and E. Majerotto, astro-ph/0603353.
100. M. Fairbairn and A. Goobar, astro-ph/0511029.
101. D. J. Eisenstein et al., *Astrophys. J.* **633**, 560 (2005).
102. D. Gorbunov, K. Koyama, and S. Sibiryakov, Phys. Rev. **D73**, 044016 (2006).
103. Ch. Charmousis, R. Gregory, N. Kaloper and A. Padilla, hep-th/0604086.
104. A. Adams, N. Arkani-Hamed, S. Dubovsky, A. Nicolis, and R. Rattazzi, hep-th/0602178.
105. P.J.E. Peebles, *Principles of Physical Cosmology.* Princeton University Press 1993.
106. J.A. Peacock, *Cosmological Physics.* Cambridge University Press 1999.
107. A.R. Liddle and D.H. Lyth, *Cosmological Inflation and Large Scale Structure.* Cambridge University Press 2000.
108. S. Dodelson, *Modern Cosmology.* Academic Press 2003.
109. G. Börner, *The Early Universe.* Springer-Verlag 2003 (4th edition).
110. N. Straumann, Helv. Phys. Acta **45**, 1089 (1972).

111. N. Straumann, *Allgemeine Relativitätstheorie und Relativistische Astrophysik*, 2. Auflage, Lecture Notes in Physics, Vol. 150, Springer-Verlag (1988); Kap. IX.
112. N. Straumann, *Kosmologie I*, Vorlesungsskript, http://web.unispital.ch/neurologie/vest/homepages/straumann/norbert
113. V.A. Belinsky, L.P. Grishchuk, I.M. Khalatnikov, and Ya.B. Zeldovich, Phys. Lett. **155B**, 232 (1985).
114. A. Linde, *Lectures on Inflationary Cosmology*, hep-th/9410082.
115. E.J. Copeland, A.R. Liddle, and D. Wands, Phys. Rev. **D57**, 4686 (1998).
116. C.Wetterich, Nucl. Phys. **B302**, 668 (1988).
117. B. Ratra and P.J.E. Peebles, Astrophys. J. Lett. **325**, L17 (1988); Phys. Rev. **D37**, 3406 (1988).
118. R.R. Caldwell, R. Dave and P.J. Steinhardt, Phys. Rev. Lett. **80**, 1582 (1998).
119. P.J. Steinhardt, L. Wang and I. Zlatev, Phys. Rev. Lett. **82**, 896 (1999); Phys. Rev. **D59**, 123504 (1999).
120. C. Armendariz-Picon, V. Mukhanov and P.J. Steinhardt, astro-ph/0004134 and astro-ph/0006373.
121. P. Binétruy, Int. J. Theor. Phys. **39**, 1859 (2000); hep-ph/0005037.
122. E. J. Copeland, M. Sami, and S. Tsujikawa, hep-th/0603057.

Appendix

K.-H. Rehren[1] and E. Seiler[2]

[1] Institut für Theoretische Physik, Universität Göttingen, 37077 Göttingen, Germany
rehren@theorie.physik.uni-goe.de
[2] Max-Planck-Institut für Physik (Werner-Heisenberg-Institut), 80805 München, Germany
ehs@mppnu.mpg.de

1 Quantum Theory

Atomic systems exhibit features that are in conflict with the laws of classical mechanics: certain quantities (bound state energies, angular momentum) can take only discrete values in individual measurements; the probabilites for the outcome of measurements of position and momentum are subject to the *uncertainty relation*, limiting their joint precision by *Planck's constant* $\hbar \approx 10^{-34}$Js; and there may occur constructive or destructive interference between the *probability amplitudes* associated with two states, as demonstrated by the "double-slit" experiment (*superposition principle*). As was experimentally confirmed in recent years, composed systems possess *entangled states*, in which measurements in the subsystems show nonlocal correlations (violation of *Bell's inequalities*) that exclude any classical description in terms of "local hidden variables" (= limited knowledge of initial conditions in deterministic local dynamics). These characteristics are suppressed when only a subsystem is observed. The related effect, that a quantum system may appear classical as a result of its unmonitored interaction with the environment, is known as *decoherence.*

Quantum theory explains these features by describing quantum states as vectors (or density matrices) in a *Hilbert space.* In quantum mechanics, these vectors are often represented as wave functions, which are interpreted as probability amplitudes in position space or in momentum space, although different but equivalent representations are possible. When there is dynamical particle production or annihilation, however, it is more convenient to abandon this description in favor of a more abstract one, e.g., in terms of asymptotic multi-particle *scattering states* characterized by representations of the Poincaré group in a relativistic setting.

Just like the time evolution of a classical state is determined by the laws of Newtonian dynamics once the forces are specified, the evolution of a

K.-H. Rehren and E. Seiler: *Appendix*, Lect. Notes Phys. **721**, 399–405 (2007)
DOI 10.1007/978-3-540-71117-9_14

quantum state is determined by the Schrödinger equation once the *Hamiltonian* operator is specified.

The conjunction of the principles of quantum mechanics with the *principle of locality* and the symmetries of special relativity, is called quantum field theory (QFT). Its application to Maxwell's theory of electrodynamics gives rise to quantum electrodynamics (*QED*). Generalizations of QED are non-abelian *gauge theories* (*Yang–Mills theory*, quantum chromodynamics (*QCD*), the standard model of elementary particles). In QFT, the observables are no longer the positions of individual particles, but rather their energy or charge densities and, as derived concepts, *cross sections* in scattering processes. Their quantitative prediction is the prominent aim of a realistic QFT model.

2 Field Theory

Originating from hydrodynamics (describing the motion of fluids and gases in terms of velocity and density fields), the classical concept of fields had its triumph in Maxwell's theory of electrodynamics. The electromagnetic fields are local agents which mediate interactions among charged particles, thus enabling the passage from Coulomb's "action at a distance" to local interactions. At the same time, fields are carriers of energy and momentum, contributing to the balance of the dynamical system including the charged particles.

The dynamics of classical fields is most efficiently described in terms of an action functional (the integral over a *Lagrangean* density) from which the equations of motion are obtained, and from which energy, momentum and charge densities are derived by canonical prescriptions.

In QFT, the quantum mechanical *superposition principle* and the *causality principle* of special relativity require that also particles are represented by fields. More precisely, while the observables (charge or energy densities) are expressed in terms of fields which are understood as "defining the model", *particles* arise as spectral features of the energy-momentum operator, or as features of *scattering states* at asymptotic times, related to the localized distribution of energy-momentum and charge. In the *perturbative approach*, the local interactions are specified by polynomial expressions (couplings) in the fields, representing elementary processes. Their precise form is strongly constrained by symmetry principles, most prominently the "minimal coupling" of gauge theories, and by the condition of renormalizability.

The latter refers to the fact that perturbative computations in quantum field theory produce divergent results due to the large quantum fluctuations at small distances (*UV singularities*). These arise formally because a QFT is treated as a system of "infinitely many coupled harmonic oscillators". *Renormalization* is a systematic treatment, extracting finite quantities from the theory, to be compared with measurement. A simple criterium ("power counting") separates models where this works, from those where it does not.

A particularly useful tool is the passage to "imaginary time" (Wick rotation) which is possible, thanks to spectral and covariance principles in flat spacetime. Taking advantage of the resulting formal similarity of the resulting *Euclidean QFT* with classical four-dimensional statistical mechanics, powerful *functional methods* (path integrals, lattice approximations, *constructive QFT*, *non-perturbative renormalization*) have been developed allowing to go far beyond perturbation theory.

3 Gauge Theory

Gauge theory is the name for (quantum) field theories with a specific type of interaction determined by the *principle of gauge invariance.* It appears to be the only consistent way to describe quantum interactions involving vector particles (gauge bosons).

Gauge symmetry was originally observed within Maxwell's theory of classical electrodynamics as an ambiguity, related to the artificial introduction of unobservable potentials in order to solve two of Maxwell's four equations. Only in Dirac's equation for *QED*, this symmetry reveals its geometric interpretation as a local complex phase ambiguity for the charged (electron) field, which does not affect observable quantities such as current densities (Gauge Invariance). Promoted to a principle, Gauge Invariance turned out to be a most fruitful and universal symmetry paradigm, determining the detailed structure of almost all couplings in the standard model of elementary particles.

For this purpose, the local complex phase has to be replaced by a space time dependent unitary matrix taking values in the model-specific gauge group ($= U(1) \times SU(2)_L \times SU(3)_c$ in the standard model). *Gauge transformations* express the absence of a global comparability of internal degrees of freedom at different spacetime points.

The geometric interpretation of gauge symmetry is most natural in mathematical terms of *vector bundles.* Charged fields are sections in a vector bundle on which the gauge group acts in a given representation. The gauge potentials define a *parallel transport* on this bundle, whose curvature can be identified with the generalized "electric" and "magnetic" fields. The associated *covariant derivative* of the charged fields gives rise to "minimal couplings", through which the gauge potentials mediate the interactions between the charged particles. Unlike *QED* whose gauge group $U(1)$ is abelian, the gauge fields exhibit a self-interaction if the gauge group is *non-abelian.* The quantum field theory of this self-interaction (without other charged particles) is called *Yang–Mills theory.*

Non-abelian gauge theory poses several challenging problems to QFT: (1) the observed mass of the intermediate vector bosons (the particles associated with the gauge potentials) of the weak interaction requires a mechanism of mass creation for these particles (presumably through a *Higgs* field); (2) in perturbation theory, the covariant quantization and elimination of redundant degrees of freedom and unphysical states requires an ingenious cohomological (*BRST*) method; and (3) the ground state of *QCD* in the confining phase is completely unknown, and the hadronic particle states must be parametrized by suitable phenomenological structure functions which so far cannot be derived from the theory.

In order to attack (3), *lattice gauge theory* takes advantage of the natural geometric interpretation in order to formulate a covariant UV and IR regularized approximation, which is accessible by computerized numerical simulations and can be used to study the behavior when the cutoffs are removed.

4 The Standard Model

The standard model of elementary particles is the state-of-the-art theory of all fundamental interactions except gravity. It combines huge experimental evidence, comprising both *particle* spectroscopy and detailed measurement of *scattering cross sections*, with ingenious theoretical modeling on the basis of symmetry principles.

Roughly speaking, the standard model comprises the *electroweak interaction* (with typical particle lifetimes $\sim 10^{-6} - 10^{-10}$ sec) and the *strong interaction* (with typical particle lifetimes $\sim 10^{-23}$ sec or less).

At particle energies well below 100 GeV, the long-ranged electromagnetic interaction with the massless photon mediating the interaction between charged particles appears independent from the short-ranged *weak interaction*, mediated by massive W (charged) and Z (neutral) bosons. The weak interaction is distinguished from all other interactions by its maximal *violation of parity* (left-right) symmetry. Above 100 GeV, the Z boson and the photon appear as different combinations of the two neutral gauge bosons of the electroweak gauge group $U(1) \times SU(2)_L$, one of which has acquired mass through the *Higgs mechanism* as follows. The (postulated) scalar Higgs field has a quartic self-interaction symmetric around zero, but its potential energy has an orbit of minima away from zero. The field fluctuates around this minimizing orbit; the fluctuations along this orbit conspire with the gauge fields to give a mass to some of them. With suitable fixing of the gauge such as required in perturbation theory, the vacuum expectation value of the Higgs field becomes non-zero, leading to what is called *spontaneous symmetry breakdown*. In this way, three of the four electroweak gauge bosons acquire a mass, without explicit mass terms that would break the gauge invariance. Likewise, massless Fermi fields (leptons and quarks) will acquire mass through *Yukawa couplings* to the Higgs field.

The *strong interaction* (QCD) has the non-abelian gauge group $SU(3)_c$ with the corresponding internal degree of freedom called "color". At the fundamental level, its gauge bosons (*gluons*) mediate the interactions among the hadronic constituents (*quarks*). However, due to the strength of the interaction, the *ground state* of QCD cannot be viewed as a "small" perturbation of that of the corresponding free particles. Instead, its (unknown) involved structure is supposed to lead to the effect of *confinement*, observed at sufficiently low energies: quarks and gluons do not exist as isolated *particles* in asymptotic *scattering states*, but manifest themselves only as color-neutral hadronic bound states or "jets". On the other hand, at high energies (corresponding to small distances), the strong interaction becomes weak due to *renormalization* effects leading to the screening of color charges by vacuum polarization. In this regime (*asymptotic freedom*, experimentally accessible in *deep inelastic scattering* processes) quarks and gluons can be treated perturbatively.

Of all the particles mentioned, only the *Higgs* particle has not yet been detected in accelerator experiments.

The model input for the standard model consists of the gauge groups along with their representations for the charged fields, and 19 free numerical parameters: two coupling constants for the electroweak sector (or equivalenty, the unit of electric charge and the Weinberg angle), the masses of the W and Higgs bosons (or equivalently, the two parameters describing the Higgs potential), the strong coupling constant along with a parameter related to chiral symmetry breaking in QCD, and finally 13 (or more, if neutrinos are massive) coefficients of Yukawa couplings (mass matrices). Besides, at least at the present time, one should also regard the empirical structure functions phenomenologically describing hadronic states of QCD as additional "free parameters".

Because this amount of freedom is often considered as unsatisfactory for a fundamental theory, *grand unified theories* (GUT) or *supersymmetric* extensions attempt to reduce it with the assumption of extended symmetries, which are broken at present energies. There is at present no experimental evidence in favor of such theories "beyond the Standard Model", while physical facts like the lifetime of the proton exclude non-supersymmetric scenarios. The next generation of experiments (LHC) is expected to give some clear signals.

5 Symmetries

Symmetries play several (distinct and related) important roles within the paradigms of modern physics. Conceptually, one distinguishes "active" symmetry transformations which relate different configurations or states of the same physical system from "passive" symmetry transformations which describe the same configuration or state in terms of different references (frames, calibrations, gauges). On the other hand, one distinguishes symmetries of space and time ("external symmetries") from symmetries of other degrees of freedom ("internal symmetries", e.g., different isospin or color values in the standard model).

Relativity in the broadest sense is the expectation or postulate that the laws of nature are invariant under (certain) active and passive transformations. The passive point of view strongly constrains the possible form of these laws ("covariance"); the active point of view allows to make predictions about the outcome of different experiments and to relate different measurable quantities among each other.

Depending on the realm of physics one has in mind, the laws of nature exhibit different external symmetries: Newtonian Mechanics is Galilei invariant, Maxwell's theory is Poincaré invariant (*special relativity*, containing Galilei invariance as the limiting case when velocities are small compared to the speed of light), and gravity is diffeomorphism invariant (*general relativity*). The most important internal symmetries are gauge symmetries, which in particular largely determine the structure of (almost) all interactions in the standard model.

Supersymmetry is a generalized symmetry concept, uniting internal and external symmetries. Because of a pertinent No-Go theorem, it cannot be realized as a group of transformations, but only as a graded Lie algebra of "infinitesimal transformations". If Supersymmetry is realized in nature, it must be broken (i.e., not realized in the physical states). Its detection in the next generation of accelerators would, among other benefits ("dark matter" in the universe?), shed new light on a possible *grand unified theory* beyond the standard model.

Symmetries may be broken in various ways. *Spontaneous symmetry breaking* is the phenomenon that the ground state may not exhibit the full symmetry of the field algebra and its dynamics. On the other hand, one speaks of explicit breaking if either the algebra itself or its dynamics are disturbed in an asymmetric way.

6 Spacetime and General Relativity

The universality of the speed of light, predicted by Maxwell's theory of electrodynamics and verified in numerous experiments, requires a revision of the Newtonian concepts of space and time. The resulting Theory of *special relativity* states that all laws of physics take the same form when referred to inertial frames of reference. This implies the relativity of length and time measurements, and that mass is a form of bound energy that is convertible into other forms (e.g., particle production). Empirically confirmed to the extent that no accelerator would work without it, special relativity is at the basis of every fundamental theory of physics.

General relativity is the accepted dynamical field theory of *Gravitation*. It comprises both the motion of massive bodies in a gravitational field, and the dynamics of the gravitational field itself. The *principle of equivalence* asserts the approximate validity of special relativity even in gravitational fields. By introducing a *curvature of spacetime*, the general theory incorporates gravity; in particular, the trajectories of free-falling massive bodies are geodesics of the curved spacetime, thus explaining the observed equality of gravitational and inertial mass.

The dynamics of the gravitational field relates the spacetime curvature to the distribution of energy and momentum of matter and radiation. The field equations obey the *principle of general relativity* ("diffeomorphism invariance"). In the absence of matter they have special solutions such as *gravitational waves* or *black holes*.

Vast simplifying symmetry assumptions (homogeneity and isotropy) about the large-scale structure of the universe reduce the dynamical degrees of freedom of the gravitational field to a single "scale parameter" $a(t)$. Its coupling to the matter content of the universe and to itself gives rise to the *Friedmann–Lemaître–Robertson–Walker cosmology*, which explains the observed *Hubble expansion* and *cosmic microwave background* as the aftermath of a " Big Bang" at the beginning of time. While widely accepted, this standard model of Cosmology still has several problems, for the solutions of which various extensions ("inflation", "dark energy") are presently under discussion.

Although the theory is not consistent with quantum theory, the physical effects of this conflict are negligible except at very small length scales (the *Planck length* $l_P \approx 10^{-35}$ m). One of the greatest challenges in fundamental physics is the formulation of a *quantum theory of gravity*, including a quantum spacetime structure. Some approaches (including loop quantum gravity) are footed on the fact that at least formally the theory of general relativity shares many aspects of gauge theories (with respect to external rather than internal symmetries). Other approaches (such as string theory) pursue more radical ideas.

Glossary

K.-H. Rehren[1], E. Seiler[2], and I.-O. Stamatescu[3,4]

[1] Institut für Theoretische Physik, Universität Göttingen, 37077 Göttingen, Germany
rehren@theorie.physik.uni-goe.de
[2] Max-Planck-Institut für Physik (Werner-Heisenberg-Institut), 80805 München, Germany
ehs@mppmu.mpg.de
[3] Forschungsstätte der Evangelischen Studiengemeinschaft (FESt), Schmeilweg 5, 69118 Heidelberg, Germany
[4] Institut für Theoretische Physik, Universität Heidelberg, Philosophenweg 16, 69120 Heidelberg, Germany
stamates@thphys.uni-heidelberg.de

Action at a distance. Instantaneous effects of distant objects onto each other not mediated by a physical support, such as Newton's law of gravity. In relativistic physics it is replaced by local interactions.

Action functional (Lagrangian). The integral over the Lagrangian density, usually the difference between kinetic and potential energy. It determines the classical equations of motion by a stationarity condition; it is also the main ingredient of the path integral (functional integral, generating functional).

Asymptotic freedom. The decrease to zero of the strength of the interaction with decreasing distance, which is found in the perturbative quantization of non-abelian gauge theories due to "antiscreening effects". Physically it means that at very high energies the quanta behave almost like free particles.

Bekenstein entropy. The entropy formally assigned to a black hole is proportional to the area of its horizon. Its derivation in terms of microscopic degrees of freedom is considered a test for every theory of quantum gravity.

Bell's inequalities. Inequalities that must hold among expectation values of localized classical quantities. Their experimentally confirmed violation (in agreement with quantum mechanics) proves that the observed correlations cannot be reproduced by a local classical theory.

Black holes. Solutions of the gravitational field equations in the vacuum, exhibiting an event horizon ("nothing, not even light, can escape"). Stellar black holes can form around collapsing stars of sufficient mass, supermassive black holes are believed to exist in the centre of galaxies.

Big Bang. Sloppy name for the singularity met when the cosmological evolution is extrapolated backwards in time. Has been coined after the Hubble's observation of cosmic expansion and became the dominant

K.-H. Rehren et al.: *Glossary*, Lect. Notes Phys. **721**, 407–414 (2007)
DOI 10.1007/978-3-540-71117-9_15

cosmological paradigm after the observation by Penzias and Wilson of the (predicted) microwave background radiation.

Bosons. Particles obeying the Bose–Einstein statistics requiring symmetry of the state vector under interchange of particles of the same type. In quantum field theory described by Bose fields which commute at spacelike separation; they necessarily have integer spin.

BRST method (after Becchi, Rouet, Stora, and Tyutin). A two-step prescription to quantize non-abelian gauge theories. One first quantizes an auxiliary theory with redundant degrees of freedom, which can be done by standard methods but introduces unphysical states. From this, one can descend to the physical theory.

Confinement. The empirical fact that quarks and gluons cannot be observed as asymptotic particles; it is suggested by the increase of the coupling with distance in quantum chromodynamics and is realized in the lattice formulation of the theory.

Constructive quantum field theory. The attempt to construct quantum field theoretical models in a mathematically rigorous form (not based on perturbation theory). In the case of gauge theories, the main approach is via a Euclidean lattice theory used as an approximation to continuum QFT.

Cosmic microwave background. The 2.7 K thermal radiation filling the Universe. It is predicted by the Big-Bang cosmology as the remnant of the hot photon gas coupled to charged matter in early cosmological times, cooled down by the Hubble expansion after decoupling following the formation of (neutral) atoms.

Covariant derivative. The prescription generalizing partial coordinate derivatives in a way that is compatible with local gauge invariance. It involves an infinitesimal parallel transport of the fields. Its use in the equations of motion leads to characteristic couplings between the fields.

Cross section. Measures the intensity of a particular scattering process in dependence of its energy, scattering angle, and possibly other characteristics such as polarizations.

Curved spacetime. The dynamical structure of space and time in general relativity. The curvature depends on the local energy (mass) and momentum densities, but may also be present in empty spacetime. The geodesics of curved spacetime define the trajectories of free-falling (pointlike) bodies.

Dark energy. Invisible form of energy accompanied by a negative pressure whose existence is inferred from an observed acceleration of the expansion of the Universe. Theories concerning its nature are highly speculative.

Dark matter. Invisible matter within and between the galaxies whose existence is inferred only from its gravitational effect manifest, e.g., in their movement. Dark matter is estimated to make up about 25% of the total energy content of the Universe (visible "baryonic" matter $< 5\%$, dark energy 70%). Candidates for non-baryonic dark matter are, e.g., weakly interacting massive particles (WIMPs).

Decoherence. The emergence of classical behaviour for a quantum system through the irreversible interaction with its environment. It allows to explain in the framework of quantum mechanics why under certain conditions the typical quantum correlations are unobservable.

Deep inelastic scattering. High energy scattering processes between leptons and hadrons with large energy–momentum transfer and the production of many secondary particles. Because strong interaction is suppressed at very high energies (asymptotic freedom), in this regime, it can be treated perturbatively.

Electroweak interaction. The dynamical theory unifying the electromagnetic and weak interactions of leptons and quarks, formulated in the standard model as a gauge theory with gauge group $U(1) \times SU(2)$. The $SU(2)$ gauge transformations act in a parity-asymmetric way. The gauge quanta are the photon and massive W (charged) and Z (neutral) bosons. At low energies, the electromagnetic and the weak interactions separate.

Entanglement. The non-local nature of generic quantum states that describe a composite system. It entails the impossibility in general of assigning a pure state to a subsystem.

Euclidean quantum field theory. An approach to QFT which exploits a formal similarity with statistical mechanics (Statistical Field Theory) if "time" is replaced by an imaginary parameter. Functional integrals become mathematically more tractable in this setting. The transition to imaginary time is justified by locality and positivity of the energy in the real-time QFT; the transition back is possible under suitable conditions via the "Osterwalder–Schrader reconstruction".

Fermions. Particles obeying Fermi–Dirac statistics requiring antisymmetry of the state vector under interchange of particles, leading to the Pauli exclusion principle. In quantum field theory described by Fermi fields which anticommute at spacelike separation; they necessarily have half-integer spin.

Friedmann–Robertson–Walker–Lemaître cosmology. Models of the long-time evolution of the Universe based on the assumptions of spatial homogeneity and isotropy. They are at the basis of the standard Cosmological Model.

Functional methods. Allow manipulations of the generating functional (correspondingly, the path integral) for the computation of scattering amplitudes, convenient to exhibit symmetries and other general structures, and to control the renormalization. Being rigorously justified in lattice field theory, they form the cornerstone of non-perturbative quantum field theory.

Gauge principle. The geometric principle underlying gauge theories, according to which internal degrees of freedom at different spacetime points cannot be directly compared, but only through the intervention of a parallel transport between the two points. The latter is described by a gauge field carrying the gauge information from point to point, generalizing the scalar and vector potentials of electrodynamics.

Gauge theory. A quantum field theory in which the interactions are determined by the gauge principle. Here typically charged particles interact through the exchange of vector bosons. All fundamental interactions of the standard model are described by gauge theories.

Gauge transformations. Redefinitions of the fields according to some representation of a (gauge) group. In a gauge symmetric theory the observables are invariant under such transformations. Local gauge transformations can act arbitrarily at each spacetime point.

General Relativity. Einstein's classical theory of gravitation, based on the local indistinguishability of inertial and gravitational forces. The gravitational field is described by the curvature of spacetime, dynamically coupled to energy and momentum of matter. General relativity is also the basis of cosmological models.

GeV. See MeV.

Gluons. The gauge bosons of quantum chromodynamics. Since they have a colour charge they also interact directly with each other (in contrast to an abelian gauge theory such as QED).

Grand Unified Theories (GUTs). Models designed to unite the electroweak and the strong interactions of the standard model into one interaction with a single coupling constant, obtained as convergence point of the running coupling constants of the standard model. GUTs are usually based on simple gauge groups containing $U(1) \times SU(2) \times SU(3)$ (the group of the standard model).

Gravitation. The extremely weak gravitational interaction dominates all other forces at macroscopic distance scales because it cannot be shielded. General relativity, the successful theory of classical gravitation and continuum spacetime, must break down at the Planck scale, where the gravitational field of quantum fluctuations of the energy would be strong enough to form a black hole and thus essentially affect the structure of spacetime.

Gravitational waves. Solutions of the gravitational field equations in the vacuum in the weak field approximation, describing small perturbations of flat spacetime propagating at the speed of light.

Ground state. A state of lowest energy in quantum theory, e.g., a non-excited particle in quantum mechanics, or the vacuum state in quantum field theory. The existence of a ground state is required to ensure the stability of a system.

Hamiltonian. The observable (usually with the meaning of "energy") that generates the time evolution of a dynamical system. Classically it can be derived from the action functional and vice versa.

Hawking radiation. A semiclassical treatment of quantum field theory in the vicinity of a black hole predicts that the latter emits thermal radiation of a temperature inversely proportional to the mass. (See Bekenstein entropy.)

Higgs mechanism. In the standard model, the gauge bosons of the weak interaction are given a mass by coupling them to the scalar Higgs field, whose potential has a non-trivial minimum away from zero. This procedure is compatible with gauge symmetry, while explicit mass terms would destroy it.

Hilbert space. The space of state vectors in quantum theory, equipped with a scalar product representing transition amplitudes. In quantum mechanics, a state vector can be given as a wave function concerning some degree of freedom of a system, while typical state vectors in quantum field theory have an interpretation in terms of asymptotic multiparticle states.

Hubble expansion. The expansion of the Universe observed by means of the flight velocity of distant galaxies. The "Hubble parameter" is given by speed over distance, its inverse is related to the age of the Universe in the Big Bang model.

Inflation. A class of models stipulating a period of extremely rapid expansion of the early Universe, required to solve problems (flatness problem, horizon problem) arising with the Friedmann–Robertson–Walker–Lemaître cosmology.

Lagrangian. See Action functional.

Lattice approximation. An approximation to quantum field theory by a theory in which Euclidean spacetime is replaced by a discrete lattice. Basis for the constructive approach and non-perturbative analysis because it allows to give a precise meaning to the path integral. Especially useful in gauge theories, because the approximation preserves gauge invariance. Allows numerical determination of hadronic properties, such as their mass spectrum and weak decay matrix elements.

MeV. Convenient unit of energy and mass (MeV/c^2) in high-energy physics. 1 MeV $= 10^6$eV where 1 eV is the energy an electron acquires when it runs through an electric potential difference of 1 V, 1 eV $=1.602 \times 10^{-19}$ J. 1 $\text{GeV}/c^2 = 1000\ \text{MeV}/c^2$ is close to the mass of the proton.

Newton's constant. The fundamental constant of nature $G \approx 6.67 \cdot 10^{-11}\text{Nm}^2/\ \text{kg}^2$ determining the strength of the gravitational interaction.

Non-abelian. A group of non-commuting transformations is called non-abelian, e.g., three-dimensional rotations. The weak and strong interactions are gauge theories with non-abelian gauge groups.

Non-perturbative renormalization. The procedure to define functional integrals, and thus to rigorously construct quantum field theories, starting from well-defined approximating measures, e.g., provided by lattice field theory.

Parallel transport. The prescription required to compare field amplitudes in a gauge theory (points in a vector bundle) at different spacetime points with each other. Necessary prerequisite to define a gauge covariant derivative.

Particles. The classical notion of a (point) particle represents a distinguishable object (mass point) possessing a well-defined trajectory. Both these properties are lost already in quantum mechanics; in quantum field theory particles manifest themselves as localized carriers of energy and momentum showing up in the asymptotics of scattering processes.

Perturbation theory. Originating from celestial mechanics, perturbation theory is the systematic approximation to an interacting theory obtained by regarding the interaction as a (small) perturbation of an exactly solved (usually: free) theory. Perturbation theory provides divergent, at best asymptotic expansions in QFT and fails when the coupling constant is large, e.g., in quantum chromodynamics at low energies.

Planck scale. The length and mass scales $l_P = (G\hbar/c^3)^{\frac{1}{2}} \approx 10^{-35}$ m, $m_P = \hbar/cl_P \approx 10^{19}$ GeV/c^2 obtained by combination of the fundamental constants of nature c (speed of light), $\hbar$ (Planck's constant), and G (Newton's constant). It roughly represents the scale at which quantum mechanical localization uncertainty becomes comparable with the gravitational black hole horizon and where, therefore, phenomena of quantum gravity are expected to show up.

Planck's constant. The fundamental unit of action $h = 2\pi\hbar \approx 6.6262 \times 10^{-34}$ Js in quantum mechanics. It sets the scale of the Heisenberg uncertainties $\Delta p \cdot \Delta x \geq \hbar/2$, of energy quanta ($E = h \cdot \nu$ for photons), and of quantized angular momentum $L \sim \hbar$.

Power counting. A simple method to decide the possibility of perturbative renormalization by determining the degree of UV singularities.

Principle of Causality. Postulates the absence of inacceptable causal paradoxa due to superluminal propagation of signals or causal influences. In quantum field theory, this principle is implemented by requiring that observables localized at spacelike distance commute; in the Lagrangian formulation this means that the interaction has to be local.

Principle of Equivalence. Transcending the empirical equality of inertial and gravitational mass, this principle asserts the local indistinguishability between inertial and gravitational forces. It provides the physical basis of general relativity.

Principle of general relativity. Various versions of postulates about the formal structure of field theories such that they comply, if coupled to gravity, with the Principle of Equivalence.

Principle of locality. See Principle of Causality.

Probability amplitude. In quantum mechanics the complex value of the wave function, whose modulus squared gives the probability per volume of finding a particle in some region of position or momentum space. In quantum field theory a complex number whose modulus squared gives the probability per phase space volume of finding a particular outgoing state in a scattering process.

QCD, Quantum Chromodynamics. The dynamical theory of quarks and gluons describing the strong interaction. Its fundamental field quanta do not arise as particles (confinement), but become almost free at very high interaction energies (asymptotic freedom). Its gauge group is SU(3).

QED, Quantum Electrodynamics. The dynamical theory of quantized electrons and photons (electromagnetic fields). Prototype of a gauge theory. Its gauge group is the (abelian) group U(1).

Quantum Gravity. Fundamental theory which accomodates the gravitational interaction into the quantum framework. It should be mainly relevant for understanding the early universe and the fate of black holes and the dynamics of gravitation near the Planck scale. This theory is still elusive.

Quantum probability. Knowledge of a quantum state does not in general predict the outcome of an individual measurement but the expectation values (averages) of observables in a large set of measurements on identically prepared systems, by providing corresponding probabilities.

Quarks. The quanta of the matter fields in quantum chromodynamics, coupled to the gluons (gauge fields). While they are not observable as isolated particles (confinement), quarks (and gluons) can in some sense be regarded as constituents of hadronic particles.

Relativity. The independence of the laws of nature on the state of motion of the system or of the observer. If this is required only for reference frames in uniform motion (inertial systems) and if the speed of light does not depend on the reference frame, it is called special relativity; if there is no such distinguished speed, Galilean relativity. In general relativity this independence extends to any reference frame by incorporating gravity.

Renormalization. The systematic treatment to express the observables of a theory with the help of physical (renormalized) parameters – couplings, masses, field strengths. In this way one can extract finite quantities from a theory that predicts divergent results in terms of its unobservable "bare" parameters, by absorbing the singular behaviour in the bare parameters themselves.

Scattering processes. Most experiments in high-energy physics proceed by scattering particles off each other, thereby producing new particles. The comparison of the ingoing and outgoing states tests the underlying dynamical theory.

Special Relativity. See Relativity.

Spontaneous symmetry breaking. Occurs when the ground state of a dynamical system does not exhibit the full symmetry of the dynamics itself (the action functional or the equations of motion).

Strong interaction. The dynamics of hadronic particles leading to the cohesion of nuclei and the decay processes with very short lifetimes ($\sim 10^{-23}$ sec or less). Described by quantum chromodynamics.

Superposition Principle. The characteristic feature of quantum states to allow linear combinations of state vectors to describe new states. It leads to constructive and destructive interference of probability amplitudes.

Supersymmetry. A generalized symmetry concept involving transformations which mix fermionic and bosonic fields. While supersymmetry is often a desirable feature in quantum field theory for theoretical reasons (renormalizability), it is not (yet?) observed in the experiment. Hence, if it is part of a fundamental theory, it must be broken by some unknown mechanism.

Ultraviolet singularities. The apparent prediction that the exchange of high-energy ("UV") quanta gives infinite contributions to an interaction.

It occurs because products of fields at the same point are mathematically ill-defined. Renormalization is designed to remedy these singularities.

Vector bundle. The geometrical notion of the space of field configurations in a gauge theory. In every point of the base space (spacetime), a vector space describes the possible values of a field at that point. Relations between vectors at different base points are specified by parallel transport or its infinitesimal version, a connection.

Violation of parity. The characteristic feature of the weak interaction is its maximal violation of parity (left-right symmetry). It is implemented in the standard model by the asymmetric ("chiral") action of $SU(2)$ gauge transformations, acting only on the fermions with left-handed helicity.

Weak interaction. The dynamics of particles responsible for "slow" processes such as radioactive β-decay (typical times from 10^3 sec (neutron) to 10^{-13}sec (τ lepton)).

Yang–Mills theory. Prototype of a non-abelian gauge theory describing only the self-interaction of gauge fields resulting from the non-trivial covariant derivative used in the kinetic term of the action.

Yukawa interaction. A model for short-range interactions mediated by a massive scalar particle. In the standard model, Yukawa couplings of the Higgs field to fermions give mass to the latter.

Index

MIX
Papier aus verantwortungsvollen Quellen
Paper from responsible sources
FSC® C105338

If you have any concerns about our products,
you can contact us on
ProductSafety@springernature.com

In case Publisher is established outside the EU,
the EU authorized representative is:
Springer Nature Customer Service Center GmbH
Europaplatz 3, 69115 Heidelberg, Germany

Printed by Libri Plureos GmbH
in Hamburg, Germany